FORENSIC ENGINEERING

PROCEEDINGS OF THE 4TH CONGRESS

October 6–9, 2006
Cleveland, Ohi

Technical Council o ... ering (TCFE) of
the American ... Civil Engineers

EDITED BY
Paul A. Bosela
Norbert J. Delatte

Published by the American Society of Civil Engineers

Library of Congress Cataloging-in-Publication Data

Forensic Engineering Congress (4th : 2006 : Cleveland, Ohio)
Forensic engineering : proceedings of the 4th congress, October 6-9, 2006, Cleveland, Ohio / sponsored by Technical Council on Forensic Engineering (TCFE) of the American Society of Civil Engineers ; edited by Paul A. Bosela, Norbert J. Delatte.
p. cm.
Includes bibliographical references and index.
ISBN 0-7844-0853-X
1. Structural failures--Investigation--Congresses. 2. Forensic engineering--Congresses. I. Bosela, Paul A. II. Delatte, Norbert J. III. Title.

TA656.F664 2006
624.1'71—dc22 2006049890

American Society of Civil Engineers
1801 Alexander Bell Drive
Reston, Virginia, 20191-4400

www.pubs.asce.org

ISBN 13: 978-0-7844-0853-7
ISBN 10: 0-7844-0853-X
Manufactured in the United States of America.

Preface

These proceedings consist of the set of papers presented at the Fourth Forensic Engineering Congress, held in Cleveland, Ohio, October 6-9, 2006. The Congress was organized by the ASCE Technical Council on Forensic Engineering (TCFE), whose mission is to apply engineering principles to investigate failures and performance problems of engineered facilities, and to develop practices and procedures to reduce the number of future failures. TCFE uses the Forensic Congress to disseminate information on failures and practices to mitigate failures, and to provide a forum to discuss ethical practice within the field of forensic engineering.

The publication of these papers is the result of many hours of volunteer labor by active members of TCFE. Each paper completed a peer review process and received a minimum of two positive reviews. All papers in this proceedings are eligible for both discussion in the Journal of Performance of Constructed Facilities and possible ASCE awards.

The editors of these proceedings would like to acknowledge the TCFE EXCOM Chairman Kevin Rens, TCFE members, authors, reviewers, session moderators, ASCE staff and other individuals who contributed to the success of this congress. In addition, the editors would like to thank Nehal Desai and Ayan Ghosh, Civil Engineering Graduate Students at Cleveland State University (CSU) for their help in providing assistance in managing the organizational process, and the faculty and staff at CSU. Finally, we would like to thank Angela Bosela and Lynn Delatte for their support while we were working on this project.

Paul A. Bosela
Professor and Chair
Department of Civil and Environmental Engineering
Cleveland State University

Norbert J. Delatte
Associate Professor
Department of Civil and Environmental Engineering
Cleveland State University

Reviewers

The Technical Council on Forensic Engineering would like to thank the following individuals who served as reviewers for the Proceedings

Benjamin Allen
Summit Engineering

Jack D. Bakos
Youngstown State University

Bruce Barnes
Knott Laboratory

Kimball Beasley
Wiss, Janey, Elstner Associates Inc.

Glenn Bell
Simpson Gumpertz & Heger

Paul A. Bosela
Cleveland State University

William C. Bracken
Bracken Enginering

Pamalee Brady
California Polytechnic State University

Keith E Brandau
Frauenhoffer & Associates P.C.

Merle E. Brander
Brander Construction Technology Inc.

Chris Cardillo
Knott Laboratory

Kenneth L. Carper
Washington State University

Shen-En Chen
University of North Carolina

Julie Mark Cohen
Structural and Forensic Engineer

James S. Davidson
University of Alabama at Birmingham

Philip H. De Groot
Cleveland State University

Norbert J. Delatte
Cleveland State University

Timothy J. Dickson
Dickson-Senffner Ltd.

Anthony M. Dolhon
Wiss, Janney, Elstner Associates, Inc.

William G. Fleck
Cleveland State University

David J. Frost
Georgia Institute of Technology

Dario Gasparini
Case Western Reserve University

Michael E. Gogel
City & County of Denver Colorado

Roberto Gori
University of Padua, Italy

Howard F. Greenspan
H.F. Greenspan Associates

O.C. Gueoelhoefer
Raths, Raths & Johnson Inc.

Decker B. Hains
United States Military Academy

John M. Hanson
North Carolina State University

Peter Marx Hausen
Higgens and Associates

Oswald Rendon-Herrero
Mississippi State University

Michael C. Higgins
Higgins and Associates

Y. Henry Huang
Los Angeles County Dept. of Public Works

Arthur Huckelbridge
Case Western Reserve University

Yung-tse Hung
Cleveland State University

Anwarul Islam
Youngstown State University

Lynn E. Johnson
University of Colorado at Denver

Harvey A. Kagan
Wiss Janney Elstner Associates Inc

Thomas A. Kallio
Scheider Corp

Lutful Khan
Cleveland State University

Walter M. Kocher
Cleveland State University

Melissa Kubischeta
Higgins and Associates

Michael P. Lester
Forensic Engineering Inc.

Nicholas B. Lehmann
PTC Forensic Consulting Services

Brian Lindsey
PTC Forensic Consulting Services

Peter Maranian
Brandow & Johnston Structural Engineers

David C. Mays
University of Colorado at Denver

Andrew T. Metzger
Case Western Reserve University

Mehdi Modares
Case Western Reserve University

J. P. Mohsen
University of Louisville

Leonard Morse-Fortier
Simpson Gumpertz & Heger Inc.

Theodore Padgett
Consulting Engineer

Shane M. Palmquist
Western Kentucky University

M. Kevin Parfitt
Penn State University

Azadeh Parvin
University of Toledo

Satinder P.S. Puri
Consulting Engineer

Leonard T. Rudick
Rudick-McClellan Associates

Sammy R. Saed
Cleveland State University

Ziad Salameh
Inspec Incorporated

Robin Shepherd
Earthquake Damage Analysis Corp

Kenneth B. Simons
Damage Consultants Inc.

Eric Stovner
Miyamoto International

Kevin G. Sutterer
Rose-Hulman Institute of Technology

Glenn Thater
LZA Technology

John J. Tomko
Cleveland State University

William Vermes
Euthenics, Inc.

Moon Won
Texas Department of Transportation

Rubin M. Zallen
Zallen Engineering

Lewis Zickel
Consulting Engineer

Plenary Session

Forensic Study of New Orleans Hurricane Protection in Katrina

Moderated by Dr. Paul F. Mlakar, U.S. Army Corps of Engineers (USACE)

Hurricane Katrina was one of the strongest storms to hit the coast of the United States in the past century. The magnitude of the destruction, the extensive damage to the hurricane protection system, and the catastrophic failure of a number of structures raised significant questions about the integrity of the flood protection system prior to the storm and the capacity of the system to provide future protection, even after repairs. Immediately following the storm, the USACE established the Interagency Performance Evaluation Task Force (IPET) to provide credible and objective answers to these technical questions. The work of IPET involved ten tasks, each of which was studied by a team led jointly by an expert from USACE and an expert from an external organization. Comprised of some 150 individuals from 50 organizations, these teams provided a diversity and depth of knowledge and experience. A continuous detailed review of IPET was provided by an external review panel under the auspices of ASCE and an independent panel of the National Research Council is providing further strategic oversight and synthesis of the findings. The IPET released its report on June 1, 2006. The session consists of three presentations on the forensic aspects of the study as follows:

The Hurricane Protection System – Dr. John J. Jaegar, USACE

The state of the hurricane protection was established through a comprehensive search of the design documents and a thorough physical inspection. This infrastructure, consisting of levees and floodwalls, was generally built as designed. Unfortunately this was done piecemeal over a period of decades rather than as an integrated system all at once. The criteria used a Standard Project Hurricane dating to 1959 that did not encompass all the information now available regarding the hurricane hazard. Some sections of the protection were lower than intended due to an inaccurate relation between the geodetic datum and mean sea level and the variable and considerable subsidence in the area.

The Storm Loading - Drs. Bruce A. Ebersole and Donald T. Resio, USACE

The hurricane loading was studied through widespread physical observation, state-of-the-art numerical analyses, and physical modeling. Hurricane Katrina generated water levels that for much of the system significantly exceeded the design criteria. Detailed hydrodynamic analyses showed that dynamic forces were a significant portion of the total forces experienced. Overtopping by waves generated very high velocities over the crest and back sides of the levees, leading to a high potential for scour and erosion. Contrary to some early speculation, the southeast trending leg of the Mississippi River Gulf Outlet had little influence on the water levels in the Inner Harbor Navigation Canal.

The Levee Response - Drs. Reed L. Mosher and Michael K. Sharp, USACE

The response of the levees was examined by limit equilibrium analyses, finite element calculations, and physical models in the centrifuge. In spite of loadings in excess of the design conditions, many sections of the protection performed well. Some 46 breaches occurred due to overtopping and erosion. Unfortunately 4 additional breaches were caused by foundation failures in floodwalls at water elevations less than the design level that were induced by the formation of a gap along the canal side of the floodwalls.

Contents

Forensics and Design Practices

Tunnels and Culverts

Indexes

Validity and Reliability of Forensic Engineering Methods and Processes

Joshua B. Kardon[1], Ph.D., S.E., Member, ASCE;
Robert G. Bea[2], Ph.D., Fellow, ASCE;
Robert Brady Williamson[3], Ph.D., P.E., Life Member, ASCE

[1]Principal Structural Engineer, Joshua B. Kardon + Company Structural Engineers, 1930 Shattuck Avenue, Suite C, Berkeley, California 94704; email: jbkse@jbkse.com
[2]Professor at Department of Civil and Environmental Engineering: Engineering & Project Management; University of California at Berkeley, Berkeley, California 94720
[3]Professor in the Graduate School, Department of Civil and Environmental Engineering: Structural Engineering, Mechanics and Materials; University of California at Berkeley, Berkeley, California 94720

Abstract

Under the Federal Rules of Evidence, and in accordance with case law, technical forensic evidence presented in court has to be valid and reliable—a judge may rule evidence inadmissable if it is shown to be invalid or unreliable. Engineering methods and processes operating in arenas other than litigation also must be valid and reliable in order that those methods and processes achieve their intended results. This paper discusses of types validity and reliability, and gives examples of expert witness evidence and of an engineering method that lacked validity and reliability.

Introduction

The Federal rule of evidence for expert testimony, Rule 702, is based on case law, such as *Daubert v. Merrill Dow Pharmaceuticals* (1993). It allows the judge to assess "whether the testimony's underlying reasoning or methodology is scientifically valid and properly can be applied to the facts at issue" (Federal, 2001). To determine that, the judge can consider whether the method "can be (and has been) tested, whether it has been subjected to peer review and publication, its known or potential error rate, the existence and maintenance of standards controlling its operation," as well as "whether it has attracted widespread acceptance within a relevant scientific community." The validity and reliability of a method generally have to do with the applicability of the method to the question asked, and the suitability of the method for the intended purpose.

Engineering methods and processes in areas other than forensics and litigation also must have validity and reliability. Engineering decisions that affect people's lives and livelihoods are made on the bases of methods and models that must be valid and reliable, or they will not result in acceptable levels of safety, durability, compatibility or serviceability.

Validity

There are two general types of validity: external and internal. External validity (Campbell & Stanley, 1963) is the extent to which the method is generalizable or transferable. A method's generalizability is the degree the results of its application to a sample population can be attributed to the larger population. A method's transferability is the degree the method's results in one arena can be applied in another similar arena.

In contrast to external validity, internal validity "is the basic minimum without which the method is uninterpretable" (Campbell & Stanley, 1963). Internal validity of a method addresses the rigor with which the method is conducted (e.g., the method's design, the care taken to conduct measurements, and decisions concerning what was and wasn't measured). There are different types of internal validity: face validity, content validity, criterion-related validity, and construct validity.

Face validity is the degree to which a method appears to be appropriate for measuring what it intends to measure (Fink, 1995). An example of face validity is the observation that a ruler appears to be an appropriate tool to measure length.

Content validity has to do with the degree to which the method measures the trait it is intended to measure. An example of a test which lacks adequate content validity is one which intends to measure a subject's mathematical ability by testing only addition (Carmines & Zeller, 1979).

Criterion-related validity has to do with the degree to which the method allows for assessment of a subject's performance in situations beyond the testing situation—in a different domain than the test. Criterion-related validity may be concurrent or predictive. That is, the test result may either be intended to assess a criterion independently measured at the same time (concurrent), or to predict achieving a criterion in the future (predictive). An example of predictive criterion-related validity is the extent to which a written driver's test accurately predicts how well the tested population will drive (Carmines & Zeller, 1979). The written driver's test does not involve physically driving a car on the road, but only involves answering several multiple choice questions. The extent to which good performance on that written driver's test correlates well with future good driving performance on the road is a measure of the test's criterion-related validity. The physical act of driving a car takes place in a different situation than the testing environment, and involves different skills and abilities. Threats to the predictive validity of the written driver's test include the possibility that a test subject can't read English but might be a good driver. That threat to criterion-related validity is addressed by having the test printed in several languages besides English. An inquiry of a test's predictive criterion-related validity asks the question, "How accurately does this test measure future performance in that setting?"

Construct validity has to do with the degree to which the results of the method can be accounted for by the explanatory constructs of a sound theory. A method's construct validity is understood by first specifying the theoretical relationships between the concepts, examining the empirical relationships between the measures of the concepts, and then interpreting how the observed evidence clarifies the concepts being measured (Carmines & Zeller, 1979). Construct validity is demonstrated when measures that are theoretically predicted to be highly interrelated are shown in practice to be highly interrelated. An inquiry of a test's construct validity focuses on the question of whether the results of the test are in fact a true measure of the construct, or theory, being tested, and not of some other phenomenon or process which might produce the same results. Such an inquiry asks, "Is this theory the best explanation for the results?"

Not all types of internal validity are applicable to any one method (Rossi & Freeman, 1979). For instance, a method may not be one which is intended to predict an outcome of a process; it may not seek to answer the question, "How accurately does the test measure future performance in a different setting?" so its criterion-related validity would not be an issue. The intended result of a method may not be something that is predicted by a theory; it may not ask, "Is the theory the best explanation for the results?" If the result of the method can't be measured and contrasted with a theoretically predicted result, its construct validity would not be an issue.

Reliability

A reliable method is one that yields consistent results upon repeated use; it is suitable for its intended purpose. However, when a reliable method is used in court by experts on two opposing sides in a dispute, the results will not necessarily be identical. After all, the reason disputes end up in court is just that there are good arguments for both sides. Expert witnesses are ethically obliged to help their attorney clients explain the case to the juries from the particular point of view of their client, within the bounds of truth (Kardon, Schroeder, & Ferrari, 2003). It is not unethical for an expert witness to explain technical aspects of the dispute from the particular point of view of their client. Issues end up in court because there are differences of opinions and shadows of doubt, and experts retained by attorneys representing both sides of a dispute often come to different and contrary opinions based on reasonable interpretations of the evidence they each review or develop. Each side's expert presents technical evidence for the purpose of aiding the trier of fact. The trier of fact is best served by the effective presentation of technical arguments from both sides of the dispute. The reliability of a method used as the basis of an opinion given as expert testimony, therefore, can not be evidenced by identical opinions being supported upon its repeated use in the dispute resolution process. Instead, the reliability of such a method will originate in the understanding that the method is suitable for supporting the opinion of the expert.

Example - Expert Testimony

An expert testified on behalf of an insurance company concerning the amount of structural movement that must have occurred in a house that was in the throes of a major remodel and seismic upgrade when it was allegedly damaged by the 1989 Loma Prieta earthquake. The expert did not observe the actual damage caused by the earthquake, but was asked by his client to determine whether the damage that was claimed by the homeowner could have been caused by the earthquake. The expert performed a computer-based analysis of the house, and relied on that analysis to come to his opinion regarding the amount of movement the house underwent during the earthquake, and therefore the amount of damage to the house that occurred as a result of the earthquake.

The structural engineer-of-record for the remodel and seismic upgrade, who performed structural observation during construction both before and after the earthquake, testified that most of the plywood on the exterior walls of the house at the time of the earthquake was attached with duplex nails (double-headed nails for easy removal) at a much wider spacing than the final shear wall nailing was to be, and the shear transfer clips and hold downs were not in place. The purpose of the plywood in place at the time of the earthquake was not to provide lateral load resistance, but to provide some jobsite security by preventing unauthorized access to the building after working hours.

The expert testified his computer model was based on the assumption that the plywood for all the shear walls was in place at the time of the earthquake, but was not nailed with all the nails specified in the design documents for the strengthening of the house. He testified he assumed the plywood was sufficiently nailed so that there was continuity between the plywood and the framing. He assumed the plywood was sufficiently attached to the framing at the time of the earthquake because that was the assumption of the computer model. He also testified his model assumed none of the hold down hardware was in place. He did not testify as to whether his model assumed the bottom plate nailing or the top plate shear transfer clips were in place.

There was no assurance the analysis by the insurance company's expert accurately replicated the actual behavior of the building in the earthquake. This was because in finite element analysis the real assembly of framing, plywood, nails, clips, hold downs, etc. is modeled using elements of assumed strength, stiffness and boundary conditions. These assumptions must be verified either by comparing the actual assembly to previously tested assemblies that have been shown to be accurately modeled, or by carrying out physical tests of the assemblies to compare their behavior with the model element's behavior.

Published models of wood-framed shear wall assemblies used in finite element analyses are based on assumptions of fully nailed walls, with competent, active shear and overturning transfer hardware in place. There was no testimony given by the insurance company's expert that he used a verified model of a wood-framed shear wall without hold downs and shear transfer hardware, and with plywood sheathing only lightly nailed at the panel edges.

In addition to the modeled wall assemblies, the interface of the house foundation and the ground, and the actual condition of the real foundation (cracks and all) must be accurately modeled in order for the analysis to be correct. There was no assurance the soil-structure interaction or the foundation were accurately modeled.

The same criticism can be brought against the modeled loads. The insurance company's expert testified that he estimated the earthquake ground motion at the subject house by examining published earthquake ground motion records from two nearby seismometer stations, and used that assumed ground motion as input for his computer analysis of his model of the house. The actual ground motions to which the building was subjected must be accurately modeled in order for the finite element analysis to produce results which reliably duplicate the actual behavior of the building. The insurance company expert relied on records of earthquake motion recorded at two stations which he stated were close to the building site, and which he stated were on similar soils. The insurance company's expert offered no testimony of any characterization of the soils at the subject site, or of the sites where the earthquake motions he used were recorded.

Interpolation of earthquake ground motion at a particular site from ground motions recorded at other sites is inaccurate. The United States Geological Survey, the source of one of the ground motion records used by the insurance company's expert, publishes maps of earthquake-induced ground shaking. They state, "ground motions and intensities typically can vary significantly over small distances, these maps are only approximate. At small scales, they should be considered unreliable." By the same token, any ground shaking at the subject house the insurance company's expert deduced from the two recording stations to which he referred must be viewed as unreliable. The only true measure of actual ground shaking at the site, absent a calibrated and functioning recording device, is the amount of damage which actually occurs at the site.

Because the model used by the insurance company's expert was not based on actual conditions at the building, and because no verification or justification of the model structure or loading was presented, the analysis was not valid or reliable. The computer model did not recreate an accurate depiction of the actual condition of the house or of the actual loads applied to the house by the Loma Prieta earthquake, it therefore lacked face validity. Because the model of the structure and of the loads was not representative of the real structure or loads, the method could not predict or describe the movement of the building, and therefore lacked criterion-related validity. Because of the absence of validity and reliability, it was argued that the expert evidence should be disallowed in the determination of damages.

Example - Engineering Method

A primary obligation of an engineer is to anticipate failure modes in the element, component, or system being engineered and then provide measures to prevent those failure modes from developing or from developing catastrophic results (Petroski 1985, 1994; Harr 1987; Wenk 1989). This obligation requires two primary elements: 1) anticipation of possible failure modes, and 2) provision of defenses in depth to prevent and/or mitigate those failure modes. The second element requires valid and reliable analytical models.

During Hurricane Katrina, a large segment of a drainage canal levee and floodwall lining the 17th Street canal in New Orleans failed catastrophically before the design water elevations were realized. The Corps of Engineers Interagency Performance Evaluation Task Force analyses (Interagency Performance Evaluation Task Force 2006) of this failure concluded that a failure mode developed that was not recognized by the designers. This finding led to the official contention that this was a "design failure." Information developed by the Independent Levee Investigation Team (2006) clearly indicates that this failure was a result, not a cause.

The failure mode involved lateral deflection of the concrete floodwall and the sheet piles that supported that floodwall. This deflection resulted in separation between the stiff supporting sheet piling and the soft soil of the levee on the flood side of the wall. Water was then able to enter the gap and exert additional lateral forces on the remaining 'half' of the levee-floodwall. Now the levee only had about 'half' of its width able to transmit the lateral forces to the underlying soils. This combination resulted in lowering the lateral resistance with a commensurate lowering of the factor of safety.

This development was incorrectly reported as "unforeseen and unforeseeable" by the Interagency Performance Evaluation Task Force (Marshall 2006; Seed and Bea 2006). In 1985, the New Orleans district of the Corps of Engineers conducted a full scale instrumented lateral load test of a 200-foot long sheet pile flood wall in the Atachafalaya basin (U.S. Army Corps of Engineers 1988a). This particular location (south of Morgan City, Louisiana) was chosen because of the close correlation of the soil conditions in the New Orleans area with those at the test location. "The foundation soils are relatively poor, consisting of soft, highly plastic clays, and would be representative of near worst case conditions in the NOD (New Orleans District)." (U.S. Army Corps of Engineers 1988a).

Test data from the highly instrumented sheet pile wall and adjacent supporting soils indicated a gapping behavior (separation of the sheet piles from the soils). The test was designed to take an eight foot height of water (above the supporting ground level) with a factor of safety of 1.25. But the wall was already in a failure condition (increasing lateral displacements with no increase in loading) when the water level reached only 8 feet instead of the calculated 10 feet. Strain gage readings on the sheet piles indicated that they were well below the steel yield point, thus the yielding had to have been developing in the supporting soils. Two very important pieces of information developed by the E-99 sheet pile tests were that there was potential soil separation from the sheet piles (allowing water to penetrate below the ground surface between the piles and the soils) and that the calculated safety factor was not reached (it was over-estimated due to unanticipated deformations in the soils).

Additional reports and professional papers further developed the experimental information and advanced analytical models that could be used to help capture such behavior (U.S. Army Corps of Engineers Waterways Experiment Station 1989). Later developments in this work were reported by Oner, Dawkins and Mosher (1997):

> *As the water level rises, the increased loading may produce separation of the soil from the pile on the flooded side (i.e., a "tension crack" develops behind the wall). Intrusion of free water into the tension crack produces additional hydrostatic pressures on the wall side of the crack and equal and opposite pressures on the soil side of the crack. Thus part of the loading is a function of system deformations.*

These developments in technology were *not* reflected in the design guidelines used (U.S. Army Corps of Engineers 1988b, 1989, 1990). A traditional method of active and passive pressures acting along the length of the sheet piles embedded in the earth levee was used to determine stresses induced in the concrete wall - sheet pile joint and in the sheet piles. This traditional method did not incorporate the information developed from the E99 floodwall test. The traditional design guideline-based method used to design and engineer the floodwall system did not possess the required attributes of validity and reliability.

A second element in this development regarded characterizations of the soils that supported the earth levee and sheet piling in the vicinity of the 17th Street canal breach. The processes used at the time of design to analyze the soil types and engineering characteristics did not capture the unique characteristics of the soils. Higher soil strengths beneath the crest of the levee were used to characterize the strengths of the soils at and beyond the toes of the levees. In addition, the spatial averaging process (vertical and lateral) did not capture the unique soil characteristics in the vicinity. Soils in Southern Louisiana and other parts of the Gulf Coast have very complex histories due to past floods, hurricanes, the rise and fall of sea level, changes in vegetation, and other events. Far from being uniform in properties or geometry, they contained complicated and rapidly varying strata of different materials with very different characteristics.

A traditional design guideline-based method of planes with a prescribed geometry was used to model the failure surfaces (U.S. Army Corps of Engineers 1990, 2000, 2003). The shear resistance along these surfaces was based on averaged (laterally and vertically) soil shear strengths for soil units that did not represent the same depositional environments. The geometry of the soil units was assumed to be horizontal. The combination of these design guidelines and practices were used to evaluate the stability of the levee-floodwall system.

In 1964 - 1965 the Corps ran a full scale levee test in the Atachafalaya basin in which advanced studies were conducted regarding characterizations of the soil strengths and performance and stability characteristics of the levee (U.S. Army Corps of Engineers 1968; Kaufman and Weaver 1967). The levee test sections were thoroughly instrumented and their performance monitored during and after construction. Various analytical methods were used to evaluate the usefulness and reliability of the various methods. These developments clearly indicated the need to understand the geologic soil depositional processes and the associated variations in soil strengths (horizontal and vertical) in order to understand the performance and stability characteristics of levees. The importance of local soil conditions to the performance of the levee was clearly pointed out. Additional reports and professional papers were published that resulted in significant advances to the engineering knowledge (Duncan 1970, Ladd et al. 1972; Edgers et al. 1973; Foott and Ladd 1973, 1977).

In-depth background on the geologic and depositional environment of vital importance to understanding the characteristics of the Mississippi Basin soils were developed in the 1950s and 1960s (Kolb and Van Lopek 1958; Krinitzsky and Smith 1969) and the Corps of Engineers lead in development of this background. Of particular importance was recognition that the marsh and swamp deposits were "treacherous" and highly variable. It was repeatedly pointed out that "careful and detailed characterization of the soil properties was required." Further the studies that the method based on traditional Corps of Engineers soil characterization and stability analyses gave factors of safety that were too large (Foott and Ladd 1977). As in the first instance, these developments were not reflected in the design guidelines and practices that were used. Again, the traditional methods used in design and engineering the levee-floodwall system did not possess the required validity and reliability.

Important failure modes in the 17th Street canal levee-floodwall system components were not recognized. The combination of methods used to perform the design was neither valid nor reliable. When the system was tested, it failed.

Conclusion

Engineering must be based on valid and reliable methods and processes in order to yield its intended results. This is true for forensic engineering performed for the purpose of resolving disputes, and for engineering for the design and construction of the built environment. Lacking validity and reliability, engineering testimony may be inadmissible, and so would not serve to support the resolution of a dispute. Lacking validity and reliability, an engineering method or process intended to protect or enhance people's lives and livelihoods can fail to do so, possibly resulting in losses on a catastrophic scale.

References

Campbell, Donald T.; Stanley, Julian C., Experimental and Quasi-Experimental Design for Research, Houghton Mifflin Company, Boston, 1963.

Carmines, Edward G.; Zeller, Richard A., Reliability and Validity Assessment, Sage University Paper Series on Quantitative Applications in the Social Sciences, John L. Sullivan, Ed., Series No. 07-017, Newbury Park, California, 1979.

Daubert v. Merrill Dow Pharmaceuticals, Inc., 509 U. S. 579, 113 S. Ct. 2786, 125 L. Ed. 2d 469 1993.

Duncan, J.M. (1970). *Strength and stress-strain behavior of Atchafalaya foundation soils.* Research Report TE70-1, Department of Civil Engineering, University of California, Berkeley.

Edgers, L. et al (1973). *Undrained creep of Atchafalaya levee foundation clays.* Research Report R73-16, Soils Publication No. 319, Dept. of Civil engineering, Massachusetts Institute of Technology, Cambridge MA.

Federal Rules of Evidence, U. S. Government Printing Office, Washington, DC, 2001.

Fink, Arlene, Editor, The Complete Survey Research Kit, Volumes 1-9, Sage Publications, Inc., Thousand Oaks, California, 1995.

Foott, R., and Ladd, C. C. (1973). *The behavior of Atchafalaya test embankments during construction.* Research report R73-27, Dept. of Civil Engineering, Massachusetts Institute of Technology, Cambridge MA.

Foott, R., and Ladd, C. C. (1977). "Behaviour of Atchafalaya levees during construction." *Geotechnique*, 27(2), 137-160.

Harr, M. E. (1987). *Reliability-Based Design in Civil Engineering*, McGraw-Hill, New York.

Independent Levee Investigation Team (2006). *Investigation of the Performance of the New Orleans Flood Protection Systems in Hurricane Katrina on August 29, 2005*, University of California Berkeley, Report No. UCB/CCRM-06/01, May

Interagency Performance Evaluation Task Force (2006). *Performance Evaluation of the New Orleans and Southeast Louisiana Hurricane Protection System,* Final Draft Report, U. S. Army Corps of Engineers, Washington, DC, May.

Kardon, Joshua B.; Schroeder, Robert A.; Ferrari, Albert J., "Ethical Dilemmas of Technical Forensic Practice," 3rd Forensic Congress, Technical Council on Forensic Engineering, American Society of Civil Engineers, San Diego, California, October, 2003.

Kaufman, R. I., and Weaver, F. J. (1967). "Stability of Atchafalaya levees." *J. Soil Mechanics and Foundations Division*, 93(4), 157-176.

Kolb, C.R., and Van Lopik, J.R. (1958). *Geology of the Mississippi River Deltaic Plain, Southern Louisiana.* Technical Report No. 3-483, U.S. Army Engineer Waterways Experiment Station, Vicksburg, MS.

Krinitzsky, E. L., and Smith, F. L (1969). *Geology of backswamp deposits in the Atchafalaya basin, Louisiana.* Technical Report S-69-8, U.S. Army Engineer Waterways Experiment Station, Vicksburg, MS.

Ladd, C. C. et al. (1972). *Engineering properties of soft foundation clays at two south Louisiana levee sites*. Research Report R72-26, Dept. of Civil Engineering, Massachusetts Institute of Technology, Cambridge MA.

Marshall, R. (2006). "Floodwall failure was foreseen, team says." *Times Picayune*, New Orleans, LA, 03/14/2006, <http://www.nola.com> (May 1, 2006).

Oner, et al (1997). "Soil-Structure Interaction Effects in Floodwalls." *Electronic Journal of Geotechnical Engineering*, <http://www.ejge.com/1997> (Jan,1, 2006).

Petroski, H. (1985). *To Engineer is Human: The Role of Failure in Successful Design*, St. Martins Press, New York, NY.

Petroski, H. (1994). *Design Paradigms, Case Histories of Error and Judgment in Engineering*, Cambridge University Press, Cambridge, UK.

Rossi, Peter H.; Freeman. Howard E., Evaluation: A Systematic Approach, Sage Publications, Inc., Newbury Park, California, 1979.

Seed, R. B., and Bea, R. G. (2006). *Initial Comments on Interim (70%) IPET Study Report*, National Science Foundation-Sponsored Independent Levee Investigation Team (ILIT), University of California, Berkeley, Mar. 12, 2006.

U.S. Army Corps of Engineers (1988a). *E-99 Sheet Pile Wall Field Load Test Report.* Technical Report No. 1, U.S. Army Engineer Division, Lower Mississippi Valley, Vicksburg, MS.

U.S. Army Corps of Engineers, New Orleans District (1968). *Field tests of levee construction, test sections I, II, and III, EABPL, Atchafalaya Basin Floodway, Louisiana.* Interim Report, New Orleans, LA.

U.S. Army Corps of Engineers (1988b). *Lake Pontchartrain, LA., and Vicinity Lake Pontchartrain High Level Plan, Design Memorandum No. 19 Orleans Avenue Outfall Canal*. Three Volumes, New Orleans District, New Orleans, LA.

U.S. Army Corps of Engineers (1989). *Development of Finite-Element-Based Design Procedure for Sheet-Pile Wall.* U.S.Army Corps of Engineers Waterways Experiment Station, Tech. Report GL-89-14, Vicksburg, MS.

U.S. Army Corps of Engineers (1990). *Lake Pontchartrain, LA., and Vicinity Lake Pontchartrain High Level Plan, Design Memorandum No. 20, General Design, Orleans Parish, Jefferson Paris, 17th. Outfall Canal (Metairie Relief)*. Two Volumes, New Orleans District, New Orleans, LA.

U.S. Army Corps of Engineers (1994). *Design of Sheet Pile Walls*. Engineer Manual EM 1110-2-2504, <http://www.usace.army.mil/inet/usace-docs> (Apr. 1, 2006).

U.S. Army Corps of Engineers (2000). *Design and Construction of Levees*. Engineer Manual EM 1110-2-1913, <http://www.usace.army.mil/inet/usace-docs> (Apr. 1, 2006).

U.S. Army Corps of Engineers (2003). *Slope Stability*. Engineer Manual EM 1110-2-1902, <http://www.usace.army.mil/inet/usace-docs> (Apr. 1, 2006).

Wenk, E., Jr. (1989). *Tradeoffs: Imperatives of Choice in a High Tech World*, The Johns Hopkins University Press, Baltimore, MD

Learning From the Past Experiences of Practicing Engineers

F. N. Rad[1] and A. M. James[2]

[1]Department of Civil and Environmental Engineering, Portland State University, Portland, Oregon 97207-0751; PH (503) 725-4205; FAX (503) 725-5950; email: franz@cecs.pdx.edu
[2]425 SW Stark Street, Second Floor, Portland, OR 97204; PH (503) 445-8694; FAX (503) 273-5696; email: art.james@nishkiandean.com

Abstract

The authors initiated a course in Forensic Structural Engineering at Portland State University two years ago. The main goal of the course is to learn from the past experiences of practicing engineers, thus leading the students to a more critical, creative, and cautious thinking process. The course introduces the students to the basic principles and approaches of forensic engineering, along with studying several case histories. A portion of each case study is devoted to outlining the ways to help minimize potential failures of similar nature, as the students embark on their journey to professional practice. The authors have observed that teaching students by examining cases presented by the forensic engineers who actually investigated the cases, is an effective way to accomplish the goals established for this course. This paper describes the authors' experience conducting the course in the past two years.

Introduction

The course in Forensic Structural Engineering is designed to introduce the students to the basic principles of forensic engineering by utilizing actual case histories. The prerequisite for the course is senior or graduate standing in Civil Engineering and knowledge of concrete, steel, and timber.

The objectives of the course are to teach the students about forensic engineering by a methodical study of several case histories, and to help them better understand ways to prevent similar failures from recurring, thus leading to improved design and construction.

The format of the course is to first present what happened in a case that involved structural failure and the evidence of failure or non-performance. The students are then asked to consider issues that could have contributed to this failure. They are encouraged to think freely, and come up with a variety of possible causes of failure. The report from the structural engineer who investigated the failure is then reviewed.

Utilizing this format of teaching has helped the students learn to think more freely and creatively as the cases are discussed, analyzed, and the structural engineers' observations are presented.

Reference Book

The reference book used for the course is: "Locomotive in The River and Other Stories", by Art James. Mr. James is also the main speaker who presented the case studies to the class. The "book" includes 22 cases of forensic investigations conducted by the author over a span of fifty years. In order to give the reader of this paper an idea on the types of failure covered in the book and discussed in class, the cases have been cataloged according to the nature of the failure, summarized below, along with a brief description of each case.

1. Roof Failures

Link Beams in Las Vegas - A roof collapses with no known cause. No known live loads. What happened and why?

Walla Walla Cold Storage Roof Collapse - Nail laminated wartime trusses fail under ice loads on the lower chords.

Plugged Roof Drains - Several different failures that resulted from the water load caused by a plugged roof drain.

Warehouse Roof Collapses Shortly After Construction - Six inches of snow fell and five roof trusses failed at Warehouse; the effect of joint eccentricity in a Howestring truss. Why load tests need to be made safely.

Cracked Roof Slabs at the Airbase - The Colonel was insistent. "Get out here and look at your cracked roof slabs". It was a roofing failure and an overload from gravel.

Bowstring Roof Trusses Collapse Under Heavy Snow Load – The collapse destroyed a lot of pleasure boats; the insurance paid, then the law suits started.

2. Problems with Columns

Chiller Tank Columns - A Puzzler, the rebar were bulged out like they failed in compression, but that was deceptive.

What Is A Squaring Post and Why Did It Fail? - A major roof support is knocked off its pedestal. What caused this near collapse?

Removing a Main Column in a Portland Hotel – The engineers were very careful and did not expect what happened when their transit tilted.

3. Failures due to Storms, High Tides, Wind, and Strong Currents

The Old Salt, the Shark, and the Loose Barges – Case investigates major damage caused when a loaded rock barge and a fuel barge break their mooring lines and ride a big ebb tide into a dock in Astoria, Oregon.

What a Big Wind Did To Sally's School - Oregon's Famous October Storm did major damage to many structures, including the author's daughter's grade school. This leads to an effort to get the state legislature to pass a law requiring wind and seismic design for schools. Sounds easy, but it was not!

Battering Ram and the Freezer Dock Collapse - A heavy log driven by strong current knocks out a post and causes a domino effect failure.

Landslide Behind House on Montgomery Drive - Slides came up to the foundation, and the insurance company was prepared to call it a total loss!

Freeze Tunnel Collapses at Modesto, California - The tunnel imploded, the author was asked to investigate the cause, and ways to fix it.

Church Steeples- Be Careful What You Pull!- The author's curiosity causes a thrill ride during his inspection at the top of a church steeple.

4. Failures of Walls

The Big Jose' - The biggest retaining wall west of the Mississippi falls five days after backfilling.

My Retaining Wall Failed, Will Your Design Last? - The owner said "I'll sue you for everything you ever earn if your wall fails!"

5. Failures due to Construction-Related Causes

Down Pipe at the Teton Dam - Author investigates a fatal construction accident.

Why Did The Boom Collapse? - The operator was lifting a marine leg out of the hold of the barge when the boom collapsed. Why?

M V PacKing Deck Crane Collapse - A log loading deck crane toppled on a new vessel with no overload. Why?

The Wilsonville Bridge Cofferdam Failures - The cofferdams failed during high water and the State bridge engineer wanted extra seal concrete and larger cells in the rebuild. The contractor had a novel approach. Would it work?

Class Schedule

The course included ten sessions carried out in ten weeks, with each session running for two hours. Four sessions contained cases that dealt with similar topics and presented as a group by Art James, as described above. In another four class sessions, cases were presented by guest speakers. The class schedule thus took the following form.

Session No. 1	Introduction to Forensic Engineering, course syllabus and format; speaker: Franz Rad, PE, SE. An example of forensic investigation: Locomotive in the River - What happened to put the switch engine in 30 feet of water? Presented by Art James, PE, SE
Session No. 2	Roof Failures; speaker: Art James, PE, SE
Session No. 3	Failures of wood structures; speaker: Don Neal, PE, SE
Session No. 4	Problems with columns; speaker: Art James, PE, SE
Session No. 5	Four cases: Salem K-Mart roof failure, Portland International Airport parking structure collapse during construction, Clark County Square Dance Center, KGW-TV transmission tower collapse; speaker: Jack Talbott, PE, SE
Session No. 6	Failures due to storms, high tides, wind, and strong currents; speaker: Art James, PE, SE
Session No. 7	Two cases of failure due to foundation settlement; speaker: Gary Peterson, PE, SE
Session No. 8	Failures of Walls, and failures due to Construction-Related Causes; speaker: Art James, PE, SE
Session No. 9	Four cases: Church building roof failure, concrete slab excessive cracking, failure of a Bowstring truss, concrete tilt-up panel connection; speaker: Ray Miller, PE, SE
Session No. 10	Case studies prepared by students; speakers: student groups. The last class session is allocated to student presentations.

In session No. 1, part of the lecture relates to the principles of forensic engineering, course format, and a discussion of a series of books and reports brought to class as examples of references. The purpose of showing the reference materials to the students is to introduce them to the wealth of literature available in the library, and to allow students to "check out" one or more books from the instructor. Also, students are encouraged to consider selecting some of the cases described in the reference books as their cases to be fully described and presented at the last class session. A sampling of the reference books is shown below.

Forensic Structural Engineering Handbook, by Robert Ratay
Construction Failures, by Jacob Feld and Kenneth Carper
Building Failures, by Thomas McKaig
Why Buildings Stand up, by Mario Salvadori
Failure Mechanisms in Building Construction, Edited by David Nicastro
To Engineer is Human, by Henry Petroski
Dsign Paradigms, by Henry Petroski
Failures in Civil Engineering: Structural, Foundation and Geoenvironmental Case Studies, edited by Robin Sheperd and David Frost
Structural and Foundation Failures, by Barry LePatner and Sidney Johnson
Lessons Learned Over Time, Learning from Earthquake Series, Volumes I, II, III, and IV, published by EERI
Proceedings of ASCE 1st, 2nd, and 3rd Forensic Congress

General Guidelines for the Speakers

The guest speakers are provided with guidelines for their presentation. The text of the guidelines is shown below.

"Please consider presenting three or four cases in each session; 15 min for presentation, plus 15 min Q/A for each case. A one-or two-page summary of the facts about each case (pre-lecture material), including sketches or photos, should be given to students a week before your session. The 'Summary of Facts' will not contain your conclusions as to what caused the failure. The students will think about the items that may have contributed to the failure before hearing your lecture. When you present the seminar, you will describe the way you went about finding what caused the failure, including your detective work, calculations, references, etc. The students will write a report on each case and submit for grade."

The Class Room

The class room used for this course is a "distance learning lab" that contains several AV equipment. The lab contains equipment and the means to show on a large screen: 35-mm sides, transparencies, hard copy pages, photographs, digital images, videos, movies, and the image on the computer screen. The authors have found that these capabilities in the class room have improved instructional effectiveness.

The Teaching Process, Class Format, and Weekly Reports

As a part of the teaching methodology employed by the authors, for each case study the students are asked to first review the "pre-lecture" materials, consider what happened when the structure failed and how many potential weaknesses could have contributed to this failure; to study the structural engineer's verbal and written reports and to determine what could be learned in order to prevent similar failures in the future. The students are directed to follow the format below when reviewing the cases to be presented, and preparing for discussions. The text of the instructions given to the students follows.

"Before coming to class, study the assigned cases (pre-lecture materials) to learn what happened. In class, listen and take notes as the Forensic Engineer describes the case. Think freely as to the potential cause(s) for the failure. Think about how you may go about the process of finding the facts about this failure and potential causes. Are perceived and/or expressed facts really facts, or half-truths, or fiction? How would you find corroborating 'perceived facts' and/or evidence to point you to discovering the most probable cause?"

The course requirements include a report to be submitted by each student on each case study. One half of the final student's grade is based on his/her reports on the case studies. The student reports on the case studies follow a specific format. The text of the instructions to the students is shown below.

"For each case, describe what happened, the location, date, and other pertinent information. Describe what, in your judgment, may have either caused or contributed significantly to this failure. Briefly describe the reasons for the failure, as observed by the Forensic Engineer and reported to you verbally and in writing. Describe what you learned from this failure. Describe how you may use the lessons learned in your future design practice. Describe any other ideas and/or information, as related to other actual failures or potential failures that may be prevented by utilizing the lessons learned."

Examples of Case Studies

The reference book contained 22 case studies, investigated by Art James, and documented in the book. About 16 other cases are presented by four guest speakers. Three examples are presented below, to familiarize the reader with the types of cases presented in class.

The Old Salt, the Shark and the Loose Barges - The barge company had a minimal view of the under-dock damage. We felt it was more serious and the international consultants from Vancouver B.C. favored our view. The energy calculation and use of a table to "bracket" the exact solution provides a valuable tool. Comparing the impact energy to a major earthquake is useful.

Locomotive in the River - The longshoreman felt the trestle collapsed first. He said he had reversed and is heading inshore. We felt the engine hit the wheel stop first and the impact sheared the bolts in the track splices then the tracks pulled the trestle down. The underwater photo showed the throttle toward the river proving which way the train was headed.

The Battering Ram and the Cold Storage Dock Collapse - Two insurers contest the issue of what failed first, and why? The dispute goes to trial. Did the dock fail from an overload? Or did a heavy moving object dislodge a post and cause a "domino effect"? The case involved a court contest with "big bucks" at stake and a structural engineering solution.

Lessons Learned

At the end of each case study, a segment of the oral and written presentations are devoted to the lessons that can be learned. Lessons learned normally refers to a series of lessons and observations useful in the future practice of young engineers. For example, in the three cases summarized above, i.e., "The Old Salt...", "Locomotive...", and "The Battering Ram..." the students learn about the application of the energy method ($\frac{1}{2} mv^2 = Fd$, where m = mass, v = velocity, F = force, and d = distance) in finding the proximity of the solution for cases that involve impact of moving objects. By assuming a range of reasonable values for the parameters, it is possible to "bracket" the solution.

The other aspects of "lessons learned" refer to the processes used for:

- Gathering evidence
- Assessing or estimating loads
- Estimating the actual properties of materials, rather than those assumed in design
- Field testing
- Structural analysis, the assumptions, and the validity of assumptions
- Assessing different views of opposing sides on the probable cause of failure

Number of Cases Presented

As mentioned earlier, there are 22 cases in the reference book, plus usually four cases that are presented by each of the four guest speakers, for a total of about 38 cases covered in nine sessions. On an average, four cases are covered in each session, making the presentation time (including Q/A) for each case to be approximately 30 minutes. With the class format adopted, and the students having a chance to read about the cases in advance (pre-lecture materials), the presentation time allowed appeared sufficient. Of course, more complex cases take a bit more time.

Another instructional approach may be to cover a lower number of cases, but allow more time to present and discuss each case. For the available number of hours for this course, which is nine sessions of two-hour length, another possibility is to cover, say only 18 cases, with one hour presentation time per case. The authors have not experimented with this format, as of yet.

Speakers' Commentary to the Students

In addition to the specific cases presented, the salient points addressed and general advice given by the speakers to the students relate to the following items.

- Structural engineers learn from past failures, and the case studies covered in the course provide examples of valuable lessons from structural failures.

- Teaching by example exposes the students to the power of deductive reasoning.

- When you examine a failure, you must identify all possible causes. Then reason your way through them and discard the ones that defy logic.

- The possible causes must be considered and analyzed with approximate computations to gauge the probability of the causes. The ultimate answer sometimes is not the early favorite. Do not close your mind too soon!

- Make sketches of the critical parts, take photos and measurements and avoid early conclusions, before all the data are known.

- Avoid statements like "it might be this or that cause". Your duty is to eliminate the alternates and determine the "proximate cause".

- Sometimes when there are several equal possibilities it may be necessary to say: I can not say for certain, but here are the most likely causes of failure.

- Why does forensic engineering fascinate us? Because we are using a combination of our engineering knowledge and deductive reasoning, and knowledge is power!

Students' Final Projects

The last class session is allocated to student final projects and presentations. Four teams of students are formed early in the term. Each team selects one or more cases to investigate, write a report, and make a presentation to the class.

Below is a summary of the cases presented in the past two terms the course was offered.

Case 1, Building a cantilevered floor for a residence/winery. This is a local project, the student is the structural engineer on the project, and it involved a cantilevered beam with inadequate inboard length. The load on the cantilevered portion raised the inboard end excessively. The student described how he went about retrofitting the structure to minimize the uplift.

Case 2, The Britannia tubular bridge, a paradigm of tunnel vision in design. This is the story of the bridge designed by Robert Stephenson, who faced the challenge of building a bridge rigid and strong enough to carry a heavy train of many carriages. This is done by making the bridge out of two long iron tubes, rectangular in shape, through which the trains would travel. The bridge opened in 1850, with many problems that followed in the following 150 years.

Case 3, Roof collapse at the shoe store. This is a local case, the building constructed with timber roof joists and partially grouted CMU walls. The roof collapsed in winter of 2004 due to plugged roof drain, ice accumulation, and a heavy HVAC unit. The students described their detective work in finding the cause, and the structural repairs.

Case 4, Sagging roof at a commercial warehouse. This is a local building, built in the 1930s, made of timber roof trusses with bolted and nailed connections. The problem was a sagging roof, what caused the sag, and how it was repaired.

Case 5, The failing of Fallingwater House. This is one of Frank Lloyd Wright's most famous houses, located in southwestern Pennsylvania. The case describes the original design, and subsequent problems with cracking and leaking, and restoration of the building completed in 2002 at a cost of $11 million.

Case 6, The Good, the Bad, and the Galloping Gertie. The famous Tacoma-Narrows bridge failure is revisited, the history of design, details of design, and theories of failure.

Case 7, Collapse of parking structure, a case included in the Proceedings of the Second Forensic Congress. The students' report included a discussion of expansion/isolation joints, and the collapse of a parking structure. The structure included two sliding isolation joints which did not allow the required movement. The lessons learned that may be applied to other similar cases were presented.

Case 8, Kansas City Hyatt Regency Hotel walkways collapse of 1981. This notorious case is re-visited, describing what happened and why.

Case 9, Roof Investigations, January 2004 Portland Snow Storm. This project investigates the effects of heavy snow storm on the roofs of several buildings, with emphasis on snow drift. The students make observations on the need to include drift in snow load calculations, and the potential structural problems associated with snow drift.

Assessment and Future Plans

The written course assessment includes the following questions, assessing the level of achievement by students, as perceived by them.

1. Acquire knowledge about what forensic engineers do.
2. Be able to do (or help with) forensic investigative work.
3. Be able to identify potential "pitfalls" in design and construction.
4. Be more cautious about my own (and other people's) assumptions regarding analysis, properties of materials, construction quality, and inspection.
5. Be a better inspector.
6. Design ways to strengthen structures.
7. Become a better engineer and to minimize potential failures in structures that include my services.
8. Learn about topics that are not commonly addressed in other courses.

Most of the students have been graduate students, with some from the practicing sector. Course assessment indicates that the course format seems to be generally working well. As for the quantitative assessment of the course, with a rating of 1 for "poor" and 5 for "excellent", the course rating is about 4.5.

The speakers and the students seem to enjoy their interaction and sharing of knowledge. The students are especially fond of the course because it is "different" from the other courses; it provides a wider scope of information, and it teaches them about numerous situations where similar types of failure can be avoided in their practice.

The current course is a 2-credit course, and our plan is to expand it to 3 credits. Moreover, we have considered assembling a compendium of forensic cases, as the practicing engineers continue to contribute to the course. Most of the cases covered in the compendium will be local cases, involving local buildings, engineers and constructors.

Observations

The authors believe that the value of a course in Forensic Engineering as part of a students' lifetime learning of Structural Engineering is substantial. By showing the students real life examples of failures and the investigations that led to conclusions, the students are enabled in several ways, including an illustration of the tug of war between opposing points of view. The authors further believe that a significant part of educating engineering students in promoting design and construction integration is by understanding the causes of structural failures. As an added benefit, the course in Forensic Structural Engineering has brought about a closer working relationship between academia (students and faculty) and design professionals, the engineers who are willing to share their experiences in forensics with students. Course assessments by students as well as speakers indicate that this course is beneficial in rounding out the students' education in structural engineering.

Benchmarking Forensic Engineering Practice – A Philosophical Discussion

S.E. Chen[1], D. Young[2], D. Weggel[3], D. Boyajian[4], J. Gergely[5] and B. Anderson[6]

[1]Dept. of Civil Engineering, University of North Carolina at Charlotte, 9201 University City Boulevard, Charlotte, NC 28223-000; PH (704)687-6655; FAX (704)687-6953;email:schen12@uncc.edu
[2]Dept. of Civil Engineering, University of North Carolina at Charlotte, 9201 University City Boulevard, Charlotte, NC 28223-000; PH (704)687-4175; FAX (704)687-6953;email:dyoung@uncc.edu
[3]Dept. of Civil Engineering, University of North Carolina at Charlotte, 9201 University City Boulevard, Charlotte, NC 28223-000; PH (704)687-6189; FAX (704)687-6953;email:dweggel@uncc.edu
[4]Dept. of Civil Engineering, University of North Carolina at Charlotte, 9201 University City Boulevard, Charlotte, NC 28223-000; PH (704)687-3038; FAX (704)687-6953;email:dboyajia@uncc.edu
[5]Dept. of Civil Engineering, University of North Carolina at Charlotte, 9201 University City Boulevard, Charlotte, NC 28223-000; PH (704)687-4166; FAX (704)687-6953;email:ggergely@uncc.edu
[6]Dept. of Civil Engineering, University of North Carolina at Charlotte, 9201 University City Boulevard, Charlotte, NC 28223-000; PH (704)687-6039; FAX (704)687-6953;email:jbanders@uncc.edu

Abstract

Recent series of unfortunate events resulted in thousands of damaged/deteriorated structures. To validate insurance claims and rehabilitation efforts and to assist disaster-worn citizens, a significant amount of forensic engineering work is currently on-going and will continue for a long time. In the midst of these activities, a significant amount of disputes/mis-judgements will occur and will cause further difficulties in settling claims and restoring normal operations. To ensure quality of forensic work, this paper attempts to address the more fundamental issue of current forensic science and engineering practices, which adopts an inverse engineering approach whereupon knowledge is accumulated from construction design and post-event observations. The reasoning process is based on pure deduction with very little in-between causality evidence. Current approach relies heavily on an engineer's interdisciplinary expertise, training, and reasoning ability, and lacks the fundamental scientific process of elimination of possibilities. It, therefore, often fails to produce complete multidisciplinary solutions to complex forensics problems. This paper attempts to establish quality quantification by suggesting forensic benchmarking such that the involved procedures can be standardized and eliminate the "guess work" still common in an otherwise rapidly developing and highly challenging field.

Introduction

Recent catastrophic events (2005 Hurricane Katrina, the 2005 Asian tsunami, 1990-2005 earthquakes and terrorist activities since 9/11) in the United States and around the world have resulted in millions of damaged or destroyed homes and structures. Hurricane Katrina alone destroyed more than 160,000 homes and generated 90 million tons of solid waste (Esworthy et al. 2005 and The White House, 2006). To assist disaster-worn citizens and other involved parties, forensic engineers are currently working to validate insurance claims and rehabilitation efforts, and this work may continue for a long time. Disputes result in difficulties in settling the claims, thereby delaying a return to everyday lives. To reduce lawsuits and pain and suffering for disaster victims, there is a need for more timely and reliable forensic practices. The ultimate goal for all forensic investigations should be to provide quality results that can substantiate a logical and valid conclusion to the case. Hence, the ability of forensic engineers to quantify forensic work qualities will be an important service to the clientele and society at large. This ability is also critical for future enhancement of engineering performance, including the increase in reliability and the reduction in forensic practice liability.

Forensic engineering is a highly versatile profession. It differs from other engineering and scientific practices in that there is very little obvious evidence available for establishing valid causality of failures. As a result, reliance on conclusions from past case studies becomes highly essential. A review of past publications in the ASCE *Journal of Performance of Constructed Facilities* shows majority papers are related to historical data or case studies. Literature published on forensic engineering pedagogy also tends to focus on the teaching of case studies (Bosela 1993, Rendon-Herrero 1993, Fowler et al. 1994, Pietroforte 1998, Rens et al. 2000 and Delatte et al. 2002). However, since failures in civil engineering studies are either rare or rarely reported, to establish a quantitative measure of failure probabilities is difficult.

To establish a reasonable measure of forensic investigation quality, a total shift of paradigm may be required, including the establishment of forensic benchmarking for generation of statistically sound causality samples. This approach would require significant experimentation efforts and instrumentation needs. With valid benchmarking, a probabilistic causality quantifier can then be established. This paper represents a first attempt in establishing forensic quality measures. It is the intent of the authors to open a dialog amongst forensic engineering professionals to determine ways to improve and moderate forensic investigations.

Rationale behind Current Forensic Practices

Forensic engineering is "the application of engineering in the jurisprudence system requiring services of legally qualified professional engineers. Forensic engineering include investigation of physical causes of accidents and other sources of claims and litigation, preparation of engineering reports, testimony at hearings in judicial proceedings, and rendition of advisory opinions to assist the resolution of disputes" (National Academy of Forensic Engineering,2005). Important elements for forensic work are professionalism, legal knowledge, and the capability to provide expert solutions to the judicial proceedings (Lewis 2003).

Forensic practice is a "fact finding mission" that provides the legal process with a doubt-free explanation of the causality for structural failures (Janney 1979). Current forensic engineering adopts an inverse approach, where knowledge is accumulated from construction design and post-event observations, instead of typical scientific approach of generating statistically reliable data to validate causal hypotheses. Current reasoning is based on pure deductive process with very little experimental support. Additionally, the field lacks "in-between data", thereby relying heavily on an engineer's interdisciplinary expertise, training, and reasoning ability. Therefore, current forensic practices often fail to produce complete multi-disciplinary solutions to complex forensics problems, resulting in unreliable forensic conclusions. The National Society of Professional Engineering has addressed this unreliability issue as a national concern (Lunch 1987). Cohen et al. (1992) indicated several areas of needs for forensic engineering that may able better practices including the collection of historical data, in-situ monitoring techniques, and experimental results.

Since forensic engineering does not rely on significant experimentation - this non-experimental approach (known as causal-comparative study, *ex post facto*) is characterized by reliance of collected structural failure observations and established causality based on past experiences (Patten 1997). Such an approach requires significant experience in deductive reasoning and complete compatibility of past and present data, including structural components, boundary conditions and failure observations. Since failures have already occurred, the causal relationship is typically established with a high level of speculation. Santamarina and Chameau (1989) showed that limitations in human cognitive capability with weak or limited evidences can induce bias.

Causality Quantification

Essential to establish reliable forensic studies is the promotion of deductive reasoning, hypotheses validation and false hypotheses elimination via reasonably-accurate experiments and modeling. *For forensic investigations, this approach means an accurate representation of the entire system including its components and boundaries.* Using a systems approach, Yao (1985) first suggested damage indices to quantify state conditions. Castaneda and Brown (1994) and Kaggwa (2005) suggested using fuzzy logic for causality investigations. However, both approaches required the causality relationships for a certain failure type are well defined from adequate failure cases. While this may not be true for structural components, failures of complex structures are definitely rare.

Assuming adequate data or case studies can be collected, test data from different measurements or observations can then be fused and integrated to explain causality. This process, though initially difficult and time-consuming, will gradually accelerate with accumulated knowledge. The results can be treated as random processes and probability quantifiers can be used to estimate causality likelihood. The probability of identifying causality can be defined as:

$$P(a_i) = P(A_{mi} > 0 | A = a_i) \quad (1)$$

where a_i is damage outcome due to specific causal relation i . A_{mi} is the measured outcome and A is the actual outcome. If a delta function is specified as

$$\delta(k_j) = \begin{cases} 1 \sim \text{when causality exists} \\ 0 \sim \text{otherwise} \end{cases} \quad (2)$$

Then probability of each causality, denoted using metric vectors a_i, where weighting coefficients b_i, are actual causalities and k_j defines unrelated causalities with weighing coefficients z_j, can be expressed as (with an error term, ε_m):

$$P(a_i)_m = \sum_{i=1}^{p} b_i a_{mi} + \sum_{j=1}^{q} Z_j \delta_m(k_j) + \varepsilon_m \tag{3}$$

To ensure positive causality, the square sum residuals, S are utilized

$$S = \sum_{m=1}^{n} \left[P(a_i)_m - \left\{ \sum_{i=1}^{p} b_i a_{mi} + \sum_{j=1}^{q} Z_j \delta_m k_j \right\} \right]^2 \tag{4}$$

The most likely causality can be established by optimization procedures:

$$\frac{\partial S}{\partial b_i} = 0 \tag{5}$$

$$\frac{\partial S}{\partial Z_j} = 0 \tag{6}$$

A set of simultaneous equations can then be assembled and decoupled into causal and non-causal variables. When adequate data have been collected, reliability of the causalities can be determined.

Adequate data sample is essential for all scientific research processes; this is because proper deductive reasoning and scientific research requires sufficient evidence to prove that a causal relationship is accurate. For most forensic engineering work, such as manufacturing and industrial engineering, a significant amount of data may be retrieved to establish the causes of failure. The investigation may start with problem definition, proceed through analysis and modeling, and conclude with testing and simulation of failure conditions. However, such databases usually do not exist for civil engineering forensic work. Databases such as the Forensic Anthropology Data Bank, where hundreds of human bodies were allowed to decompose for scientific study at the University of Tennessee at Knoxville, are impossible in civil engineering work if only because of the size of the evidence (The Forensic Anthropology Data Bank, 2005).

Forensic Benchmarking

In the cases, where limited data are available, additional data can be generated using extensive structural modeling and numerical simulation. If standard procedures can be established to ensure consistent data generation, then positive causality can be established for a wide variety of problems - **Forensic Benchmarking Experimentation**. Forensic benchmarking can be used as a forensic investigation procedure by generating extensive causal relationship data. The proper scientific procedures for a true experiment-based investigation typically involve the following steps (Blaxter et al. 2001):

- Observe some aspect of the structural failures;
- Establish a set of structural parameters for reproduction;
- Establish a set of hypotheses that is consistent with what has been observed about the causes of failure.
- Construct structural replicas;

- Use the hypothesis to make predictions.
- Test those predictions by experiments on the structural models ;
- Repeat the above steps until there are no discrepancies between theory and experiment and/or observation.

Such experimentation requires skilled structure model construction and extensive testing/instrumentation setups. Numerical simulation, especially for nonlinear phenomena, also requires extensive detailed modeling of realistic physical parameters. Experimentation-based forensic benchmarking differs from current forensic investigation methodologies in two notable ways:

1. **Unknown variables:** Typical engineering investigation focuses on only one or possibly a few specific variables, where other variables are eliminated or simplified. Forensic benchmarking allows consideration of a wide range of variables.

2. **Multiple processes**: Typical engineering investigations focus on specific damage process, whereas forensic benchmarking targets at capturing damage sequences in order to establish comprehensive explanations of failure mechanisms.

Figure 1 compares current forensic practices and a possible experiment-based forensic investigation. The objective of the experimentation is to generate databases that can be used in statistical inference to generate reliable causal relations. To accomplish such a premise using a structural model, certain design criteria must be established. It should be noted that depending on the application, a perfectly scaled-down model is not always essential; other small-scale relaxed models may be developed (Harris and Sabnis 1999).

Benchmarking Sensing and Nondestructive Testing

To ensure proper forensic benchmarking, it is essential that consistent measurements are made. Today, structural inspection is done mostly through visual assessment; however, the reliability of inspector skills remains an unsolvable issue (Phares et al. 2000). To address the causality issue of a failed structure, it is critical to ensure that collected data accurately capture the entire failure process. Such ambitious monitoring can be achieved only by using several sensors that measure different physical parameters, such as temperature, strain, deformation, ground motion, etc. Recent vigorous research and developments of Nondestructive Testing (NDT) techniques enabled the adoption of these techniques for forensic studies (Chase 1997, Rens et al. 1997, Davidson et al. 1998 and Washer 2000). In particular, nondestructive testing techniques such as X-Ray, Infrared thermography, ultrasound, ground penetration radars, impact-echo, fiber-optic strain sensing, ferromagnetic and geophysical testing methods can provide detailed spatial description of damage and are gaining increasing popularity as common testing methods (Azacedo et al. 1996, Cumming et al. 1997, Gucunski et al. 2000, Fu et al. 1996, Ballard, 1996, Chen et al. 1997, Kalinski 1997, Poston et al. 1997, Rens et al. 1998 and Sansalone et al. 1997).

It should be noted that NDT techniques can be differentiated into local and global methods. Methods such as acoustic emission, ultrasonics, impact echo, x-ray and radar are focused on damage studies on localized defects (Mannings 1985). Techniques, such as dynamic characterization of a structure, allow global damage quantification (Adams et al. 1975). By capturing the vibration signatures of a structure, determination of the deteriorating state of a structure can be assessed (Cawley et al. 1979, Mazurek et al. 1990, Hearn et al. 1991 and Aktan et al. 1997). Maser (1988) described the various physical parameters required during structural assessment to include: 1) inventory data, 2) condition data and 3) performance data. For material failure measurements such as corrosion, the range of interest could be only a fraction of a millimeter (Carper 1989, Day 1999 and Noon 2000).

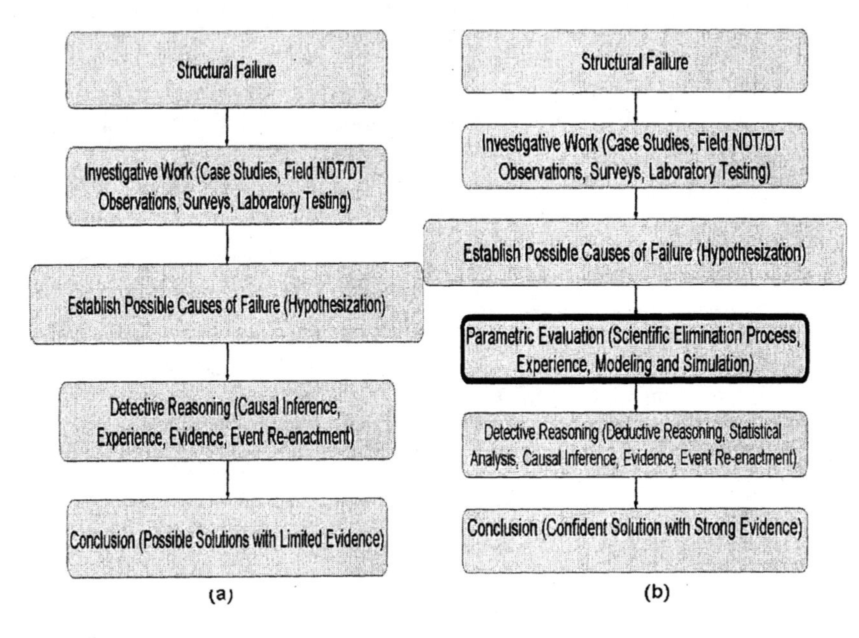

Figure 1 Forensic Investigation Processes: a) Current and b) Experiment-Based

Discussion

Good forensic engineering practices demand a valid measure of the relationship between failure and its causes. A true scientific-based investigation would require the generation of statistically sound sample population of failure cases. While real life data are hard to come by, using structural modeling, forensic benchmark studies can be conducted. To establish a valid structural model or numerical model, certain issues need to be addressed, including:

- Structural dimensions and in-place material verification - for many existing structures, complete as-built plans may not be available; even if plans are

available, they may not be correct or may not reflect errors or changes during construction;
- Damage mapping such as fatigue cracking; and
- Environmental effect determination and delineation from common load effects – some stresses are due to load and some are due to changes in moisture and temperature.

Since most of the above criteria are difficult to establish, forensic benchmarking is extremely difficult to establish for historical structures. Additional reasons why experiment-based studies have not been popular in forensic engineering include:
- Most structural systems are difficult to define (due to changes in surrounding conditions and challenges in scaling and reproducing structural aging and repetitive loading history);
- Few actual structures will allow extensive and repetitive experiments;
- There is currently no comprehensive instrumentation that allows synchronized multiple-level sensing and automated data fusion and self-calibration; and
- Instrumentation for such studies is too expensive for most educational institutions.

Another difficulty that faces a valid forensic benchmarking is that all measured observations should be synchronized: all collected data should be time/location-stamped, such that a chronological sequence may be plotted out. Generally, output data from a sensor may have one of several possible formats:
- Continuous or sampled waveform complete with amplitude, frequency and phase information;
- Two-dimensional imagery with spatial amplitude or spectral data;
- Vector consisting of parametric positional data, state vector, target identity or characteristics; or
- Other useful outputs such as temperature, health index and background noise thresholds (Hall 1992).

Critical to forensic engineering is the quality of information that can be extracted from sensor measurements. Thus sensor reliability has a direct impact on the quality of the forensic work being performed (Joshi et al. 1992 and 1998). Faulty sensors need to be identified to ensure that bad data do not continue to be collected. Currently sensor reliability issues are not commonly considered when data acquisition systems are designed and developed, although the reliability of collected information is highly dependent on the sensors used. In fact, most of the existing schemes used in industry for fault detection are confined to simple strategies for pre-determined small fault-sets.

Example

The causality calculation is demonstrated using an example where a truss bridge (the Haupt bridge) with 21 members is used (Chen et al. 2002). The West Point Bridge Designer (Ressler 2000) is used to simulate the truss bridge behavior under a standard AASHTO H20-44 truckload. To illustrate the effect of individual member stiffness as a random variable for a structure, static analysis was conducted to determine if the bridge will fail in the load test using the program. The span of the bridge is 24 m long. The original design consists of members of 120 mm ×120 mm cross-section, except for members No. 11 and 21, which are 140 mm×140 mm cross-sections. A Monte Carlo simulation is then conducted using randomly generated members and sections. The cross-sections are limited to 10 mm × 10 mm to 140 mm × 140 mm range. The computation of causality quantification is then conducted. Figure 2 shows the WPBD program with the standard truck crossing over the bridge and caused bridge failure. The members under high stresses are shown in red color. A total of 207 data is generated for the example problem. Figure 3 shows the bridge member numbers. Table 1 shows the outcome probabilities for each of the members.

In this example, causality is defined as a function of the bridge members. Without considering the measurement techniques, the weighting coefficient b_i is considered 1 for all damage cases. For the non-causal terms, coefficients Z_j and errors ε_m are considered zero. The maximum S for the problem is at members 6 and 14, indicating the most probable failure cases. However, the closeness of the percentage S values for all members indicating that they have equal likelihood of failure, which is expected for the simple problem. The accuracy of the causality study will improve with more data sets.

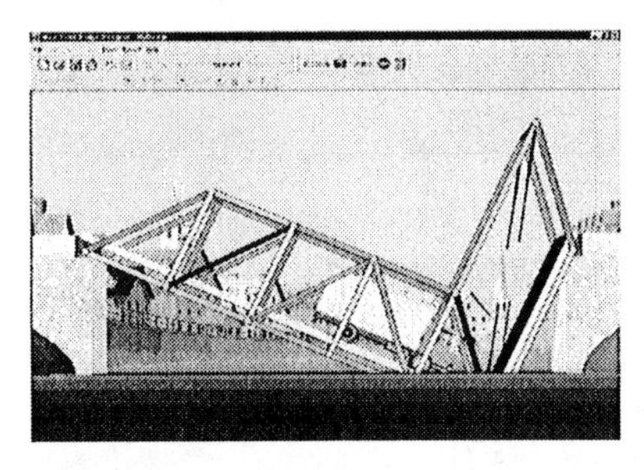

Figure 2 WPBD showing failed bridge during truck crossing

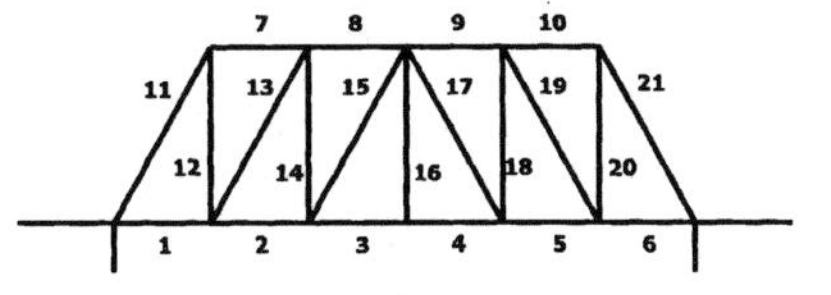

Figure 3 Member numbering system

Conclusion

This paper suggests the possibility of improving forensic causalities by massive generation of case data using structural modeling and forensic benchmark studies. This approach provides the in-between data that can help create direct conclusions of causal relationships between failures and their causes. However, such approach requires significant capabilities in structural modeling as well as extensive instrumentation and nondestructive sensing. To establish such approach may require an industry-wide joint effort and sharing of information.

A simple probabilistic model is suggested in order to identify the most likely causality to a failure, which is illustrated by a simple truss problem. A total of 207 failure cases have been generated to demonstrate that under the same test scenarios, all members should have the same likelihood of failures.

References

Adams, R.D, Walton, D., Flitcroft, J.E.and Short, D. (1975). "Vibration Testing as a Non-Destructive Test Tool for Composite Materials," *Composite Reliability*, ASTM STP 580, 159-175.

Aktan, A.E., Farhey, D.N., Helmicki, A.J., Brown, D.L., Hunt, V.J., Lee, K.L. and Levi, A. (1997) "Structural Identification for Condition Assessment: Experimental Arts." *J. Struct. Engrg.*, 123(12), 1674-1685.

Azacedo, S.G., Mast, J.E., Nelson, S.D., Rosenbury, E.T., Jones, H.E., McEwin, T.E., Mullenhoff, D.J., Hugenburger, R.E., Stever, R.D., Warhus, J.P. and Wieting, M.G. (1996). "HERMES: A high-speed radar imaging system for inspection of bridge decks," *Nondestruc. Eval. of Bridges and Highways*, SPIE # 2946, 195-205.

Ballard, C.M. and Chen, S.S. (1996). "Automated Remote Monitoring of Structural Behavior via the Internet," *Smart Sys. for Bridges, Struct. and Highways*, SPIE # 2719, 90-101.

Blaxter, L., Hughes, C. and Tight, M. (2001). *How to Research*, 2nd Ed., Open University Press, Buckingham, UK.

Bosela, P. (1993). "Failure of Engineered Facilities: Academia Responds to the Challenge," *J. Perform. Constr. Facil.*, 7(2), 140-14.

Carper, K.L. (1989). *Forensic Engineering*, Elsevier Science, New York, N.Y.

Castaneda, D. and Brown, C. (1994). "Methodology for Forensic Investigations of Seismic Damage," *J. Perform. Constr. Facil.*, 120(12), 3506-3524.

Cawley, P. and R.D. Adams (1979). "The location of defects in structures from measurements of natural frequencies," *J. Strain Anal.*, 14(2), 49-57.

Chase, S. and Washer, G. (1997). "Nondestructive Evaluation for Bridge Management in The Next Century," *Public Roads*, 61(1), 16-25.

Chen, S., Pong, W., Chen, P. and Nishihama, Y. (2002) "Teaching Reliability to Civil Engineers," *Proc. Amer. Soc. Engrg. Educ. Southeast Section Annual Conference*, Gainesville, Fl., 2002.

Chen, Z., Cudney, H.H., Giurgiutiu, V., Rogers, C.A., Quattrone, R. and Berman, J. (1997). "Full-scale ferromagnetic active tagging trdting of C-Channel Composite Elements," *Smart Sys. for Bridges, Structures, and Highways,* SPIE # 3043, 169-180.

Cohen, J.M., Corley, W.G., Wong, P.K. and Hanson, J.M. (1992). "Research Needs Related to Forensic Engineering of Constructed Facilities," *J. Perform. Constr. Facil.*, 6(1), 3-9.

Cumming, N.A. and Ooi, O.S. (1997) "Locating Delaminations and Other Defects in Concrete Silo Walls Using the Impact-Echo Procedure," in Pessiki, S., and Olson, L., Ed., *Innovations in Nondestructive Testing of Concrete,* SP-168, ACI, Chicago, IL.

Day, R.W. (1999). *Forensic Geotechnical and Foundation Engineering*, McGraw-Hill, New York.

Davidson, N.C. and Chase, S.B. (1998). "Radar Tomography of Bridge Decks," *Proc. Structural Materials Technology III: An NDT Conference*, SPIE v.3400, San Antonio, TX, NJ, 250-256.

Delatte, N. and Rens, K.L. (2002). "Forensics and Case Studies in Civil Engineering Education: State of the Art," *J. Perform. Constr. Facil.*, 16(3), 98-109.

Esworthy, R., Schierow, L.J., Copeland, C. and Luther, L. (2005). *Cleanup after Hurricane Katrina: Environmental Considerations,* Congressional Research Service Report for Congress, RL33115, Washington DC.

Fowler, D. W., and Delatte, N. J. (1994). "Graduate Course in Forensic Engineering," *Proc. ASCE Texas Section Spring Meeting*, Corpus Christi, Texas.

Fu, X., and Chung, D.D.L. (1996). "Self-Monitoring Concrete," *Smart Sys. for Bridges, Structures and Highways*, SPIE # 2719, 62-68.

Gucunski, N., Vitillo, N. and Maher, A. (2000). "Pavement Condition Monitoring by Seismic Pavement Analyzer (SPA)", *Proc. Structural Materials Technology IV – An NDT Conference*, Feb. 28-Mar. 3, Atlantic City, NJ, 337-342.

Hall, D.L. (1992). *Mathematical techniques in Multisensor Data Fusion*, Artech House, London, UK.

Harris, H.G. and Sabnis, G.M. (1999). *Structural Modeling and Experimental Techniques*, 2nd edition, CRC Press, Boca Raton, FL.

Hearn, G., and Testa, R. B. (1991). "Modal Analysis for Damage Detection in Structures." *J. Struct. Engrg.*, 117 (10), 3042-3063.

Janney, J.R. (1979). *Guide to Investigation of Structural Failures*, ASCE, New York, NY.

Joshi, B. and Seyed H. (1992). "Reliability Analysis of Self-Diagnosable Multiple Processor Systems," *Proc. 23rd Pittsburgh Conf. Model. & Simul.*, Pittsburgh, PA, 1993-1999.

Joshi, B. and Hosseini, S. (1998). "Diagnosis Algorithms for Multiprocessor Systems," *Proc., IEEE Workshop on Embedded Fault-Tolerant Systems*, Boston, MA, 112-116.

Kaggwa, W.S. (2005). "Probability-Based Diagnosis of Defective Geotechnical Engineering Structures," *J. Perform. Constr. Facil.*, 19(4), 308-315.

Kalinski, M.J., (1997) "Nondestructive Characterization of Damaged and Repaired Areas of a Concrete Beam Using the SASW Method," Pessiki, S. and Olson, L., Ed., *Innovations in Nondestructive Testing of Concrete,* SP-168, ACI, Chicago, IL.

Lewis, G.L. (2003). *Guidelines for Forensic Engineering Practices*, ASCE Pub., Reston, VA.

Lunch, M.F. (1987). "Liability Crisis – Where do We Go from Here?" *J. Perform. Constr. Facil.,* 1(1), 30-33.

Manning, D.G. (1985). *Detection of Defects and Deteriorations in Highway Structures*, NCHRP No.118, TRB.

Maser, K.R. (1987). "Sensors for Infrastructure Assessment," *J. Perform. Constr. Facil.*, 2(4), 226-241.

Mazurek, D. F., and DeWolf, J. T. (1990). "Experimental study of bridge monitoring technique." *J. Struct. Engrg.*, 116 (9), 2532-2549.

National Academy of Forensic Engineering (2005). http://www.nafe.org/NafeMainDef.htm>12/25/05.

Noon, R. (2000). *Forensic Engineering Investigation*, CRC Press, Boca Raton, FL.

Patten, M.L. (1997). *Understanding Research Methods – An Overview of the Essentials*, Pyrczak Pub., Los Angeles, CA.

Phares, B.M., Rolander, D.D., Graybeal, B.A., Washer, G.A. and Moore, M. (2000). "Visual Inspection Reliability Study," *Proc. Structural Materials Technology IV – An NDT Conference*, Feb. 28-Mar. 3 Atlantic City, NJ, 14-22.

Pietroforte, R. (1998). "Civil Engineering Education Through Case Studies of Failures," *J. Perform. Constr. Facil.*, 12(2), 51-55.

Poston, R., and Sansalone, M. (1997). "Detecting Cracks in the Beams and Columns of a Post-Tensioned Parking Garage Using the Impact-Echo Method," in Pessiki, S., and Olson, L., Ed. *Innovations in Nondestructive Testing of Concrete,* SP-168, American Concrete Institute, Chicago, IL.

Rendon-Herrero, O. (1993). "Too Many Failures: What can Educators do?" *J. Perform. Constr. Facil.*, 7(2), 133-139.

Rens, K.L., Rendon-Herrero, O. and Clark, M.J. (2000). "Failure of Constructed Facilities in Civil Engineering Curricula," *J. Perform. Constr. Facil.*, 14(1), 27-37.

Rens, K.L. and Transue, D.J. (1998). "Recent Trends in Nondestructive Inspections in State Highway Agencies," *J. Perform. Constr. Facil.,* 12(2), 94-96.

Rens, K.L., Wipf, T.J. and Klaiber, F.W. (1997). "Review of Nondestructive Evaluation Techniques of Civil Infrastructure," *J. Perform. Constr. Facil.*, 11(4), 157-160.
Ressler, S.J. (2000). *West Point Bridge Designer, Dept. Civ. and Mech. Engrg.*, US Military Academy, West Point, NY, v.4.06.
Sansalone, M., and Streett, W. (1997). *Impact-Echo: Nondestructive Evaluation of Concrete and Masonry*, Bullbrier Press, Ithaca, NY.
Santamarina, J.C. and Chameau, J.L. (1989). "Limitations in Decision Making and System Performance," *J. Perform. Constr. Facil.*, 3(2), 78-86.
The White House (2006). *The Federal Response to Hurricane Katrina: Lessons Learned*, The White House, Washington DC.
The Forensic Anthropology Data Bank (2005). http://web.utk.edu/~anthrop/FACdatabank.html, last accessed July 2005.
Washer, G. (2000). "Developing NDE Technologies for Infrastructure Assessment," *Public Roads*, 63(4), 44 – 50.
Yao, J.T.P. (1985). *Safety and Reliability of Existing Structures*, Pitman Advanced Pub., Marshfield, MA.

Table 1 The failure probability for each bridge member for 207 cases

Member number	Occurrences	Failed cases	Causality probability	Square sum residuals S	% ratio to total S
1	11	4	0.364	105.5	4.89%
2	13	7	0.538	102.0	4.73%
3	14	4	0.286	107.1	4.97%
4	10	5	0.5	102.8	4.76%
5	5	1	0.2	108.9	5.05%
6	8	1	0.125	110.5	5.12%
7	5	4	0.8	96.8	4.49%
8	11	8	0.727	98.2	4.55%
9	10	5	0.5	102.8	4.77%
10	7	2	0.286	107.1	4.97%
11	11	10	0.909	94.6	4.39%
12	18	9	0.5	102.8	4.77%
13	10	8	0.8	96.8	4.49%
14	8	1	0.125	110.5	5.12%
15	8	7	0.875	95.3	4.42%
16	10	4	0.4	104.8	4.86%
17	12	5	0.417	104.4	4.84%
18	21	5	0.238	108.1	5.01%
19	7	5	0.714	98.5	4.57%
20	3	1	0.333	106.2	4.92%
21	5	5	1	92.9	4.31%

RETURNING BUILDINGS DAMAGED BY HURRICANES OR EARTHQUAKES TO PRE-DAMAGE CONDITION

Gary C. Hart[1], Anurag Jain[2] and Stephanie A. King[3]

[1] Emeritus Professor, Department of Civil Engineering, University of California, Los Angeles, and Principal, Weidlinger Associates, Inc., 2525 Michigan Avenue, Bergamot Station D2-3, Santa Monica, CA 90404; PH (310) 998-9154; FAX (310) 998-9254; email: hart@wai.com
[2] Associate Principal, Weidlinger Associates, Inc., 2525 Michigan Avenue, Bergamot Station D2-3, Santa Monica, CA 90404; PH (310) 998-9154; FAX (310) 998-9254; email: jain@wai.com
[3] Associate Principal, Weidlinger Associates, Inc., 399 W. El Camino Real, Suite 200, Mountain View, CA 94040; PH (650) 230-0210; FAX (650) 230-0209; email: sking@wai.com

Keywords: loss estimation, earthquake, hurricane, repair, rehabilitation, building damage, Bayesian

Abstract

In this age of high technology, the expert consultant usually must state an opinion on what must be done to a building damaged by an earthquake or hurricane to bring it back to its pre-earthquake or pre-hurricane condition. This paper presents an approach for defining the pre-earthquake or pre-hurricane condition of the building in terms of the estimated expected damage to a natural hazard event. With this definition, it is possible to then determine what is required to return the building to its current post-event condition to that same level of expected damage, i.e., its pre-event condition.

Introduction

This paper addresses a topic that is very important in post-hurricane or post-earthquake building repair: the determination of the dollar cost necessary to bring the building back to its pre-earthquake or pre-hurricane condition. Stated in terms of the common request asked of the structural engineer, it is: "Please recommend the damage repair to the building that is required to restore the building to it pre-hurricane or pre-earthquake condition." This request requires a definition of what is the measure that is to be used to define the pre-earthquake or pre-hurricane condition of the building. It is here that the structural risk and reliability research over the last few decades in calibration of building designs, performance-based design, and damage estimation is very important and of great value. This paper offers an approach for determining what is required to return the building to its pre-earthquake or pre-hurricane condition. Due to space constraints in this paper, we limit the discussion to earthquake-damaged buildings, but the same approach can be applied for hurricane-damaged buildings, or for any other type of natural hazard for which a well-developed analytical method exists for computing expected damage to individual buildings.

The Proposed Approach

The approach relies on the use of a well-developed and comprehensive earthquake damage and loss estimation methodology. In this paper, we use the publicly-available and widely-used FEMA HAZUS [1] methodology and computer software that was developed in the mid-1990's by the National Institute of Building Sciences. It should be noted, however, that any other comprehensive analytical methodology (publicly available or privately held, e.g., ATC-13 [2]) that will provide a quantified estimate of the expected damage to a given building can be used as well.

The pre-earthquake condition of the building is determined using the fragility and loss models provided in the HAZUS methodology. In this methodology, for a defined level of earthquake ground motion, e.g. a 10% in 50-year hazard level, models are provided for determining the earthquake ground motion at the building site, called the demand curve. Then the capacity of the building is determined by developing a load versus deflection – or capacity (pushover) curve – using the parameters supplied with the methodology that were developed from nonlinear structural analysis methods. Lastly, a fragility curve is used to estimate the expected damage based on the building displacement level computed from the analysis of the earthquake demand curve and building capacity curve.

In HAZUS, the fragility curves incorporate the uncertainty in the ground motion, uncertainty in the capacity of the structure, and uncertainty in the expected damage. The result of the use of the HAZUS methodology for a given building is the probability that the building will respond to the selected earthquake (e.g. the 10% in 50-year hazard level) in each of five defined damage states. The five damage states in order from least to worst damage are: None, Slight, Moderate, Extensive and Complete. For example, the analysis of the pre-earthquake building could result in the following 5%, 20%, 40%, 25% and 10% probability of being in the None, Slight, Moderate, Extensive and Complete Damage States, respectively, for a 10% in 50-year earthquake.

This HAZUS analysis provides a quantified measure of the pre-earthquake condition of the building. The building in its post-earthquake condition is then analyzed using the same HAZUS procedure (with a modified capacity curve to account for the damage to the building) and the probability of being in each of the five damage states is calculated (e.g. it could be 2%, 10%, 50%, 30%, and 8% probability of being in the None, Slight, Moderate, Extensive and Complete Damage States, respectively, for a 10% in 50-year earthquake). To bring the building back to its pre-earthquake condition these probabilities must be brought back to their pre-earthquake values. Figure 1 illustrates this approach, where the damage state probabilities are the equivalency parameters.

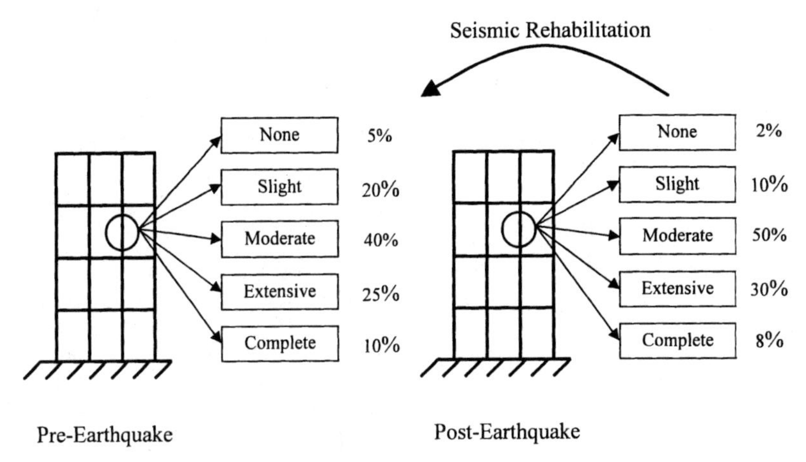

Figure 1. Seismic rehabilitation to prior condition damage state probability equivalence.

Incorporation of More Advanced Technologies

The fragility curves in the HAZUS methodology directly incorporate the following three types of uncertainty in determining the uncertainty in the probability of the building being in one of the five structural damage states:

(1) The uncertainty in the building's capacity curve;
(2) The uncertainty in the estimation of the earthquake ground motion at the building site; and
(3) The uncertainty in the estimate of the median value of the threshold of structural damage.

It is the beauty of this proposed approach that the tools of modern risk and reliability analysis are used to define the uncertainties associated with the different steps in the damage estimation process. This enables a direct rewarding for the transfer of technology into this part of structural engineering design. Of course, it is to be expected that the approach will be resisted by those that either do not really want to try to estimate the real performance of a real building during future earthquakes or do not recognize that uncertainty exists and can be quantified using the analytical tools of structural risk and reliability theory. Engineers or public officials in this group are uncomfortable with accepting the realities of structural engineering: that when design deviates from published fixed formulas and computerized design equations and attempts to address reality, excellent structural engineers can often differ on parameter values (e.g. mean, standard deviation and probability density functions of structural variables – for example, damping) in their analyses. Therefore, two excellent structural engineers can obtain different – and equally acceptable – answers to the same problem. Those resistant to probability are also often uncomfortable using Bayesian methods of analysis, where limited experimental data exists and professional judgment is directly and scientifically incorporated into design. This head-in-the-sand, anti-probabilistic-methods mentality is hopefully vanishing as we face, and more rationally and analytically address, uncertainties in many areas of our lives.

The Earthquake Ground Motion Used For Equivalency

As previously discussed, there is some flexibility in the selection of the earthquake ground motion used to determine the probabilities that the structure will respond to the ground motion in the five HAZUS damage states. For example, the selected ground motion could be the recorded, or a best estimate, earthquake ground motion at the site of the building under consideration. If this is the case, the repair would be to bring the expected building performance back to what it was prior to experiencing the earthquake.

This concept can be extended to include the pre-earthquake condition of the building for several different levels of earthquake hazard (e.g. 2%, 10%, and 50% in 50-years). It could also be included with a Monte Carlo simulation analysis where many (e.g. 200) future 50-year scenarios of annual earthquake ground motions at the building site are simulated and the results used to provide the pre-earthquake quantification of the building condition. A Monte Carlo simulation approach opens up the use of a broad range of possible probability density functions to define the input variables, and is also often more understandable to some engineers.

Equivalency In Terms Of Dollars

Continuing with this approach for quantifying the pre-earthquake condition of the building with a methodology such as HAZUS, the next step is to convert the probability of being in each of the five damage states to a dollar loss value for the building. For example, the analysis might calculate the expected pre-earthquake damage in the five damage states to be $200K, $300K, $600K, $400K, and $100K for the None, Slight, Moderate, Extensive and Complete Damage States, respectively, for a 10% in 50-year earthquake. This is a quantification of the pre-earthquake condition of the building. Then the building is analyzed in its post-earthquake condition (again by modifying the capacity curve in the HAZUS methodology) and for example, the expected damage in each damage state might be $300K, $400K, $650K, $500K, and $300K for the None, Slight, Moderate, Extensive and Complete Damage States, respectively, for the same 10% in 50-year earthquake. The building is then rehabilitated to bring it back to its pre-earthquake condition, now quantified as expected dollar loss for each of the five HAZUS damage states.

Illustrative Example

The following example illustrates the use of the proposed approach. Consider a wood-frame residential building, subjected to an earthquake that causes damage to the building. The structural engineer is then asked to provide a repair that brings the building back to its pre-earthquake condition.

The structural engineer first obtains, for the earthquake that occurred, an estimate of the earthquake ground motion at the building site. Either by direct measurement or the use of earthquake ground simulation methods, the earthquake ground motion can be estimated. Figure 2 shows one form of such an estimate which is a plot of the Spectral Acceleration (S_a) versus Spectral Displacement (S_d) – essentially the capacity curve used in the HAZUS methodology, denoted as the Earthquake Ground Motion in Figure 2.

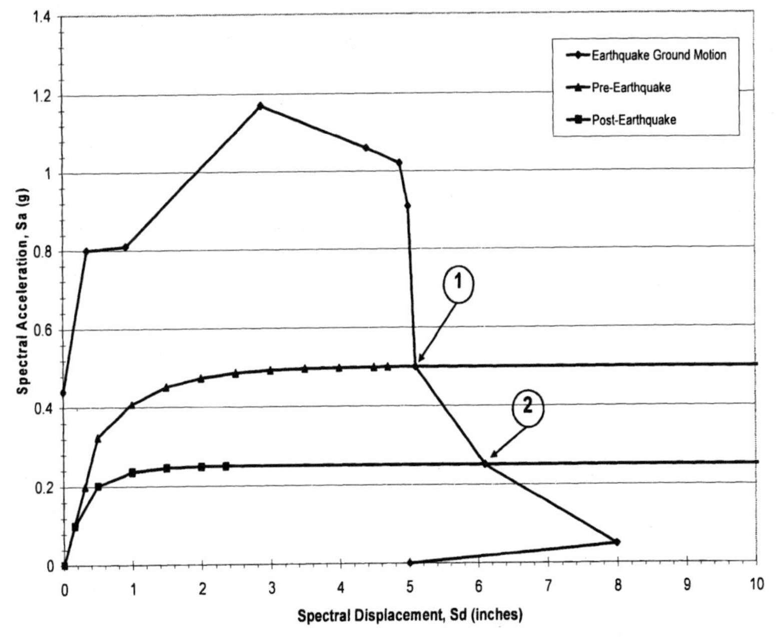

Figure 2. HAZUS earthquake ground motion and building capacity curves.

The next step in the method is to develop the building capacity curve for the pre-earthquake and the post-earthquake conditions. Figure 2 shows these two capacity curves. For this specific example ground motion and capacity curve, the value of the spectral displacement at the building roof for the pre-earthquake condition, Point 1 in Figure 2, is 5.1 inches and for the post-earthquake condition, Point 2 in Figure 2, is 6.1 inches.

The HAZUS procedure defines default values for the three types of uncertainty stated in the previous section. Therefore, the structural engineer can use these default values or select, based on additional analysis, more accurate values. Table 1 provides, for this example, the values used for the lognormal standard deviation that describes the total variability for a damage limit state. Note that in general the uncertainty for the post-earthquake building exceeds the pre-earthquake building.

Table 1. Lognormal Standard Deviation for HAZUS Damage States

Damage State	Pre-Earthquake Building	Post-Earthquake Building
Slight	0.89	1.04
Moderate	0.95	0.97
Extensive	0.95	0.90
Complete	0.92	0.99

Figure 3 shows plots of the boundaries of the different HAZUS damage states for the building in its pre-earthquake condition. The pre-earthquake building roof displacement from Figure 2 is 5.1 inches. At this 5.1-inch displacement level, Point B is on the boundary between the Extensive and Complete damage states. Point C is on the boundary between the Moderate and the Extensive damage states. The difference in the values on the vertical axis between Points C and B define the probability that the building in its pre-earthquake condition will experience damage in the Extensive damage state. The probability of the building being in any one of the five damage states for the pre-earthquake and post-earthquake building is given in Table 2.

Table 2. Probability of Experiencing Different HAZUS Damage States

Damage State	Pre-Earthquake Building (%)	Post-Earthquake Building (%)
None	3	2
Slight	15	7
Moderate	44	35
Extensive	28	34
Complete	10	22

Figure 4 shows how the boundary between the Extensive and Moderate damage states changes when the building is damaged. The pre-earthquake Point C has now moved to location C^1 for the roof displacement value of 6.1 inches for the damaged building. It then follows from this figure that the probability that the building experienced either Extensive or Complete damage has increased from 38% for the pre-earthquake condition to 56% for the post-earthquake condition. Figure 5 shows all of the fragility curves for this example pre- and post-earthquake building.

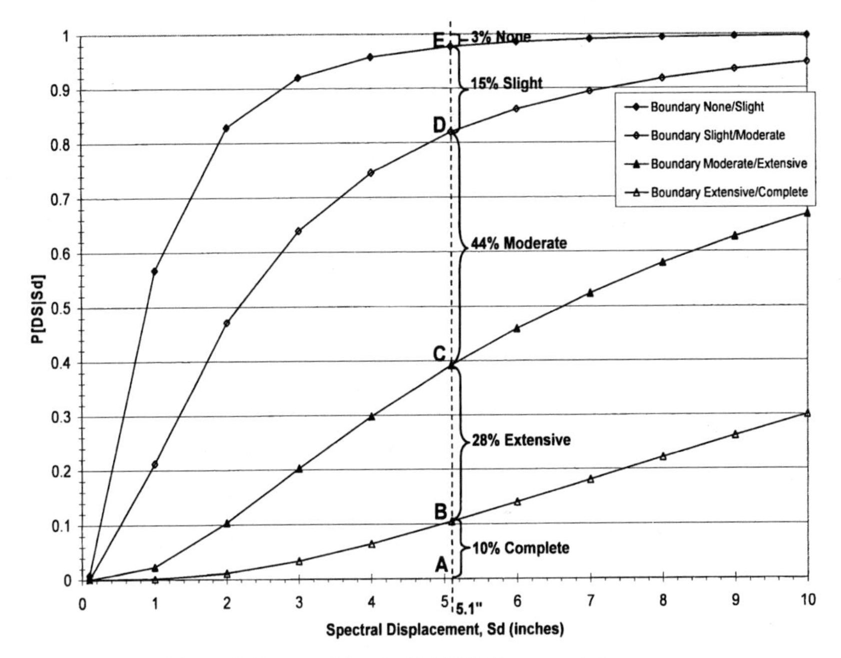

Figure 3. Pre-earthquake HAZUS damage states.

It is interesting to note that if the owner desires to spend more money and provides the fees for the engineer to develop a more accurate estimate of the ground motion, the capacity curve, and the threshold of damage for this specific building, then even if the methodology produces the same roof displacement estimate, there will be a reduction in uncertainty in this estimate. In this case, the total variability in the damage state estimate will decrease. If the only damage state that is important to the owner is the Complete damage state, and if this uncertainty is reduced by a factor of two on the post-earthquake building damage estimate, then the probability of being in the complete damage state will reduce from 22% to 6%. Table 3 shows the corresponding damage state probability values for all of the building damage states for this example case.

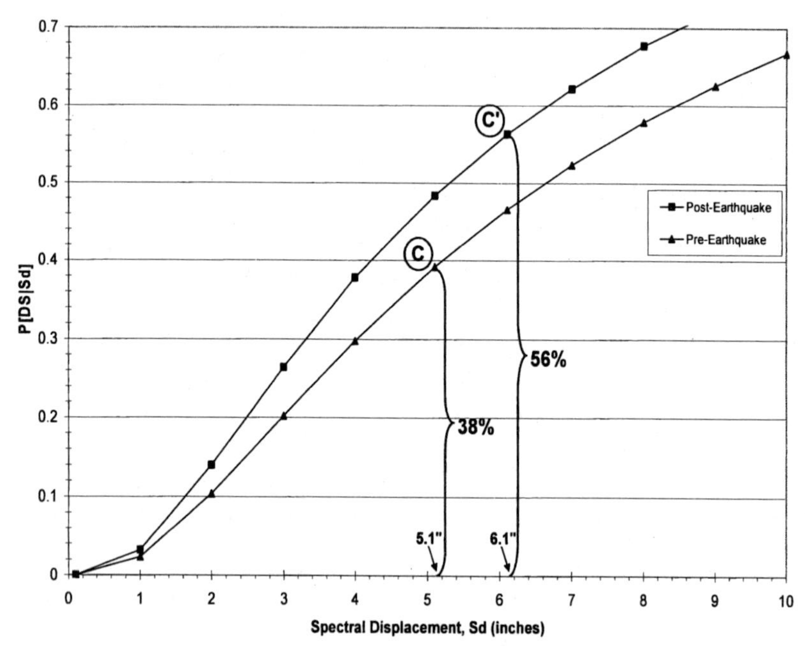

Figure 4. Pre- and Post-earthquake boundary curves between Moderate and Extensive HAZUS damage states.

Table 3. Probability of Experiencing Different HAZUS Damage States after Advanced Analysis and Uncertainty Reduction

Damage State	Pre-Earthquake Building (%)	Post-Earthquake Building (%)
None	0	0
Slight	3	0
Moderate	68	38
Extensive	28	56
Complete	1	6

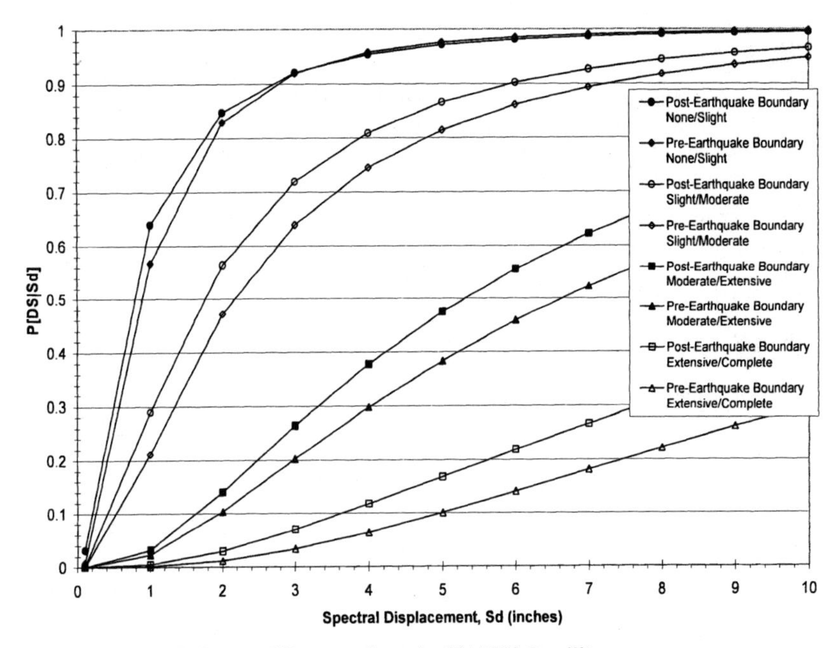

Figure 5. Pre- and Post-earthquake HAZUS fragility curves.

Conclusions

The use of probabilistic loss estimation methods provides a rational method for evaluating the pre-earthquake and repaired conditions of a building. This paper presents an approach for doing this using the publicly-available and widely-used HAZUS methodology and computer software developed by the federal government. The approach is not limited by the loss estimation methodology used – the same approach can be used with any other comprehensive analytical methodology (publicly available or privately held) that will provide a quantified measure of the expected damage to a given building. The discussion of the approach in this paper is limited to earthquake-damaged buildings, but it can also be used for estimating repair for other types of damage such as hurricanes.

Acknowledgements

The authors wish to thank Dr. Chukwuma Ekwueme and Dr. Can Simsir for their assistance on this paper.

References

1. Federal Emergency Management Agency (FEMA). "HAZUS99 Earthquake Loss Estimation Methodology, Technical Manual." Prepared by the National Institute of Building Sciences for the Federal Emergency Management Agency, 1999.

2. Applied Technology Council (ATC). "Earthquake Damage Evaluation Data for California." ATC-13 Report. Redwood City, CA: ATC, 1985.

Lessons Learned from Hurricane Katrina

James W. Jordan, SE, PE, M.ASCE[1]
Saul L. Paulius, SE, PE, M.ASCE[2]

Abstract: Hurricane Katrina has been estimated to be the costliest storm in U.S. history, with total losses exceeding $100 billion. Hurricane Katrina was at Category 5 strength while in the Gulf of Mexico, and reportedly diminished to a strong Category 3 when it struck the Louisiana and Mississippi coast on August 29, 2005. The hurricane surge forces breached protective levees in New Orleans, and resulted in catastrophic flooding in the City. Although the flooding catastrophe in New Orleans drew world-wide attention and the media's focus during the reporting of the hurricane aftermath, the strongest winds and highest storm surge from Hurricane Katrina ravaged Mississippi coastal communities. The Mississippi coast was vulnerable to the destructive forces developed on the "right side" of the hurricane, and had little or no protection from levees, barrier reefs, or breakwater structures. Given the severity of this storm and resultant damages to the Mississippi coastal communities, the structural engineering profession has a unique opportunity to evaluate our approaches to hurricane damage assessment.

The purpose of this paper is to provide structural engineers with lessons learned from building damages that occurred along the Mississippi coastline during Hurricane Katrina. The authors personally inspected and/or supervised the damage assessment of over 300 buildings and structures in Mississippi as a result of Hurricane Katrina. This paper presents an approach to assessing hurricane damages from this storm in order to provide answers to society so as to rebuild from this catastrophe. The insurance industry requires that specific questions be answered by structural engineers who assess hurricane damages, and we will address those issues in this paper.

[1]Branch Manager, Rimkus Consulting Group, Inc., 8910 Purdue Road, Suite 170, Indianapolis, Indiana 46268; PH (317) 510-6484; FAX (317) 510-6488; email: jwjordan@rimkus.com.

[2]Senior Consultant, Rimkus Consulting Group, Inc., 999 Oakmont Plaza Drive, Suite 550, Westmont, Illinois 60559; PH (630) 321-1846; FAX (630) 321-1847; email: slpaulius@rimkus.com.

Introduction

Hurricane Katrina was a very powerful storm that struck the Mississippi coastline on August 29, 2005. As reported by the National Oceanographic Atmospheric Administration (NOAA), many of the key weather stations failed during the hurricane making it difficult to quantify information on the storm. NOAA and other forensic meteorologists published their findings and analyses several weeks after the hurricane event [see Figure 1]. In some cases, computer modeling was performed to estimate the hurricane characteristics as it struck the coastline.

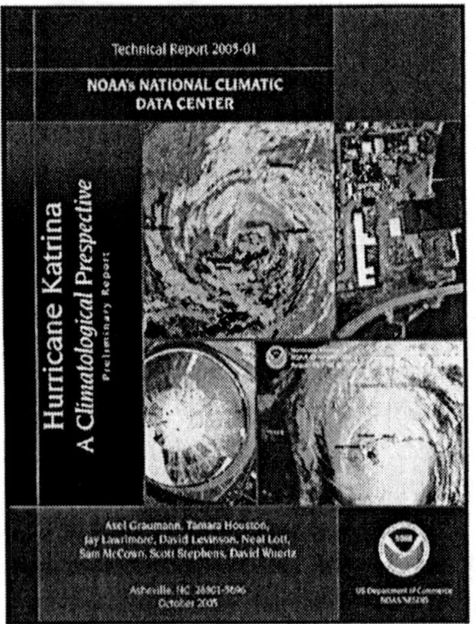

Figure 1: NOAA preliminary report[4] issued 10/05

The insurance industry also found Hurricane Katrina to be very destructive. In fact, the Insurance Information Institute estimated Hurricane Katrina to be the costliest hurricane in U.S. history, with claims expected to approach $40 billion [see Figure 2].

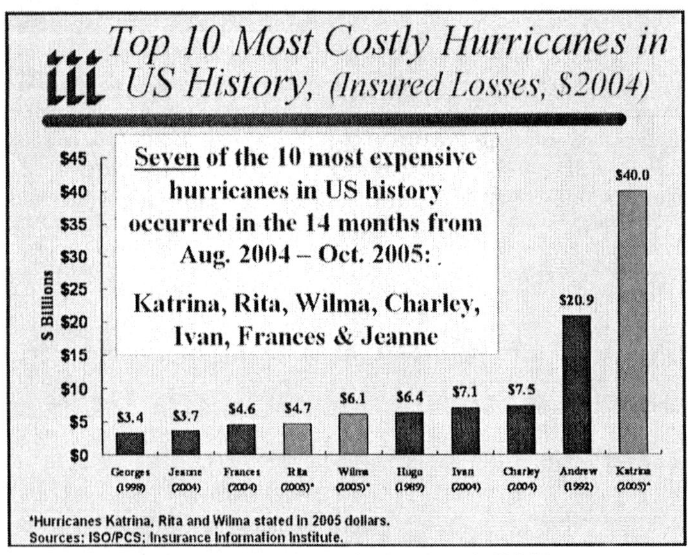

Figure 2: Hurricane Katrina was the costliest hurricane in U.S. history

The Structural Engineer's Role in Assessing Hurricane Damage

Due to the extensive nature of damages to buildings along the Mississippi coastline from Hurricane Katrina, structural engineers played a pivotal role in assessing the extent of damages and reparability of the damaged structures. The insurance industry responded almost immediately to the damages incurred, and sent hundreds (if not thousands) of adjusters to Mississippi coastal communities after the hurricane struck. Most of these adjusters required the services of structural engineers so as to produce accurate assessments of the monetary cost of damages to insured buildings and structures.

First and foremost, structural engineers were charged with determining the extent of damages to buildings and structures ravaged by the hurricane. Armed with the understanding of load paths of various structural systems, behaviors of building materials, and general construction of residential and commercial buildings, structural engineers offered valuable insights into the evaluation of buildings exposed to the hurricane. In many instances, apparent damages to buildings pre-existed Hurricane Katrina. The structural engineer determined which damages were caused by the hurricane, and which pre-existed the storm or were caused by other means.

Secondly, structural engineers were typically asked to determine which of the hurricane damages were caused by high winds versus damages caused by the associated storm surge and waves. This was of particular interest to the insurance industry, as certain policies were triggered by wind damage versus other policies that covered only flood damage. Generally, the insurance industry considered wind damages to result from direct air movement against a building or structure, or from wind-blown materials that impacted the building or structure. Damages that resulted from water (other than wind-driven rain), including tidal or wave forces, and fast or slow moving water were typically covered by flood policies.

Finally, structural engineers were asked to determine if a building or structure could be feasibly repaired. During the initial damage assessments, this determination was typically limited to an opinion based upon the structural engineer's visual observations at the site.

Initially, the quantity of licensed engineers in Mississippi qualified to evaluate structural damage was limited. As such, many structural engineers supervised field technicians who gathered data on damaged buildings for them. These technicians required instruction on recording the damages, and communicating this information to the structural engineers.

Challenges for the Structural Engineer Inspecting Katrina Damage

The first step for supervising structural engineers headed to Mississippi to inspect building damage was to get properly licensed. Mississippi required structural engineers in responsible charge to be licensed as professional engineers in the State of Mississippi. It was unlawful for out-of-state engineers to offer their services as structural engineers without first obtaining licensure in Mississippi. Inspectors not licensed as engineers in Mississippi required supervision from Mississippi professional engineers. State government officials recognized the need for out-of-state help, and expedited the licensure process by reviewing applications weekly (rather than monthly). These officials helped in-state engineers displaced by the hurricane by temporarily postponing their continuing education requirements.

The next challenge for structural engineers headed to the hurricane ravaged areas was to acquire appropriate protection for their health and safety. The Center for Disease Control provided a listing of required immunizations and other recommendations for those entering the hurricane-damaged areas [see Figure 3]. Furthermore, on-site security was of concern to structural engineers inspecting properties where reports of looting occurred. Hard hats, protective gloves, boots and appropriate clothing were a necessity, and highly visible vests were recommended to properly identify inspectors from trespassers. Many neighborhoods within a few blocks of the beach were under martial law after the hurricane, and only those who obtained special authorization from the local municipality officials were permitted to access these areas.

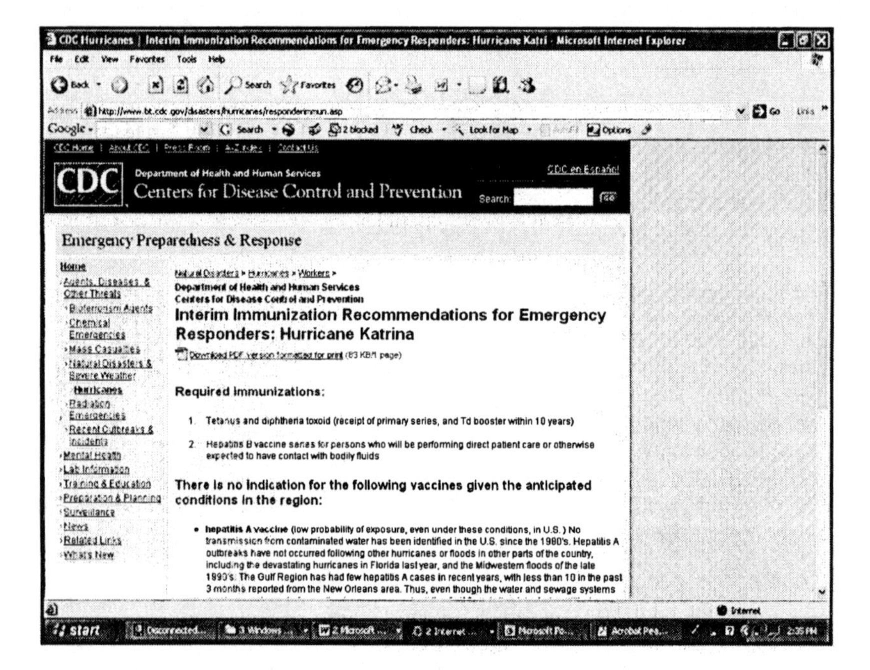

Figure 3: Immunization requirements from the CDC[2]

Assessing Damage to Structures – Wind Versus Surge

A tremendous amount of information became available for those assessing Hurricane Katrina damage. Of particular value were aerial photographs of the hurricane damaged areas published by NOAA[3] [see Figure 4]. Other valuable resources included weather data from a variety of sources, United States Geological Survey (USGS) topographic maps[5], and aerial photographs taken prior to the hurricane. The use of these resources to supplement on-site observations provided structural engineers with the means to determine damages caused by high winds versus the storm surge. These resources were available free of charge on the internet. Forensic meteorologists were also available to provide weather information and opinions on the chronology of the wind and surge components of the hurricane.

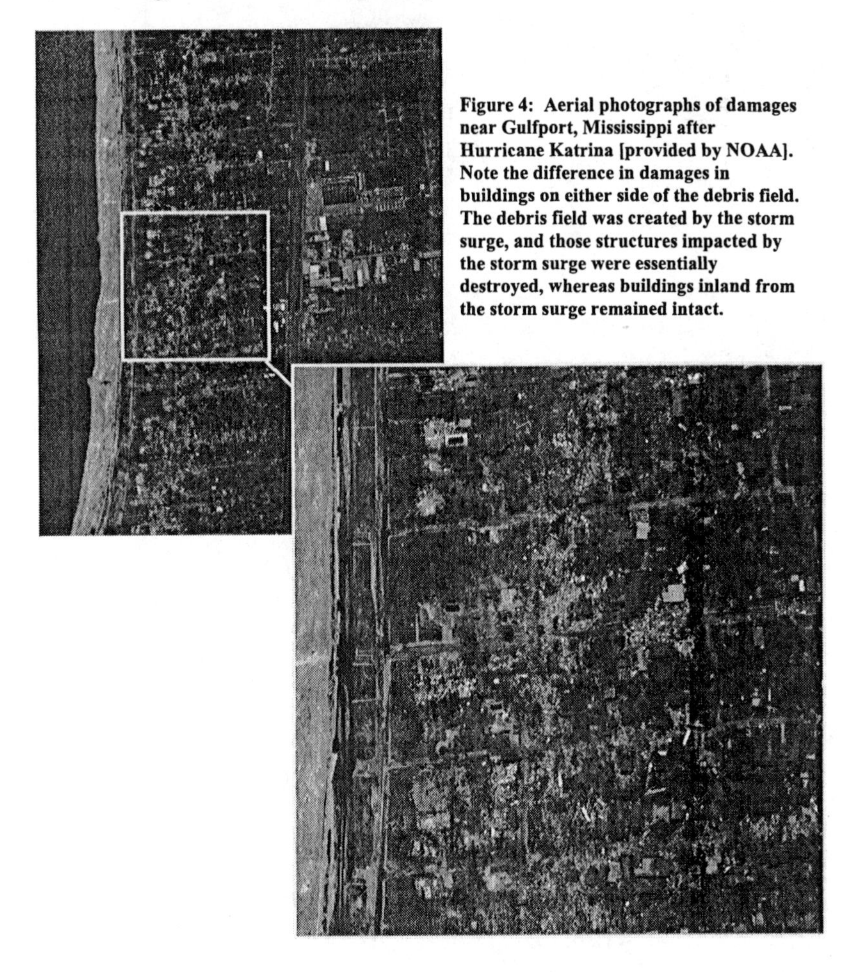

Figure 4: Aerial photographs of damages near Gulfport, Mississippi after Hurricane Katrina [provided by NOAA]. Note the difference in damages in buildings on either side of the debris field. The debris field was created by the storm surge, and those structures impacted by the storm surge were essentially destroyed, whereas buildings inland from the storm surge remained intact.

In most areas along the Mississippi coast line, a distinct debris field was evident near the beachfront. By comparing the outer edges of the debris field with topographic maps, one could estimate the storm surge height – or at least provide an approximate confirmation of reported surge estimates [see Figure 5]. In addition, a Federal Emergency Management Agency (FEMA) website[8] provided reference maps that showed the estimated inland extent of the storm surge.

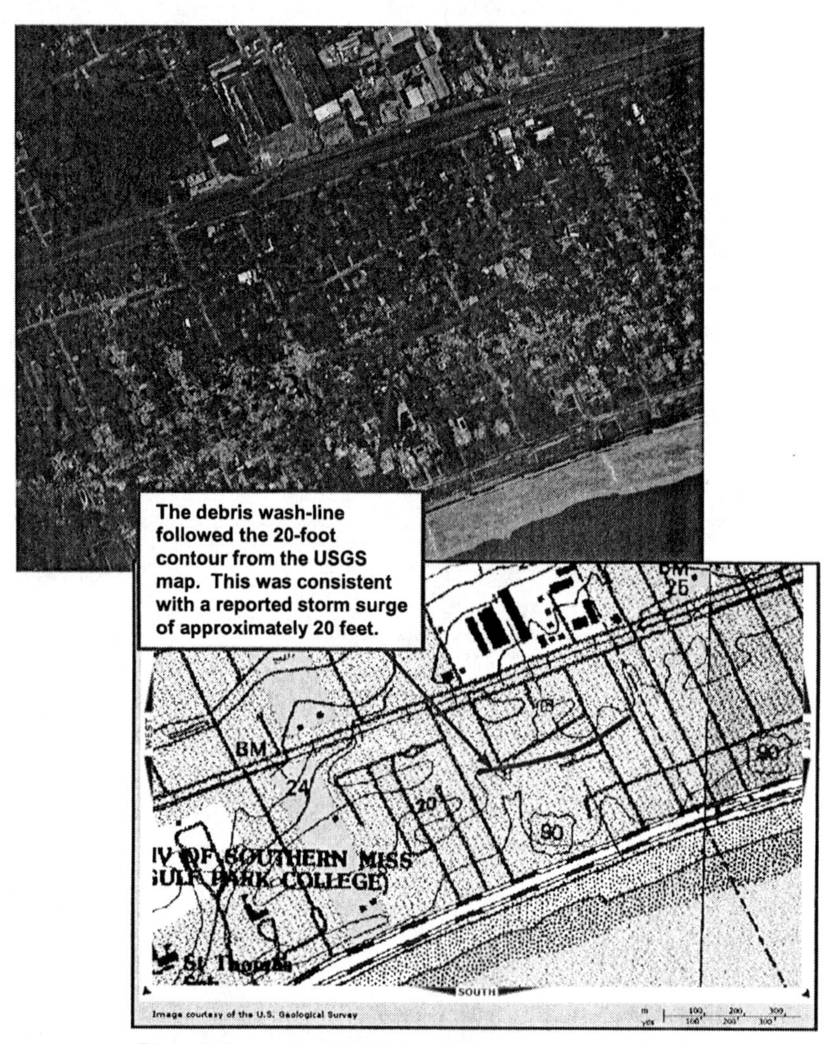

Figure 5: By orienting the NOAA aerial damage photograph from Figure 5 and comparing to a USGS topographic map of the area, the storm surge height may be estimated by examining the location of the debris wash-line.

In some coastal neighborhoods, all of the visible building superstructures were destroyed. A structural engineer assessing a distinction of damages caused by high winds before the storm surge swept through the site had a very daunting task. Without the aid of the aforementioned resources, an opinion regarding the extent of the likely wind damages was difficult (if not impossible) to substantiate. Figure 6 shows some of the typical views along the areas with the most severe damage.

Figure 6: Typical views of devastation along the Mississippi coast caused by Hurricane Katrina. Determining the extent of wind damages prior to the storm surge sweeping through these sites was a daunting task for the structural engineer.

In some instances, there was enough of the superstructure left on-site for a structural engineer to determine the distinction between wind and surge damages. It is vital that structural engineers understand that the forces from surge water current typically exceed wind forces. For example, a 1-foot tall seawater current moving at 10 mph will produce a total lateral force equivalent to a 310 mph wind force acting over a 1-foot height (including dynamic and hydrostatic conditions):

$$\text{Force} = \frac{mv^2}{2} + \frac{Ph^2}{2}$$

Hydrostatic Force Component* (arrow to $\frac{Ph^2}{2}$)

Dynamic Force Component (arrow to $\frac{mv^2}{2}$)

$$\text{Force of 1' water @ 10 mph} = \frac{64.0\,\text{pcf}}{2 \times 32.2\,\text{ft/sec}^2} \times (10\,\text{mph})^2 \times \left(\frac{5280\,\text{ft/mile}}{3600\,\text{sec/hr}}\right)^2 + 64.0\,\text{pcf} \times (1\,\text{ft})^2 / 2$$

$$= 245.8\ \text{lbs-force}$$

$$\text{Wind Pressure}^1 = 0.00256 \times V^2$$

$$245.8\,\text{lbs} = 0.00256 \times V^2$$

* We included the effects of hydrostatic force in this calculation, since video documentation taken during Hurricane Katrina showed a flood height outside a building to be 2 to 4 feet above the water depth inside the building as the surge rose.

Solve for Equivalent Wind Speed, V = 310 mph

Maximum wind gusts reported along the Mississippi coast was 140 to 150 mph, much less than in our example calculation shown above. The purpose of this exercise is to show that surge forces were more likely to cause destructive forces than wind forces during the hurricane. This example is very conservative. In reality, it is likely that water current during the surge exceeded 10 mph. Furthermore, it is likely that the water had a greater density than clean seawater, since sand and debris was churned up into the flood waters. With this in mind, damages due to the storm surge will be greater at lower elevations on a structure, while wind forces are greater at higher elevations on a structure. Typical damages due to high winds include torn and missing roofing and siding, and localized removal of wall and roof sheathing. Typical damages due to storm surge (other than water damages due to flooding) include displaced walls, columns, and piers; and scouring/undermining of foundations. The demolition of superstructures was consistent with the powerful forces of Katrina's storm surge, while localized damages to roof coverings was consistent with damages from the hurricane winds.

We will now examine several case examples to show the distinction between damages caused by high wind forces versus damages caused by the storm surge from Hurricane Katrina.

Case Study 1: Wood-Framed Church in Pascagoula

Slow-rising flood waters from the storm surge reached a height of 30-inches above the first floor, causing extensive water damage to interior floor and wall coverings. High winds damaged the roof covering, caused wind-blown debris to penetrate the roof sheathing, and removed roof and wall sheathing panels from the windward gable end. Wind-driven rain entered breaches in the roof covering and gable wall openings causing water damages to the ceilings and partial second story.

Figure 7: Series of photographs that documented damages to the church structure. In this example, the forces from wind caused the majority of structural damages, while the surge caused extensive water damages to the interior. This is a good example of typical wind damages caused by Hurricane Katrina.

Case Study 2: Wood-Framed, Single-Family House in Long Beach

The storm surge reached a height of about 10-feet at this site, and caused the destruction of the south exterior wall (facing the beach). The severity of damage was greatest at the first story, consistent with the destructive forces of the storm surge. Relatively minor damage to the roof covering was the extent of wind damages found at this structure.

Figure 8: Series of photographs that documented damages to the house. Note that minimal damage was present to the roof covering, while severe damages were on the lower part of the structure. This was a good example of typical damages caused by the storm surge from

Case Study 3: Wood-and-Steel-Framed, Two-Story Restaurant in Diamondhead

The storm surge reached a height of about 20-feet at this site (near the roof eave), and caused extensive damage to the first two stories. High wind forces caused localized damages to the roof structure. All surrounding wood-framed residences were demolished by the hurricane such that only some of the piers and beams remained.

Figure 9: Series of photographs that documented damages to the restaurant building and surrounding residences. In this example, the extent of wind damages was evident by the remaining roof structure, while the devastation caused by the storm surge was readily apparent. The remaining portions of the restaurant structure supported the likelihood that the destruction of surrounding wood-framed houses was caused by the storm surge, and that probable damages from high winds would have been relatively localized.

Case Study 4: Steel and Precast Concrete Shopping Mall in Gulfport

The storm surge did not reach this site, thus the storm damage was caused by high winds. Wind gusts of 132 mph were estimated in Gulfport. Damages to these buildings included localized portions of missing metal roof deck, removal of about half of the membrane roof covering, and the collapse of an elevated roof structure.

Figure 10: Series of photographs that documented damages to the shopping mall. The partial collapse of the elevated roof structure was attributed to construction defects (inadequate bracing / unstable construction) whereas the removal of localized portions of roof deck occurred at the perimeter due to the failure of puddle welds.

Case Study 5: Steel and Concrete Casino in Biloxi

The storm surge reached a height of about 25-feet at the site, and caused severe damages to all of the first and second story walls and finishes. Waves from the storm surge caused a floating casino structure to break from its moorings and batter the building as it drifted ashore. There was finish damage caused by the wind to the portions of the structure above the second floor.

Figure 11: Series of photographs that documented damages at the casino structure. Note the severity of destruction to the lower levels of the building. Also note the final resting position of a floating casino barge that drifted ashore after impacting the casino building.

Regarding Ethics Issues

The extent of damages along the Mississippi coast exceeded practically everyone's worst predictions. In particular, flooding from the storm surge reached areas that were thought to be relatively risk-free (above the floodplain) causing damages further inland than anticipated. Since many property owners had no flood insurance on their buildings damaged by the hurricane, the determination of wind versus water damages by structural engineers could bring financial ruin to individuals or families. How do structural engineers deal with the pressures created by this unfortunate circumstance? Ethics! Ethic standards, in particular – the ASCE *Code of Ethics*[6], provide help to structural engineers in this situation.

The key factors in dealing with ethical pressures placed upon the inspecting structural engineers are honesty, integrity, and objectivity. The inspecting structural engineer must uncover the facts so as to form substantiated opinions. If opinions cannot be substantiated based upon available resources, then the structural engineer must explain that specific answers to their client's questions cannot be determined.

Another problem that many structural engineers faced was the tremendous volume of buildings to inspect. Over 250,000 people were displaced by Hurricane Katrina – their homes damaged or destroyed during the storm. Rebuilding efforts were not likely to start until insurance issues were resolved regarding these damaged buildings. These insurance issues more than likely hinged on the conclusions needed from an inspecting structural engineer. In this situation, structural engineers must give due consideration to each building they inspect. Practically every building is unique in its ability to resist lateral forces from a hurricane. Factors that differ among buildings include age/deterioration, building materials, over/under-designed structural systems, construction defects, over/under-loaded conditions, and surrounding terrain features. Thus, we cannot simply assume that conclusions reached for a particular building will also apply to all of the surrounding buildings.

Considering the huge number of damaged buildings to inspect and the finite number of licensed structural engineers to examine them, it should come as no surprise that there have been reports of individuals not properly qualified as structural engineers performing or supervising damage assessments of buildings. For guidance on this issue, consider Cannon 2 of the ASCE *Code of Ethics*[6], "Engineers shall perform services only in areas of their competence." Should you come across someone known to be misrepresenting themselves as structural engineers, consider Cannon 1: "Engineers shall hold paramount the safety, health and welfare of the public and shall strive to comply with the principles of sustainable development in the performance of their professional duties"; and subpart d: "Engineers who have knowledge or reason to believe that another person or firm may be in violation of any of the provisions of Canon 1 shall present such information to the proper authority in writing and shall cooperate with the proper authority in furnishing such further information or assistance as may be required.") The Mississippi Board of Licensure for Professional Engineers and Surveyors requires that complaints be submitted in writing by the person or persons making them. A recent newsletter from the Board explained that they typically cannot pursue anonymous complaints[7].

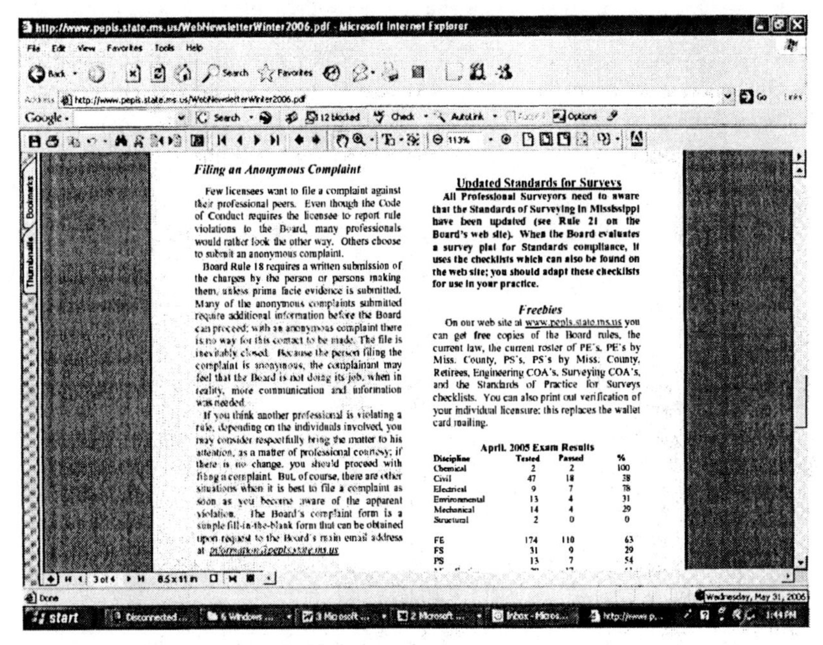

Filing an Anonymous Complaint

Few licensees want to file a complaint against their professional peers. Even though the Code of Conduct requires the licensee to report rule violations to the Board, many professionals would rather look the other way. Others choose to submit an anonymous complaint.

Board Rule 18 requires a written submission of the charges by the person or persons making them, unless prima facie evidence is submitted. Many of the anonymous complaints submitted require additional information before the Board can proceed; with an anonymous complaint there is no way for this contact to be made. The file is inevitably closed. Because the person filing the complaint is anonymous, the complainant may feel that the Board is not doing its job, when in reality, more communication and information was needed.

If you think another professional is violating a rule, depending on the individuals involved, you may consider respectfully bring the matter to his attention, as a matter of professional courtesy; if there is no change, you should proceed with filing a complaint. But, of course, there are other situations when it is best to file a complaint as soon as you become aware of the apparent violation. The Board's complaint form is a simple fill-in-the-blank form that can be obtained upon request to the Board's main email address at information@pepls.state.ms.us

Updated Standards for Surveys

All Professional Surveyors need to aware that the Standards of Surveying in Mississippi have been updated (see Rule 21 on the Board's web site). When the Board evaluates a survey plat for Standards compliance, it uses the checklists which can also be found on the web site; you should adapt these checklists for use in your practice.

Freebies

On our web site at www.pepls.state.ms.us you can get **free** copies of the Board rules, the current law, the current roster of PE's, PE's by Miss. County, PS's, PS's by Miss. County, Retirees, Engineering COA's, Surveying COA's, and the Standards of Practice for Surveys checklists. You can also print out verification of your individual licensure; this replaces the wallet card mailing.

April, 2005 Exam Results

Discipline	Tested	Passed	%
Chemical	2	2	100
Civil	47	18	38
Electrical	9	7	78
Environmental	13	4	31
Mechanical	14	4	29
Structural	2	0	0
FE	174	110	63
FS	31	9	29
PS	13	7	54

Figure 12: Article from Mississippi Board newsletter

Conclusions

Hurricane Katrina imposed unique challenges to structural engineers charged with the task of assessing damages to buildings and structures. Although most of the attention for the aftermath of the hurricane was focused on the levee failures and resultant flooding in New Orleans, the strongest wind and surge forces that made landfall from the hurricane were imposed on the Mississippi coastal communities. The resultant damages to the Mississippi coastline from Hurricane Katrina were the most expensive from any storm in U.S. history. As the insurance industry mobilized to help property owners recover during the aftermath of the hurricane, the services of structural engineers were needed to uncover facts for adjusters so that they could determine which insurance policies applied. Specifically, structural engineers had the challenge of determining a distinction between damages caused by high winds versus damages caused by the associated storm surge and waves. In this paper, we reviewed techniques and discussed available resources for structural engineers to formulate substantiated opinions regarding Hurricane Katrina damage. We also discussed unique challenges for structural engineers involved in assessing damage to the Mississippi coast, including health and safety issues, licensure / legal aspects, and ethical issues.

A note regarding SI units:

Data from the referenced publications and weather sources were provided in English units. In order to maintain continuity and to minimize confusion for the reader, who may wish to review these references, the sample calculations provided in this paper are also provided in English units. We offer the following basic conversions for those who wish to consider SI units:

$1\ kg/m^3$	=	$0.0624\ pcf$	$1\ cm$	=	$2.54\ in$
$1\ m$	=	$3.281\ ft$	$1\ km$	=	$0.621\ miles$
$1\ m^3$	=	$35.3147\ ft^3\ (cf)$	$1\ m^2$	=	$10.764\ ft^2\ (sf)$

References

[1]American Society of Civil Engineers (ASCE). 1995. *Minimum Design Loads for Buildings and Other Structures.*

[2]Centers for Disease Control and Prevention (CDC). 2005. *Interim Immunization Recommendations for Emergency Responders: Hurricane Katrina,* http://www.bt.cdc.gov/disasters/hurricanes/responderimmun.asp

[3]National Oceanographic and Atmospheric Administration (NOAA). 2005. *Hurricane Katrina Images,* http://ngs.woc.noaa.gov/katrina/

[4]National Oceanographic and Atmospheric Administration (NOAA). 2005. *Hurricane Katrina – A Climatological Perspective, Preliminary Report.*

[5]Unites States Geological Survey (USGS). *2005. Topographic Maps (general library),* http://www.terraserver.microsoft.com/

[6]American Society of Civil Engineers (ASCE). 1914 (Amended 1996). *Code of Ethics, http://www.asce.org/inside/codeofethics.cfm#note3*

[7]Mississippi Board of Licensure for Professional Engineers and Surveyors. 2005. *Winter, 2006 Newsletter,* http://www.pepls.state.ms.us/WebNewsletterWinter2006.pdf

[8]Federal Emergency Management Agency (FEMA). 2006. *Hurricane Katrina Flood Recover Maps,* http://www.fema.gov/hazard/flood/recoverydata/katrina/katrina_ms_maps.shtm

Strategies for Damage Assessments and Emergency Response

David B. Peraza, PE, M. ASCE

Exponent, 420 Lexington Avenue, Suite 1740,
New York, NY, PH (212) 895-8103,
email: **dperaza@exponent.com**

Civil and structural engineers can provide crucial expertise following natural disasters, man-made catastrophes, and other accidents. In recent years we have experienced major terrorist attacks on buildings and on public transportation, devastating storms, and singular collapses of public structures.

Even highly qualified engineers, when faced with a large disaster often feel overwhelmed and have initial difficulty dealing with the situation and knowing where to start. There is very little training on this topic that is targeted at the engineer.

This paper discusses strategies from the author's experience. It is hoped that this will assist other engineers in coping with emergency situations, so that they can best apply their skill and knowledge for the public good.

Be Prepared

Calls regarding emergencies typically come at awkward times. They come at 5pm on a Friday afternoon, when you are out of town, and when you are at a family event. If you are part of a team that from time to time receives emergency calls, then a little bit of preparation can go a long way.

Redundancy, redundancy, redundancy! Just as redundancy is good for structures, it is good for engineers. Following the 9/11 attacks in New York, mobile phones were unreliable, and many businesses had no internet access for days. It is prudent to have key information or equipment in more than one place, and to have backup systems. For example, have a list of key phone numbers on a piece of paper in your wallet or purse, also have them programmed into your cell phone or PDA, and also have them on your company intranet. This way if one of these is not available for some reason, you have a backup. Sole reliance on internet-based solutions is not wise, since in major disasters the internet or electricity may not be available.

Keeping an up-to-date list of emergency phone numbers is an ongoing task. This might include mobile and home phone numbers for staff, as well as for key contractors. The list should be updated at least once per year, or preferably twice per year.

Ideally, when an engineer first arrives at the site, he/she will ideally have some basic items such as a hard hat, camera, safety shoes, measuring tape. Proper identification should not be overlooked. Probably the most useful piece of equipment that an engineer can have is a fully-charged mobile phone, preferably with email capabilities. This will allow the engineer to communicate with other staff, brief the client, and generally coordinate activities. Consider giving key staff a duplicate set of basic equipment to keep at home in case they need to respond directly from home.

PUBLIC SAFETY ISSUES

One of the issues that the structural engineer may be asked to assess is whether the situation continues to pose a threat. Is there, for example, a possibility of further collapse?

If the answer is "yes"--or even "maybe"--further consideration is needed. Is the threat aligned toward the general public, or toward a limited population such as rescue workers or site personnel? Are important public utilities, such as mass transit lines, communications facilities, or power transmission lines affected? Is private property threatened?

Once the potential threat has been assessed, possible responses can be developed and evaluated. They may include:

- Restrict public access: This may mean the temporary evacuation of residences and businesses, closing of streets, or the cordoning off of large areas or neighborhoods. The displacement of families and of businesses obviously entails myriad difficulties.
- Stop construction: If the incident occurs on a construction site, construction will usually be halted until the situation is under control. There are situations, however, where it would be prudent to continue work, if that work will bring added stability. For example, if a building shifts while a contractor is installing

underpinning for its foundation, it may not be wise to stop the work until the underpinning is complete.

- Stabilization: Shoring or bracing may be needed in order to prevent further movement or further collapse of a structure. In the case of retaining wall failures or landslides, stabilization might involve regrading the slope, removing soil, and possibly "hardening" the slope with rip rap.
- Demolition: If shoring or bracing cannot be safely installed, demolition of the structure, or of a portion, may be the only remedy.
- Protection: If the hazard cannot be immediately removed, the installation of soft and hard barriers to protect the public or personnel may be an option. Netting can provide a useful means of containing loose debris, or of catching lightweight debris. Sidewalk sheds are often used as hard barriers to shield pedestrians from loose façade material. In the case of the 1997 partial façade collapse of 540 Madison Avenue in New York City, a "sidewalk" shed was even installed across Madison Avenue so that it could be reopened to traffic!

So Much to Learn; So Little Time

Very often in emergency situations, the engineer is called upon to make important decisions, with incomplete information and with very little time. For example, a fire chief may ask if it's safe for his men to enter a building that has been structurally compromised. The engineer may have just arrived at the site, there are no drawings available, and it is dark.

This is an uncomfortable position for most engineers, since it is contrary to our training and nature. We are taught to first gather all of the information, and then carefully analyze it, and finally make our own independent calculations before rendering an engineering opinion. But the nature of emergency situations requires a different type of response, and a few principles may be helpful.

First, it may be helpful to realize that a precise answer may not be needed. Fuzzy answers may suffice. An acceptable answer might be, "I would suggest staying on the left side, under the girder. And of course not staying there any longer than absolutely necessary, and not having more people there than is needed."

The reward should justify the risk. What is the risk, both in terms of probability and of consequence? Obviously a risk that may injure a rescuer or the public is weighed differently than one that may cause additional property damage. And, if successful, what will the reward be? If, for example, if there is a reasonable potential that someone's life may be saved, a high level of risk is justified. Or, if the goal is to recover victim's bodies, a moderate level of risk is acceptable. If there is no rescue or recovery involved, then it will be hard to justify unusual risks, and the normal safety measures that are in-place on a typical construction site would probably be appropriate.

Consider monitoring. It is hard to accept, but some conditions defy analysis. For these conditions the best course of action may be to monitor the questionable structure. The engineer can assist in setting the criteria for monitoring, such as the frequency, the number of points watched, the type of equipment used, and the thresholds.

We must err on the side of safety. This may mean that, lacking information and lacking time, that a very conservative decision is made. This is completely appropriate under those circumstances. The professional should remain open to revising his/her opinion as additional information is received and as additional analysis is performed.

SPEED MATTERS!

In most emergency situations, time is a precious luxury. Often, it is not possible to go through the normal procedures that are used on typical projects. The engineer must look for ways to shorten the amount of time needed to reach each particular goal.

One of the ways to do this is to make the best use of the materials that are already on hand. For example, after the 1998 collapse of the scaffold-like hoist structure at Four Times Square in Manhattan, ways were sought to stabilize the upper part of the structure so that it did not fall. The contractor had on hand extras of the components that part of the hoist structure. After reviewing this components, were able to use mast sections for the hoist as horizontal cantilevers on several floor to support the weight of the hoist structure. And we were able to use lengths of the hoist cables as "slings" that snugged the hoist structure against the building.

Another example is from the World Trade Center Recovery in New York, following the 9/11 terrorist attacks. When a platform was needed for an enormous 800 ton crane that was "on its way", there was no time for a conventional solution. We looked at what we had on hand. Ironically, many of the tower's core columns, which were built-up boxes with 3" plates, were in near-perfect condition in spite of the collapse. Working closely with the contractors, we were able to locate and extract a sufficient quantity of suitable segments in the debris, and we designed a platform using these as the basis.

VOLUNTEERS VERSUS SUBCONSULTANTS

Following the 9/11 attacks on the World Trade Center in New York, the author, then with the firm LZA/Thornton-Tomasetti, directed the engineering associated with the rescue and recovery on behalf of the New York City Department of Design and Construction. This effort continued 24/7 for about nine months, during which time thirty nine engineering firms came under our direction. Were these firms or individuals "volunteers?" Absolutely not. These firms and individuals were engaged as subconsultants and from the beginning were told that they would be compensated. But it is important to understand the multiple meanings that the term "to volunteer" has.

One meaning of "to volunteer" is to perform work without pay. For example, someone may volunteer to work at the local hospital. Typically the duties assigned to a volunteer would be of a routine nature. Brain surgery is not usually performed by a volunteer!

"To volunteer" also means to offer to perform a service of one's own free will. In the military, a volunteer is a person who enlisted for service, as opposed to someone who was drafted. Also, volunteers are sometimes sought in situations where a paid employee is needed to perform a particularly dangerous or distasteful task. We do not normally refer to employees as "volunteers."

There are numerous reasons why it is generally not a good idea for unpaid "volunteers" to provide engineering services. One of these reasons relates to liability and insurance.

Most states have some form of a Good Samaritan Law, as it applies to first aid or emergency medical assistance. For example, as defined in New York State, a Good Samaritan is "...any person who voluntarily and without expectation of monetary compensation renders first aid or emergency treatment at the scene of an accident or other emergency..." [emphasis added]. The purpose of a Good Samaritan law is to keep people from being reluctant to help a stranger in need for fear of legal repercussions if they make some mistake in treatment.

In most states, Good Samaritan laws do not protect engineers. This is needed legislation, and engineering societies are working to persuade lawmakers. Sometimes, however, there is the implication that the hundreds of engineers who "volunteered" in search and rescue efforts at the World Trade Center became financially exposed by so doing. In reality, these engineers were not volunteers; they were sub consultants that were retained from the very beginning. As such, they are covered by an insurance policy made possible by legislation passed by the U.S. Congress, directing the Federal Emergency Management Agency (FEMA) to provide the City of New York with up to $1 billion in coverage for the City and its contractors for claims arising from debris removal performed after the collapse of the World Trade Center buildings. This insurance policy specifically excludes any firm or person who was acting as a volunteer.

Another difficulty with unpaid volunteers is that they tend to be transitory. Except for the most trivial and short-lived of assignments, services are typically required on an on-going basis. It is simply not economically feasible for most firms to be able to provide pro bono services on a sustained basis. Consequently, the type of work assigned to an unpaid volunteer must be carefully selected. It should be of short duration, it should require minimal training, and it should result in a well defined deliverable. For example, rapid assessment of large numbers of residences following an earthquake is a task that is ideally suited for volunteer engineers.

And finally, the adage "you get what you pay for" comes to mind. Although a professional would provide service to the same standard, whether or not he/she was being compensated, it might *appear* to other parties that a substandard level of service was being provided if it was provided gratuitously.

The engineering firms and individual engineers who worked at the WTC served of their own free will, and they served under difficult and sometimes dangerous conditions. From beginning to end, they provided highly professional services that helped deliver a project that was lauded by all for its speed, budget, and remarkable safety. These firms and individuals were not volunteers. They were heroes.

INTERACTION WITH PUBLIC AGENCIES

Most emergencies that affect public welfare will trigger the response of public agencies. For each incident, there will be an organization that serves as the incident commander. This may be the local fire department, or possibly special emergency management agency that provides a coordinating function. Public agencies normally adhere strictly to the "chain of command" concept. To private firms, the chain of command on a particular project may not be immediately obvious and an effort should be made to determine it. The chain of command may also change as a project develops.

For large emergencies, there will usually be a command post at the site. Some municipalities have mobile command centers that have communications equipment, video monitoring capabilities, and meeting space. Regular meetings will often be scheduled, perhaps twice per day at set times, and impromptu meetings may also be held to address specific developments. Private consultants and/or contractors may be invited to these, particularly if they have some special expertise or knowledge regarding the situation.

Public agencies normally will rely on the advice of another public agency over that of a "free-lance" private consultant. Thus, the fire department may seek advice from the local building department regarding the stability of a building. The building department may have the in-house expertise and staffing to serve the project, or it may engage outside consultants as needed. Opinions provided by a consultant that has been engaged by a public agency will carry typically more authority.

If the incident occurred at a construction site, the Occupational Safety and Health Administration (OSHA) will play a role. OSHA will perform an investigation to determine whether safety regulations were not followed by those parties having control over the construction site. Typically, OSHA will exert some degree of control after the rescue and recovery portion has been completed. In the author's experience, the degree of control that OSHA exerts has varied tremendously over the years and from project to project. On L'Ambiance Plaza for example, OSHA took total control of the site and did not seek cooperation with any of the private firms. In other cases, such as the Miller Park Crane Accident, OSHA allowed the private firms to develop a testing program for critical items of evidence and to execute it.

ROLE OF ENGINEERING SOCIETIES

Engineering societies can play a crucial role in a number of ways.

Engineering societies can help provide trustworthy and unbiased information to the media. Often after a newsworthy event, news media reaches out for experts who can comment on the situation. Often the media has difficulty locating someone with the appropriate qualifications and authority. The local engineering society can help by identifying those individuals. Also, the engineering society may consider preparing an informational press release

Engineering societies can sponsor programs that help train their constituents. These program might include training in the use of ATC 20 for the rapid evaluation of buildings following a seismic event, Condition Assessments of damaged structures, and rescue and recovery.

Engineering societies can develop their own emergency response plan. Shortly after the 9/11 attacks, the National Council of Structural Engineers Association (NCSEA) formed a committee that eventually produced a model document titled "SEERPlan Manual (Structural Engineers Emergency Response Plan). Local societies can use as a basis for developing their own plan.

24/7 OPERATIONS—MAINTAINING CONTINUITY

Most emergency situations initially are around-the-clock operations. This is usually the case if the rescue and victim recovery operations are underway, or if there is a threat of additional collapse, or if there is a pressing need to restore a public service. Although some public agencies, like police and fire departments, are organized on this basis, most engineering firms are not. Therefore, engineering firms will be challenged to institute new practices in order to successfully operate in this environment.

Logistically, consideration will have to be given to whether the engineering staff will have an on-site office, or will use another space as their center of operations. The amount of time that the effort will be needed, the location, and other factors will play a part in this decision.

A 24/7 operation will obviously require shifts. Typically, 12 hour shifts and 8 hour shifts are the only options. 12 hour shifts have some definite advantages. First, 12 hour shifts require less of coordination effort, which is important. In an emergency situation, communication lines are often strained, and people may be difficult to reach. Any procedure that minimizes the amount of coordination needed is preferable. Secondly, it makes more effective use of staff. Prior to the emergency, office staff was probably busy with other projects. To suddenly add a project that consumes vast amounts of additional staff time is difficult. The people assigned to the emergency project will need to be using overtime immediately. Staff not assigned to the emergency project will likewise have to be working overtime, to cover the other projects.

Continuity over the various shifts will be important. Consider scheduling briefings at each shift change, where the persons leaving inform the new staff as to new developments and priorities. Also consider requiring a short written report from each team leader at the end of each shift that gets handed to the replacement person and a copy filed. This system was successfully implemented at the World Trade Center.

Conclusion

Natural and man-made catastrophes will continue to occur, possibly with increasing frequency. Structural engineers can play an important role in minimizing loss of life and property damage, and in assisting with rapid recovery. Engineering societies can assist by providing appropriate training to their members.

INVESTIGATION INTO BRICK MASONRY AND CONCRETE FOUNDATION WALL DISTRESS OF A SINGLE-FAMILY RESIDENCE

Timothy J. Dickson, P.E., S.E., Member ASCE[1]

ABSTRACT:

This paper describes the work performed to determine the cause of out-of-plumb masonry veneer and horizontal cracks in a concrete foundation wall of an approximately ten-year-old single family residence. The investigation included a visual inspection of the property, measurement of the lateral movement of the brick veneer and concrete foundation wall and determination of masonry brick ties locations. The investigation was performed on behalf of the homeowner to resolve a construction defect dispute between the homeowner and builder. Based on the findings of the investigation, it was concluded the outward movement of the brick veneer was the result of an insufficient number of brick ties installed to secure the brick veneer to the building. In addition, the bricks used to construct the house were highly expansive. The horizontal crack in the east foundation wall was due to the unreinforced concrete wall not having sufficient strength to resist the applied earth load. In addition, the construction of the foundation of the porch adjacent to the east foundation wall on uncompacted fill material may have contributed to the stresses in the foundation wall as an eccentric load was applied to the foundation wall as the two-story brick enclosure was constructed on the porch foundation.

BACKGROUND:

The subject property is a two-story, wood-framed, single-family residence, built over an unfinished full basement (Fig. 1). The house was constructed in 1994. The exterior of the house is comprised of a four inch-thick brick veneer around the entire house, except for lap siding applied to the second floor level on the north, west, and south sides of the house. The front entry to the house consisted of a two-story brick enclosure. No construction drawings were available for review.

The homeowner stated the apparent movement of the brick veneer was first observed four years after the house was completed, and the movement has continued since that time. In addition, the homeowner stated that a horizontal crack was observed in the east concrete foundation wall when he took possession of the house. The general contractor filled the horizontal crack with a cementitious repair mortar at that time.

[1]Senior Project Manager; American Consulting, Inc., 7260 Shadeland Station, Indianapolis, IN 46256; (317) 547-5580, (317) 543-0270; tdickson@amercons.com

Fig. 1 East elevation of the house.

FINDINGS:

The investigation included a visual inspection of the distressed areas of the building, along with photographic documentation. In addition, a vertically-oriented spinning laser was used to determine the magnitude of movement of both the bowed east brick veneer wall and cracked east foundation wall. In addition, a metal detector was used to locate ties attaching the east veneer to the building and determine if vertical steel reinforcing bars had been placed on the interior face of the east foundation wall.

East Brick Veneer Wall:

An approximately 3/4-inch-wide horizontal gap was observed between the exterior brick veneer and the window frame at the second story window of the master bedroom (Fig. 2). In addition, there was an approximately 3/4-inch-wide vertical gap between the window frame and the brick veneer at the curved head of the window (Fig. 3).

Similar horizontal and vertical gaps were observed between the brick veneer at the front entry door and at the curved-head window above the door. The location of maximum movement appeared to be near the top of the curved window head, and the width of the gap diminished toward the base of the front door. A four-foot-long carpenter's level was placed vertically on the outside face of the window frame above the front entry door. The window frame appeared to be plumb.

Gaps were also observed at the masonry openings for the gable vent above the front entry, the two windows of the east exterior wall of the first floor and the second floor window to the north of the front entry.

Fig. 2: Gap between window frame and brick veneer

Fig. 3: Gap between curved-head window frame and brick veneer

At the north jamb of the master bedroom window, it was possible to view between the brick veneer and the oriented stand board (OSB) sheathing. A corrugated galvanized steel brick tie was visible (Fig. 4). The vertical leg of the corrugated brick tie appeared to have pulled out away from the wall. No building wrap was observed over the OSB sheathing, which appeared water damaged.

A metal detector was used to locate brick ties securing the brick veneer the east exterior wall, and the brick tie locations were marked with chalk on the exterior face of the brick veneer. Since a few brick ties were visible through the gap between the master bedroom window frame and the brick veneer, the accuracy of the metal detector for locating the brick ties was verified. The frequency of brick tie locations appeared to diminish toward the upper portion of the east wall. Seven brick ties were found in an area of approximate 30 sq. ft. between the master bedroom window on the second floor and the first floor window to the south of the front porch, resulting in

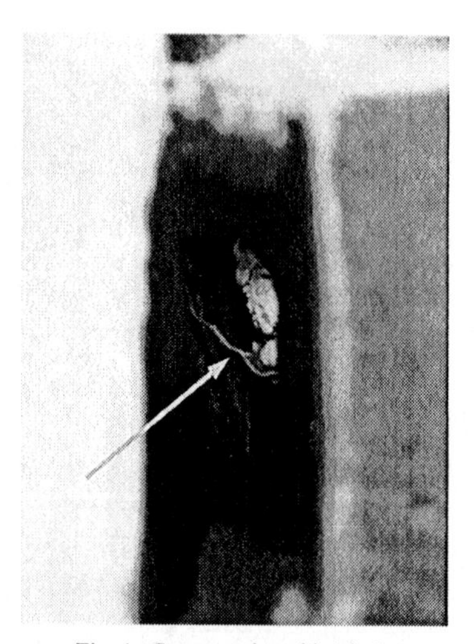

Fig. 4: Corrugated steel brick tie

an average spacing of approximately one tie for every 4.25 sq. ft. of wall. Seven brick ties were also found in an area of approximately 40 sq. ft. above and to the north of the master bedroom window, giving an average spacing of approximately one tie for every 5.7 sq. ft. of wall. At the front entry, only four brick ties were found above the front entry door head elevation, which had a wall area of approximately 50 sq. ft. This equates to one brick tie for every 12.5 sq. ft. of wall.

A vertically-oriented spinning laser was used to determine the deformed shape of the east brick veneer of the house on the south side of the porch and around the front entry and curved-head window within the porch. Using the average measurements at the base of the wall to establish a vertical plane for the theoretical wall position, the measurement indicated the brick veneer at the north jamb of the master bedroom window had moved outward from the building approximately 3/8 in. The masonry at this location was approximately 1-1/4 in. further out from the building than the masonry at the top of the veneer wall. At the curved-head window above the front entry door, the maximum horizontal movement of the brick veneer was determined to be approximately 1-1/4 in. when compared to the average measurements obtained at the base of the wall.

No cracking in the brick or mortar joints were observed in the east wall brick veneer that had experienced the horizontal and vertical movement. No cracks were observed between the porch structure and the east wall of the house.

East Concrete Foundation Wall:

The cast-in-place concrete foundation wall on the east side of the house was approximately 35 feet long and 8.5 feet from the top of concrete slab-on-grade to the top of foundation wall. Wood floor joists bear on the east foundation wall at a spacing of approximately 16 in. on-center.

A horizontal crack was located in the east foundation wall approximately five feet above the floor and extended horizontally for approximately 25 ft. (Fig. 5). The crack extended up and down on both ends and had a maximum crack width of approximately 1/16 in.

Fig. 5: Cracked east foundation wall

A vertically-oriented spinning laser was used to measure the deformed shape of the east foundation wall. Measurements were taken on an approximately seven ft. by five ft. grid, with seven points of measurement horizontally and five points of measurement vertically, for a total of 35 points. Measurements were also taken along the crack. The maximum horizontal movement measure of the wall was found to be approximately 0.75 in. at the mid-point of the wall (Fig. 6). This amount of horizontal movement was also measured at the south end of the horizontal crack, which was determined by using the average of the measurements at the base of the wall to establish a vertical plane for the theoretical location of the non-deformed foundation wall. The distance from this vertical plane to the face of the wall was calculated by subtracting the measurement from the location of the non-deformed wall. The measurements also showed that the top of the east foundation wall had moved horizontally approximately 0.55 in. inward at the mid-length of the wall.

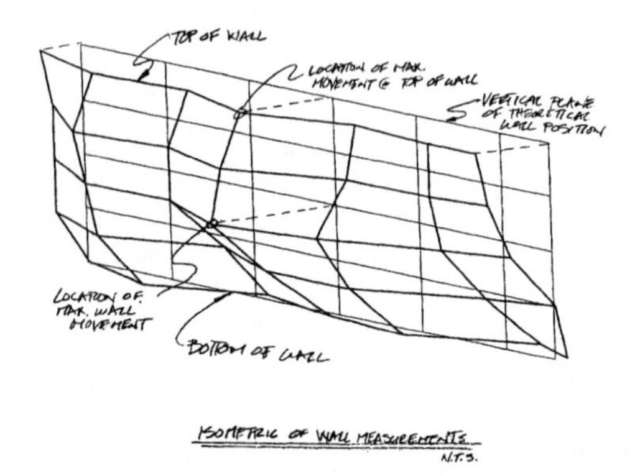

Fig. 6: Isometric diagram of wall configuration

To determine the thickness of the wall, a hammer drill was used to drill a 3/8-in.-diameter hole through the east foundation wall approximately one foot from the top near the mid-length of the wall. Drilling was immediately stopped when the drill bit reached the outside face of the wall. The depth of the drill bit extended into the wall was then measured. A rigid borescope with a 90-degree viewing mirror was also inserted in the hole to locate the outside face of the foundation wall and verify the thickness of the wall. The measurements revealed the east foundation wall was 8 in. thick.

Using a metal detector, an attempt was made to locate vertical reinforcing bars on the inside face of the concrete foundation wall. No reinforcing bars were detected.

DISCUSSION:

East Brick Veneer Wall:

According to Section R-503.4 of the 1990 edition of the *Indiana One and Two Family Dwelling Code* one corrugated steel anchor must be provided for each 3.25 sq. ft. of wall area. The field measurements found areas, particularly at the upper portions of the east wall that did not meet this code requirement. The fact that the number of brick ties diminished toward the upper portion of the wall also verified the veneer was not sufficiently anchored. More movement, both vertical and horizontal, was observed at the upper portion of the veneer than the lower portion.

According to the Brick Industry Association's *Technical Note 18 – Volume Changes and Effects of Movement, Part 1*, bricks expand slowly over time when exposed to moisture or humid air. However, when a brick looses the gained moisture, it does not return to its original size. A brick unit is at its smallest size when it cools after being removed from the kiln. It increases in size due to moisture expansion from that time, with most of the expansion occurring within the first few months, but expansion will continue for several years. The moisture expansion characteristics of a brick unit depend on the raw materials and the firing temperatures used. Bricks made from the same raw materials fired at a lower temperature will expand more due to moisture than those fired at a higher temperature. For this particular case, the observed expansion of the brick material was greater than the typical expansion expected.

This expansion, along with the fact that an insufficient number of brick ties were installed, apparently has allowed the brick veneer to move outward, away from the wood framing, creating the gaps around the windows. In addition, the type of brick ties used for this project does not appear to be effective since a deformed brick tie was observed. Since corrugated steel brick ties are nailed to the exterior wall studs and then bent into an "L"-shape, it is possible for the brick veneer to move laterally before the brick tie becomes engaged and provides lateral restraint, as was observed near the north jamb of the master bedroom window. The vertical leg of the bent brick ties shows evidence of lateral movement as it was pulled by the expanding brick veneer. Therefore, the brick ties failed to adequately restrain the horizontal movement of the brick veneer. Other type of brick ties are available that are configured such that this horizontal movement will not occur prior to the tie becoming engaged.

Another item that was not in compliance with the building code was that fact the no building wrap or water-resistant barrier was observed behind the brick veneer. Sections R-503.5, R-503.7 and Table No. R-503.6 of the 1990 edition of the *Indiana One and Two Family Dwelling Code*[1] indicate that a weather-resistant barrier is required to be installed over non-water repellant sheathing behind brick veneer. This barrier is required to both protect the wood sheathing from deterioration due to exposure to moisture, as well as to direct water in the cavity to the flashing and weeps at the base of the wall.

East Concrete Foundation Wall:

Structural calculations were prepared to determine the ability of the 8-in.-thick concrete wall to resist code-prescribed loads. For the analysis, an equivalent fluid pressure of 60 pounds per cubic foot (pcf) for the "at-rest" soil pressure on the wall, and a 28-day compressive strength of the concrete of 3,000 pounds per square inch (psi) were assumed. The analysis was performed in accordance the American Concrete Institute's *Building Code Requirements for Reinforced Concrete* (ACI 318-89). The results of the structural analysis indicated that the 8-in.-thick concrete wall was overstressed and therefore insufficient for a wall height of 8.5 ft.

The depth of the concrete foundation for the brick porch enclosure was determined to be approximately 32 in. below grade by others during previous work on the property. The depth of the east foundation wall was most likely approximately 9.5 ft. below grade, and the porch foundation appears to have been anchored to the east foundation wall. Therefore, any settlement of the porch foundation would impose an eccentric force at the top of the east foundation wall (Fig 7). Since it is likely that the porch foundation was constructed on backfill material placed against the east foundation wall, it is likely the porch foundation has experienced some settlement. This eccentricity would create a moment at the top portion of the wall developing increased tensile stress on the inside face of the foundation wall. The fact that the horizontal crack terminates with vertical cracks rather than diagonal cracks that extend to the corners of the foundation wall also indicates that the eccentricity of the porch structure contributed to the formation of the crack.

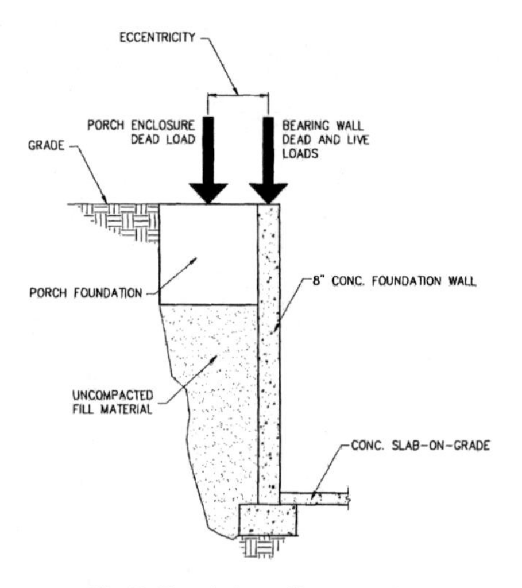

Fig. 7: Foundation wall cross-section

If the porch footings had been constructed on undisturbed soil at the same depth as the basement footing, the crack would not have developed since the likelihood of settlement of the porch would have been greatly reduced and the porch foundation would not have applied an eccentric load to the east basement wall. However, the 8-in.-thick foundation wall still would not have met the thickness requirement prescribed in the building code for plain concrete foundation walls.

The analysis did not take into account the effect the foundation of the porch at the front of the house had on the forces applied to the east foundation wall.

Since the horizontal crack was present when the homeowner took possession of the house, it is likely that the crack developed during construction. The horizontal crack may have developed as the masonry was being placed for the porch enclosure. The application of the dead load of the bricks would have been a slow process. As the enclosure was built, the porch foundation could have settled, and the masons would have unknowingly filled any potential cracks with fresh mortar.

A review of the *Indiana One and Two Family Dwelling Code* (1990 Edition) regarding requirements for unreinforced concrete foundation walls was performed. Table No. R-304.3a indicated that for an 8-in.-thick unreinforced concrete foundation wall, the maximum unbalanced backfill is 7 ft. Since the existing grade was observed at the top of the foundation wall, the unbalanced backfill is 8.5 ft for this particular case. Therefore, the 8-in.-thick concrete foundation wall did not meet the building code requirements.

The fact that the horizontal crack in the concrete foundation wall was repaired 1994 and has reopened indicates the foundation wall is still experiencing some lateral movement.

CONCLUSIONS:

East Brick Veneer Wall:

The outward and vertical movement of the brick veneer was believed to be the result of two factors. First, the bricks appear to be a highly expansive material, a statement supported by the amount of both horizontal and vertical movement observed at the second floor level of the house. As the brick veneer expanded due to thermal and moisture changes, the outer face of the wall wants to grow longer. This difference in expansion between the inside and outside face of the veneer causes the wall to bow outward. Vertical movement was observed at the top of the curved window head, most likely due to a cumulative effect of the brick expansion. Secondly, the type and number of brick ties installed were insufficient to adequately secure the brick veneer to the wood-framed structure. The brick ties were observed to have deformed, which allowed the veneer to move outward, and the number of brick ties installed did not meet the building code requirement.

East Concrete Foundation Wall:

The investigation into the cause of the horizontal crack in the east foundation revealed the crack was due to the fact that the foundation wall was improperly constructed and the eccentricity created at the top of the wall as the porch foundation apparently settled. The unreinforced 8-in.-thick concrete wall was too thin to resist the earth pressure, and the added stresses of the porch foundation eccentricity compounded the problem. The settlement of the porch foundation is believed to have occurred slowly as the brick porch enclosure was constructed. If it would have occurred after the house was completed, vertical cracks would likely have developed between the east brick veneer wall and the perpendicular walls of the porch enclosure.

These two issues are not believed to be interrelated, as was stated by other engineers that inspected the property. Since there was no mechanical connection between the concrete foundation wall and the base of the brick veneer wall, rotational movement at the top of the foundation wall will not be transferred into the brick veneer. If this movement was transferred, it would have been localized at the base of the veneer. Movement of the top of the foundation wall would not cause movement of the brick veneer at the second floor level where the greatest movement of the veneer was observed.

Based on the conclusions of the investigation, recommendations for remediation work to correct the problems with the exterior brick veneer and concrete foundation walls were developed. Using that information, a general contractor was able to develop anticipated construction costs to perform the repairs. The homeowner used that information to negotiate a settlement with the builder of the home during mediation proceedings.

ACKNOWLEDGEMENT:

We wish to thank Mr. Mike Mathioudakis for his assistance on this project.

REFERENCES:

1. State of Indiana (1990), *Indiana One and Two Family Dwelling Code* (1990), Indianapolis, Indiana.

2. American Concrete Institute (1989), *Building Code Requirements for Reinforced Concrete* (ACI 318-89), Detroit, Michigan.

3. Brick Industry Association (1991), *Technical Note 18 – Volume Changes and Effects of Movement, Part 1*, Reston, Virginia.

FORENSIC INVESTIGATION OF A PATIO FAILURE

By Andrew Halter, A.M. ASCE[1]

[1]NRC Engineering, PO Box 657, Delaware, OH 43015; PH (740) 363-6021; FAX (740) 363-9419; email: andy@nrcengineering.com

Abstract

A patio under construction failed catastrophically, resulting in a back yard full of concrete blocks and rubble fused together with sand and cement. My assignment was to determine the cause of failure.

A homeowner had laid concrete unit masonry blocks around the perimeter of the patio. A combination of flowable fill (a mixture of sand, cement, fly ash, and water) and debris were used in the center as a substrate for the slab which was to be poured at a later time. The patio failed during the final pour of flowable fill. The patio failed because the moment created by the fluid pressure of the flowable fill against the wall overcame the adhesive force of the mortar between the blocks

Introduction

A homeowner who has a historic house with a rear entry that is elevated above grade decides to install an elevated patio. To preserve the architectural characteristic of the foundation line, he plans to construct the perimeter foundation of the patio with 9.2 cm concrete blocks clad with stone to match the existing building exterior.

After a concrete foundation is poured to form the perimeter of the patio, the 9.2 cm blocks are laid to an elevation of 1.2 m. The blocks are not steel-reinforced and have no lateral reinforcement. The interior of the patio is partially filled with construction rubble. The homeowner has a friend who works at a local concrete plant whom he has enlisted to help pour the top slab. The friend suggests that the interior of the patio be finished off with flowable fill, which is sturdy and self-leveling, to provide an excellent base for the slab. The homeowner agrees that this is a reasonable plan, and the friend begins to bring in truckloads of the material to top off the interior of the patio.

During the last pour of flowable fill, the structure fails catastrophically, destroying most of the walls and causing the interior of the patio to cement together into one solid mass of concrete (Figure 1).

Figure 1. Failed patio.

As a forensic engineer, my assignment was to investigate this case and determine the cause of failure.

Discussion

Solid and *liquid* are terms of convenience for discrete materials. The human mind seems configured to understand these categorically. Instinctively, we feel that something which is solid (or having properties of a solid) will not exert lateral pressure. Isaac Newton understood that lateral forces come from a solid object only in response to the forces applied to it; in the absence of external forces, the solid object is content to just sit where it is. Also instinctively, we feel that a solid should not readily change its shape, or flow. For example, we assume that the giant boulder at the park will not become more bottom-heavy over time as gravity causes moving material to accumulate toward the bottom of the rock (antique glass mythology not withstanding) (Gibbs, 1). A liquid is commonly defined as something that has no intrinsic shape and so takes the shape of its container.

Yet few things in nature are either discrete or homogenous. For example, is the earth a solid or a liquid? Most natural things defy these easy categorizations. Yet, civil engineers on a regular confront such "grey areas". One strategy is to use empirical information rather than theories. I have long since given up trying to find a solution for the Navier-Stokes equations. That is why Reynolds number scaling works well; if I can determine how a small, easy-to-handle model reacts, Reynolds allows me to scale a pond up to an ocean.

Figure 2. Fill flowed in between pieces of rubble.

For irregular mixtures of materials, sometimes the only descriptor that an engineer can grasp is the angle of repose, which describes how much slope there will be if the "stuff" is made into a pile, and the edges examined. Such heterogeneous irregular mixtures of materials are constantly surprising us with their properties. New, counterintuitive states for "stuff" are still being discovered today.

Some rudimentary guidelines for irregular mixtures can be formulated. Mixtures with large, rough aggregate tend to have a high angle of repose, which is to say they behave more like solids than like liquids. Mixtures with small and/or smooth aggregate and high percentages of water tend to have low angles of repose, which is to say they behave more like liquids. This does not always hold true, as traditional concrete contains a large amount of water, yet after it is placed, a worker can form control joints in its surface which easily hold their shape.

Flowable fill, according to the U.S. Environmental Protection Agency (EPA), is a mixture of sand, cement, fly ash, and water. The agency specifies flowable as having a slump of 20.3 cm or higher (EPA, 1). Traditional structural concrete mixtures have slumps in the range of 10cm. In this case, when the homeowner and concrete contractor brought in the flowable fill material, they likely thought that it would behave like traditional concrete. Unfortunately, the design requirement for a wall that must resist the lateral forces of a liquid is vastly different from the design requirement for a wall which is not subjected to lateral forces.

Soil is one material that falls somewhere between a solid and a liquid. The Uniform Building Code provides some guidelines for minimum designs that may be necessary for a sub-grade wall. However, it has been my experience that most contemporary builders vastly underestimate the power of lateral soil forces. And why wouldn't they? When foundations and basements are excavated, the soils do not (hopefully) immediately flow in. It is easy to think instinctively that soil resembles a solid more than a liquid.

Returning to the case in question, the homeowner filled in the center of his patio with miscellaneous debris. The construction debris was solid enough on a piece-by-piece basis; it was mostly chunks of concrete. In fact, it even passed the homeowner's own informal angle of repose criteria: he had to fill in the gap between the central pile and the block wall itself with small pieces. In the homeowner's estimation, very little lateral pressure was being transferred to the wall. In fact, the debris used by this homeowner has similar properties to typical granular fill used as substrates for concrete slabs.

Masonry performs well in compression, but an order of magnitude less in tension. In fact, in the design of steel-reinforced concrete beams, the tensile strength of concrete is traditionally ignored as being non-contributory. The resistance of lateral forces in concrete block walls almost always entails installing steel in the assembly to resist the tensile forces, which this homeowner failed to do.

When the flowable fill was placed on top of the loose fill material, it began to fill in the gaps between pieces of rubble (Figure 2). More importantly, it also began to do exactly what the contractor envisioned: accumulate on top of the rubble and form a pond. It acted as a fluid in that it flowed to the edges of the patio and formed a liquid layer on the top, which began to increase in thickness. Conventional substrates for concrete slabs require measuring, working, and compacting to achieve a level, substantial surface suitable for pouring over. Flowable fill would seem to bypass all of these steps and consolidate the base by infiltrating the voids beneath and, being a fluid, leveling itself.

Figure 3. Fill and blocks in yard.

The one factor that the contractor overlooked was that a fluid will exert great lateral pressures on the surfaces it contacts. With a weight of about 2000 kg/m^3, the flowable fill is twice as heavy as water. The concrete masonry units which made up the patio perimeter, which were never intended to function as a formwork, unintentionally became a formwork. Even though the fluid formed a relatively thin layer, and the lateral stresses themselves were within the range of what the blocks could withstand, the height of the walls allowed the resultant force to generate a large moment at the base of the wall. The wall was overwhelmed and toppled outward, resulting in a yard full of concrete blocks (Figure 3) and thousands of pounds of rubble fused together with sand and cement.

Analysis

To find the moment at the base (hinge) of the wall caused by the flowable fill, I integrated from the bottom of the flowable fill to the top of the pour

$$M_B = \sum M_B = \int_{h-t}^{h} fx\,dx = \tfrac{1}{2}\gamma h x^2 - \tfrac{1}{3}\gamma x^3 \Big|_{h-t}^{h}$$

where f is the forcing function of the fluid pressure at the elevation x such that $f = \gamma(h-x)$ where γ is the unit weight of the flowable fill. h, t are the height of the wall and the thickness of the pour, respectively. The distance x assumes that the wall has no thickness, for conservative simplification.

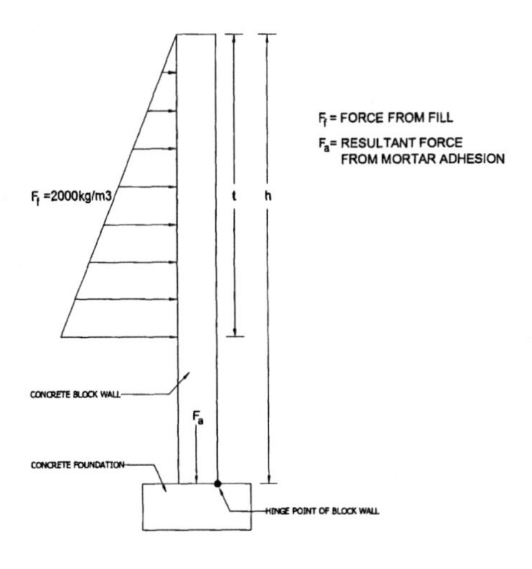

Figure 4. FBD for masonry wall.

To find the moment at the base (hinge) of the wall caused by the adhesion of the mortar to the block, I summed the contributing moments from each element of the block where mortar is in contact

$$M_B = \sum Fx = 3F_1x_1 + F_4x_4 + F_5x_5$$

where F is found by multiplying the area of the element by a, which is the adhesion of the mortar to the brick. The average adhesive force of type N mortar on masonry prisms was assumed to be about .72MPa at a 90-day cure (Tahal and Shrive). The distance x is the farthest point from the hinge on the individual element being analyzed, for conservative simplification. The width, length, and web thickness of the brick is assumed to be 9.2cm, 39.7cm, and 2cm respectively.

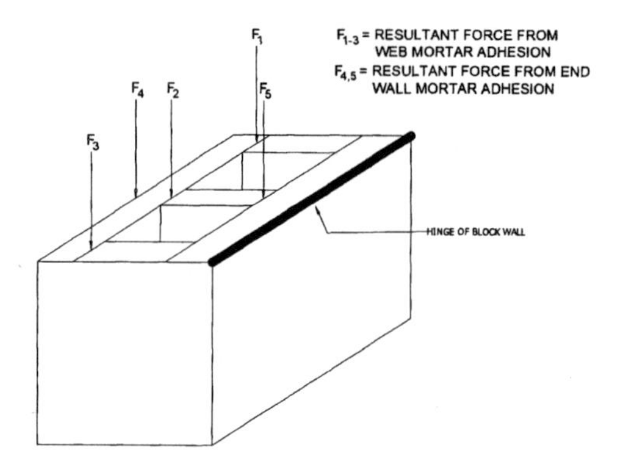

Figure 5. FBD for mortar adhesion to masonry.

Conclusion

The patio failed because the moment created by the fluid pressure of the flowable fill against the wall overcame the adhesive force of the mortar between the blocks. The maximum thickness of flowable fill that these concrete block walls would be able to resist is .48 m.

I have not seen flowable fill used prior to this encounter. In fact, the world of engineering is constantly seeing new materials and methods being introduced as old ones are phased out. The key to making proper design decisions for a structure is to always understand the nature of the materials involved and envision the potential perils.

References

Gibbs, Philip. (1996). *Is Glass Liquid or Solid?*
http://math.ucr.edu/home/baez/physics/General/Glass/glass.html

National Ready Mixed Concrete Association. (2006). *Flowable Fill,*
http://www.flowablefill.org

Shrive, N.G., and Tahal, M.M. Reda. (2001). *The Use of Pozzolans to Improve Bond and Bond Strength,* Proceedings of the 9th Canadian Masonry Symposium,
http://www.unm.edu/~mrtaha/Publications/Fredericton-2-2001.PDF

U.S. Environmental Protection Agency (EPA). (2006). *Flowable Fill Content Mixtures and Specifications,* http://www.epa.gov/cpg/products/flow-ast.htm

The Leaking Basement Epidemic – Causes, Cures and Consequences

Stuart Edwards, P.E., M.A.S.C.E[1]

[1]Verdant Energi and Environment, LLC, 5815 Crabtree Lane, Cincinnati, OH 45243
PH (513) 271-0623; FAX (513) 271-0623; stuart_edwards@verdantenv.com

Abstract

One need only consult the local Yellow Pages under 'Waterproofing Contractors' to appreciate the scale of the leaking basement epidemic in the United States. This paper draws from a detailed case history to look at its causes, financial ramifications, and ideas on mitigation. It concludes that most often the failures are systemic involving multiple factors, but that very significant among these is the role of the downspout drain, its frequent deterioration over time and consequent loss of functionality. This is compounded by poor design decisions that have resulted in walls unable to support the lateral loads that are created when hydrostatic pressures develop. Solutions range from proactive maintenance to strengthening of walls. However, there must first be more public awareness of the causes and liabilities involved, and this may best be accomplished through more focused attention when property transfers occur.

Introduction

The 'Waterproofing Contractors' section of many local Yellow Pages is a strong indicator of the scale of the leaking basement epidemic in the United States. In Cincinnati, Ohio, for example, there are 12 pages of them representing some 34 contractors who serve a population of about 2 million (Cincinnati Bell, 2005). What has happened to the housing stock that makes it necessary for the 'Basement Doctor', a Superman look-a-like and 'Gotta Crack Call Jack' to exist?

In attempting to answer this question, the paper draws primarily from a detailed case history where a home with seriously distressed basement walls was purchased for investment. Evidence that there is indeed an epidemic became obvious from screening a large number of additional investment prospects. From an engineering stand-point, the scale of the problem was puzzling and an effort was made to understand the causes and financial ramifications. Ideas have been developed on how to mitigate the problem and the huge financial burden that it represents.

Case History

Equity Erosion: The home is a colonial style, two-story, brick, single-family residence built in 1951. In May 2003 it was offered for sale at a price of $790,000, and shortly thereafter an offer was accepted contingent upon the usual inspections. Removal of some paneling in the basement revealed that the walls not only leaked, but that the west wall was seriously cracked. That deal fell through and the house was returned to the market at a price of $649,000. Foundation problems tend to have a chilling effect on property value, and the eventual purchase price was $555,000; close to the value of the 1.25 acre lot.

This may be an extreme case, but clearly a cracked basement wall is not just a nuisance, it is potentially a huge financial liability. The seller lost 30 percent (%) of the value of a premium property, and arguably 100% of the value of the structure. There was a large equity and the owner could absorb the loss, but if the purchase price had been below the remaining mortgage balance, default and foreclosure would have been likely. It should be clear then, that this is not just a problem for homeowners, but for mortgage lenders too.

Property Condition: The house is a two-story brick structure with a tile roof. Except for the garage and laundry room that have slab-on-grade foundations, there is a complete basement with 8-inch [200 millimeter (mm)], non-reinforced concrete walls. The cracked west basement wall beneath the living room had bowed inwards 1-3/4 inches (45 mm) and the larger cracks were open about 3/8 inches (9 mm). Leakage occurred along some of the cracks during and after rainfall.

Historically, previous owners had taken measures to mitigate leakage by installing an interior perimeter interception and drainage system along the westerly basement wall. These measures were reasonably effective at controlling water once it entered the basement. A remedial system in the non-paneled portion of the basement included the use of large fiber-glass sheets riveted to the wall that allowed water to run unobserved behind them into the sub-floor drainage system. A detrimental aspect of this system was that cracks could no longer be seen either, and so were 'out of sight, out of mind'.

The performance history of the walls was pieced together roughly from available records of a) efforts made by previous owners to alleviate their water problem and b) periodic remodeling initiatives. A major remodeling appears to have taken place in about 1963 that included finishing a family recreation room in the portion of the basement beneath the living room. While not conclusive, it suggests that the owner was not overly concerned about the moisture conditions or the state of the west wall at that time. The work involved the attachment of 1 inch x 2 inch (25 mm x 50 mm) furring strips to the concrete walls and installation of rough finished 1/2 inch (12 mm) pine paneling. This remodeling is significant in that it undoubtedly prevented direct observation of the condition of the west wall in this part of the basement in the ensuing years.

In 1988, an effort was made to control water seepage through two cracks that had developed in the northern part of the west basement wall, beneath the dining room / first floor hall. This remediation consisted of a trench drain leading to a sump, and fiber glass panels attached to the wall that directed water from the cracks to the drain. The sump appears to have been designed to discharge into the drainage system provided for the roof drain down spouts. The floor drain portion of this system was still operational in 2003 and appears to have successfully controlled the flow of seepage water within this part of the basement.

In June 2000, a contractor proposed to conduct remedial work on the recreation room drainage system. A new catch basin and metal grate were constructed at the outside door to the room, and new discharge piping installed.

May 2003 saw further remedial drainage works related to the recreation room. A drain was installed on the outside of the west basement wall in an apparent effort to relieve hydrostatic pressure on the wall and redirect surface water flows. The wall was also reportedly waterproofed. Two cracks in the east wall of the furnace room were 'injected', although no performance warranty was offered. Significantly, both allowed water into the basement during and after periods of heavy rain on several occasions in 2003 and 2004. The remediation in 2003 was an apparently failed attempt to deal with the drainage issues in preparation for sale of the property.

In hind-sight, it might have been more cost effective to demolish this structure and rebuild a modern style house. However, it was a strikingly good looking and potentially functional home. This, together with the perceived environmental benefits of 'recycling' it argued in favor of renovation. Naturally, a key element of this was to stabilize the structure and deal with the water. Fortunately, the brick shell of the house was perfectly intact and beyond the basement walls themselves, cracked plaster was the only damage to the building envelope.

Stabilizing the Basement Walls and Eliminating the Leakage: A completely dry and structurally sound basement was essential for an upgraded family room that would be a valuable feature of the home. To accomplish this it was first necessary to answer some fundamental questions.

If the walls had performed satisfactorily for a decade or two after the house was completed, what happened to change things? The walls didn't 'wear out' like a shingle roof. Concrete continues to gain strength (slightly) indefinitely and so should have improved over time rather than deteriorating. The only other variable is the loading condition. In addition to the weight of the house, the major load on the west basement wall is the pressure created by soil backfill acting laterally, together with hydrostatic pressure (again a lateral load) caused by water that happens to become dammed up against it. The pressure created by the backfill is unlikely to change significantly as long as the moisture conditions remain fairly constant. This leaves hydrostatic. The key issue here is not whether hydrostatic pressure can create a critical load condition, but why did it increase over time?

The answers to this question appeared to be obvious with the first heavy rains when the gutters started to overflow, flooding the area adjacent to the house and above the basement wall. It was assumed that the gutters and/or down-spouts were blocked. Upon further investigation, it was found that the down-spouts were clear, but had a significant head of water in them. The down-spout drains were blocked. Further investigation of several of the offending pipes revealed that they were blocked solid, from end to end, with roots and decaying organic matter.

Another observation during times of rain was that surface water pooled for long periods of time in the front yard adjacent to the house. The lot slopes down from the street towards the house which tends to impede the natural drainage causing ponding at several locations along the west wall. It is hard to believe that the original builder paid so little attention to surface drainage. Even if it was marginal to start with, most likely it was functional. So an additional factor was needed to explain changes that could account for subtle grade modifications in the front yard near the house.

The answer was not obvious until some months later during installation of an in-ground utility sink in the laundry room. Having cut a rectangle out the concrete slab-on-grade it was obvious that the soils beneath it had subsided about 12 – 15 inches (300 – 375 mm). This would have been backfill behind the adjacent basement wall (furnace room west wall) and the surface of the soil showed the clear imprint of the concrete slab that had been in direct contact at the time it was poured. With periodic, prolonged periods of saturation caused by overflowing gutters and ponding, it is reasonable to surmise that poorly compacted clay backfill had settled significantly over time. With only modest surface drainage grades to start with, it is easy to see how backfill-related subsidence adjacent to the house, could have resulted in enough grade changes to disrupt drainage and cause ponding.

The problem was now clearly defined and there were two actions that had to be taken to eliminated the source of the water, and help reduce the loads on the basement wall. First, all the down-spout drains had to be replaced, and second, the front yard near the house had to be regraded to provide proper drainage away from the structure. Regrading involved creating a swale the length of the house about 15 feet (about 5 meters) from it. All the front landscaping was sacrificed and had to be replaced, but the result was a remarkable and immediate improvement. Watching a torrent of water several inches deep in the swale during a storm drove home just how serious the stress on the building envelope had been when much of this had just dammed up against the house. While the regrading was in progress, new 6-inch (150 mm) diameter drains were installed and connected to the down spouts. Large diameter piping was selected after seeing the non-functional state of the original 4-inch (100 mm) lines. The pipes were laid so that they daylighted in the side yard, and discharged onto a small rock apron protected by dry stone wing walls. This will allow future inspection and maintenance of the pipes, a feature that was not present previously when disposal of roof run-off relied on tile drains.

There were three reasonably feasible options for strengthening the basement walls:

- Complete replacement
- Soil anchors
- Reinforcement
  - a) steel
  - b) externally bonded fiber reinforced polymer (FRP)

Bonded FRP was selected primarily because it is minimally disruptive (compared to total replacement), unobtrusive when complete (compared to most steel reinforcement and anchoring systems) and overall costs were reasonable. As a secondary factor, contractors that are familiar with this technology also tend to be knowledgeable about epoxy crack injection for both leak abatement and structural restoration. This was important for the other objective, the dry basement. All cracks were injected with epoxy to restore waterproofing. Several of the larger cracks were injected with structural epoxy to restore strength to the distressed concrete section.

The FRP product consisted of 6-inch wide (152 mm) kevlar/carbon-fiber woven 'straps' bonded vertically to the surface of the concrete with epoxy every four feet (1220 mm) along the wall. This creates the equivalent of tensile reinforcement on the bowed face of the wall. Initially, straps were applied only to the recreation room west wall since it was in the worst condition. Subsequently, the decision was made to reinforce the whole of the west basement wall. This was based on the notion that if stress conditions could develop that would cause serious distress at the south end of the wall, there may in the future be nothing to prevent those same conditions developing at the north end, with similar results.

The net result of all this work was that the basement has remained completely dry and stable for more than a year.

The installation contractor provided a lifetime warranty on the straps. However, the approach to design seemed simplistic and could most generously be described as 'empirical'. Straps are always placed on 4-foot centers (1220 mm). That is the design. To evaluate this further, structural analysis of the wall was conducted for both before and after remediation scenarios. An American Concrete Institute design and construction guide (ACI 440.2R-02) provides a valuable resource for information on the application of bonded FRP, and the British Cement Association has published a PC-based software package for the design of simple reinforced-concrete structures including basement walls (Goodchild, 1999). The generalized loading conditions are shown in Figure 1.

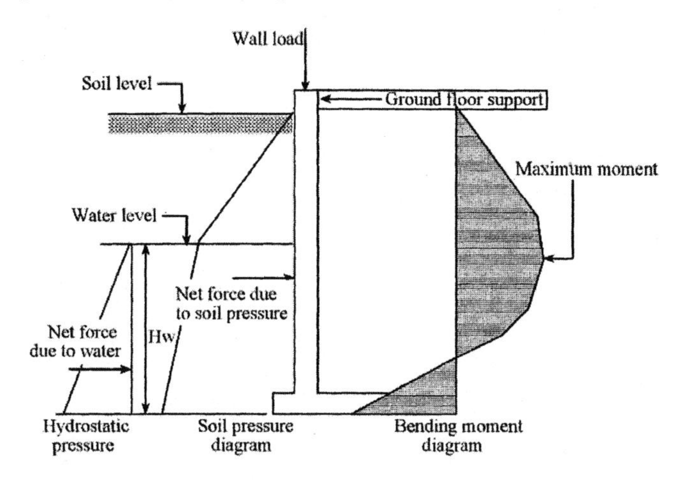

Figure 1. Cross Section to Show Basement Wall and Primary Imposed Loads (not to scale)

Input data based on field measurements of the geometry of the structure, laboratory strength testing of a concrete core recovered when forming a penetration for new HVAC duct-work, and estimates of soil properties based on field observation were input to the spread-sheet based model. The results of interest are the maximum bending moment creating tension on the inside face of the wall, and the moment resistance of the concrete section.

Lateral loads were assumed to be those imposed due to the 'at rest' earth pressure of the silty clay backfill. This was likely the load condition for the first decade or more, and the wall appears to have successfully withstood it. Once the drainage situation started to deteriorate, hydrostatic pressures of ever-increasing magnitude could be assumed to develop. At some point during this process, the wall cracked and bowed. The soil pressures were probably reduced a little as some strength was mobilized and were probably closer to 'active' soil pressures. At the same time the water now had a drain since it could flow into the basement through the cracks, and the hydrostatic pressures should have been lowered somewhat. Following the 2004 remedial drainage work, the potential for high hydrostatic pressures was substantially reduced. On the other hand, the improved integrity of the wall following its repair eliminated pressure relief through the cracks, and high pressures are again theoretically possible if there is a future failure of the drainage system.

The retained soil is silty clay and, in the absence of strength test results, a range of possible friction angles were examined. The graph below (Figure 2) shows the maximum moment in the wall for soil friction angles between 20 and 40 degrees and a range of ground water levels (Hw) from full depth (water at the ground surface) to no water. The soil at the site is assumed to have a friction angle towards the bottom of the range; say 20 – 30 degrees.

Figure 2. Bending Moment Imposed on Basement Wall Due to Soil Pressure and Hydrostatic Loading

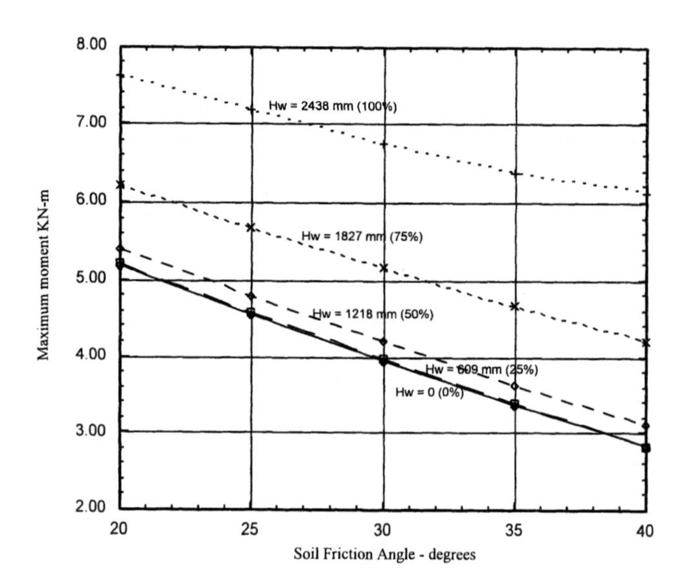

The impact of a rising water table is clear. At low water levels, (up to about 50% of the height of the wall) the presence of the water does not significantly increase the maximum moment. However, as the water level rises beyond this, it increases quite dramatically. For a 25-degree friction angle soil, the initial 'dry conditions' moment (4.55 KN-m) increases by about 5% for the first 4 feet (1209 mm) of the water level increase and by 50% to 7.18 KN-m for the next 4 feet (1209 mm).

The calculated moment resistance of the wall once the straps have been applied is 8.5 KN-m. This is computed using material factors for the concrete and FRP of 1.5 and 1.17 respectively, and exceeds the maximum moment that would be anticipated under extreme hydrostatic loading conditions, thus validating the overall design proposed by the contractor.

The moment resistance of the wall prior to cracking (assuming nominal water pressure and 'at rest' soil pressure) must have been between 4 – 5 KN-m, although it is difficult to show by calculation that a value this high is feasible in a non-reinforced concrete wall. The calculated moment of resistance of a non-reinforced wall section 8-inches in thickness is on the order of 0.6 KN-m. This may imply that the typically assumed value for the tensile strength of concrete (10 – 15% of compressive strength) is conservatively low, or that the analysis of the wall as a two-dimensional 'strip' of the overall structure is considerably conservative because it ignores three-dimensional support effects. Alternatively, the soil loading conditions immediately following construction may have been temporarily reduced. In any event, the longer term loading conditions were clearly untenable.

Costs: This was an expensive repair, although compared to the loss of equity that the former owner suffered, it is almost trivial. The costs are typical of those that must be incurred (scaled as appropriate for the size of the property) to properly stabilize a structure when the basement has reached an advanced state of disrepair.

Carbon fiber/Kevlar straps	\$8,000
Crack injection (structural and waterproofing combined)	\$6,000
Regrading and downspout drain replacement, reseeding	\$5,000
Landscaping and walkway replacement	\$5,000
Total (2004)	\$24,000

The portion of the basement wall that was strengthened measured about 100 linear feet (30.5 m) or about \$80/foot (\$263/m). About 170 linear feet (51.8 m) of cracks were injected at \$35/ foot (\$116/m) of crack (\$60/foot (\$197/m) of linear wall). The total concrete related repair costs were, therefore, on the order of \$140 per linear foot of wall (\$460/m). Regrading required the creation of 90 feet (27 m) of swale, about the same amount of buried 6" diameter pipe and two head walls. The affected area that was hydroseeded was on the order of 2,000-2,500 square feet (186 – 232 square meters).

Discussion

Each element of this remedial program, could be viewed separately: the strengthening of the wall, the leak-proofing of the wall, the provision of new drains and the modification of the surface grades to promote drainage. Indeed, all are quite independent activities and this is often how they are sold. But this misses the bigger point and, perhaps, provides a clue as to why the leaking basement problem is widespread. At its root was a systemic failure, the seeds of which were planted before the builder left the job site. It involved every phase in the life cycle of the building from conception to (very nearly) its destruction.

First was a failure of the design process and, more importantly, of that type of design that relied on simple building code 'engineering'. The wall should have been designed to withstand much higher lateral loads. Granular backfill and a drain at the heel of the wall would have been helpful too as the above graph shows. However, in 1951 all the code required was that a non-reinforced concrete wall be at least 8 inches thick (Department of Building Inspector, 1950). In defense of the code it must be said that this was followed by a statement putting responsibility squarely back on the designer's shoulders:

> "When the stresses due to earth pressure, surcharges and superimposed loads exceed the maximum working stress permitted by this Code for the material used, and the additional stresses are not otherwise provided for, the wall thickness shall be increased to bring the stresses within the required limits."

A little reinforcing steel might have been more useful; but at the least some extra concrete would have helped. The International Residential Code would, today, require #6 bars 24 inches (20mm diameter, 610 mm) on center in this wall (ICC 2003).

Severe subsidence of the backfill around the basement was a construction failure attributable to the builder. This outcome was reasonably predictable and suggest that high levels of construction quality control (by today's standards) are not likely to have been employed during this era of home construction.

Tile drains for disposal of roof run-off are bound to fail eventually. No matter how effective the leaf guards on the gutter, some solid matter is going to be washed into it. This will eventually clog the drain and it will cease to function as anything more than a small reservoir. Drains that discharge to the surface stand a better chance of long term survival, but it is not unusual for their discharge points to become blocked with debris and for roots to invade them. This speaks to a failure of maintenance. Once the drain fails, the roof run-off can only overflow the gutter, discharging on top of the poorly compacted basement backfill.

And finally, there are some institutional failures that are contributing to the problem. When the renovated house was sold, the new owner commissioned a detailed inspection. The inspector was, overall, very thorough. However the section dealing with the drains was painfully weak:

> "Gutters: Aluminum; Downspouts: Aluminum; Leader/Extension: Storm water runoff from the downspouts is to the drain tile. The underground storm drains were not inspected as part of this evaluation."

He had been told that the drains had been replaced and now discharged to surface, but that is not really the point. In his summing up where he suggested a plethora of other inspections (termites, radon etc) nothing more was said about the leaders/extensions that were not (by his own admission) inspected.

Is this Really an Epidemic?

In the introduction, it was suggested that the number of basement waterproofing contractors was a strong indication that there is a leaky basement epidemic. A lot more study could be done to decide what constitutes an epidemic and whether this is one or not. It is interesting to note that on the street where the case history came from at least two other houses had basement leakage problems out of a total of eight; almost 40%. Qualitatively the writer has observed that about 50% of the open houses visited have or have had leaking basements, or show signs that they soon will. On one street in a community east of Cincinnati, there are four houses in a row that are either for sale or being renovated. All seem to be infected.

House prices in that community have been rising at about 3.5% per year for the past few years, so while it is not exactly San Diego Ca., it is not in terminal decline either. It tends to be a place where first time buyers can find a small house for less than $100,000 (median home price in this part of the community was $95,000 in 2005) that is comfortable and close to the city. The four houses in question were built between 1946 and 1953 and are located adjacent to undeveloped woodland on the lower slopes of a hill. The following table shows the recent changes in value that were triggered, at least in part, by deterioration of a basement wall.

Table 1. Equity Loss in Properties with Distressed Basements

Street Number	County Valuation	Recent Sale price	Current Listing	Equity Loss %
6926	$88,400 (2002)	$65,000 (1993)	$34,650	61%
6924	$83,000 (2005)	$40,000 (2006)	Not currently for sale	52%
6920	$82,800 (2002)	$54,000 (2004)	$89,900	35%
6916	$86,500 (2005)	$54,000 (2005)	$69,900	38%

The three houses that are for sale all have distressed basements, and based on exterior conditions, it appears that the forth one does too. Distressed here is defined as meaning that there is horizontal cracking at the mid point of the concrete block basement walls on the 'up-hill' sides of the property, and evidence of leakage. Damage to the above ground portion of the structures varies from minor to serious with significant cracking of the brick around door and window penetrations at 6920. Ironically, that is the house where the owner expects the highest price.

The houses are in various states of repair: considerable remodeling work has been done at 6920 while 6926 is in need of a total rehab. A common factor is that in at least three of the four cases there has been a foreclosure within the past two years (Hamilton County, 2006). It seems likely that in these cases, the cost of repairing the structures may have contributed to the owner's decision to default on their mortgage and walk away. Using the prices developed above, and assuming that two walls of the typically 38 x 23 foot (11.6 x 7.0 m) basements need to be stabilized, an owner would be looking at an expense of more than $8,000. Replacement of downspout drains and improving surface drainage could easily add another $1,000 - $2,000. This is a large sum for a low or moderate-income family to be faced with for repairs and it is easy to see why abandonment could be seen as the better option. Even if an owner elects to sell the house, the condition of the basement is unlikely to escape an appraiser's eye, and any chance of financing is diminished until repairs are completed. The financial result is probably the same – foreclosure.

This is, admittedly, a small sample. However, the following statistics are striking. The median house age in the Cincinnati MSA was 36 years in 1998 (the most recent data) (US Dept of Commerce, 2004). This probably means that it is about 44 years now and one can be sure that the more than 200,000 down-spout drainage systems that are older than this are not improving with age. Many in the 20 – 40 year age bracket will also be starting to fail. So if this is not an epidemic, it will be soon.

Solutions

Once a basement wall has failed, nothing short of a major remedial project will cure the situation. The emphasis must therefore be on prevention. Since one of the most serious culprits seems to be the downspout drains and their progressive deterioration, this is where most attention needs to be directed. It will always be difficult to convince people to replace blocked drains promptly, but there is one point at which outside influence can be brought to bear: during a property transaction. The lender has the option of making the loan or not and it should be contingent on a more rigorous inspection regime when it comes to this issue.

This raises the question of what an appropriate inspection regime would look like. Here, the engineering profession could usefully apply some thought and develop ideas on how to assess the condition of a roof water disposal system. Perhaps a simple hydraulic load test or the equivalent of a 'perc test', such as is used to develop design parameters for on-site sewage disposal systems, could be developed.

Or perhaps it should always be assumed that the drains have failed if the house is more than 20, 30, or 40 years old, and should be replaced. This puts the cost burden on the current owner who, after all, is the party that 'enjoyed' the drains while they were still functioning. Assuming that they could be replaced for an average cost of $2,500, this would seem to be a prudent expenditure in light of the cost of repairing the basement – or worse yet, loosing the value of the house entirely.

Another approach would be to retrofit walls that were designed using now obsolete standards to more modern specifications. Externally bonded FRP is an excellent choice for this and can add substantial bending stress resistance to non-reinforced walls.

FRP and epoxy injection combine to form a powerful tool for remediation once damage has been done to a basement. Development of this technology should continue with a view to reducing the cost and refining the design process.

Conclusions

Evidence presented in this paper suggests that there is indeed an epidemic of leaking basements. Because of the vast number of older houses it is fair to say that it is far from over. Most often the failures are systemic involving multiple factors. Very significant among these is the role of the downspout drain, its frequent deterioration over time and consequent loss of functionality. Combined with this, poor design decisions have resulted in a large number of basement walls that are unable to support the kind of lateral loads that can be expected if high hydrostatic pressures develop.

These two factors combine with predictable results and often cause a significant loss of property value. This may be so serious that it can influence the decision of an owner to abandon the home so creating liability for lending institutions.

There are solutions to the problem that range from proactive maintenance to strengthening of the walls. However, there must first be more public awareness of the causes and liabilities involved. This can best be accomplished through more focused attention during the inspection and appraisal process when property transfers occur.

References

American Concrete Institute (ACI). (2002). *ACI 440.2R-02. "Guide for the Design and Construction of Externally Bonded FRP Systems for Strengthening Concrete Structures."*

Cincinnati Bell. (2005). *"Greater Cincinnati Edition. Cincinnati Bell Directory 2005. The Real Yellow Pages."*

Department of Building Inspector, Cincinnati, Ohio. (1950). *Hamilton County, Ohio Building Code.*

Goodchild, C.H. and Webster, R.M. (1999). "Spreadsheets for Concrete Design to BS 8110 and EC2. - RCC61 Basement Wall.xls [Reinforced Concrete Council (RCC).]." British Cement Association Publication 97.370. http://www.civl.port.ac.uk/rcc2000/spreadsheets/userguid/userguid.pdf (Feb. 12, 2006).

Hamilton County Auditor Dusty Rhodes. (2006). County Auditor On-Line. *"Property Search – 6926 Vinewood, 6924 Vinewood, 6920 Vinewood and 6916 Vinewood, Parcel Info Summary, Transfer Summary."* www.hamiltoncountyauditor.org/realestate. (Feb. 14, 2006).

International Code Council (ICC). (2003). *International Residential Code for One- and Two-Family Dwellings.* Country Club Hills, Illinois.

U.S. Department of Commerce, U.S. Census Bureau. (2004). *"American Housing Survey for the United States: 200.3"* Current Housing Reports H150/03. http://www.census.gov/prod/2004pubs/H150-03.pdf. (Feb. 12, 2006)

John Hancock Center Scaffold Collapse

Alec S. Zimmer, P.E.[1] and Glenn R. Bell, S.E.[2]

Abstract

On Saturday, 9 March 2002 at approximately 1:45 p.m., a 100-ft long, suspended scaffold platform fell from the west face of the 100-story John Hancock Center in Chicago, dropping debris along an arc from the northwest corner of the building to the building's south face. The scaffold wreckage killed three motorists on East Chestnut Street and severely injured several other passers-by.

In early 2000, the building's commercial and residential owners embarked upon a major facade restoration project. They engaged a prime contractor who contracted with a scaffold vendor to design and furnish the scaffold system. The scaffold rig involved in the accident consisted of a 100-ft long, aluminum truss work platform suspended from two outriggers on the roof of the building. On the day of the accident, the contractor determined that it was too windy to work from the platform, and, as had become their custom during off-hours, the contractor's workers moored the work platform on the building's west face at the 42nd floor.

We investigated both the technical and procedural causes of the failure. We determined that under the platform's self-weight and down-draft wind loads acting on it, the cam followers (uplift rollers) and wire rope lashing holding the north outrigger to the building's roof track failed. Without this support, the outrigger overturned, and the scaffold platform dropped. As is the case in many catastrophes of this magnitude, there were many opportunities for the project participants to avert the failure. Many of those parties failed to meet their obligations to the project, and their shortcomings contributed to the accident.

Background

The John Hancock Center (JHC) is a 1,100 ft tall, 100-story, mixed commercial and residential building completed in 1970. It is among the tallest buildings in Chicago, and its footprint covers an entire city block. The commercial portion of the building, which encompasses floors 1 through 43, floors 93 through 100, and the roof, is owned by a commercial property entity. Floors 44 through 92

---

[1] Staff Engineer, Simpson Gumpertz & Heger Inc., 41 Seyon Street, Building 1, Suite 500, Waltham, MA 02453; PH (781) 907-9000; FAX (781) 907-9009; email: azimmer@sgh.com

[2] Senior Principal, Simpson Gumpertz & Heger Inc., 41 Seyon Street, Building 1, Suite 500, Waltham, MA 02453; PH (781) 907-9000; FAX (781) 907-9009; email: grbell@sgh.com

consist of residential condominium units. The condominium unit owners are collectively represented by a separate homeowners association. (In this paper we refer to the commercial and residential owners collectively as the Owners.)

In the fall of 1999, the Owners elected to replace all of the window sealants on the building and clean the anodized aluminum cladding panels. The Owners retained an architectural firm (the Architect) to prepare drawings and specifications and to provide construction administration services for the facade restoration project. As part of the preparation for the project, the Architect investigated the means of access to the JHC curtain wall and determined that a particular scaffold supplier (the Scaffold Supplier) was well suited to provide equipment to access the exterior of the building. The Owners paid the Scaffold Supplier an advance fee to begin preliminary design work on the scaffold system.

In March 2000, the Owners engaged a restoration contractor (the Contractor) initially through letters of intent and later through formal contracts to perform the curtain wall restoration. The Contractor executed an agreement with the Scaffold Supplier to lease the suspended scaffold equipment. The Scaffold Supplier contracted with a Scaffold Specialist to assist with assembly and maintenance of the equipment and to provide the Contractor with on-site technical assistance with the equipment.

The Scaffold Supplier engaged two professional engineers, one registered in New York State and the other, pursuant to contract requirements, in Illinois, to review and seal documentation on the suspended scaffold equipment. (In this paper, we refer to these engineers as the Sealing Engineers). To assist in reviewing the suspended scaffold submittals, the Architect also engaged an Illinois-registered engineer (the Architect's Engineer).

The Owners also retained a second architecture firm to act as their advisor and representative (the Owner's Representative) to monitor the schedule, budget, and progress of the project and to provide other professional services. (We refer to the Owners and Owner's Representative collectively as the Ownership.) Figure 1 is an organization chart showing the relationships between the various parties.

The Scaffold Supplier developed a system of long, aluminum, work platforms, each of which would be suspended from a pair of outriggers on the JHC's roof. The perimeter of the building's roof is encircled by a pair of steel tracks on which the JHC's permanent window-washing rig travels, and the Scaffold Supplier designed the outriggers to roll on these tracks. In May 2000 the Scaffold Supplier delivered the first 100 ft long platform and roof-top outriggers (Rig No. 1) and installed them on the east face of the building. Figures 2 and 3 are, respectively, labeled diagrams of a scaffold platform and an outrigger.

The Scaffold Supplier provided the Contractor with an operations manual describing the equipment in detail and providing guidance on how to properly operate it. The operations manual requires that "the scaffold shall be moved (raised) to the parapet position, or to the ground (overhead pedestrian protection) whenever the equipment shall remain unmanned for an extended period of time." The operations manual also indicates that "the platform shall be moored up to the parapet...or lowered to the ground when severe weather is forecast. Contractor judgment and experience must determine what is severe."

The Scaffold Supplier installed a second scaffold rig (Rig No. 2) on the south facade in March 2001. At the Contractor's request, the Scaffold Supplier moved Rig No. 2 to the west facade in September 2001. As the Contractor worked from the top of the west facade toward the bottom in the autumn of 2001, workers routinely moored the platform at the 42nd floor of the JHC when they were not working in spite of the operations manual's requirements. Doors in the facade at this floor provided the workers with ready access to the platform. When working near the mid-height of the building, mooring the platform at the 42nd floor reduced the amount of time the workers had to spend in transit at the beginning and end of the work day.

The Contractor last worked from Rig No. 2 on 7 March 2002 and moored it at the 42nd floor access doors. That day, local weather forecasts for 8 March called for sustained winds of 20 to 30 mph. On 8 March, the Contractor determined that it was too windy to work. That afternoon, local television stations issued a "wind advisory" for 9 March. The winds did not diminish during the night, and on the morning of 9 March, the winds shifted from the south to the west-southwest with sustained wind speeds of 35 mph and gusts of 56 mph measured near ground level. Again, the Contractor did not work.

As these strong westerly winds hit the JHC's west facade, the north outrigger of Rig No. 2 broke free from the roof track and overturned at approximately 1:45 p.m. The north outrigger fell from the roof and landed on the sidewalk near the southeast corner of the intersection of North Michigan Avenue and East Delaware Place. Without the north outrigger to support it, the north end of the scaffold platform swung downward and around the southwest corner of the JHC where it hit the building's south facade. Upon impact, the platform's aluminum frame partially disintegrated. Platform debris and broken window glass rained down onto East Chestnut Street below. The wreckage of the scaffold platform killed three motorists driving along East Chestnut Street and severely injured several other passers-by.

In May of 2002 attorneys for several plaintiffs in the ensuing litigation retained Simpson Gumpertz & Heger Inc. (SGH) to investigate the collapse.

The Scaffold System

The platform of Rig No. 2 was approximately 100 ft long and 7 ft wide. As shown in Figure 2, the platform was an aluminum truss assembled from three-foot-high modular pyramids with a plywood deck. The main platform was about 5 ft 6 in. wide, but it could be extended toward the building face with moveable platform extensions shown on the inset diagram in Figure 2. The work platform had a fiberglass panel roof supported on a light aluminum frame.

The platform was suspended from the outriggers by 9/16 in. wire rope suspension lines. The suspension lines engaged the scaffold platform 20 ft from each end at walk-through stabilizer frames shown in Figure 2. Sheaves atop the stabilizer frames reduced the hoisting force needed to raise the platform. The platform was also attached to three of the building facade's mullion tracks by friction clamp and roller assemblies spaced at 25 ft along the platform's length. The upper and lower rollers shown in the inset in Figure 2 restrained the platform in the two horizontal directions but allowed it to ride freely up and down the mullion tracks. The friction clamps could be tightened onto the mullion tracks to restrain the platform in the vertical direction when it was parked. As originally conceived and built, each 100 ft long platform had three pairs (total of six) friction clamps that could resist upward and downward motion. However, when the Scaffold Supplier moved Rig No. 2 to the west face, they installed only three friction clamps that restrained only upward motion.

Figure 3 shows the outrigger assembly used at Rig No. 2's north and south suspension lines. The outriggers had inboard and outboard steel A-frames supported on roller carriage assemblies that rolled on the inboard and outboard roof tracks. Each roller carriage had bearing rollers on top of the track flange, side rollers to restrain it laterally, and under-flange rollers to resist uplift. When the Scaffold Supplier's workers assembled the outriggers, they discovered that the large under-flange rollers shown on the scaffold design drawings (Figure 3, left inset) interfered with the splice plates connecting the track segments, so they substituted smaller "cam followers" (Figure 3, right inset) for the large under-flange rollers. This substitution did not have the benefit of engineering review or approval. Additionally, the cam followers were installed with their axles in substantial bending, contrary to the cam follower manufacturer's literature. Along with this substitution, the Scaffold Supplier added wire-rope lashing between the inboard A-frame and the inboard roof track to supplement the outrigger's overturning resistance.

The upper portion of the outriggers consisted of a cantilever aluminum pipe trussed with wire rope. Steel counterweights in a steel basket at the inboard end of this pipe provided some overturning resistance. The scaffold design drawings called for one continuous truss cable on each outrigger, but the Scaffold Supplier provided two-part cables spliced between the outriggers with fist-grip clips. Our investigation revealed that the fist-grip splices were not installed per manufacturer's recommendations and did not develop the strength of the truss cables. The outriggers were also equipped with hoist motors, cable winding spools, load limiters, and emergency "block stops" to arrest the hoisting cable in the event of a mechanical malfunction.

Codes and Regulations

When the project began in the spring of 2000, the Chicago Building Code did not require that scaffold operators obtain a permit for the equipment; it required only that scaffolding be constructed to insure the safety of persons working on, or passing under or by the scaffold. The code did, however, reference American National Standard ANSI A10.8-1977 "Safety Requirements for Scaffolding" for accepted engineering practice.

ANSI A10.8-1977 contains requirements for live loading and stability of scaffolds but has no specific requirements for wind load design for suspended scaffolds. It requires that "scaffolds and their components shall be capable of supporting without failure at least four times the maximum intended load." In 1988 ANSI A10.8-1988 superseded the 1977 document, and although it was not referenced by the Chicago Building Code, it was in wide use throughout the scaffold industry when the scaffold system was designed. Like the older standard, it is silent on explicit wind load design requirements for suspended scaffolds. It requires that the outriggers be "capable of sustaining four times the rated load of the hoist, and shall be secured against movement." It also requires a safety factor of 4:1 against overturning. Additionally, suspended platforms with overhead protection canopies like the one at the JHC must have "additional independent support lines equivalent in strength to the suspension ropes to support the units if the primary suspension system fails."

OSHA Part 1926 Subpart L "Safety and Health Regulations for Construction: Scaffolds" applied to the JHC restoration project. It provides specific requirements for live loads and overturning resistance but does not explicitly address design wind loads. It, too, requires independent support lines tied back to the building structure that can stop the fall of the scaffold should one of the main support lines fail. It refines the language in ANSI A10.8-1988 by requiring that the outriggers be capable of resisting the greater of four times the tipping moment imposed by the scaffold operating at the rated load of the hoist or 1.5 times the tipping moment imposed by the scaffold operating at the stall load of the hoist.

Field Investigation

Shortly after the collapse, the wreckage from the street as well as scaffold components that remained on the roof and suspended from the side of the building were taken to a warehouse near Chicago controlled by one of the defendant's experts. Over the course of more than two years we made many visits to observe and examine this debris. We organized several "cooperative" laboratory sessions with the other engineering experts during which we sorted, cataloged, and weighed debris. We performed metallurgical, microscopic, and mechanical testing on certain pieces. We also reviewed the conditions of the outrigger track on the building's roof.

Sorting through the debris, we were able to locate, identify, and arrange most of the scaffold platform and outrigger components from Rig No. 2. Figure 4 is an overall view of the north outrigger parts re-positioned in the warehouse. We carefully examined the condition of each piece, noting the degree of conformance with Scaffold Supplier's design drawings, the quality of fabrication and assembly, and identification of fractures and deformations. Based on the accumulated weight of the platform components, with adjustments for broken or missing pieces where appropriate, we estimate the dead load of the scaffold platform was 9,397 lb at the time of the collapse excluding the hoisting wire rope and friction clamps.

Examination of deformations and fractures of the outrigger components and roller assemblies was useful for establishing the sequence and mode of failure. Although detailed documentation of these findings is beyond the scope of this paper, the positions and configurations of the fist grip clips at the north outrigger's cable truss splice are particularly important. Our observations led us to conclude that the splice slipped during the failure.

In examining the north outrigger's roller carriages, we noted that all sixteen of the original under-flange rollers on the four A-frame legs had been replaced with the smaller cam followers, and all but two of the cam followers had broken free from the carriages. The two cam followers that remained intact were on the outboard side of one of the outboard A-frame legs. Figure 5 shows one of these remaining cam followers along with one of the fourteen broken cam followers. This observation is consistent with overturning of the outrigger about the outer roof track.

We also had an opportunity to examine the south outrigger's components. Figure 6 is a photograph of the south outrigger taken shortly after the accident. Although the Scaffold Supplier substituted smaller cam followers for fourteen of the original, larger under-flange rollers, one larger under-flange roller remained on each of the two inboard (uplift) roller carriages.

By examining scuff markings on the roof track beams we were able to confirm the position of the north and south outriggers just prior to failure.

Wind Tunnel Testing

For our investigation, plaintiffs' counsel retained Rowan Williams Davies & Irwin Inc. (RWDI) of Guelph, Ontario, to determine the wind forces acting on the Rig No. 2 platform at the time of its collapse. RWDI obtained wind speed and direction data from O'Hare and Midway airports for the date and time of the accident and tested a 1-to-500 scale model of the JHC and the surrounding portion of downtown Chicago in a boundary layer wind tunnel to determine the wind pressures acting on the platform. To obtain appropriate drag coefficients for the scaffold platform, RWDI also tested a larger model of the platform in the boundary layer wind tunnel.

RWDI determined that at the time of the accident, the mean wind load acting on the platform at the 42nd floor was 2,800 lbs downward with a transient load component of 9,000 lbs (maximum) acting upward or downward. Thus, the maximum wind load on the scaffold platform was 11,800 lbs acting downward. RWDI prepared a power spectrum of the wind force on the platform that we subsequently used in a dynamic analysis of the scaffold system.

Testing of Cam Followers

Early in our investigation, we identified the cam followers on the north outrigger as critical elements in the collapse—the outrigger would not have overturned if the cam followers had not failed. We obtained exemplar cam followers from a local vendor and verified that the exemplar cam followers had the same chemical and mechanical properties as those recovered at the JHC.

We simulated the conditions at the JHC by threading the cam followers into a 3/4 in. x 2 in. steel bar, leaving the axle exposed. In this case the cam follower axle loaded is primarily in bending. We loaded the cam follower's cylindrical bearing surface radially to cause flexural failure of the cam follower axle. To better simulate the conditions at the time of the collapse, we chilled the cam followers to 33°F and loaded the cam followers at a rate of approximately 40 in./min. In our tests, the cam followers failed at an average load of approximately 2,360 lb.

Analysis of the Adequacy of the Outrigger Design

Both ANSI A10.8-1977 and OSHA 1926 Subpart L require that all scaffold components be capable of withstanding four times the sum of the dead and live loads to which one may reasonably anticipate they would be subjected. Using the outrigger component weights and measurements and scaffold platform weights we obtained during our field investigation, we determined that the outrigger weighed 5,329 lb including its counterweights and winder cart. With the platform near the base of the building, we determined that the platform's dead load (including the hoisting wire rope) was 11,560 lb, shared equally by the outriggers. According to the Scaffold Supplier's submittals and operations manual, the system's design live load capacity was 2,650 lb shared equally by the outriggers.

We determined the distribution of forces in the outrigger structure using a RISA 3-D finite element analysis model. We based the model's geometry, section properties, and material properties primarily on the Scaffold Supplier's design submittals supplemented by our field and laboratory observations. We assumed that the cable truss (Part 47 in Figure 1) was installed correctly. We also assumed that the original 3-1/2 in. diameter under-flange rollers were installed as shown in the left inset of Figure 3. This design check revealed that many components of the outrigger were substantially overstressed under the design loads. We calculated demand-to-capacity (D/C) ratios for several outrigger elements. A D/C ratio in excess of 1.0 indicates an overstress or design deficiency, and we found D/C ratios as high as 2.59 in some of the outrigger A-frame leg elements. Bearing stresses at several of the aluminum outrigger pipe connections had even higher D/C ratios. Thus, there were many design deficiencies in the outrigger frame.

For stability, OSHA 1926 Subpart L requires that the outriggers be capable of resisting four times the tipping moment imposed by the scaffold operating at the rated load of the hoist. This tipping moment is 218,000 ft-lb. In two tests, the Scaffold Supplier determined that the ultimate uplift capacity of a carriage assembly with four, 3-1/2 in. diameter under-flange rollers is at least 48,000 lbs as originally designed. We determined that the D/C ratio for the as-designed carriage assembly subjected to this load combination is 0.35.

Failure Analysis

We used the weights and measurements that we obtained during our field investigation and the results of RWDI's wind tunnel studies to determine the loads acting on the scaffold system at the time of the collapse. With the platform moored at the 42nd floor, we determined that the platform weight including the hoisting wire rope was 10,564 lb.

The wind forces on the platform consist of a static component and a dynamic component. To assess the dynamic effects on the outrigger, we prepared a NASTRAN finite element analysis model of the outrigger, scaffold platform, and roof track structure, and using the power spectrum of the force on the platform provided by RWDI, we determined that the dynamic component of the wind forces on the scaffold platform amplifies the total wind load acting on the outrigger by ten percent. Thus, the maximum equivalent static wind load on the scaffold platform is 11,800 x 1.1 = 12,970 lb. The overturning moment on the outrigger at the time of the failure was approximately 75,800 ft-lb. See Figure 7 for a diagram of the loads.

Again, we used a RISA-3D finite element analysis model to determine the forces in the outrigger's members. For this analysis, we considered the smaller cam followers that had been substituted for the larger, 3-1/2 in. diameter under flange rollers, and our model included the 0.435 in. gap that existed between the underside of the inboard roof track flange and the top of the inboard cam followers' bearing surface. We accounted for our observation that the fist grip clips connecting the two segments of the cable truss slipped, causing the cable truss to go slack prior to overturning. We also omitted the wire rope lashing between the inboard roof track and the outrigger because our analysis of the lashing's stiffness indicated that it was ineffective until after the cam followers failed. We compared the deflected shape of our model under failure loads to the deflected shape of the south outrigger following the collapse (Figure 6), and we found that they were very similar.

The following table summarizes the D/C ratios of critical outrigger components using expected material strengths. (In this table a D/C ratio in excess of 1.0 indicates that the member's ultimate strength is exceeded.)

Table 1
Demand-to-Capacity Ratios of Outrigger Components at Time of Collapse

Load Combination	Dead + Dynamic Wind Demand-to-Capacity Ratio
Bearing on 8 in. Alum. Pipe at Outboard Splice	1.31 (on top of splice)*
Bearing on 8 in. Alum. Pipe at Inboard Frame	1.19*
Bearing on 8 in. Alum. Pipe at Outboard Frame	1.82*
Peak Axial / Flexure Interaction in 8 in. Aluminum Outrigger Pipe	0.70
Outboard Upper Leg Brace	0.92
Outboard Upper Leg	0.39
Outboard Lower Leg	0.80
Inboard Lower Leg	0.79
Cam Followers	0.98

* Bearing "failures" at these connections would cause local deformation but not gross fracture of the connection.

Technical Causes of the Failure

The overall location and condition of the scaffold debris following the collapse indicates that the north outrigger overturned from the roof and fell to the ground. All of the major north outrigger's components, including the winder cart and counter weights, were on the ground near the northwest corner of the building after the collapse. The south outrigger, though deformed, remained on the roof track.

The north outrigger overturned under combined the action of dead load and downdraft wind loads on the platform. Both the static and dynamic components of wind were significant contributors to the failure. At initiation of the collapse, the fist-grip splice of the outrigger's cable truss slipped, compromising the effectiveness of the cable truss and placing substantial uplift loads on the inboard track roller carriages. Our static and dynamic structural analyses of the north outrigger under conditions acting at the time of failure showed that the loads in the uplift cam followers were essentially equal to the ultimate strength of the cam followers as tested in our laboratory. Other outrigger components, such as certain A-frame legs, were also near failure as shown in Table 1. The wire rope lashing between the outrigger and roof track was too flexible to work effectively in parallel with the roller carriage's uplift resistance. This lashing failed following fracture of the cam followers.

While the downward wind loads were very large, they would have been adequately resisted by a scaffold installation that complied with the applicable codes and standards for strength and overturning resistance. Our analysis shows that the failure would also likely have been averted if one or more of the following conditions had been met:

- The original undercarriage rollers had been left in place
- The inboard A-frames of the outriggers had been lashed more substantially to the roof track
- The platform had been raised to the roof or lowered to near the ground and properly stowed as required in the Scaffold Supplier's operations manual
- Independent support lines required by OSHA 1926 had been used
- The Contractor had secured the scaffold using supplementary friction clamps supplied by the Scaffold Supplier for use in windy conditions

The south outrigger, which resisted loads similar to the north outrigger just prior to collapse, deformed significantly in regions where our analysis showed high D/C ratios, but it did not overturn. The only significant difference between the north outrigger and its southern counterpart was that the Scaffold Supplier retained two of the original, large under-flange rollers on the inboard side of the south outrigger's inboard A-frame, whereas all of the large under-flange rollers were replaced with the smaller cam followers on the north outrigger.

Procedural Causes of the Failure

In our investigation, we examined thousands of documents and other materials related to the procedural responsibilities for this failure. These documents included industry codes and standards, project manuals, operations manuals, design drawings and calculations, shop drawings and other submittals, photos and videotapes, weather records and forecasts, contracts, project correspondence, minutes of construction progress meetings, reports by the Owners' representatives, and deposition transcripts. While a full analysis of this material is beyond the scope of this paper, this section offers a summary of our findings and conclusions regarding procedural issues.

The Scaffold Supplier had no engineers on staff appropriately qualified to take responsibility for the design of a scaffold system as complex as the one they supplied for the JHC project. There were numerous errors in the design concept and calculations for the scaffold, including problems as basic as a severe underestimation of the scaffold's dead load. The Scaffold Supplier retained two Sealing Engineers, but the scope of work, responsibility, and communication between these engineers and the Scaffold Supplier's in-house "designers" was confused and, in some cases, non-existent.

The project submittals for the scaffold system were problematic and were never completed to the satisfaction of the Architect. Moreover, the Scaffold Supplier made changes to the scaffold assembly after initial submittal approval that were not authorized through subsequent engineering review and approval. Some of these changes proved fatal, including the substitution of the smaller cam followers for the larger under-flange on the outrigger roller carriages. In another unauthorized change, the single length of wire rope for the cable truss was modified such that it consisted of two segments of wire rope joined with improperly tightened fist grip clips.

OSHA regulations require that scaffolds of this type have independent support lines, redundant to the primary support lines. The scaffold design did not provide for such redundancy. In addition, early installations of the scaffold outriggers had independent tie-back lines securing the outrigger frames to a separate anchor point on the building's structure. These tie-backs were eliminated in the installation of Rig No. 2 on the west face.

There were also numerous performance problems with the scaffold, indicating that moving forward with the project without further study was not prudent. Among these, all of the parties involved were aware that the inboard outrigger A-frames had well-documented history of lifting off the inboard roof track. The Architect repeatedly reported to Ownership that the scaffold submittals were incomplete.

Lines of communication and authority were complex, as shown in Figure 1 but, more importantly, they were confused. The Scaffold Supplier had two Sealing Engineers. The Architect retained outside engineers to consult on scaffold issues on its behalf. Both the Architect and Ownership involved themselves in actions traditionally considered part of the Contractor's means and methods. In the face of problems with scaffolding submittals and performance as well as delays in the project's completion date, Ownership pressed the Architect to reduce its scope, fee, and scrutiny.

The Contractor, who had the ultimate responsibility for the scaffold's day-to-day operation, did not operate and moor the scaffold in accordance with the Scaffold Supplier's instructions, in spite of the forecasts of high winds in the days before the failure. Even on the morning of the failure, passers-by questioned the building's security staff about the safety of the scaffolding. Such warnings did not result in any action to avert this disaster.

Conclusions

The scaffold collapsed under the forces of dead load and down-draft wind loads on the scaffold platform. The failure initiated when the splice in the outrigger's cable truss slipped, the cam followers on the inboard (uplift) track roller carriages fractured, and the wire rope lashing the outrigger to the roof track failed. Without its anchorage to the building, the north outrigger overturned. On loss of its north suspension lines, the scaffold platform swung down and around the southwest corner of the building where it collided with the south face of the building, raining debris and broken window glass onto the street below.

The collapse occurred through the failure of many parties to meet their obligations on the project. These failures ranged from detailed errors in execution of the scaffold's design and construction to broad shortcomings in establishment and management of the project team. The Scaffold Supplier's design did not meet required standards for strength and overturning resistance. The Contractor failed to provide complete and accurate submittals and permitted the Scaffold Supplier to substitute scaffold components at variance with the design calculations and submittals. The Scaffold Supplier did not provide clear, written directions for securing the scaffold when it was not in use, and the Contractor failed to moor the scaffold in accordance with the instructions they were given. The project came under severe time and budget constraints, and several entities involved themselves in the Contractor's customary role of means and methods, confusing lines of responsibility and communication. From the start of the project until the day of the collapse, the parties involved with the project missed many opportunities to avert the failure.

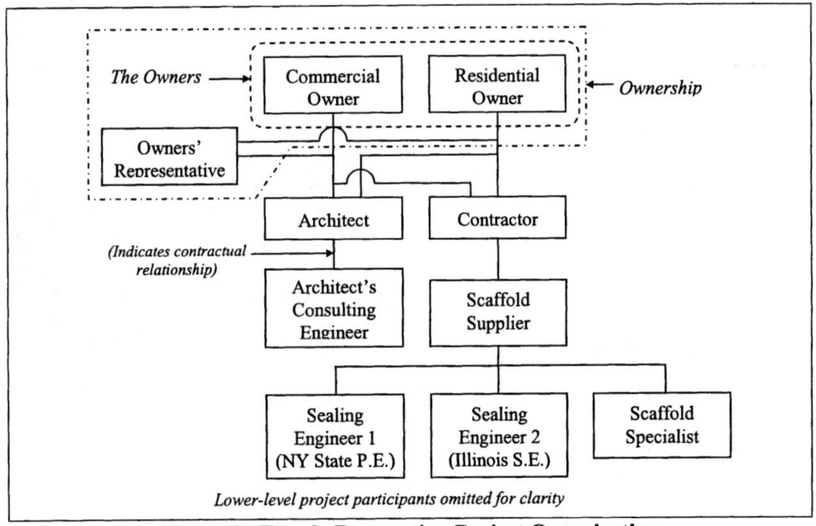

Figure 1. Facade Restoration Project Organization

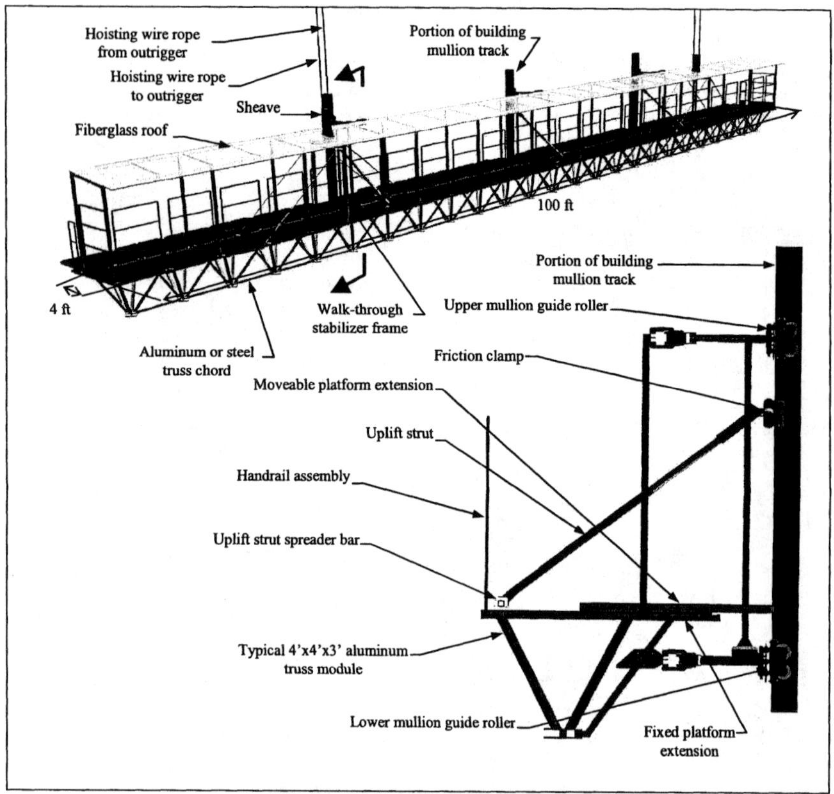

Figure 2. Scaffold Platform

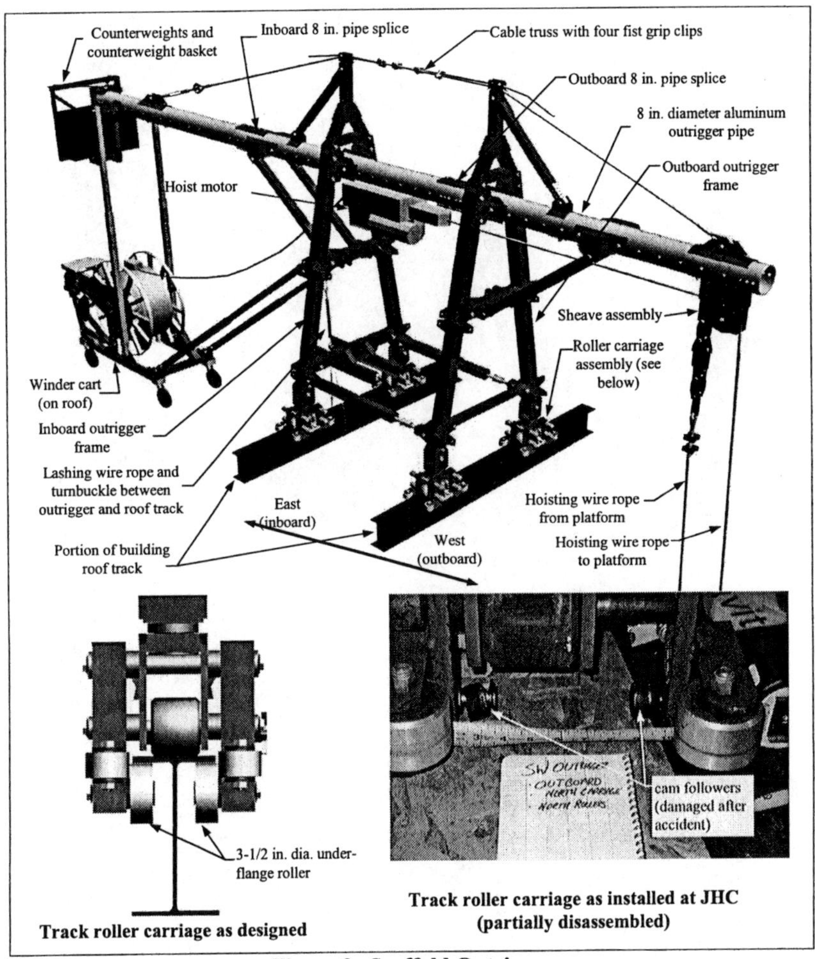

Figure 3. Scaffold Outrigger

Figure 4. Layout of North Outrigger Components

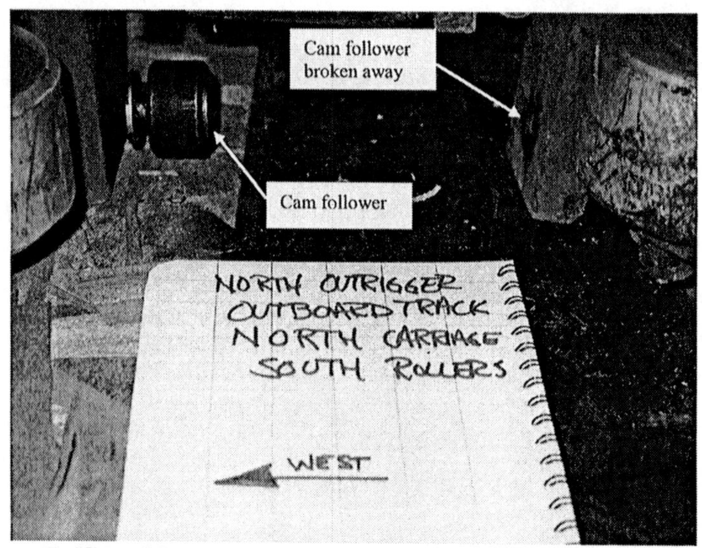

Figure 5. View of Remaining and Broken Cam Followers on North Outrigger

Figure 6. South Outrigger Following Collapse

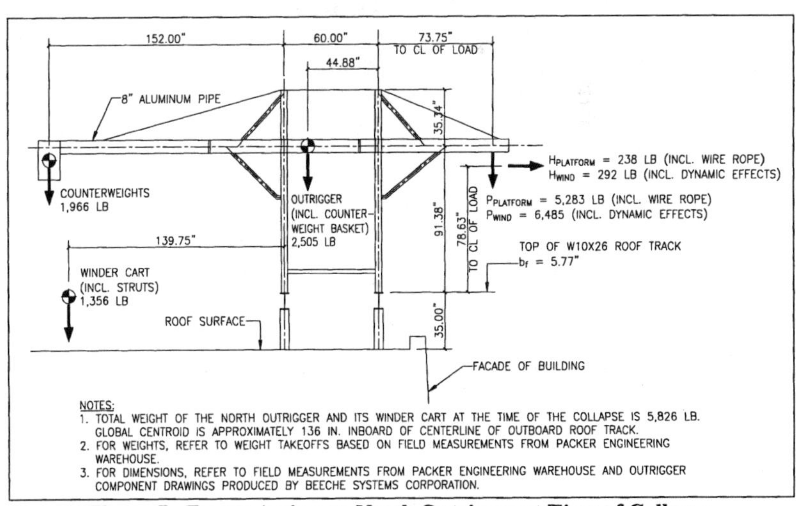

Figure 7. Forces Acting on North Outrigger at Time of Collapse

Nonlinear Static Analysis of Reinforced Concrete Residential Structure with Non-fixed Connections

José Ramón Arroyo[1], Ph.D., ASCE Member., and Drianfel Vazquez[2], Ph.D.

[1] Associate Professor, General Engineering Department, University of Puerto Rico, PO Box 9044 Mayagüez, P.R. 00681, PH (787) 832 – 4040 x 3789; FAX: (787) 265 – 3816; email: jrarroyo@uprm.edu
[2] Assistant Professor, Department of Engineering, University of Puerto Rico at Ponce, PO Box 7186 Ponce PR, 00732, PH (787) 844 – 8181 x 2411 email: drian@uprp.edu

Abstract

This study presents the evaluation of reinforced concrete residential structures with a non-fixed connection between the first and second floor. In Puerto Rico, the construction of concrete second floor over the existing first floor is a common practice due to the high cost of the land and the increasing necessity of houses. The lack of appropriate connection between the existing structure and the new one produces the non-fixed connection. The structure consists of two floors with all rigid connections except the connection of the base of the columns of the second floor. This connection can be assumed as pinned or semi-rigid. A field survey was made to identify what engineers are doing in practice to connect the upper and lower stories. Finally, a vulnerability analysis was performed to typical reinforced concrete frames. It was concluded that the structural integrity of almost all the residence frames analyzed is compromised when subjected to loads specified by the UBC 1997.

Introduction

It is quite common in Puerto Rico to find residential structures with more than one story that were originally designed and constructed as single-story units. Increasing real state costs and continuous population growth frequently force the owners of residential structures to construct new facilities over the existing ones. Engineers in Puerto Rico design these additional levels for all the load combinations specified in the design and construction codes currently in use, the UBC 1997.

However, the boundary conditions at the new created connections are not well defined. In general, these connections are modeled as a fixed support, that is, the second-story columns are fixed to the slab of the existing structure. Actually, the proper way to model this connection is to use an intermediate condition, between a fixed support and a pinned connection. The problem is that with a non-fixed connection the natural periods of the structure increase and the displacements of the structure due to the occurrence of a seismic event also increase. It is a well-known fact that the magnitude of the displacement that the structure undergoes during an earthquake provides a good measure of the damage caused by this phenomenon. Therefore, this kind of connection may provide the perfect environment to induce seismic damage in this type of two-story residential structures.

This paper present this problem and a methodology required to predict the behavior of these structures. Two-dimensional finite-element models, were used to simulate the seismic response of these structures. The Acceleration-Displacement Spectrum Method is used to determine the seismic behavior of these structures with non-fixed connection. Figures 1 and 2 present two constructions of second stories over the existing structure. In these cases, the appropriate anchorage length of the reinforcing bars was not achieved. This would suggest that the connection between the new structural elements and the existing ones should be modeled as a pin connection, instead of a fixed-support condition, which is the condition typically assumed for these cases by most professional engineers. In many cases, the resulting anchorage depth is not enough to guarantee a fixed connection. Sometimes, new columns are erected over the slab without any supporting beams immediately below. In most cases, the slab has not been originally designed to support any additional concentrated loads. This kind of connection reduces the overall stiffness of the structure, rendering a system with a lower natural frequency and a higher natural period of vibration. Under earthquake loading, structures with higher natural periods will develop lower accelerations but higher displacements. The increased displacement will produce higher levels of structural damage, as it has been observed in past seismic events. Therefore, we can conclude that if a destructive earthquake occurs, such as the one that hit the island of Puerto Rico in 1918, these residential structures could be in imminent danger.

Field Survey

The first step in the evaluation of the residences is to obtain representative physical parameters of the residences to be evaluated. A field survey was performed to obtain the dimensions of the residences and to examine the mechanical properties of the materials used. The important parameters are the spans length, the height of the columns, the number of spans, steel reinforcement, dimensions of elements, and slab thickness. This parameter describes the residences and its parameters obtained from the survey made at the Permits and Regulation Administration. Four of eight areas of this state government agency were visited, as shown in Figure 3. These areas are Arecibo, Mayagüez, Ponce, and San Juan. These zones represent a 42% of the total municipalities of Puerto Rico.

Figure 1. New second and third floor over an existing structure

Figure 2. Inadequate anchorage length for columns

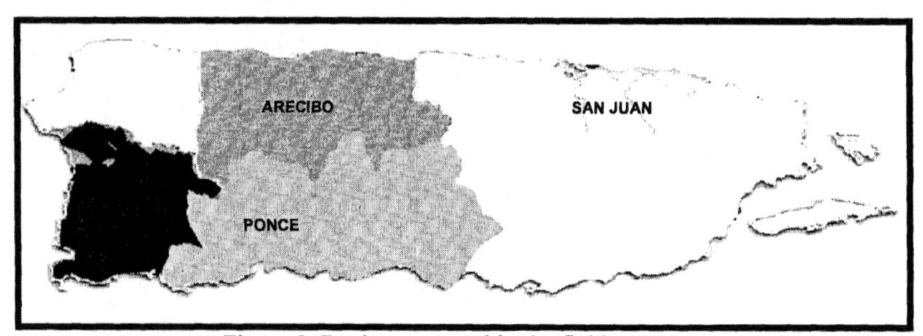

Figure 3. Regions covered in the field survey

A total of 45 residences with additional stories were found in the Agency search. The information was determined by the information presented in the construction plans. The obtained data for each of the 45 reinforced concrete plane frame residences are the following:

1. Span dimension
2. Height of each floor
3. Yield stress of the reinforcing bars
4. Strength of concrete at 28 days.
5. Slab thickness
6. Dimensions of the structural elements (beams and columns)
7. Reinforcing bars dimension and distribution

Result of the Field Survey

After a careful evaluation of the plane frames, nine structural parameters were detected and studied. To observe the distribution of the variables, the amount in terms of percent of the field study cases was calculated. The first parameter analyzed was the frame distribution. That is, this variable described the amount of frames in each direction (north-south, east-west frames) of the structure. This is also a measured of the number of bays per frame in each direction. Table 1 shows the frame distribution in terms of percent.

Table 1. Frame distribution

FRAME DISTRIBUTION	1 X 1	2 X 2	2 X 3	2 X 4	3 X 3	4 X 4
%	4.44	15.56	53.33	17.78	6.67	2.22

The second parameter analyzed was the span of the frame. That is, this variable described the distance between columns in each frame. This is also a measured of the length of the beams. Table 2 shows the span distribution in terms of percent.

Table 2. Span dimension (1 ft = 0.3048 m)

SPAN DIMENSION (ft)	5' to 9'	10' to 12'	13' to 16'	17' to 18'	Over 18'
%	15.18	59.82	21.88	0.89	2.23

The third parameter analyzed was the columns dimensions. Table 2.49 shows the columns dimensions in terms of percent.

Table 3. Columns dimension (1 in = 25.4 mm)

COLUMN (inches)	12" X 6"	18" X 6"	24" X 6"	12" X 12"	16" X 16"
%	37.99	38.69	10.60	4.77	7.95

The fourth parameter analyzed was the beams dimensions. That is, this variable described the dimension of the cross sectional area of the beams. Table 4 shows the beams dimensions in terms of percent.

Table 4. Beam dimensions (1 in = 25.4 mm)

BEAMS	14 in X 6 in	16 in X 6 in	17 in X 6 in	18 in X 6 in
%	2.22	26.67	48.89	22.22

The fifth parameter analyzed was the height of the stories. That is, this variable described the height of the frames. Table 5 shows the frame height in terms of percent.

Table 5. Frame height (1 ft = 0.3048 m)

HEIGHT	8 ft	9 ft	10 ft	11 ft
%	11.11	31.11	55.56	2.22

The sixth parameter analyzed was the thickness of the slab. Table 6 shows the slab thickness in terms of percent.

Table 6. Frame height (1 in = 25.4 mm)

SLAB THICKNESS	4 in	5 in	6 in
%	28.89	60	11.11

The seventh parameter analyzed was the reinforcing bars diameters frequently used in the construction of residential structures. Table 7 shows the slab thickness in terms of percent.

Table 7. Reinforcing bars dimensions (1 in = 25.4 mm)

BARS #	#4 (1/2 in)	#5 (5/8 in)
%	19.15	80.85

The eight parameter analyzed was the strength of the concrete at 28 days specified in the structural plans found in the field review. Table 8 shows the resulting compressive strength of concrete in terms of percent.

Table 8. Compressive strength of concrete at 28 days (1 ksi = 6.8947 MPa)

f'c	3 ksi	3.5 ksi
%	91.11	8.89

The last parameter analyzed was the strength of the reinforcing bars specified in the structural plans found in the field review. Table 9 show the yield strength specified in the structural plans in terms of percent.

Table 9. Yield stress of reinforcing bars (1 ksi = 6.8947 MPa)

Fy	40 ksi	60 ksi
%	40	60

The specific values of the parameters used for each analysis are specified in the following section for the Non-linear Static Analysis (Pushover Analysis).

Vulnerability Evaluation of Residences

Nonlinear Static Pushover analyses were performed on the stiffest residences, the medium stiffness residences, and the most flexible residences in order to evaluate the vulnerability to potential earthquake loads. A total of twenty four (24) frames (twelve with fixed connection and twelve with the second story pinned connection) were analyzed. Their behavior was examined when they were subjected to the design spectrum as the one defined in the current building code adopted in Puerto Rico, The Uniform Building Code 1997 (UBC-97). The Capacity Spectrum Method, (Badoux, 1998, Mahaney, Paret, Kehoe, and Freeman, 1993), was implemented as the evaluation tool to study the seismic performance of the structures in a nonlinear response basis.

Selection of the Residences Parameters

It was observed from the field survey that the story heights of the residences remains constant with nine feet (2.74 m) of clearance and the span length inspected varied from nine feet (2.74 m) to seventeen feet (5.18 m). Also it was noticed that the bar size or diameter of longitudinal steel reinforcement varied from bars # 4 to # 5 for the structural elements. The structural elements were basically two cross sections for the columns depending on the span length of the structures and one section for the beams. These two sections for the columns are a section of 152.4 mm wide by 304.8 mm height and a section of 152.4 mm width by 457.2 mm height while the predominant sections of the beam was 152.4 mm wide by 304.8 mm height. However, 127 mm were added to the beam sections to include the effect of the "built in" concrete slab for a final section of 152.4 mm wide by 431.8 mm height for the beams.

From this information, three extreme cases of residences were identified to obtain models that vary from the stiffest to the most flexible system. The stiffest residences are those with the residence analyzed with their structural elements acting in their strong axis direction, whereas the most flexible ones are those with the opposite characteristics, specifically loaded in their weak axis direction. Based on these considerations, three types of residential constructions were established. The stiffer residences, with three different span length, loading and element sections, will be identified with a K prefix. The most flexible residences will be the same span length, loading and element section of the K residences, but loaded in the weak direction of the structural elements. They will be referred to with the F prefix which stands for flexible. To obtain a preliminary vulnerability analysis, structural systems consisting of two dimensional plane frames with two and three spans length, two stories height and with the cross sections mentioned before were defined and modeled. A typical structural system used in the analysis is shown in Figure 4.

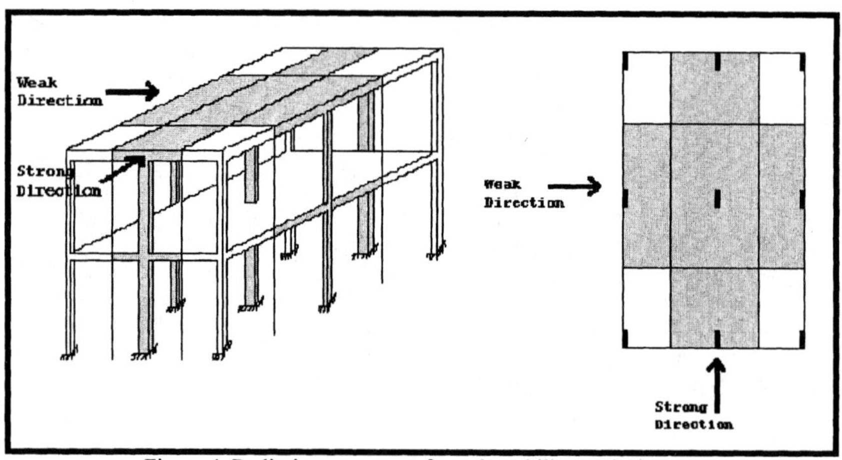

Figure 4. Preliminary systems for vulnerability analysis

Tables 10 and 11 present a summary of the geometric properties of the two spans frame, span lengths, column height and reinforcement analyzed as extreme cases in the vulnerability analyses. The nomenclature used to described the frames consists of three variables, i.e. K21 and F21, where K and F represents the strong (stiffer) and weaker direction of the structural elements respectively, the second number corresponds to the number of spans and the third number correspond to three cases (i.e. 1, 2 and 3) that implies variation in the span length. The numbers in the third letter are 1, 2, and 3 that represents a span length of ten (10), thirteen (13) and sixteen and half (16.5) feet of span measured between columns center to center, respectively.

Table 10. Parameters of the two span residences (1 in = 25.4 mm)

				Cross Section		Reinforcement	
Residence	**Story**	**Height [ft]**	**Span [ft]**	**Beams [in]**	**Columns [in]**	**Beams**	**Columns**
F21	1	9	11	6X17	12X6	4#5	6#5
	2	9	11	6X17	12X6	4#5	6#5
F22	1	9	13	6X17	12X6	4#5	6#5
	2	9	13	6X17	12X6	4#5	6#5
F23	1	9	16.5	6X17	18X6	4#5	6#5
	2	9	16.5	6X17	18X6	4#5	6#5
K21	1	9	11	6X17	6X12	4#5	6#5
	2	9	11	6X17	6X12	4#5	6#5
K22	1	9	13	6X17	6X12	4#5	6#5
	2	9	13	6X17	6X12	4#5	6#5
K23	1	9	16.5	6X17	6X18	4#5	6#5
	2	9	16.5	6X17	6X18	4#5	6#5

Table 11. Parameters for the three span residences (1 in = 25.4 mm)

Residence	Story	Height [ft]	Span [ft]	Cross Section		Reinforcement	
				Beams [in]	Columns [in]	Beams	Columns
F31	1	9	11	6X17	12X6	4#5	6#5
	2	9	11	6X17	12X6	4#5	6#5
F32	1	9	13	6X17	12X6	4#5	6#5
	2	9	13	6X17	12X6	4#5	6#5
F33	1	9	16.5	6X17	18X6	4#5	6#5
	2	9	16.5	6X17	18X6	4#5	6#5
K31	1	9	11	6X17	6X12	4#5	6#5
	2	9	11	6X17	6X12	4#5	6#5
K32	1	9	13	6X17	6X12	4#5	6#5
	2	9	13	6X17	6X12	4#5	6#5
K33	1	9	16.5	6X17	6X18	4#5	6#5
	2	9	16.5	6X17	6X18	4#5	6#5

Nonlinear Static Pushover Analysis

The modeling and nonlinear static pushover analyses were performed using the computer program SAP2000 Nonlinear Version (Computers and Structures, 1994). The first step in the generation of the model is to define the geometry of the residence or structure. The geometries for all of the residences analyzed were presented in Tables 10 and 11 for the two and three spans residences, respectively. After the geometry of the structure is defined, the next step is to assign the element sections to the beams and columns as listed in Tables 10 and 11. The boundary conditions or restraints are also defined in this step.

To perform a Nonlinear Static Pushover Analysis in SAP2000 it is necessary to perform the following steps. The applied loads (i.e. Dead Load, Live Load, and Earthquake Load) of the model must be defined. It is also necessary to assign the joint masses to the system before computing the natural frequencies and vibrational modes of the structure. The last step, one of the most important, is to assign the plastic hinges properties and their locations in the structural elements.

In the present study, the plastic hinges properties assigned to the columns were different from the plastic hinges properties of the beams. The plastic hinges properties provided by SAP2000 are typically based on the documents FEMA–273, (FEMA, 1997) and ATC–40, (ATC 40, 1996). For the columns the PMM (flexural and axial) hinge relation was used and the Concrete Moment M3 (flexural) hinge relation was used for the beams. The difference between them is the inclusion of the axial load in the PMM relation. The hinges were located at a distance from each end of the frame elements equal to 10% of the element length.

The next step after defining and assigning the plastics hinges is to define the pushover load case. In this study, the lateral loads applied to each story of the systems were defined following the vertical force distribution of the static force procedure of the Uniform Building Code, 1997. To carry out a nonlinear static pushover it is necessary first to run a static analysis and a modal analysis. The static analysis is needed to obtain the initial stiffness matrix and the modal analysis is needed to obtain the structure's period and certain coefficients and factors used in the capacity spectrum methodology. After performing these analyses, one is ready to run the Static Pushover analysis. It is important to depicts that actually two static nonlinear pushover analyses were performed, one for gravity loads and one for the lateral loads. SAP 2000 is also capable of including the nonlinearity due to the geometry of the structure. This feature, known as the P-Delta effect or Frame Instability effect, was also included in these analyses.

Nonlinear Static Analysis Results

For this study, we are interested in obtaining the base reaction versus the monitored displacement curve and with the demand spectrum in an ADRS format superimposed to it. The program automatically calculates the plots in the ADRS format. These curves allow us to observe the behavior of the structure and compared it to the demand spectrum. The vulnerability analyses of the frames as presented in Tables 10 and 11 were performed and are summarized in this section using the Capacity Spectrum Method in ADRS format for both Sub Cases.

All of the results produced by these analyses are presented from Figures 5 to 16. Examining the capacity demand curve of residences, it can be observed that there is no intersection or Performance Point in Cases F21, F22, F23, K22, K23, F31, F32, F33, K31, and K32 (10 out of 12) which means that the frames of these residences are not capable of withstand the ground motion in the nonlinear range. It is important to recall that the spectrum was based on rock (UBC S_b Soil Type) and the effect of soft soils was not included in the analyses. Including the soft soil condition will increase the demand spectrum and the Performance point will not be reached. Furthermore, notice that the two models that resist the demand imposed are the K21 and K33 which are the same residences as F21 and F33 but loaded in the weak direction. This means that these models that resist the demand in one direction are not capable of resisting the same earthquake in the other direction. In summary none of the models would resists the earthquake demand because the main frames could collapse in one or the other direction.

We can conclude from these nonlinear static analyses, that the structural integrity of almost all the residences studied will be compromised when subjected to the design spectrum considered in this study, that is, the spectrum presented in the UBC-97 with S_b soil type. Moreover, they will confront even more problems if the effects of a softer soil types are included in the analysis.

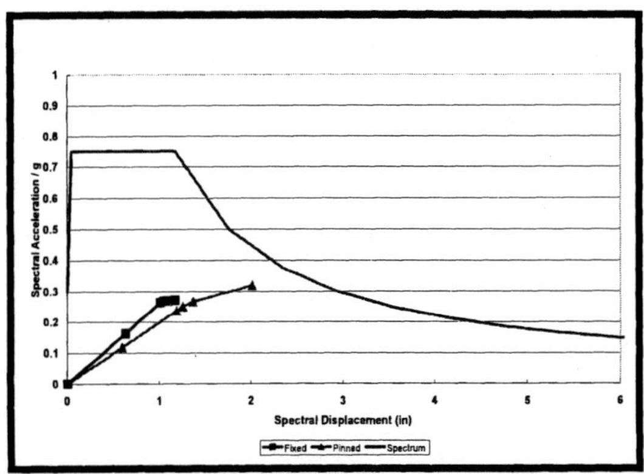

Figure 5. Capacity Spectrum Methodology for Case F21 (1 in = 25.4 mm)

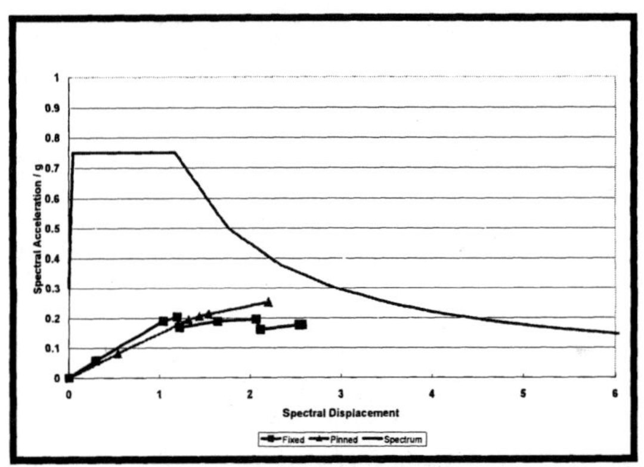

Figure 6. Capacity Spectrum Methodology for Case F22 (1 in = 25.4 mm)

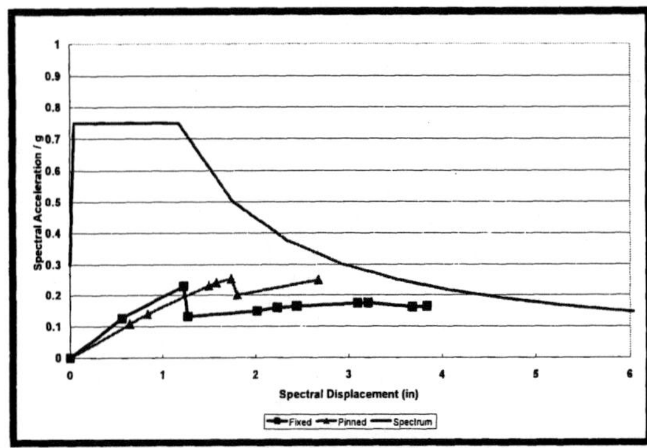

Figure 7. Capacity Spectrum Methodology for Case F23 (1 in = 25.4 mm)

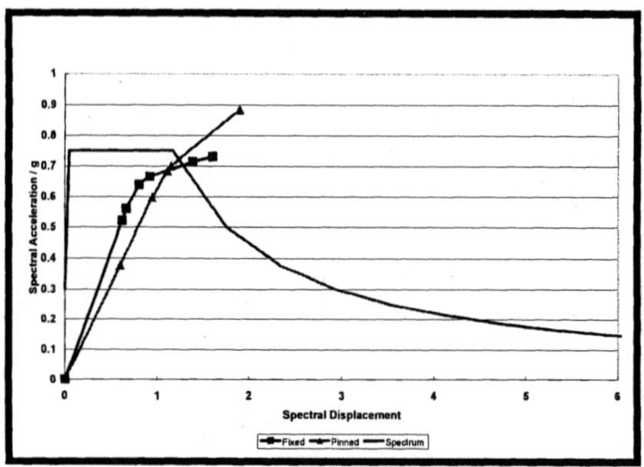

Figure 8. Capacity Spectrum Methodology for Case K21 (1 in = 25.4 mm)

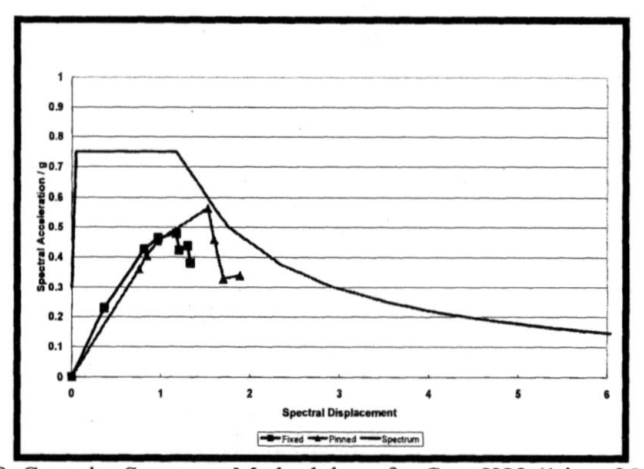

Figure 9. Capacity Spectrum Methodology for Case K22 (1 in = 25.4 mm)

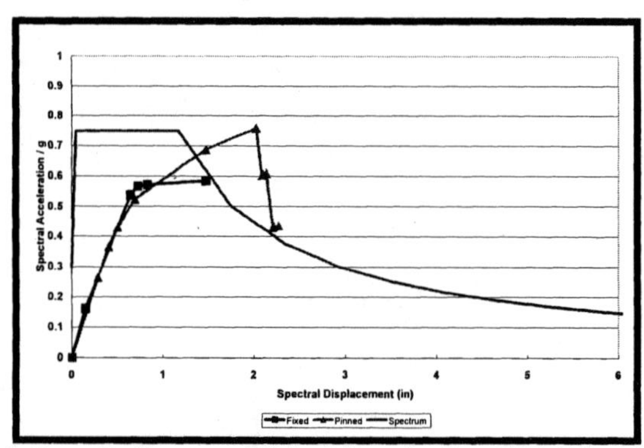

Figure 10. Capacity Spectrum Methodology for Case K23 (1 in = 25.4 mm)

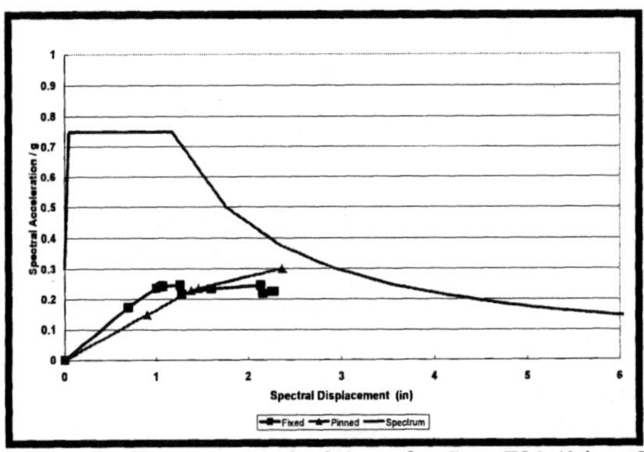

Figure 11. Capacity Spectrum Methodology for Case F31 (1 in = 25.4 mm)

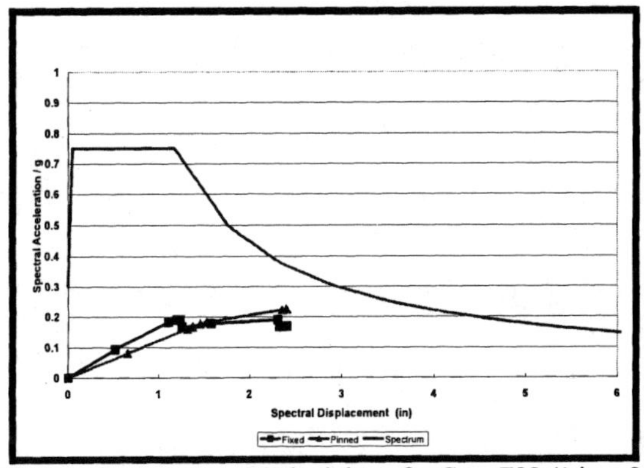

Figure 12. Capacity Spectrum Methodology for Case F32 (1 in = 25.4 mm)

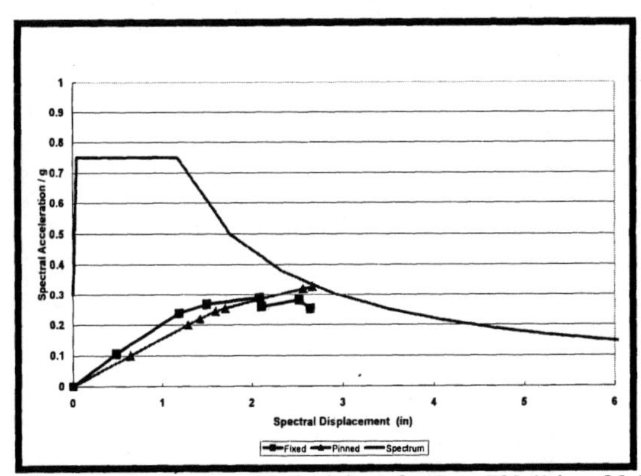

Figure 13. Capacity Spectrum Methodology for Case F33 (1 in = 25.4 mm)

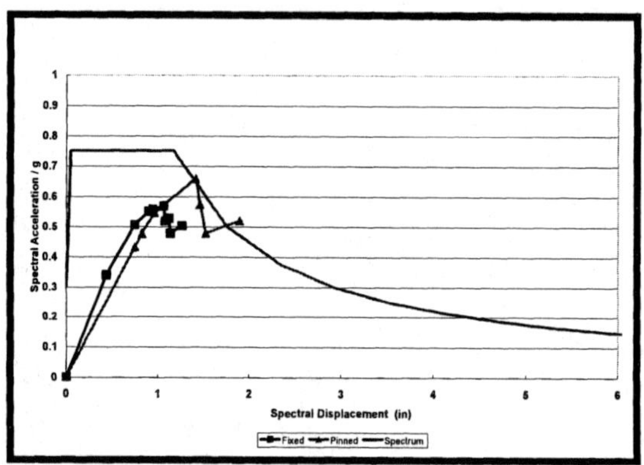

Figure 14. Capacity Spectrum Methodology for Case K31 (1 in = 25.4 mm)

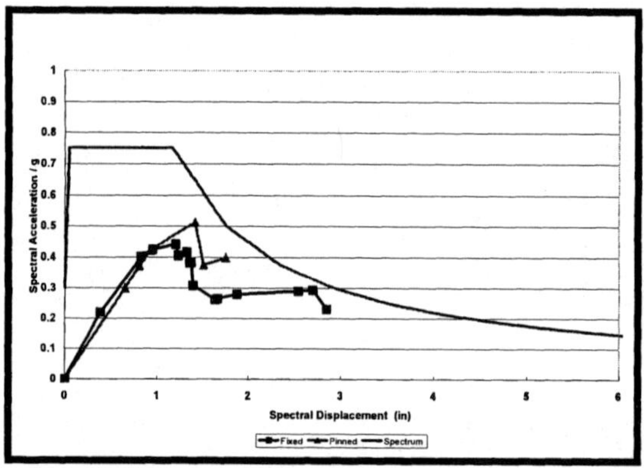

Figure 15. Capacity Spectrum Methodology for Case K32 (1 in = 25.4 mm)

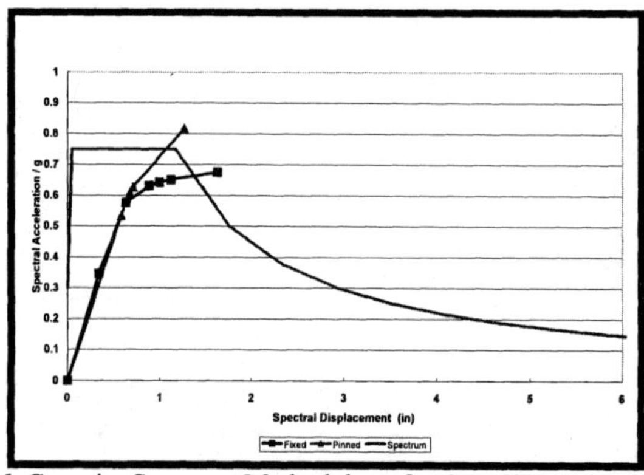

Figure 16. Capacity Spectrum Methodology for Case K33 (1 in = 25.4 mm)

Conclusions

It is common in Puerto Rico to find residential structures with more than one story that were originally designed and constructed as single-story units. Increasing real state costs and continuous population growth frequently force the owners of residential structures to construct new facilities over the existing ones. Professional engineers design these second floors for all the load combinations specified in the construction codes. However, the boundary conditions at the newly created connections (bottom of second floor columns) are not well defined. In general, these connections are modeled as a fixed support, that is, the second-story columns are fixed to the slab of the existing structure. The proper way to model this connection is to use an intermediate condition, between a fixed support and a pinned connection. Actually, these connections behave more like pinned connections than fixed connections. The problem is that with a non-fixed connection the natural periods of the structure increase and the displacements of the structure due to the occurrence of a seismic event also increase. It is well-known that the magnitude of the displacement that the structure undergoes during an earthquake provides a good measure of the damage caused by this phenomenon. Therefore, this kind of connection may provide the perfect environment to induce seismic damage in this type of two-story residential structures.

A field survey was done and typical construction plans, were obtained from the Regulation and Permits Administration. From these typical structural plans obtained from almost all the Island regions, a series of reinforced concrete moment resisting frames were adopted as the most common concrete frames used in two-story houses. The models selected represent the most commonly used in the field, and covers almost all the important variables that enter in the structural design. After the analyses performed, it can be concluded that the structural integrity of almost all the residences will be compromised when subjected to strong motion similar to the ones considered in this study, the UBC-97 S_b Soil Type (the minimum expected to the island). Moreover, they will confront even more problems if the effect of soft soil is included in the analysis.

Finally, to comply with the moral and ethical point of view, the principal author is preparing a set of articles related to this theme to be submitted to the magazine of the College of Professional Engineers and Land Surveyors of Puerto Rico. This magazine reaches all the professional engineers in the Island. In this way, professional engineers can adopt the information produced in this research and can be implemented in the projects to reduce risk and bring safety to the public. Also, during the time the research was done, the State Agencies related to the construction permits approval were contacted and the problem of residences with non rigid connections was presented and explained. They help in the field survey made during the research.

Acknowledgements

This research project was made possible due to the support provided by the Federal Emergency Management Agency (FEMA) who provided the financial support for the development of this research project under the Hazard Mitigation Grant Program 1372 – DR – PR – Project - PR – 0006.

References

1. American Concrete Institute, (1994) "Guide for Evaluation of Concrete Structures Prior to Rehabilitation", ACI364.1R-94.
2. ATC 40, (1996) "Seismic Evaluation and Retrofit of Concrete Buildings", Applied Technology Council, California Seismic Safety Commission.
3. Badoux, M. (1998), "Comparison of Seismic Retrofitting Strategies with the Capacity Spectrum Method", 11th European Conference on Earthquake, Balkema, Rotterdam pp 1 - 8.
4. FEMA (1997), "FEMA-273-NEHRP Guidelines for the Seismic Rehabilitation of Buildings", Washington, D.C.
5. International Conference of Building Officials (ICBO), (1997), Uniform Building Code. Vol. 2. Whittier, California.
6. Mahaney, J. A., Paret, T.F., Kehoe B.E., and Freeman A. (1993), "The Capacity Spectrum Method for the Evaluating Structural Response During Loma Prieta Earthquake. National Earthquake Conference, USA.

Case Study of Stone Veneer Failure

Deepak Ahuja, M.S., P.E., M. ASCE
Stewart M. Verhulst, M.S., P.E., M. ASCE
Andrew M. Noble III, B.Arch., R.A.

Abstract

The primary concerns for veneer systems are typically cosmetic (appearance) and the function as part of the building envelope, especially regarding moisture issues. However, the selection of the veneer system and the execution of the veneer support can become issues of public safety. This case study outlines issues of veneer-related safety on two public buildings with similar construction. Improper selection of the veneer materials during the design phase of the project and improper execution during the construction phase caused several safety concerns around the entire building perimeters. The safety issues were not caused by special or unique conditions imposed on the structure. Consideration and execution of fundamental issues related to exterior veneer material selection and the installation of a stone veneer system would have prevented all of the unsafe conditions.

Introduction

Two public buildings located in San Antonio, Texas, were experiencing failure of the stone veneer system. For the purposes of this discussion, "failure" is defined as a loss of function or serviceability, including conditions which render a building unsafe. Generally speaking, the failure observed was in the form of stone veneer (both pieces of stone and also entire stones) which had fallen from the façades or were in danger of falling. The buildings were of similar construction with the same types of veneer systems and materials. The buildings were so similar that the topics discussed apply to both; however, where discussed separately they will be referred to as Building A and Building B.

The authors performed an investigation of the cause of failure of the veneer. The structures were approximately six years old at the time of investigation (they were built in the late 1990's); however, stone veneer failure had been ongoing for some time prior. Both Buildings A and B are five-story buildings with conventional reinforced concrete framing. Refer to Figures 1 and 2 for representative views of each building exterior.

The total height of the dimension stone veneer varies from three stories to five stories at each building; with porcelain tile at the remaining portions of the building façades at the fourth and fifth floors. The lighter veneer in Figures 1 and 2 is indicative of the limestone, and the darker-colored veneer is indicative of the sandstone (also, the porcelain tile is visible and is darker in appearance than the sandstone).

Figure 1. General view of Building A exterior.

Figure 2. General view of Building B exterior.

Documents made available during the investigation include architectural and structural plans, the project manual (specifications), and some submittals for the stone, which included pre-construction test data.

Dimension Stone Veneer

General. Dimension Stone is defined as, "a natural building stone that has been cut and finished to specifications" (MIA, 1999). Two main types of dimension stone veneer were used in the veneer for the buildings, limestone with a natural rock face and red sandstone with a natural rock face. All of the stone veneer on the buildings has been observed to have the rift or grain in a vertical orientation. Therefore, the forces from the stones' self-weight are in-line with the direction of the layering within the stone (refer to Figure 3). The specifications indicate a dimensional stone depth of 2 1/2"-3" for the veneer system; and Type N mortar with Type S lime for the mortar in the dimension stone veneer system. A typical dimension stone in the veneer system measures 24" wide x 20" high and has an estimated weight of 115 lbs. The specifications indicate stainless steel anchors, spaced at 18" on-center, for the stone veneer anchors. The design documents indicate a steel lintel support for the veneer at each floor level.

Figure 3. View of stone veneer – note vertical orientation of the grain.

Exterior stone systems must be designed to withstand stresses affecting the building such as: self-weight of the stone veneer, pressure from wind loading, structural movements, thermal expansion and contraction, moisture migration and condensation, wet/dry cycles, and freeze/thaw cycles. To ensure that stone veneer systems are properly designed, material characteristics of the stone must be tested prior to the stones' application on a building.

The primary concerns with the veneer system included flaking of the stone veneer (delamination from the stone surface) falling off the building onto the ground below, some areas of large separations at mortar joints (reportedly increasing in size despite some efforts at previous repair), and progressive movement and deflection of stone veneer above openings. There were no reports of the sandstone delaminating at the structures and only limited amounts of delaminated sandstone were observed at the site in comparison with the limestone. The areas around the buildings are public spaces, some of which are gathering spaces and walkway areas which experience relatively high traffic volume. The delamination of the limestone was of sufficient size and frequency to be a potential danger to the public health, safety, and welfare.

Laboratory Testing. Laboratory testing for the dimension stone (limestone and sandstone) was performed prior to construction, and some additional testing was performed during the investigation.

The project specifications indicate that both the limestone and sandstone are to be Classification II type stone. Considering limestone, the minimum compressive strength is indicated as 12,000 psi and the maximum absorption rate is listed as 5% in the specifications, although ASTM C 568 *Standard Specification for Limestone Dimension Stone* specifies a minimum compression strength of 4,000 psi and a maximum absorption rate of 7.5% for Classification II stone.

The specifications for the sandstone were unclear due to conflicting stone types indicated in the drawings and specifications. Due to the limited occurrence of sandstone delamination, the testing performed during the investigation was focused primarily on the limestone material.

Some of the pre-construction test data was available for review during the investigation of the stone failure. Pre-construction testing of the limestone indicated absorption rates of up to 6.3% and compressive strengths as low as 8,480 psi for the limestone. These test results indicate that the limestone does not meet the project specifications for compressive strength and absorption rate; however, the values from the testing do comply with the requirements of Classification II stone from ASTM C 568.

Pre-construction testing for the sandstone indicated that the compressive strengths and absorption rates of the sandstone material complied with the requirements of a Classification II stone from ASTM C 616 *Standard Specification for Quartz-Based Dimension Stone*.

As part of the investigation of the failed stone, samples of the sandstone and limestone were taken from the site and submitted to a testing laboratory for additional evaluation and testing. Samples of both limestone and sandstone were examined petrographically.

During petrographic examination, the sandstone was determined to be of good quality. This testing also indicated that the internal layering is not pronounced and there does not appear to be a large percentage of impurities present in the sandstone. Based on this evaluation and the general lack of sandstone distress, further testing was not performed on the sandstone.

The petrographic examination of the limestone indicated organic matter, carbon, iron-containing minerals, and clay minerals distributed in the bedding planes. It was also determined that large iron-ore inclusions discovered in the stone could have a negative impact on the durability of the limestone.

The limestone samples were tested for compressive strength and for absorption rate. The compressive testing was performed in accordance with C170 *Standard Test Method for Compressive Strength of Dimension Stone*. The results of the testing indicated that the limestone complied with the requirements of a Classification II stone per ASTM C 568 for both compressive strength and for absorption rate, although some of the compressive strength results were below the 12,000 psi limit of the project specifications. ASTM C 568 also states that, "*Limestone shall be sound, durable, and free of spalls, cracks, open seams, pits, or other defects that are likely to impair its structural integrity in its intended use.*" The limestone observed at the buildings does not comply with this requirement, as several stones were observed to have separations and spalls at the bedding planes (Figures 3-6).

Delamination of Stone. Pieces of delaminated stone were noted all along the perimeter of the building (Figures 4 and 5). The stone debris at the surrounding ground was of various sizes, although it should be noted that the size of debris is misleading, as the stone shatters when it falls from the wall above and impacts the ground. The largest debris pieces observed were up to 7" in length (Figure 5) and were generally about 1/4" thick. All of the limestone veneer on the buildings has been observed to have the rift (or grain) in a vertical orientation.

Figure 4. Pieces of delaminated limestone veneer on the ground.

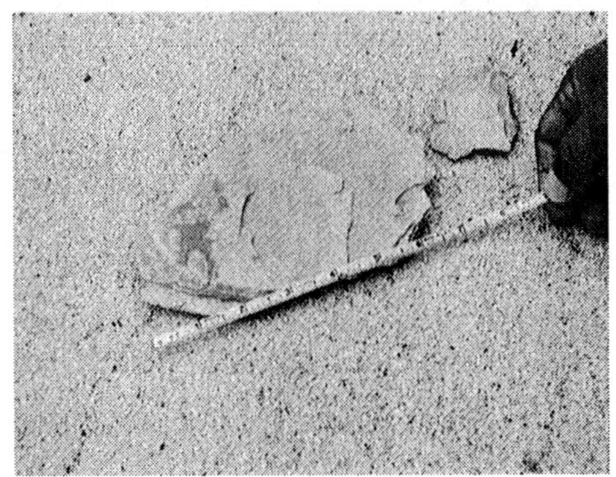

Figure 5. Piece of delaminated limestone measures 7" in length.

As noted above, the dimension stone veneer was specified to be 2 1/2" – 3" in thickness. Stone thicknesses outside of this range were observed at the subject buildings, with stones as thin as 2".

Active delamination was observed at the limestone veneer for the buildings. Figure 6 indicates a piece of limestone which has delaminated from the surface of a veneer stone, but has not yet fallen. This condition exists throughout the exterior veneer walls of both buildings.

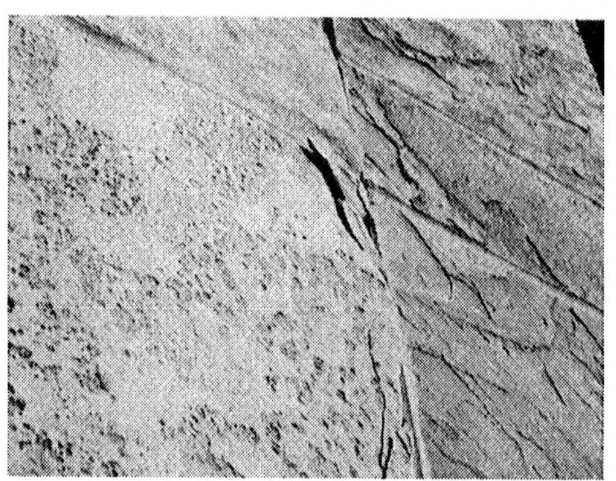

Figure 6. Delaminated limestone veneer at Building A.

The delamination and distress at the sandstone was not as significant as at the limestone. Generally, the sandstone was performing adequately; however, other issues, such as a reduced bearing area due to the lack of a full mortar bed could contribute to future distress of the stone. It is our opinion that the full depth of the mortar joints is paramount to preserving the integrity of both the limestone and the sandstone. The mortar bedding is addressed below in the "Mortar" section.

Expansion Joints. Another issue regarding the dimension stone veneer was expansion joint width. Horizontal expansion joints are indicated on the plans in the exterior walls at each floor level. The expansion joints occur where steel relief angles support the load of the veneer from floor-to-floor (approximately 16'). The horizontal expansion joints occur under the veneer course that rests directly on the relief angle and serve to separate the two fields of stone veneer and transfer the veneer weight to the concrete structure of the building.

The second floor expansion joint is indicated in the design documents for both structures to be 7/8" in width and the expansion joints at the third and fourth floors are indicated to be 3/4" wide. However, a subsequent revision note to the drawings indicates that horizontal expansion joints at all levels are to be 1/2" wide.

During evaluation of the buildings, the horizontal expansion joints were measured to have joint widths typically less than that indicated on the plans. Some expansion joints were completely closed, that is, the stones on either side of the joint were in direct contact (Figure 7). Also, expansion joint sealant material appeared compressed and was observed to be bulging at many joints. Delamination of the stones was observed at the corners of the joints where the stones were in contact.

Figure 7. Delaminated stone pieces with no expansion joint.

Some expansion joints were probed to determine if the joint extends the full depth of the stone veneer. Locations were observed where a knife probe could penetrate only 1/8" inward from the outer surface of the expansion joint sealant. Based on the observed conditions at other joints, it was likely that the stones were in contact behind the sealant, compressing the sealant out of the expansion joint and transmitting expansion stresses between the stone surfaces along the length of the joint. Alternatively, the expansion joints may be filled with mortar behind the sealant.

Measurements of the expansion joints were commonly only 1/4" wide, and several joints were determined to not extend for the full depth of the stone. Also, at some expansion joints, stones were observed to be "buckling" (i.e. deflecting outward at the joint). The lack of proper expansion joints was contributing to the delamination of the stones at the corners. The lack of joints also causes increased stresses in the stone veneer when it experiences movement or expansion, such as related to temperature or moisture.

Mortar. Mortar comprises only a small portion of a stone veneer wall; however, the influence of the mortar on the performance of the veneer wall is disproportionately large. Therefore, degradation or improper installation of the mortar can significantly impact the performance of a veneer wall. Masonry mortar is generally composed of cement, lime, sand, and water. Type I cement is the most common type used. Lime aids the workability and bonding characteristics of mortar and the only lime suitable for masonry work is Type S lime. The mortar specified for the site was Type N mortar with Type S lime.

Considering masonry construction, the most important property of a masonry mortar is bond strength. Workmanship is an important factor in establishing the bond between the mortar and adjacent stones, and full mortar joints must be provided at all contact surfaces for proper bonding (Beall, 1993).

Various instances of loose, displaced mortar were noted along the east side of the veneer clad beams and columns. Mortar was observed missing at joints between the coping stones as well. Generally, the observed mortar bed depth was not in compliance with the depth required by the construction documents. Mortar joint separations measured up to 0.06" in width, and several locations were observed where the mortar had fallen completely out of the joint. At some locations, the pattern of the mortar displacement was indicative of differential movement between the coping stones (pre-cast stone) and veneer stone.

Some samples of the loose mortar were collected and were typically observed to be only of partial depth of the full joint. Close observation of mortar samples indicated a typical intimacy of contact of 80%-90%; however, the intimacy of contact was as low as 60% at some locations.

The most significant sections of missing mortar were observed at the corners of columns, however, large sections without mortar were observed at window openings and building corners also. Based on our observations, the primary causes of mortar displacement are the physical movement of the veneer stones in conjunction with insufficient mortar depth at the joints.

The Project Specifications for both buildings stated that, the stone veneer should be set, *"... in full bed of mortar with vertical joints slushed full, unless otherwise indicated."*

Based on measurements from mortar samples taken from the site, the bed depth of the samples ranged from 1" to 3", and averaged 2" on one building and 1 5/8" on the other. Of the samples from both buildings, 64% had measured mortar depths of 2" or less. Figure 8 indicates a typical piece of mortar extracted from the joints, measuring only 1" in depth.

The lack of full mortar joint depth indicates improper construction and contributes to a loss of strength in the overall veneer wall system. Incomplete horizontal mortar beds would also cause increased bearing stresses at the ends of the stone veneer. This could cause stress concentrations at the stone surface and could contribute to the failure of the stone along lamination planes.

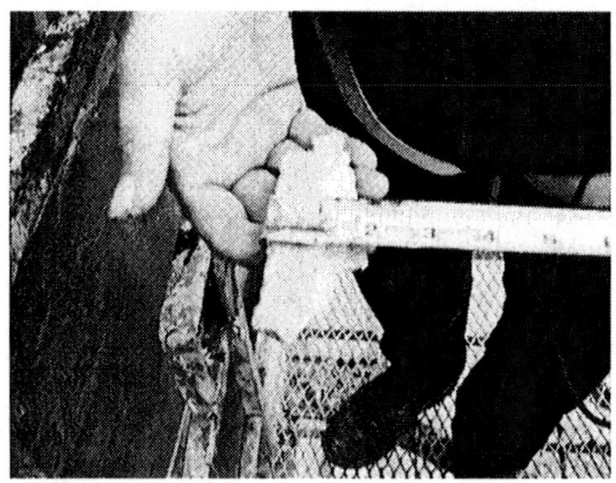

Figure 8. Depth only 1" at extracted mortar.

Veneer Anchorage and Support

Separations and evidence of movement and rotation were observed at the coping stones and windowsill stones throughout both buildings. The material for the coping stones and windowsills was precast stone. Some coping stones could be easily displaced by prying with a screwdriver in the separated joint.

The pattern and type of separations at some of the stone veneer were such that even a brief, non-intrusive evaluation raised concern over safety, especially at high-traffic areas. Further investigation was performed by removing stones or by observation with a scope at selected areas. Some of these are discussed below:

- Building maintenance personnel reported that there had been previous repairs to windowsill stones due to excessive movement. One area had reportedly moved 3"-4" outward from the building necessitating repair. No anchors were discovered during this repair and some remedial anchor clips were provided afterward. At other windowsill stones removed previously, the dowel pins called for in the architectural details were not observed during the repairs and no under-sill anchorage existed. Also, at some windowsill stones, evidence of further movement at repaired stones indicated that the remedial anchor clips were not performing adequately to anchor the stone.

- At the façade of one building, there was a veneer-clad wing wall above a fourth floor terrace area. This wall was topped with coping stones, with the top of the wall angled downward toward the exterior. There was a concrete beam running along the top of the wall (below the coping and veneer). A coping stone was removed from this wall to observe the anchorage of the coping and the condition at the wall cavity. The coping stone was easily removed and was determined to have no anchorage to the structure. Some of the other coping stones were scanned with a metal detector to locate veneer anchors into the concrete beam and veneer; no fasteners were detected.

- At the corner of the fourth floor terrace (at the front façade and located over a high-traffic courtyard area), stone coping and angled veneer were present along the wall. The coping at this corner was observed up close from the terrace above and also with the use of an articulating boom man-lift. Separations and displaced mortar were noted at the joints between the angled veneer stones and between the coping stones. The veneer is angled 45° at this location (see Figure 9 to view the angled stone veneer).

  At the first stone from the corner, mortar was missing from most of the vertical joint and the second coping stone from the corner had apparently rotated outward. Evidence of residual moisture was observed at the joints between the coping and the angled veneer and also at the joints along the bottom of the angled veneer. These joints and the stone adjacent to the joints appeared to be saturated with water.

  A detail in the architectural plans illustrates a section through the sloped veneer and the coping. Per this detail, the angled stone is anchored to a metal stud wall at the bottom by a stainless steel anchor and is anchored into the coping stone at the top by a stainless steel anchor pin. In turn, the coping stone is indicated to be set flush against the concrete frame and anchored with a stainless steel anchor fastening the stone to the concrete. Also, the coping stones are indicated to be supported vertically at the concrete wall by a relief angle. There is no indicated anchorage to the relief angle to prevent lateral displacement. This detail occurred at other areas on both buildings, raising concerns around the building perimeters. Some mortar displacement was noted at the angled veneer at other areas, indicating possible movement.

  At the northeast corner, the stone at the corner was separated 1/4" and the horizontal mortar joint below had fractured, apparently from differential movement. This corner stone and the two stones adjacent were loose. They could be rotated slightly at the outside edge by hand. These stones, especially the corner stone, exhibited a range of rotational movement which was not consistent with the attachment detail indicated in the drawings for the building.

  A metal detector was used to locate the anchors for the coping stones and for the angled veneer. Generally, it was observed that there were two anchors along both the top and bottom of the angled veneer. However, at the corner coping stone, no anchor attachment was noted.

  Because the loose and rotating coping stones at the northeast corner were a concern, a block of the stone veneer at the east side of the corner was removed to allow visibility into the wall cavity and behind the coping. This removed 15" x 7" block was not anchored to the wall at the top or bottom and was only supported by the mortar joints at the bottom and the sides and by the caulk material in the joint along the top of the stone. From this opening, the condition along the back side of the coping stones was visible. Metal shelf angles for the stone railing and for the coping at the edge were noted. Corrosion was observed at visible sections of the angles. Also, the coping was not set flush with the concrete frame and there was no anchor attachment visible from the coping into the concrete. The lack of a coping-to-concrete anchor was noted along the entire visible length at the back side of the coping.

The conditions at this area of the building represented a danger to public health, safety, and welfare.

Figure 9. Angled stone veneer at fourth floor. There is no mechanical horizontal anchorage to the concrete structure.

- The front fourth floor terrace of one structure had a stone balustrade, which separated from a storefront window system for a conference room. Large separations were observed at the vertical mortar joints in the wall below. A vertical column of three spacer stones bridged the gap between the balustrade and the storefront framing. Mortar was missing at the base of the stone assembly, indicating a lack of bearing support. The stones were easily moved and rotated by hand. Further investigation was performed from the building exterior, using the lift. It was noted that the stones were set on a relief angle below the balustrade. The stones were observed to bear on crumbling mortar, a significant amount of which was missing from the outside edge of the support. The gap between the stone and the relief angle was measured as 1/2" at the outer edge.

  Due to the obvious safety concern, the loose stones were removed immediately and were set onto the fourth floor balcony. As these stones were removed, the mortar application at the stone joints was determined to be inconsistent (i.e., uneven and incomplete "buttering" of the stones at the joints). Also, there was no anchorage from the spacer stones to the adjacent baluster stones.

A mortar sample was taken from the joints around the stones after they were removed. The mortar at this location was determined to be Type O mortar and was moderately soft, eroded severely with a high water-to-cement ratio, and of poorer quality than other mortar samples taken from the veneer walls. The quality of the mortar varied greatly even with in the relatively small sample.

- The veneer support was observable at a typical door header condition by using a boroscope to view the wall cavity. A steel angle lintel was observed behind the header stone. Observations in the cavity indicated that the angle was not set into the notch at the back side of the stone. Rather, a metal extension was apparently welded onto the edge of the angle and supported the stone at a notch in the back side of the header stone. This indicates a lack of coordination and/or workmanship during the installation of the veneer system.

- At the main ground level entry to one of the structures, separated and displaced dimension stone veneer was observed. The mortar joints in the stone veneer above the center opening were separated, and a downward movement of the veneer was observed in the approximate shape of an arch starting from the header course above the opening. Figure 10 below shows the displacement of the veneer above the entry and the loose mortar. The separations were measured, as 15/16" at horizontal joints and the mortar was loose and easily removed by hand. Some repairs had been done previously; however, the separations re-opened and the stone continued to displace. Although the support could not be seen from the wall cavity due to excessive mortar, similar details at other openings had a lack of full support for the stone. At one location above another entry door, a gap between the stone and the steel support lintel was observed.

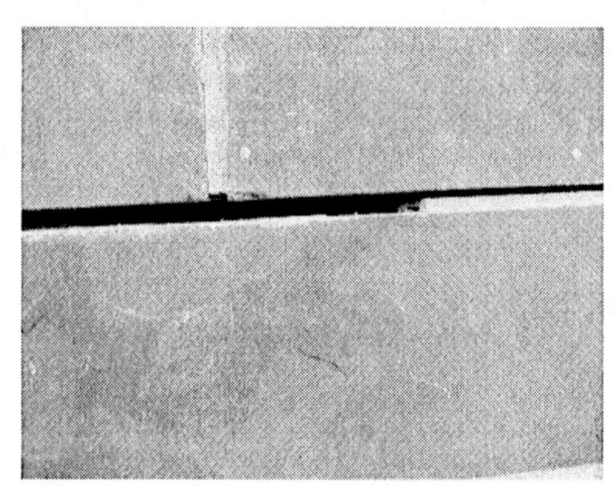

Figure 10. Veneer displacement and loose mortar above entry.

Conclusions

The damage observed at the veneer systems for the subject buildings is related to both design issues and construction issues. The contributing factors to the stone delamination included the selection of the limestone material to be edge-set in the veneer system, the lack of full-depth mortar, and the lack of proper expansion joints.

In addition to the delamination of the dimension stone, differential movement and displacement was observed at veneer which was not properly supported or anchored. The anchors specified at the coping and windowsill stones were not installed during the original construction of the buildings. Also, the dimension stone was visibly displaced at several openings due to a lack of proper vertical support.

Differential movements in the dimension stone due to improper or non-existent anchorage and the lack of proper mortar depth has also caused mortar to be completely displaced from a significant number of mortar joints. Proper anchorage of the stones and proper installation of full depth mortar would have prevented this problem.

The conditions at the dimension stone and coping at the veneer system created conditions which were potential dangers to the public health, safety, and welfare. This case study illustrates how design decisions and execution of support for heavy stone veneers can affect the safety of a building. Consideration and execution of fundamental issues related to exterior veneer material selection and the installation of a stone veneer system would have prevented all of the unsafe conditions.

References

Amrhein, James E. and Merrigan, Michael W. (1989). *Marble and Stone Slab Veneer*, Second Edition, Masonry Institute of America, Los Angeles, California.

American Society for Testing and Materials (ASTM). (1999). *ASTM C 170-90 Standard Test Method for Compression Strength of Dimension Stone*, Reapproved 1999, West Conshohocken, Pennslyvania.

American Society for Testing and Materials (ASTM). (1999). *ASTM C 568-99 Standard Specification for Limestone Dimension Stone*, West Conshohocken, Pennslyvania.

American Society for Testing and Materials (ASTM). (1999). *ASTM C 616-99 Standard Specification for Quartz-Based Dimension Stone*, West Conshohocken, Pennslyvania.

Beall, Christine. (1993). *Masonry Design and Detailing For Architects, Engineers, and Contractors,* Third Edition, McGraw-Hill, Inc.

Bortz, Seymour A. and Wonneberger, Bernhard. (1997). "Laboratory Evaluation of Building Stone Weathering." *Degradation of Natural Building Stone*, Geotechnical Special Publication No. 72, American Society of Civil Engineers, Reston Virginia, 85-104.

Labuz, J.F. (editor) (1997). *Degradation of Natural Building Stone,* Geotechnical Special Publication No. 72, American Society of Civil Engineers, Reston Virginia.

Marble Institute of America (MIA). (1999). *Design Manual, Marble and Stone Slab Veneer*, Marble Institute of America, Cleveland Ohio.

Risk Assessment and Treatment in Slope Stability Forensic Engineering

Mihail E. Popescu, Ph.D., P.E., Eur.Ing.
Wang Engineering, Inc. / Illinois Institute of Technology
6117 River Bend Drive, Lisle, IL 60532, PH: (630) 960-2673, email: mepopescu@usa.com

Abstract

Slope instability phenomena are frequently responsible for considerable losses of both money and lives. In view of above consideration, it is not surprising that slope instability phenomena are rapidly becoming the focus of major scientific research, engineering study and practices, and land-use policy throughout the world.

The paramount importance of slope instability management is by and large recognized. Herein lies the guiding principle of the current paper; i.e., to describe slope instability hazard assessment and methods to mitigate the associated risks in an appropriate and effective way.

Back analysis of failed slopes is one of the most effective ways to advance our knowledge in the field of slope stability engineering. This paper discusses procedures to back calculate soil shear strength parameters from slope failures and illustrates how these parameters can be subsequently used to design slope remedial works.

Slope Instability Risk Management

Landslides and related slope instability phenomena plague many parts of the world. Japan leads other nations in landslide severity with projected combined direct and indirect losses of $4 billion annually (Schuster, 1996). United States, Italy, and India follow Japan, with an estimated annual cost ranging between $1 billion to $2 billion. Landslide disasters are also common in developing countries and economical losses sometimes equal or exceed their gross national products.

International cooperation among various individuals concerned with the fields of geology, geomorphology, soil and rock mechanics have been contributing to improve our understanding of landslides, notably in the framework of the United Nations International Decade for Natural Disaster Reduction (1990-2000). Evidently, this provided the environment for establishing the International Geotechnical Societies UNESCO Working Party on World Landslide Inventory (abbreviated WP/WLI) which in 1994 became the IUGS Working Group on Landslides (abbreviated WG/L).

The risk management process comprises two components: risk assessment and risk treatment. Slope stability engineering has always involved some form of risk management, although it was seldom formally recognized as such. This informal type of risk management was essentially the exercise of engineering judgment by experienced engineers and geologists. In simple form, the process involves answering the following questions:

- What might happen?
- How likely is it?
- What damage or injury may result?
- How important is it?
- What can be done about it?

Slope instability hazard identification requires an understanding of the slope processes and the relationship of those processes to geomorphology, geology, hydrogeology, climate and vegetation.

Before embarking on a regional landslide hazard assessment, the following preparatory steps are to be taken (Hutchinson, 2001):

- Identify the user and purpose of the proposed assessment. Involve the user in all phases of the program.
- Define the area to be mapped and decide the appropriate scale of mapping. This may range from 1: 100,000 or smaller to 1:5,000 or larger.
- Obtain, or prepare, a good topographical base map of the area, preferably contoured.
- Construct a detailed database of the geology (solid and superficial), geomorphology, hydrogeology, pedology, meteorology, mining and other human interference, history and all other relevant factors within the area, and of all known mass movements including all published work, newspaper articles and the results of interviewing the local population.
- Obtain all available air photo cover, satellite imagery and ground photography of the area. Photography of various dates can be particularly valuable, both because of what can be revealed by differing lighting and vegetation conditions and to delineate changes in the man-made and natural conditions, including slide development.

By and large, the elements at risk involve property, people, services, such as water supply or drainage or electricity supply, roads and communication facilities, and vehicles on roads. The consequences may not, however, be limited to property damage and injury/loss of life. Other factors include public outrage, political effects, loss of business confidence, effect on reputation, social upheaval, consequential costs, such as litigation (Einstein, Karam, 2001).

Many of these may not be readily quantifiable and will require considerable judgment if they are to be included in the assessment. Consideration of such consequences may constitute part of the risk evaluation process by the client/owner/regulator (Popescu, Zoghi, 2005).

Slope Instability Causal Factors

"The processes involved in slope movements comprise a continuous series of events from cause to effect" (Varnes, 1978). When assessing slope instability hazard for a particular site, of primary importance is the recognition of the conditions which caused the slope to become unstable and the processes which triggered that movement. Only an accurate diagnosis makes it possible to properly understand the landslide mechanisms and thence to propose effective remedial measures.

Table 1. A brief list of landslide causal factors

1. GROUND CONDITIONS
(1)Plastic weak material (2)Sensitive material (3)Collapsible material (4)Weathered material (5)Sheared material (6)Jointed or fissured material (7)Adversely oriented mass discontinuities (including bedding, schistosity, cleavage) (8)Adversely oriented structural discontinuities (including faults, unconformities, flexural shears, sedimentary contacts) (9)Contrast in permeability and its effects on ground water (10)Contrast in stiffness (stiff, dense material over plastic material)
2. GEOMORPHOLOGICAL PROCESSES
(1)Tectonic uplift (2)Volcanic uplift (3)Glacial rebound (4)Fluvial erosion of the slope toe (5)Wave erosion of the slope toe (6)Glacial erosion of the slope toe (7)Erosion of the lateral margins (8)Subterranean erosion (solution, piping) (9)Deposition loading of the slope or its crest (10)Vegetation removal (by erosion, forest fire, drought)
3. PHYSICAL PROCESSES
(1)Intense, short period rainfall (2)Rapid melt of deep snow (3)Prolonged high precipitation (4)Rapid drawdown following floods, high tides or breaching of natural dams (5)Earthquake (6)Volcanic eruption (7)Breaching of crater lakes (8)Thawing of permafrost (9)Freeze and thaw weathering (10)Shrink and swell weathering of expansive soils
4. MAN-MADE PROCESSES
(1)Excavation of the slope or its toe (2)Loading of the slope or its crest (3)Drawdown (of reservoirs) (4)Irrigation (5)Defective maintenance of drainage systems (6)Water leakage from services (water supplies, sewers, stormwater drains) (7)Vegetation removal (deforestation) (8)Mining and quarrying (open pits or underground galleries) (9)Creation of dumps of very loose waste (10)Artificial vibration (including traffic, pile driving, heavy machinery)

The great variety of slope movements reflects the diversity of conditions that cause the slope to become unstable and the processes that trigger the movement. It is more appropriate to discuss causal factors (including both "conditions" and "processes") than "causes" per se alone. Seldom, if ever, can a landslide be attributed to a single causal factor. The process leading to the development of the slide has its beginning with the formation of the rock itself, when its basic properties are determined and includes all the subsequent events of crustal movement, erosion and weathering.

The computed value of the factor of safety is a clear and simple distinction between stable and unstable slopes. However, from the physical point of view, it is better to visualize slopes existing in one of the following three stages: stable, marginally stable and actively unstable.

The three stability stages provide a useful framework for understanding the causal factors of landslides and classifying them into two groups on the basis of their function:

1.*Preparatory causal factors* which make the slope susceptible to movement without actually initiating it and thereby tending to place the slope in a marginally stable state.
2.*Triggering causal factors* which initiate movement. The causal factors shift the slope from a marginally stable to an actively unstable state.

A particular causal factor may inflict either or both functions, depending on its degree of activity and the margin of stability. Although it may be possible to identify a single triggering process, an explanation of ultimate causes of a landslide invariably involves a number of preparatory conditions and processes.

When assessing landslide causes it is necessary to make a distinction between ground conditions and processes. Ground conditions are the specification of the slope system, the setting on which a process can act to prepare or trigger a failure.

The operational approach to classification of slope instability causal factors proposed by the WP/WLI (Popescu, 1994) is intended to cover the majority of landslides. It involves the consideration of the available data from simple site investigation and information furnished by other site observations. Slope instability causal factors are grouped according to their effect (preparatory or triggering) and their origin (ground conditions and geomorphological, physical or man-made processes). Ground conditions may not have a triggering function, while any ground condition or process may have a preparatory function. Table 1 is a short practical checklist of landslide causal factors arranged according with the tools and procedures necessary for documentation.

Slope Instability Treatment

Risk treatment is the final stage of the risk management process and provides the methodology of controlling the risk. At the end of the evaluation procedure, it is up to the client or policy makers to decide whether to accept the risk or not, or to decide that more detailed study is required. The landslide risk analyst can provide background data or normally acceptable limits as guidance to the decision maker but should not be making the decision. Part of the specialist's advice may be to identify the options and methods for treating the risk. Typical options would include (AGS, 2000): accept the risk, avoid the risk, reduce the likelihood, reduce the consequences, transfer the risk, postpone the decision.

The relative costs and benefits of various options need to be considered so that the most cost effective solutions, consistent with the overall needs of the client, owner and regulator, can be identified. Combinations of options or alternatives may be appropriate, particularly where relatively large reductions in risk can be achieved for relatively small expenditure. Prioritization of alternative options is likely to assist with selection.

Correction of an existing landslide or the prevention of a pending landslide is a function of a reduction in the driving forces or an increase in the available resisting forces. Any remedial measure used must involve one or both of the above parameters.

IUGS WG/L (Popescu, 2001) has prepared a short checklist of landslide remedial measures arranged in four practical groups, namely: modification of slope geometry, drainage, retaining structures and internal slope reinforcement Table 2). The flow diagram presented in Fig. 1 illustrates the sequence of various phases involved in the planning, design, construction and monitoring of remedial works (Kelly, Martin, 1986).

As many of the geological features, like the sheared discontinuities, are not well known in advance, it is better to put remedial measures in hand on a "design as you go basis". That is the design has to be flexible enough for changes during or subsequent construction of remedial works.

During the early part of the post-war period, slope instability processes were generally seen to be "engineering problems" requiring "engineering solutions" involving correction by the use of structural techniques. This structural approach initially focused on retaining walls but has subsequently been diversified to include a wide range of more sophisticated techniques

Table 2. A brief list of landslide remedial measures

1. MODIFICATION OF SLOPE GEOMETRY
1.1. Removing material from the area driving the landslide (with possible substitution by lightweight fill)
1.2. Adding material to the area maintaining stability (counterweight berm or fill)
1.3. Reducing general slope angle
2. DRAINAGE
2.1. Surface drains to divert water from flowing onto the slide area (collecting ditches and pipes)
2.2. Shallow or deep trench drains filled with free-draining geomaterials (coarse granular fills and geosynthetics)
2.3. Buttress counterforts of coarse-grained materials (hydrological effect)
2.4. Vertical (small diameter) boreholes with pumping or self draining
2.5. Vertical (large diameter) wells with gravity draining
2.6. Subhorizontal or subvertical boreholes
2.7. Drainage tunnels, galleries or adits
2.8. Vacuum dewatering
2.9. Drainage by siphoning
2.10. Electroosmotic dewatering
2.11. Vegetation planting (hydrological effect)
3. RETAINING STRUCTURES
3.1. Gravity retaining walls
3.2. Crib-block walls
3.3. Gabion walls
3.4. Passive piles, piers and caissons
3.5. Cast-in situ reinforced concrete walls
3.6. Reinforced earth retaining structures with strip/ sheet - polymer/metallic reinforcement elements
3.7. Buttress counterforts of coarse-grained material (mechanical effect)
3.8. Retention nets for rock slope faces
3.9. Rockfall attenuation or stopping systems (rocktrap ditches, benches,fences and walls)
3.10. Protective rock/concrete blocks against erosion
4. INTERNAL SLOPE REINFORCEMENT
4.1. Rock bolts
4.2. Micropiles
4.3. Soil nailing
4.4. Anchors (prestressed or not)
4.5. Grouting
4.6. Stone or lime/cement columns
4.7. Heat treatment
4.8. Freezing
4.9. Electroosmotic anchors
4.10. Vegetation planting (root strength mechanical effect)

including passive piles and piers, cast-in-situ reinforced concrete walls and reinforced earth retaining structures.

Over the last several decades, there has been a notable shift towards "soft engineering", non-structural solutions including classical methods such as drainage and modification of slope geometry but also some novel methods such as lime/cement stabilization, grouting or soil nailing (Popescu, 1996). The cost of non-structural remedial measures is considerably lower when compared with the cost of structural solutions.

Environmental considerations have increasingly become an important factor in the choice of suitable remedial measures, particularly issues such as visual intrusion in scenic areas or the impact on nature or geological conservation interests. An example of "soft engineering" solution, more compatible with the environment, is the stabilization of slopes by the combined use of vegetation and man-made structural elements working together in an integrated manner known as biotechnical slope stabilization (Kelly, Martin, 1986).

Terzaghi (1950) stated that, "if a slope has started to move, the means for stopping movement must be adapted to the processes which started the slide". For example, if erosion is a causal process of the slide, an efficient remediation technique would involve armoring the slope against erosion, or removing the source of erosion. An erosive spring can be made non-erosive by either blanketing with filter materials or drying up the spring with horizontal drains etc.

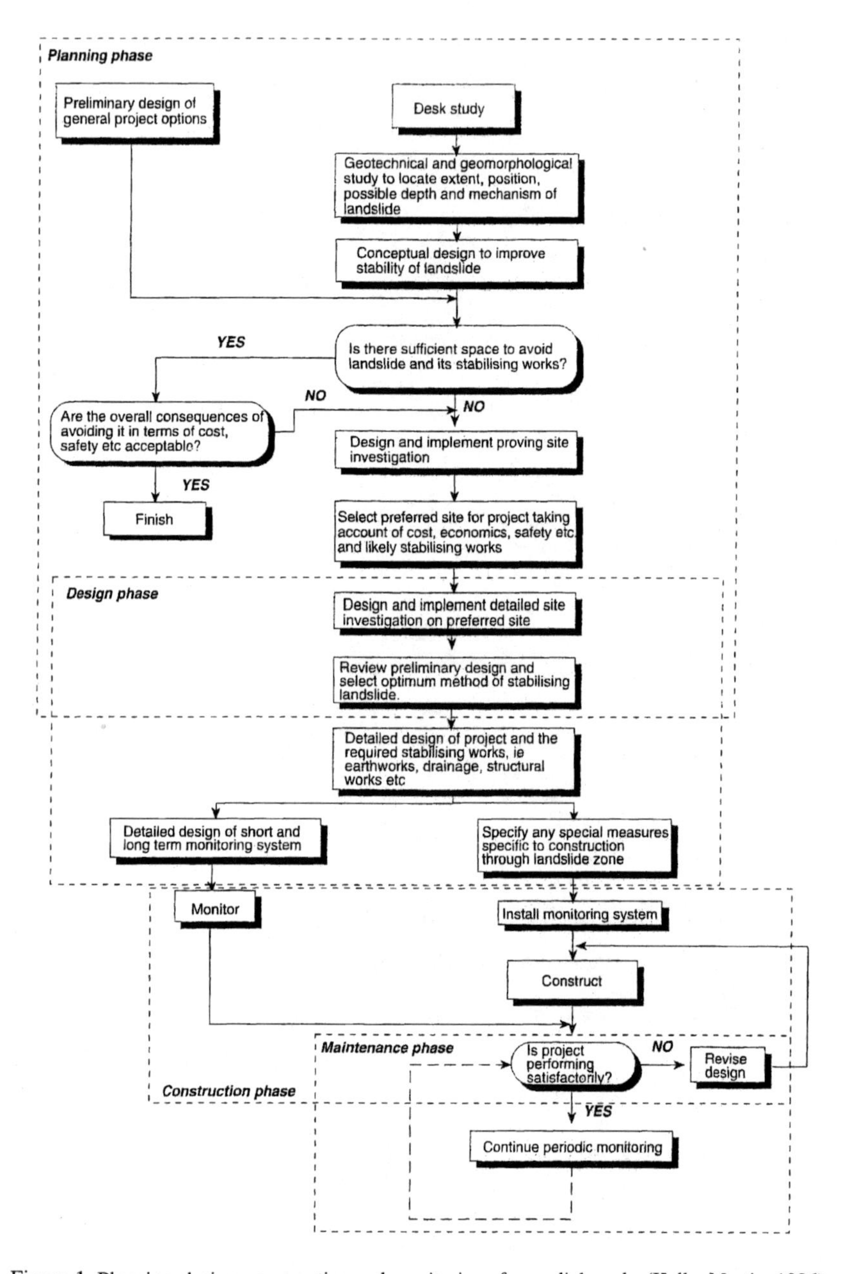

Figure 1. Planning, design, construction and monitoring of remedial works (Kelly, Martin, 1986)

Just as there are a number of available remedial measures, so are there a number of levels of effectiveness and levels of acceptability that may be applied in the use of these measures. We may have a landslide, for example, that we simply choose to live with; one that poses no significant hazard to the public, whereas it will require periodic maintenance for example, through removal, due to occasional encroachment onto the shoulder of a roadway.

Most landslides, however, must usually be dealt with sooner or later. How they are handled depends on the processes that prepared and precipitated the movement, the landslide type, the kinds of materials involved, the size and location of the landslide, the place or components affected by or the situation created as a result of the landslide, available resources, etc. The technical solution must be in harmony with the natural system, otherwise the remedial work will be either short lived or excessively expensive. In fact, landslides are so varied in type and size, and in most instances, so dependent upon special local circumstances, that for a given landslide problem there is more than one method of prevention or correction that can be successfully applied. The success of each measure depends, to a large extent, on the degree to which the specific soil and groundwater conditions are prudently recognized in an investigation and incorporated in design.

Back Analysis of Failed Slopes to Design Remedial Measures

Failure Envelope Parameters

A slope failure can reasonably be considered as a full scale shear test capable to give a measure of the strength mobilized at failure along the slip surface. The back calculated shear strength parameters which are intended to be closely matched with the observed real-life performance of the slope, can then be used in further limit equilibrium analyses to design remedial works.

The limit equilibrium methods forming the framework of slope stability/instability analysis generally accept the Mohr-Coulomb failure criterion:

$$\tau_f = c' + \sigma' \tan \phi' \quad (1)$$

where τ_f and σ' are the shear stress and effective normal stress respectively on the failure surface and c' and ϕ' are parameters assumed approximately constant for a particular soil.

A significant limitation in the use of this criterion is that the constant of proportionality is not really a constant when wide range of stress is under consideration. There is now considerable experimental evidence to show that the Mohr failure envelope exhibits significant curvature for many different types of soil and compacted rockfill. Therefore, if the assumption of a linear failure envelope is adopted, it is important to know what range of stress is appropriate to a particular slope instability problem.

In order to avoid this difficulty a more convenient approach is to adopt an analytical equation of the nonlinear Mohr failure envelope and incorporate it into the existing limit equilibrium methods to back calculate the equation parameters from the data provided by the failed slopes.

The curved failure envelope can be approximated by the following power law equation:

$$\tau_f = A\,(\sigma')^b \tag{2}$$

which was initially suggested by De Mello (1977) for compacted rockfills and subsequently found appropriate for soils (Atkinson, Farrar, 1985).

Procedures for Back Analysis of Slope Failures

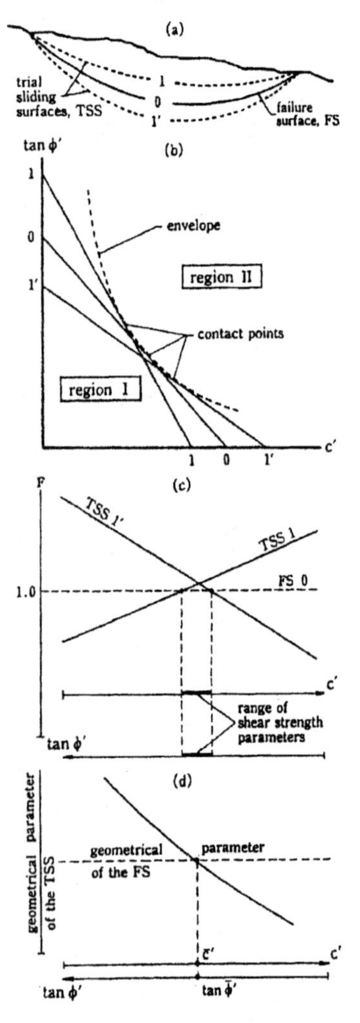

Figure 2. Shear strength back analysis methods

Shear strength parameters obtained by back analysis ensure more reliability than those obtained by laboratory or in-situ testing when used to design remedial measures.

In many cases, back analysis is an effective tool, and sometimes the only tool, for investigating the strength features of a soil deposit. However one has to be aware of the many pitfalls of the back analysis approach that involves a number of basic assumptions regarding soil homogeneity, slope and slip surface geometry and pore pressure conditions along the failure surface. A position of total confidence in all these assumptions is rarely if ever achieved.

Indeed, in some cases, because the large extension of a landslide, various soils with different properties are involved. In other cases the presence of cracks, joints, thin intercalations and anisotropies can control the geometry of the slip surface. Moreover progressive failure or softening resulting in strength reductions that are different from a point to another, can render heterogeneous even deposits before homogeneous.

While the topographical profile can generally be determined with enough accuracy, the slip surface is almost always known in only few points and interpolations with a considerable degree of subjectivity are necessary. Errors in the position of the slip surface result in errors in back calculated shear strength parameters. If the slip surface used in back analysis is deeper than the actual one, c' is overestimated and ϕ' is underestimated and vice-versa.

The data concerning the pore pressure on the slip surface are generally few and imprecise. More exactly, the pore pressure at failure is almost always unknown. If the assumed pore pressures are higher than the actual ones, the shear strength is overestimated. As a consequence, a conservative assessment of the shear strength is obtainable only by underestimating the pore pressures.

Procedures to determine the magnitude of both shear strength parameters or the relationship between them by considering the position of the actual slip surface within a slope are discussed by Popescu and Yamagami (1994). The two unknowns - i.e. the shear strength parameters c' and ϕ' - can be simultaneously determined from the following two requirements:

(a) F = 1 for the given failure surface. That means the back calculated strength parameters have to satisfy the c'-tan ϕ' limit equilibrium relationship;
(b) F = minimum for the given failure surface and the slope under consideration. That means the factors of safety for slip surfaces slightly inside and slightly outside the actual slip surface should be greater than one (Fig.2a).

Based on the above mentioned requirements, Saito (1980) developed a semi-graphical procedure using trial and error to determine unique values of c' and tan ϕ' by back analysis (Fig.2b). An envelope of the limit equilibrium lines c' - tan ϕ', corresponding to different trial sliding surfaces, is drawn and the unique values c' and tan ϕ' are found as the coordinates of the contact point held in common by the envelope and the limit equilibrium line corresponding to the actual failure surface.

A more systematic procedure to find the very narrow range of back calculated shear strength parameters based on the same requirements is illustrated in Fig.2c.

The procedures discussed above to back calculate the linear strength envelope parameters, c' and ϕ' in equation (1) can be equally applied to back calculate the nonlinear strength envelope parameters, A and b in equation (2) (Popescu, Yamagami, Stefanescu, 1995).

The fundamental problem involved is always one of data quality and consequently the back analysis approach must be applied with care and the results interpreted with caution.

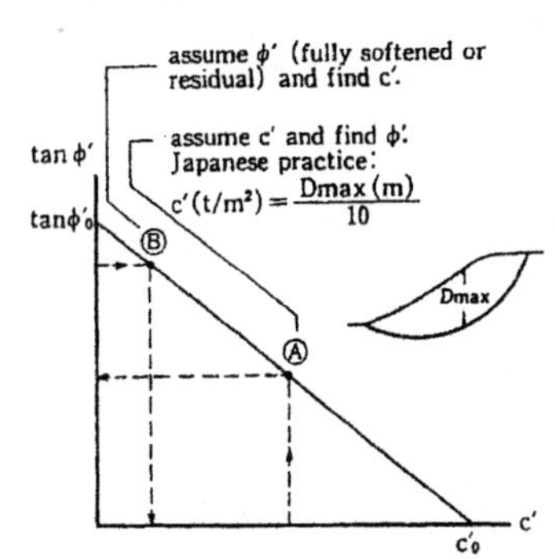

Figure 3. Assume one of the shear strength parameters and determine the other one

Back analysis is of use only if the soil conditions at failure are unaffected by the failure. For example back calculated parameters for a first-time slide in a stiff overconsolidated clay could not be used to predict subsequent stability of the sliding mass, since the shear strength parameters will have been reduced to their residual values by the failure.

It is also to be pointed out that if the three-dimensional geometrical effects are important for the failed slope under consideration and a two-dimensional back analysis is performed, the back calculated shear strength will be too high and thus unsafe.

Design of Remedial Measures Based on Back Analysis Results

Although the principle of the back analysis methods discussed above is correct, Duncan and Stark (1992) have shown that in practice, as a result of progressive failure and the fact that the position of the rupture surface may be controlled by strong or weak layers within the slope, the shear strength parameters cannot be uniquely determined through back analysis.

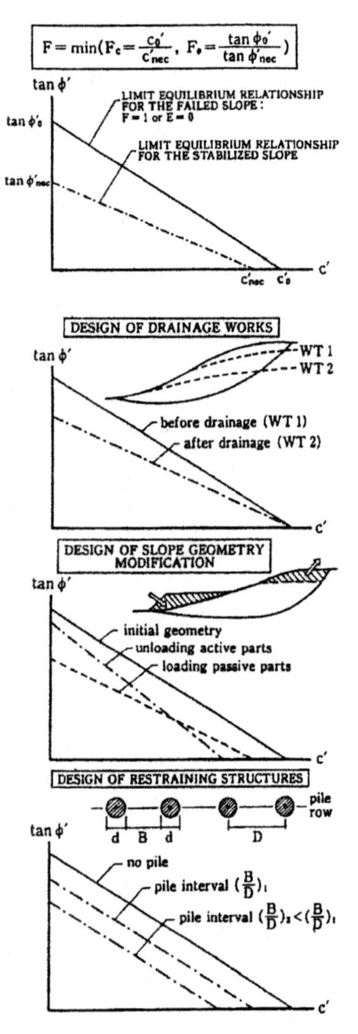

Figure 4. Limit equilibrium relationship and design of slope remedial works

The alternative is to assume one of the shear strength parameters and determine the other one that corresponds to a factor of safety equal to unity (Fig. 3). Duncan and Stark (1992) proposed to assume the value of ϕ', using previous information and good judgment, and to calculate the value of c' that corresponds to F=1. They recommended to assume fully softened strength where no sliding has occurred previously, and residual strength where there has been sufficient relative shearing deformation along a pre-existing sliding surface.

In order to avoid the questionable problem of the representativeness of the back calculated unique set of shear strength parameters a method for designing remedial works based on the limit equilibrium relationship c' - ϕ' rather than a unique set of shear strength parameters can be used (Popescu, 1991).

The method principle is shown in Fig.4. It is considered that a slope failure provides a single piece of information which results in a linear limit equilibrium relationship between shear strength parameters. That piece of information is that the factor of safety is equal to unity (F=1) or the horizontal force at the slope toe is equal to zero (E=0) for the conditions prevailing at failure. Each of the two conditions (F=1 or E=0) results in the same relationship c'-tan ϕ' which for any practical purpose might be considered linear.

The linear relationship c'-tan ϕ' can be obtained using a standard computer software for slope stability limit equilibrium analysis by manipulations of trial values of c' and tan ϕ' and corresponding factor of safety value. It is simple to show that in an analysis using arbitrary ϕ' alone (c'=0) to yield a non-unity factor of safety, F_ϕ*, the intercept of the c'-tan ϕ' line (corresponding to F=1) on the tan ϕ' axis results as:

$$\tan \phi_0' = \tan \phi' / F_\phi^* \tag{3}$$

Similarly the intercept of the c'-tan φ' line (corresponding to F=1) on the c' axis can be found assuming φ'=0 and an arbitrary c' value which yield to a non-unity factor of safety, F_c^*:

$$c_0' = c' / F_c^* \tag{4}$$

Using the concept of limit equilibrium linear relationship c'-tan φ', the effect of any remedial measure (drainage, modification of slope geometry, restraining structures) can easily be evaluated by considering the intercepts of the c'-tan φ' lines for the failed slope (c_0', $\tan \phi_0'$) and for the same slope after installing some remedial works (c'_{nec}, $\tan \phi'_{nec}$), respectively (Fig. 4). The safety factor of the stabilized slope is:

$$F = \min\left(F_c = \frac{c_0'}{c'_{nec}}, F_\phi = \frac{\tan \phi_0'}{\tan \phi'_{nec}} \right) \tag{5}$$

Errors included in back calculation of a given slope failure will be offset by applying the same results, in the form of c' - tan φ' relationship, to the design of remedial measures.

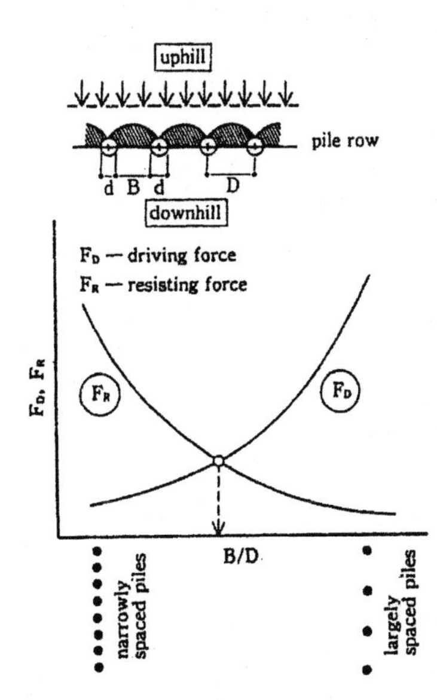

Figure 5. Driving vs. resisting force for stabilizing piles

The above outlined procedure was used to design piles to stabilize landslides (Popescu, 1991) taking into account both driving and resisting force. The principle of the proposed approach is illustrated in Fig.5 which gives the driving and resisting force acting on each pile in a row as a function of the non-dimensional pile interval ratio B/D. The driving force, F_D, is the total horizontal force exerted by the sliding mass corresponding to a prescribed increase in the safety factor along the given failure surface. The resisting force, F_R, is the lateral force corresponding to soil yield, adjacent to piles, in the hatched area shown in Fig.5. F_D increases with the pile interval while F_R decreases with the same interval. The intersection point of the two curves which represent the two forces gives the pile interval ratio satisfying the equality between driving and resisting force.

The accurate estimation of the lateral force on pile is an important parameter for the stability analysis because its effects on both the pile-and slope stability are conflicting. That is, safe assumptions for the stability of slope are unsafe assumptions for the pile stability, and vice-versa. Consequently in order to obtain an economic and safe design it is necessary to avoid excessive safety factors.

The problem is clearly three-dimensional and some simplification must be accepted in order to develop a two-dimensional analysis method based on the principles outlined above. However the only simplicity to be accepted and trusted is the simplicity that lies beyond the problem complexity and makes all details and difficulties simple by a sound and profound understanding.

Observation of natural phenomena and model studies could improve very much our capacity to understand, foresee and prevent landslides. In this respect, in-situ tests of well instrumented slopes which are forced to failure (Petley et al., 1991) provide a most valuable source of information to calibrate our back analysis techniques.

Concluding Remarks

Assessing landslide hazard is a most important step in landslide risk management. Once that has been done, it is feasible to assess the number, size and vulnerability of the fixed elements at risk (structures, roads, railways, pipelines, etc) and thence the damage they will suffer. The various risks have to be combined to arrive at a total risk in financial terms. Comparison of this with, for instance, cost-benefit studies of the cost of relocation of populations and facilities, or mitigation of the hazard by countermeasures, provides a very useful tool for management and decision-making.

Much progress has been made in developing techniques to minimize the impact of landslides, although new, more efficient, quicker and cheaper methods could well emerge in the future. There are a number of levels of effectiveness and levels of acceptability that may be applied in the use of these measures, for while one slide may require an immediate and absolute long-term correction, another may only require minimal control for a short period.

Whatever the measure chosen, and whatever the level of effectiveness required, the geotechnical engineer and engineering geologist have to combine their talents and energies to solve the problem. Solving landslide related problems is changing from what has been predominantly an art to what may be termed an art-science. The continual collaboration and sharing of experience by engineers and geologists will no doubt move the field as a whole closer toward the science end of the art-science spectrum than it is at present.

References

Atkinson, J.H., Farrar, D.M. (1985). *Stress path tests to measure soils strength parameters for shallow landslips.* Proc.11th Intern. Conf. Soil Mech. Foundation Eng., San Francisco, vol.2, p. 983-986.

De Mello, V.F.B. (1977). *Reflections on design decisions of practical significance to embankment dams*, Geotechnique, vol.27, no.3, p. 281-354.

Duncan, J.M., Stark, T.D. (1992). *Soil strength from back analysis of slope failures*, Proc. ASCE Geotechnical Conf. on Slopes and Embankments, Berkeley, 890-904.

Einstein, H.H., Karam, K.S. (2001). *Risk Assessment and Uncertainties*, Proceedings of the International Conference on Landslides – Causes, Impacts and Countermeasures," Davos, Switzerland.

Hutchinson J. N. (2001). *Landslide risk - To know, to foresee, to prevent.* Journal of Technical and Environmental Geology, n. 3, pp. 3 - 22.

Kelly, J.M.H., and Martin, P.L. (1986). *Construction Works On Or Near Landslides,* Proc. Symposium of Landslides in South Wales Coalfield, Polytechnic of Wales, 85-103.

Petley, D. J., Bromhead, E.N., Cooper, M.R., Grant, D. I. (1991). *Full-scale slope failure at Selborne, U.K.* In: Develop. in Geot. Aspects of Embankments, Excavations and Buried Structures, Balkema, Rotterdam, 209-223.

Popescu, M.E. (1991). *Landslide control by means of a row of piles,* Keynote paper. Proc. Int. Conf. on Slope Stability Engineering, Isle of Wight, Thomas Telford, 389-394.

Popescu M.E., Yamagami T. (1994). *Back analysis of slope failures - a possibility or a challenge?,* Proc.7th Congress Intern.Assoc.Engg.Geology, Lisbon, (6), p.4737-4744.

Popescu M.E., Yamagami T., Stefanescu S. (1995). *Nonlinear strength envelope parameters from slope failures,* Proc.11th ECSMFE, Copenhagen,.(1), p. 211-216.

Popescu, M.E. (1996). *From Landslide Causes to Landslide Remediation,* Special Lecture." *Proc. 7th Int. Symp. on Landslides,* Trondheim, 1:75-96.

Popescu M.E. (2001). *A Suggested Method for Reporting Landslide Remedial Measures, IAEG Bulletin,* 60, 1:69-74

Popescu M.E., Zoghi M. (2005). *Landslide Risk Assessment and Remediation,* Chapter in Monograph on Natural, Accidental, and Deliberate Hazards, edited by Craig Taylor and Erik VanMarke, American Society of Civil Engineering, Council on Disaster Risk Management, Monograph No.1, Reston, VA, p. 161-193.

Saito M. (1980). *Reverse calculation method to obtain c and ϕ on a slip surface.* Proc. Int. Symp. Landslides, New Delhi, (1), p.281-284.

Schuster, R.L. (1996). *Socioeconomic Significance of Landslides,* Chapter 2 in Landslides – Investigation and Mitigation, Special Report 247, Transportation Research Board, National Research Council, Turner, A.K. and Schuster, R.L., Editors, National Academy Press, Washington, D.C.

Terzaghi, K. (1950). *Mechanisms of Landslides,* Geological Society of America, Berkley, 83-123.

Varnes, D.J. (1978). *Slope Movements And Types And Processes,* in Landslides Analysis and Control, Transportation Research Board Special Report, 176:11-33.

WP/WLI: International Geotechnical Societies' UNESCO Working Party on World Landslide Inventory. Working Group on Landslide Causes - Popescu, M.E., Chairman. (1994). *A suggested method for reporting landslide causes,* Bulletin IAEG, 50:71-74.

An "Invisible Menace" – The Impact of Pyrite Induced Expansive Forces on Long-Term Building Failure – A Case Study

Paul G. Carr, Ph.D., P.E., Member ASCE[1]
Joseph L. Thesier, P.E. Member ASCE[2]
Mark B. Kimball, P.E. Member ASCE[2]

Abstract—There is a dearth of information available in the literature and published engineering texts addressing the impacts of pyrite expansion in buildings (Mitchell and Soga 2005). The texts have largely been silent, and the intersection of engineering geology, geotechnical engineering and foundation design has been lacking, with the exception of limited investigations related to highway engineering. Typically pyrite has been associated with shale materials when used as fill. Granite, as well as other rocks and soil can also contain sufficient pyrite to initiate the destructive forces associated with pyrite oxidation. In this case study paper, the insidious and destructive forces of pyrite expansion are presented. The investigation offered the opportunity to study the fill material manufactured and placed in 1998 in both an unaffected, or pristine condition, along with material subjected to varying states of sulfation. The long-term implications to the facilities, and strategies for remediation of the buildings are considered.

Index Terms—Pyrite Oxidation, induced heave, expansive forces, foundation failure.

I. INTRODUCTION

OUR involvement with the case began in the summer of 2002, when we were retained to investigate numerous failures throughout multiple buildings in a school district in upstate New York, which had undergone a recent capital construction project. The primary concern of the owner at that time was the movement within one of the elementary schools that had received an addition.

The building is a reinforced load-bearing masonry structure with lateral forces resisted by masonry shear walls. The building had wracked under a lateral load, which had created a ¼ to ½ inch crack between the interior shear walls and the interior and exterior perpendicular walls.

As a result of the investigation it was discovered that there were several sources of the failure of the lateral resistance system. The structural systems were inadequate to resist the loads experienced by the facility. The building was evacuated in May 2003, and temporary structural repairs were completed in the summer of that year.

During the investigation other structural errors were discovered and remediation of

[1]Paul G. Carr, PhD, P.E. is an adjunct Associate Professor with the Department of Civil and Environmental Engineering, Cornell University, Ithaca, NY 14853 USA (e-mail: pgc3@cornell.edu)
[2] Joseph L. Thesier, P.E. and Mark S. Kimball, P.E. are Partners at The Bernier-Carr Group of Companies, Watertown, NY 13601

those problems were also accomplished at the same time; summer 2003.

One situation observed in the investigation was the movement in the slab-on-grade floor at the elementary school. It was discovered that there had been differential movement between the interior slab and the exterior foundation wall. As would be expected, the first impression was that the foundation had settled due to the concentrated forces on the exterior walls, transmitted to the foundation, while the floating slab inside the building was lightly loaded. In addition, at this time the theory of slab curling was also promoted. This movement condition in 2003 is presented in Figure 1.

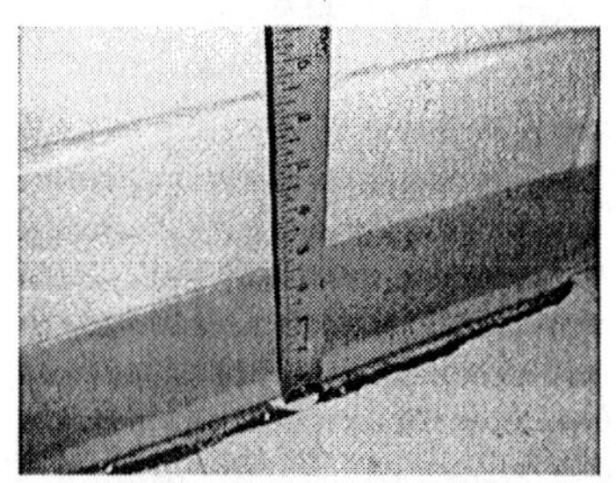

Figure 1 Differential Floor / Wall Movement

Although by the summer of 2003 it had been clearly established that the above ground structural system at this school was in need of remediation, the slab movement remained unresolved. There were meetings with the designers and constructors of the building, and a general defensive theme relative to the foundation/slab movement was that the building was settling, the slab was not rising up. A debate on the cause and responsibility ensued, and continued for the next three years during which time various theories of failure were pursued.

II. BACKGROUND

A. General

In January of 2003 a survey of the school was completed to establish a baseline of elevation. There was no 'as-built' survey performed in 1999 at the time of construction completion. It was measured that the slab within the building was higher in the center of the classrooms than at the edge, and the survey established that the edge of the slab was higher than the top of the exterior foundation wall. The cross-section in Figure 2 provides a graphical presentation of the survey results.

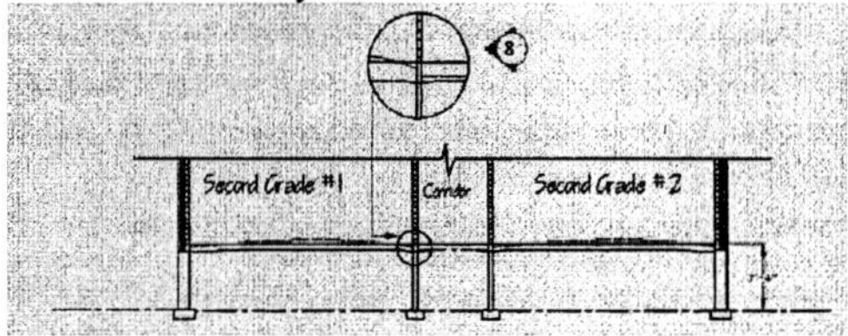

Figure 2 - Cross-Section of Elementary School

Furthermore, and critical to our investigation, it was established that the bottom of the brick veneer, on the outside of the building was essentially level along the perimeter of the building addition, matching the brick of the original building. The survey showed that the centerline elevation of the corridor was also level from the point of connection to the existing school to the end of the new addition. Yet, even though all data pointed to heave in the classroom, no proof of the cause was available, thus the original project team's defense continued that this was a foundation problem that should be addressed with the school district's original geotechnical engineer who provided the Architect/Engineer (A/E) team with foundation design recommendations.

The design had generally followed the recommendations of the geotechnical report when designing and specifying the foundation systems for the project. The A/E had called for an "engineered structural fill" for the fill within the building area. This was specified through a note on the drawings (Figure 3). The contractor however had deviated from the specified design when he requested to substitute the material used for the fill within the foundation of the structure.

12. ENGINEERED STRUCTURAL FILL SHALL BE PROVIDED AS FILL WITHIN THE BUILDING AREA INCLUDING A DISTANCE OF 5 FEET BEYOND THE PERIMETER. IT SHALL CONSIST OF A CLEAN, SCREENED, CRUSHED OR BANK-RUN GRAVEL CONFORMING TO THE FOLLOWING GRADATIONS.

ENGINEERED STRUCTURAL FILL

SIEVE SIZE	PERCENT PASSING
3"	100
1"	80-95
1/2"	45-75
#4	30-60
#40	10-40
#200	0-7

THIS MATERIAL WILL NEED TO BE OBTAINED FROM AN OFF-SITE SOURCE.

Figure 3 - Drawing Note Requiring Engineered Fill within the Building

The substituted fill at the elementary school ranged in depth from approximately 1-foot to over 6-feet. The fill had been specified to be an engineered structural fill, yet the contractor requested to use stone dust as a substitute for this material. Figure 4 shows the first lift of the substitute stone dust installation.

Figure 4 - Elementary School Foundation with Initial Lift of Substitute Material

In July of 2003, as part of the investigation, we retained the services of a geotechnical consultant, different from the owner's original geotechnical consultant. This consultant would assist in the investigation. As part of this effort, cores were drilled through the concrete floor slab at three locations in the most affected classrooms, and fill samples were collected for analysis. The fill material retrieved from these cores did not have the appearance of the material that had been placed within the foundation, as seen in the construction photos. The samples collected appeared to be sand, not stone dust as had been reported, and had been evidenced in the project photos. The material collected was strikingly brown in color, in contrast to the construction photos of the gray stone dust being placed.

The appearance of the "sand like" material collected by the geotechnical engineer, further confounded the investigation. In addition, the slab cored adjacent to the exterior wall in the end classroom was found to have a ¾ to 1-inch space below the concrete floor slab and the top of the fill material. This suggested that the foundation and adjacent backfill material under the floor slab might have settled, thus creating this space. It could also have indicated that the center of the slab was lifting, raising the edge of the slab with it. Without as-built survey data, and with such a slow movement, there was no answer here.

The investigation's geotechnical engineer was tasked with determining the physical characteristics of the soils, and to test the collected material for swell [expansion] potential. These tests were completed, and the tests indicated that the material was physically similar to the material submitted as a substitute for the structural fill, even though its visual appearance (color) was significantly changed.

Laboratory Swell Test Report

Samples obtained from Core Location C-1 and C-3, LN's 5194 and 5197

	LN-5194	LN-5197
Compacted Wet Density (pcf)	141.9	141.2
Compacted Dry Density (pcf)	127.5	128.5
Compacted Moisture Content (%)	13.8	10.6
Final Moisture Content (%)	11.3	9.9
Maximum Indication of Swell (in.)	0.003	0.004
Inundation Duration (hours)	87.0	87.0
Final Indication after draining (in.)	-0.001	+0.001
Molded Height of Specimen (in.)	4.585	4.585
Max Indication of Swell (%)	0.07	0.09
Final Indication of Swell (%)	0.00	0.00

Figure 5 Swell Test Results

As seen in Figure 5 – the Final Indication of Swell is at 0.00%. Even with the greatest measured swell at 0.09% this would result in less than 1/16th of an inch swell in a five-foot thick layer. Thus with these test results the theory of material swell was abandoned in 2003. In the sample collected in 2003, the test reports misidentified the material collected from beneath the slab as a sand material. This confounding identification led the investigation team away from immediate attempts to collect a pristine sample of the material manufactured and placed in 1998, as there was none at the elementary school site.

B. Alternative Theories

At that time alternative theories were developed, with significant consideration given to the possibility that the exterior foundation had in fact settled. It was agreed that the school would retain a surveyor to continue measurement the building, and monitor its movement. In addition, school personnel placed markings on the walls to track the relative movement of the building.

1) Water Pressure

In the winter/spring of 2004 it was reported by the school that the building had continued to move and we again mobilized for further field studies. It was posited that the spring runoff from the hillside above the school might contribute to the lifting of the slab through some form of underground spring, causing an upward pressure on the floor slab within the building foundation area. Test gages were purchased, installed and monitored. No pressure differential was recorded.

2) Frost Heave

In February 2005 we directed the installation of additional movement gages at strategic locations within the 1998 addition to track any minute movement more precisely, even though there had been no reports from the district's surveyor of any measurable change. It was during this visit that the building's continued movement was clearly evident. We requested that the school district's survey firm again do a field check. The survey results reported that the building slab had indeed risen up more. In March 2005 we meet with school's original geotechnical engineer, and jointly visited the site. The concept was to involve the original soils engineer again, and investigate the potential for the movement due to frost heave.

By the end of March the building had moved again. The movement gage installed on February 23, 2005 showed a 1mm movement in a one-month period.

Various efforts were made by the original geotechnical firm to measure ground water and to investigate the potential for frost heave. The school representatives believed that water was a contributing factor to the problem at the school. We could not disagree with the assessment, yet still we had no firm cause and effect relationship established. It was believed that frost penetration might be a contributing factor to the problem.

The original geotechnical investigation performed in 1995 determined the type and character of the in-situ soils found on the site at the location of the planned addition. One item the investigation evaluated was the density of the soil in each core boring, across different vertical locations. The material characteristics under the footing were identified by relatively high "blow counts" [100+/-]. The blow counts of 100 in this zone indicate extremely dense material. It is on this dense underlying soil layer that the footings were placed.

The implication is that as the surface water sheds from the building, it passes through the porous sand exterior backfill, downward through the soil. It will then either be captured in a footing drain [of which there was none designed, nor installed] or it will continue downward into the native soil mantle. With this underlying soil quite dense, the ability of the water to penetrate into the native soil is limited; therefore it will be captured at the base of the foundation wall. The footing, once in place, and without a footing drain, has created a moat around the building that allows water to be captured and to stand in place without the opportunity to freely drain away. This is shown in Figure 6.

Figure 6 - Water Trapped Adjacent to Footing

The implication of this moat adjacent to the foundation is that there is now a source of water to add to the deleterious effects associated with the fill materials.

We know that the material that was substituted for use as a structural fill is frost susceptible. The quantity of fines in the material [defined by the amount passing the 200 sieve] is enough to induce a capillary flow of water into the soil. The substituted stone dust material with 11 to 15% < 200 sieve material will not only draw the water, but will hold the water for long periods of time within the enclosed foundation wall system.

If we had frost susceptible fill material, with significant fines content, a water source and cold temperatures, then the three necessary ingredients for frost development existed, thus our frost investigation continued.

The computed frost potential and the movement of the building over time were evaluated. It was found that the building had in fact continued to move since its construction. The building's rate of movement was increasing with time and that at the edge of the building, where the interior shear wall tied to the exterior wall, had a differential movement of over 22 mm.

3) Footing drains

The school was advised concerning the concept of the installation of footing drains around the perimeter of the building. We reported to the Board on the potential for frost heave, and that while the control of water adjacent to the footing was a reasonable prophylactic measure, we did not believe that frost was the problem; and therefore the drains would likely not be the final solution. The footer drain project moved forward and was completed in the summer of 2005. We continued monitoring the site throughout 2005.

The crack monitoring gages were reset to zero after the drainage was installed. The site was visited in August, when the movement monitors had been in place for about one month, and when there was no frost present. The movement gage showed continued movement of 1 mm upward and ½ mm outward within the month of August, even though there was no frost penetration.

It was now evident that there were additional factors moving the building's floor other than the potential impact of frost penetration. Our investigation took yet another direction.

III. Alternative Theory and a New Direction

In March of 2005 we consulted several of my colleagues at Cornell University to solicit their input. Dr. Fred Kulhawy and Dr. Thomas O'Rourke observed the soil samples that had been collected [and retained] from the summer of 2003 during the core boring investigation (O'Rourke and Kulhawy 2005). It was seen that the samples had dried out, and upon a detailed inspection it was noted that the material was in fact, not the sand reported within the 2003 geotechnical consultant's investigation, but that it was actually the crushed granite stone dust. It was the same material substituted and approved for placement at the project, yet it was now brown.

The crushed granite fines had undergone some form of phase change where the fines had oxidized and turned brown, yet when wet appeared to be a sand product. Upon discussion with these colleagues we then returned to the theory of material expansion due to swell, a theory previously abandoned after the 2003 original swell tests had reported a negative potential.

The investigation had now returned to the theory that some constituent within the granite fines could be contributing to the expansion of the structural fill. One such material known to exhibit these characteristics is pyrite. The stone supplier assured us that there was little or no pyrite in the granite at the site of their operation, however this theory was followed.

IV. Site Conditions and Specified System

The footing drain installation in the summer of 2005 allowed us to observe and collect additional material for analysis. The footer drain was installed at the bottom of the footing elevation, and the foundation wall was exposed for the drain installation. Core drilling of the concrete foundation wall allowed us to collect additional samples of the structural fill under the slab-on-grade. The material collected had the same "wet sand' appearance as the samples collected in 2003. This is shown in Figure 7.

Figure 7 Fill material collected from below the floor slab

When constructed, the exterior of the building's foundation was backfilled with a sand material. This sand was observed to be dry during the foundation drain excavation, yet the material collected through the core samples, within the foundation wall perimeter under the building, was saturated.

V. SUBSTITUTION OF SPECIFIED MATERIAL

As previously indicted, changes were made during in the construction of this project. The material used for structural fill was physically very different from the material that had been specified.

The material that had been specified for use was an engineered structural fill, with less than 7% minus 200-sieve material. This fill would be a well-drained, stable material.

The material substituted for the engineered structural fill material contained double the amount of fines, and had 100% less than ½ inch material. In fact, upon observation it can be noted that 100% of the material was less than ¼ inch. The specified material was to have no less than 75% less than ½ inch.

While public projects in New York State require the acceptance of equivalent materials, the process of acceptance or rejection of proposed substitutes varies, but is generally subject to the judgment of the A/E. In this case the substitute material was allowed for installation.

VI. BUILDING HEAVE CAUSE AND EFFECT

The material in the project photos being placed within the building (Figure 4) is the material that had been proposed as a substitute by the contractor – however the in-situ samples collected by the geotechnical engineer in 2003, and by us in 2005 appeared different [the fines had changed to a brown color]. The two materials are virtually identical from a physical (gradation) standpoint. The change of the crushed granite from gray to brown indicated a phase change, likely due to oxidation. It was theorized that this change might have been the result of pyrite present in the source material.

A. Cause

Pyrite, [FeS_2], when exposed to water and oxygen undergoes a chemical reaction called sulfation – this releases insoluble iron oxides and sulfuric acid, allowing acidic conditions to form, which in turn creates expansive forces (Bryant, 2003, Mitchell, 2004). It was now theorized that these expansive forces, resulting from the physical/chemical reactions, contributed to the lifting of the slab.

The chemical reactions are:

$$2FeS_2 + 2\ H_2O + 7O_2 \rightarrow 2FeSO_4 + 2H_2SO_4$$

In the continuation of the chemical reaction, and in the presence of Calcium (Ca), the sulfuric acid and water create gypsum ($CaSO_4$) with its attendant volume increases.

$$H_2SO_4 + CaCO_3 + H_2O \rightarrow CaSO_4 + 2H_2O + CO_2$$

In addition, H_2SO_4 can react with feldspar, a component of the granite found in the quarried stone. This reaction forms halloysite, which can also create expansive forces.

One problem is that not all pyrite is visible to the naked eye. Therefore its detection can be difficult at times, and the reality is that it takes a very small amount of pyrite to create significant expansion forces within a fill material. As little at $1/10^{th}$ of 1 percent of the material in the fill material can initiate expansion (Belgeri and Siegel, 1998).

Given the nature of pyrite relative to other minerals known to be in the supply quarry this small percentage of pyrite can present a problem. Bryant quotes Cripps, et al in her work. The implication of the term invisible menace becomes clear. *"Unfortunately, while these micro-crystalline sulfide materials are the most problematic, they are also the most difficult to identify visually due to their small grain size. As a result, pyrite has been called the "invisible menace" since pyrite is most dangerous to engineered systems when it is too fine-grained to be visible" (Bryant, 2003).*

Once the pyrite is in the finest of particles, the surface area is much greater; therefore the oxidation process will increase in time and completeness.

The forces created by the oxidation of pyrite are reported to be substantial. Belgeri reports the magnitude of these forces as 70 to 75 kPa. Since one kPa is equal to 0.145 psi, pressures of over 10 psi within the fill material are possible.

The question then arises as to what forces would be required to raise the floor slab. The floor itself was cored and found to be approximately 6-inches thick. This would then weigh ~ 75 pounds per square foot, or ~3.58 kPa. The fill material immediately under the slab at 125 pounds per cubic foot would then add a resistance of ~ 8.97 kPa. The structural fill at 142 pounds per cubic foot would add ~40.69 kPa for a six foot depth. This all adds to less than the expansive forces that could be experienced even at the lowest level of the fill material. Therefore the floor will be forced upward, and quite possibly the bowing of the foundation walls outward.

Figure 8 is a photo of the differential movement of the floor at the classroom / corridor interface in 2003.

Figure 8 - The Effect of Pyrite Expansion

B. Effect

In addition to the conditions above, the lifting of the slab has in 2005 resulted in the structural impact to the building. The interior wall adjacent to the end classroom is a shear wall. This is a wall that transfers the horizontal wind load on the building from the walls to the foundation. The location of shear walls is shown in the Figure 9.

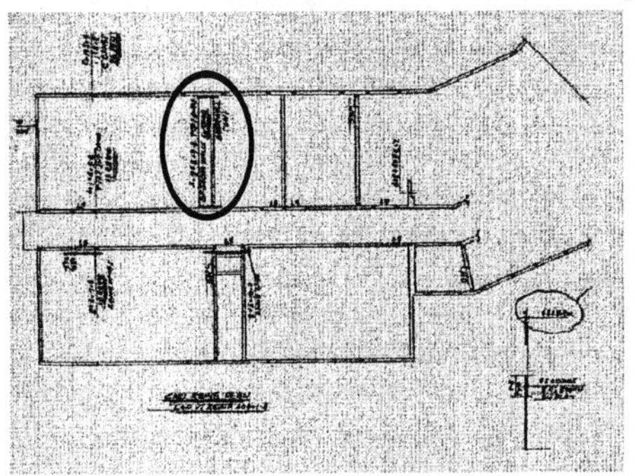

Figure 9 - Floor Plan

One such location, where the wind forces are transferred into the shear wall is at the location depicted in the circle. Without this connection, the forces imposed by the wind would have to be resisted by the exterior wall alone, and from our earlier investigation of the building's super-structure we know these walls alone are incapable of resisting these forces.

C. Impact

The impact of the expansion of the underlying soil continues to lift not only the slab, but also the shear wall resting on the slab. The wall is connected to the structural bar joist through a series of plate connections. This is how the exterior wall and the roof diaphragm forces are delivered from the wind into the joist and then into foundation.

As the interior wall lifted, the travel available in the slotted connections diminished. In

August of 2005 it was discovered that the connecting bolt had been moved to the top of the slotted connection.

The wall, which rests on the upward moving floor slab, was now lifting the structural joist. The August 2005 discovery that the joists adjacent to the two shear walls had been loaded upward to a point where it had broken free of the exterior wall. It was also noted that the joist located approximately four-feet on either side of the shear wall were also disconnected from the exterior wall. The joist's functioning as a load-transmitting member to resist the forces of wind load on the building was in serious question. Remediation was essential, and the attachment shown in Figure 10 was effected.

Figure 10 - Remediation Measure

Our attention immediately focused on what magnitude of force would be required to raise the roof sufficiently to cause the pullout of three bearing plate anchors. Each plate had a ½ - inch diameter stud anchor with 6-inches of embedment. The force required to uplift the bearing plate was calculated at approximately 3000-pounds. Considering the weight of the shear wall and the supporting slab, the uplift force would only need 40 kPa or approximately 57% of the reported potential 70-kPa force caused by pyrite expansion.

The impact of the fill material's expansion has had, and continues to have a significant deleterious impact on the building.

VII. Quantification and Results

The situation at the elementary school posed a significant challenge to the school district. The reality is that if there were no water / air fluctuations within the soil matrix to initiate the oxidation process, the expansion of the fill material may have been avoided. That is however speculative at this time since there was no footing drain designed, and none installed in the original construction.

The material that had been installed under the slab was 'crusher dust' produced at the local quarry. It is believed that the material was produced in 1998. Upon the initiation of our investigation, we had collected various samples. We had the core samples retrieved in 2003. We had the quarry samples collected in 2003. In 2005 we added the samples from the core penetrations when the footing drain was installed. Each of these represents either, a condition when the material had undergone certain reactions already (2003 core and 2005 core samples), or was produced at a time other than that during which the stone dust was installed (1998 core samples and 2003 quarry sample). We needed a "pristine" sample from 1998, which had not undergone the phase change to the degree of the core samples.

The same material installed at the elementary school was also installed in the basement area of the middle school, under the new gymnasium. In the adjacent crawl space stone dust from 1998 was found. The material was relatively dry, remained gray in color, and was unchanged by the oxidation process. A sample was collected from this location for analysis. This sample would represent the closest we could find to a pristine sample of the fill installed in 1998.

The samples were analyzed to determine the presence of sulfur in the material. The presence of pyrite would be a source of this sulfur. The test conducted was the hydrogen peroxide oxidation test as described below.

Hydrogen peroxide oxidation test

Hydrogen peroxide (H_2O_2) oxidation testing is commonly used in soil science and mining discipline to determine the acid-producing potential of a geo-material. This test method is based upon the use of hydrogen peroxide (30% H_2O_2 in de-ionized water) to rapidly oxidize pyrite, producing sulfuric acid (H_2SO_4) and iron hydroxide ($Fe(OH)_3$). The net amount of sulfuric acid produced by oxidation with hydrogen peroxide is then determined by titration with sodium hydroxide.

Recalling that minute quantities (less than 0.1%) of sulfur could cause the swell reaction and expansive forces to begin, we tested the material for sulfur content (Belgeri and Siegel, 1998). The test results for the samples are presented in Table 1.

Table 1 Test Results

Sample Analysis		NaOH (0.01M)	Meq. H+	%S
1a	2003 Quarry Sample	32.4	16.20	0.22
1b	Stone dust colected	33.4	16.70	0.23
1c	from stock pile	34.3	17.15	0.24
3a	Under floor slab	11.0	5.50	0.02
3b	Installed 1998	13.0	6.50	0.04
3c	sample July 2003	12.7	6.35	0.04
4a	North Footing	13.8	6.90	0.05
4b	Installed 1998	13.2	6.60	0.04
4c	sample July 2005	13.5	6.75	0.04
5a	South Footing	16.4	8.20	0.07
5b	Installed 1998	16.4	8.20	0.07
5c	sample July 2005	15.7	7.85	0.06
6a	Middle School	44.1	22.05	0.33
6b	Crawl Space	43.8	21.90	0.32
6c	Installed 1998	43.3	21.65	0.32
	collected August 2005			
* 2	Sample 2 destroyed			

Accepting that the material collected in the middle school crawl space is representative of the elementary school fill material the test results show that with 0.32% sulfur there was an adequate amount of sulfur present to initiate the reaction for heave. Referring to the work of Bryant, the overall potential for swell suggests that upwards of 5 inches of movement might be expected, agreeing closely with the actual movement at the school.

A. District-wide Problem

Once it was confirmed that there was indeed sulfur material present, and that the distress at the elementary school was the likely result of these expansive forces, the other schools where the material had been placed were investigated thoroughly. Up until this time it had been thought that the slab problem was isolated to the elementary school. Upon inspection it was found that the same conditions existed in the middle school and high school.

The extensive nature of the problem now encompasses the new gymnasium (~1,990 square meters) at the middle school and the new auditorium and adjacent areas (~2,390 square meters) at the high school.

While the chemical process has not yet reached the intensity of that of the elementary school, all indications are that, with time it will. The pyrite oxidation process has begun.

VIII. Conclusions

Once the oxidation process has begun, the ability to halt its action is essentially non-existent. The rate of movement, and building failure may be expected to increase with time. This is supported by the tracking of building rate of movement over time as presented in Figure 11.

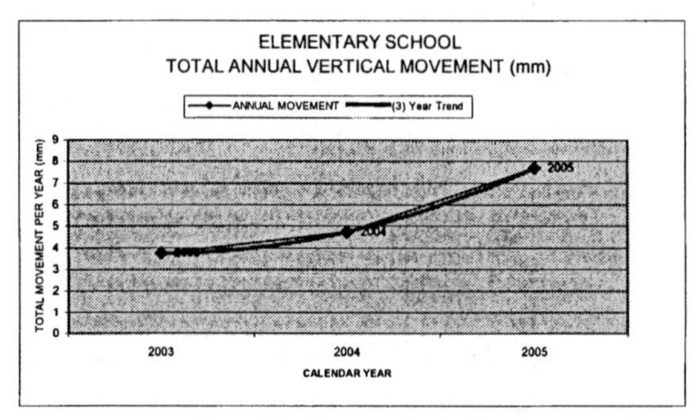

Figure 11 – Movement over Time

The solution to the problem at the elementary school will likely be dramatic. While the recently installed footing drains will remove water from the space adjacent to the foundation wall, and will allow the soils within the foundation to dry out, the long-term implications of the pyrite expansion are not good.

Bryant outlines the time-rate implications of pyrite oxidation:

"While time-rate relationships help with engineering design and expected structure life, the weathering of sulfidic material is a continuing process that builds upon itself. The rate of expansion will likely increase as the rock becomes more fractured, allowing more aerated water to interact with the un-weathered pyritic material" (Bryant 2003).

The implication here is that as the expansion takes place, the previously compacted material expands, thus becoming less dense and allowing more water to penetrate into the material, advancing the process.

Further, we know from the chemical test results that even at the elementary school, where much of the oxidation has already taken place, there remains un-oxidized pyritic material present.

The photo of the movement gage installed on February 23, 2005 is shown in Figure 12. This picture shows that the movement in eight month's time has now been in the vertical and the horizontal direction.

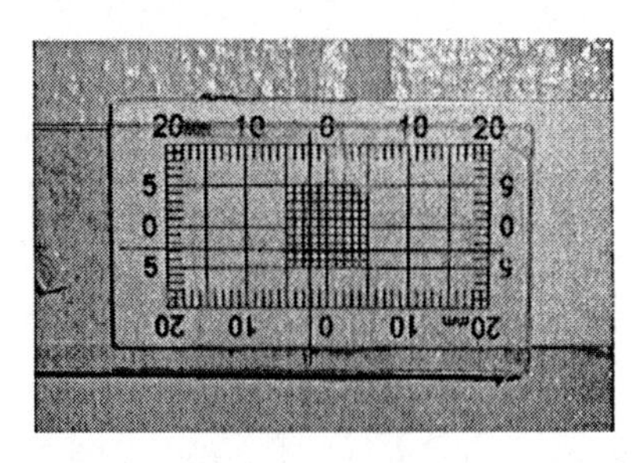

Figure 12 - Movement Gage installed July 2005 - Observed November 10, 2005

The continued movement of the building can only be attributed to the force of the pyritic expansion. We know that the building is continuing to move, and we know from the breaking of the tail of the roof joist from the exterior wall in the summer of 2005 that this movement creates a condition where the structural integrity of the building is in question.

Evidence of the expansive forces now moving the foundation of the structure in the horizontal direction is shown in the movement gage and in the exterior wall being pushed outward from the interior shear wall. The movement is greater in the lower portion of the wall than the upper wall area, where the roof joist restrains it. This creates an untenable situation.

As a result of the movement, and the evidence of the potential for a dramatic structural failure, the elementary school building was closed in January 2006, following our recommendation.

The middle school and high school continue to undergo monitoring to ensure the structural safety of those facilities.

It has become clear that over time these facilities will require extensive remedial work. The pyritic material will be removed and replaced with a non-expansive material, and the structure reconstructed to its original intended condition.

REFERENCES

[1] Belgeri, J. J., and Siegel, T. C., Design and Performance of Foundations on Expansive Shale, Ohio River Valley Soils Seminar, XXIX, Louisville, KY. 1998

[2] Bryant, Lee Davis, Master of Science in Civil Engineering Dissertation, "*Geotechnical Problems with Pyritic Rock and Soil*", Virginia Polytechnic Institute and State University, May 2003
[3] Kulhawy, Fred, P.E., G.E., Hon. M.ASCE Professor, Cornell University, Personal Communication, 2005
[4] O'Rourke, Thomas D., NAE, Professor, Thomas R. Briggs Professor of Engineering, Cornell University, Personal Conversation 2005
[5] Mitchell, James K., Soga, Kenichi, *Fundamentals of Soil Behavior,* Third Edition, John Wiley and Sons, Hoboken, New Jersey, 2004

Restoration of Distressed Secondary Monitoring System at a Hazardous Waste Landfill

J. J. Parsons, P.E.[1], J. Lyang, Ph.D., P.E.[2], K. Durnen, P.E.[3]

[1]James J. Parsons, P.E., Principal Engineer, NTH Consultants, Ltd, 38955 Hills Tech Drive, Farmington Hills, MI 48331; PH: 248-324-5329; FAX 248-324-5179; e-mail: jjparsons@nthconsultants.com

[2]Jenghwa Lyang, Ph.D., P.E., Sr. Project Engineer, NTH Consultants, Ltd, 38955 Hills Tech Drive, Farmington Hills, MI 48331; PH: 248-324-5312; FAX 248-324-5179; e-mail: jlyang@nthconsultants.com

[3]Kerry Durnen, P.E., Director of Operations, Wayne Disposal, Inc. 49350 North I-94 Service Drive, Belleville MI, 48111; PH 734-699-6265; FAX 734-697-9886; e-mail: Kerry.Durnen@eqonline.com

Abstract

Distressed secondary riser pipes were discovered at a hazardous waste disposal facility during routine sampling. The riser pipes extended from a secondary sump, up a 10-foot high (vertical) intracell berm, turn 45 degrees through the primary clay and 80 mil geomembrane liner, and extend vertically through approximately 120 feet of hazardous waste to the surface. Video inspection of the 8-inch to 12-inch diameter riser pipes revealed that at four riser locations, the field-fabricated elbows had partially buckled. At one riser location, the vertical portion of the pipe buckled at two points. Investigation of the distressed riser pipes led to a unique and challenging repair approach.

Introduction

Wayne Disposal, Inc., a subsidiary of EQ The Environmental Quality Company, owns and operates the Site No.2 disposal facility in Belleville, Michigan. The site comprises approximately 400 acres and has landfilled municipal solid waste, industrial hazardous waste, and commingled waste. Current operations include disposal of hazardous waste and waste regulated by the Toxic Substances Control Act (TSCA). The balance of the disposal areas is closed. The facility is currently licensed to landfill 11 million cubic yards of hazardous/TSCA waste.

Master Cell VI is the current active disposal area. The 34-acre cell consists of five subcells designated as A-North, A-South, B, C and D. These cells were constructed during the late 1980's and early 1990's with a typical double composite liner system that includes from the bottom up: a 60 mil secondary HDPE geomembrane liner, one to three layers of secondary drainage net, geotextile, 5-foot compacted clay liner, an 80 mil primary HDPE geomembrane liner, 12-inch peastone leachate storage layer, geotextile, 12-inch granular drainage layer and a geotextile separation layer.

A network of perforated 6-inch diameter SDR 7.3 HDPE pipe comprises the primary leachate collection system.

The secondary leak detection system for subcell D includes a sump and an SDR 17 HDPE riser pipe that extends along, and is fully supported by, the sideslope of the cell to ground surface. Each of the remaining subcells incorporates a secondary sump and an SDR 17 HDPE riser pipe (Fig.1) that extends approximately 10 vertical feet along a 2 vertical to 1 horizontal slope, then turns within the primary compacted clay layer and

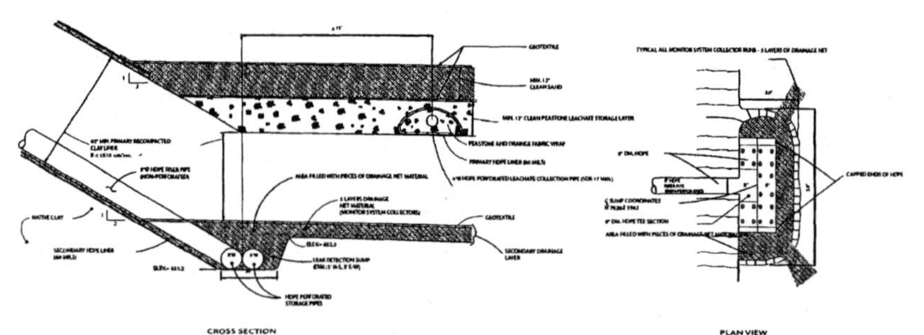

Figure 1. - Typical secondary sump and riser

extends vertically through approximately 120 feet of waste. The vertical portion of the riser pipe is sleeved with a second pipe at a point approximately 10-feet above the primary liner (Fig. 2). The sleeve was designed to reduce downdrag forces imposed upon the riser pipe as waste settlement occurred.

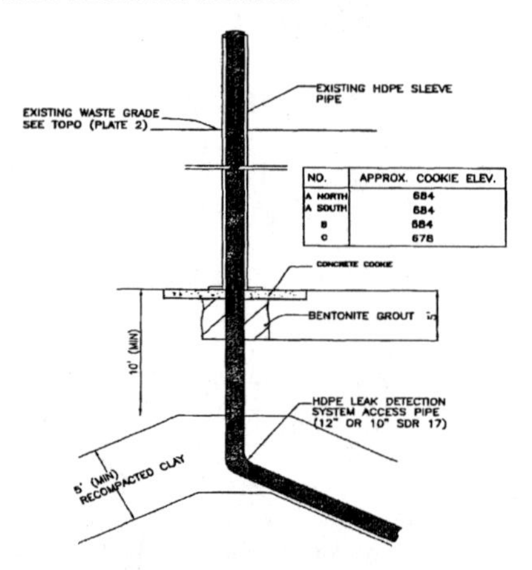

Figure 2. - Protective outer sleeve for riser pipe

Because decomposition of the waste mass is not expected, no gas collection system was necessary for the hazardous waste disposal areas.

In 2002, a vertical expansion of Master Cell VI was constructed. The vertical expansion, designated as subcell E, overfills the existing subcells and extends to the north over the adjacent closed Master Cell V.

The problem

The operating permit for the facility requires liquid in the secondary sump to be sampled and pumped dry on a quarterly basis. Water accumulating in the sump is primarily consolidation water from the compacted clay liner and has historically been free of contaminants that may indicate a leak in the primary liner system. Early in 2003, during routine sampling of the secondary sump in subcell B, the pump became stuck during extraction. The pump was eventually retrieved, however the outer pump shield remained inside the riser pipe. In addition, difficulty was encountered extracting the pump from the A-North riser pipe.

Camera inspection of the riser at subcell B was initiated to locate and determine means of retrieving the outer pump shield. A 4-inch diameter pan-and-tilt camera was lowered into the riser pipe and revealed significant deformation of the pipe wall in the field fabricated elbow within the primary compacted clay liner (Fig. 3). The magnitude of the deformations prevented camera inspection past the elbow.

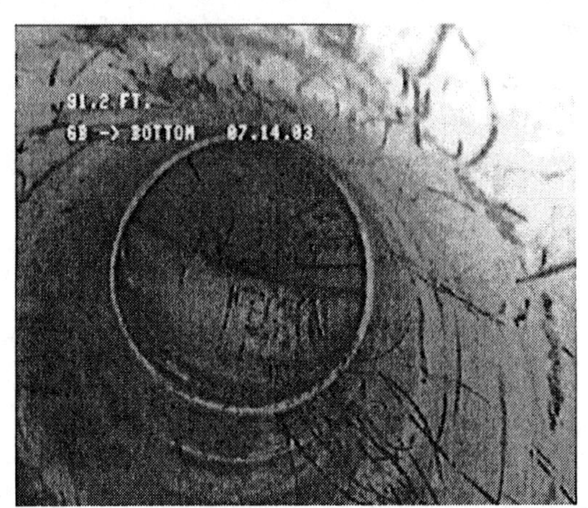

Figure 3. - Distressed elbow in Subcell B riser

Because difficulty was encountered during extraction of the pump from the riser at A-North, this pipe was also inspected using the 4-inch camera. At a depth of approximately 57 feet below the waste surface, the vertical portion of the riser pipe was severely buckled (Fig. 4). Again, this condition prevented camera inspection below the buckled section.

All of the risers were subsequently camera inspected using a 1-inch diameter push camera to depths of at least 120 feet. At A-North, a second buckled section was observed approximately five feet below the first. Deformation of the elbow was also noted. At subcells A-South, B and C, deformation within the elbows was noted but no buckling of the vertical riser pipe was observed. The riser pipe for subcell D, which is fully supported on the cell sideslope, was undamaged.

Fig. 4 – Vertical riser at A-North

The impact to the facility operations of potential total failure of the secondary risers could be enormous. Without the ability to sample and test the liquids in the secondary sump, the facility would loose the ability to demonstrate that the liner system has no leaks. Compliance and regulatory constraints could force the facility to limit or cease operations.

Evaluation of Distressed Pipe

Evaluation of the secondary leak detection riser pipes in Master Cell VI was initiated. The evaluation included a detailed survey of the damaged risers; collection of record photographs, survey notes, inspection reports and other data; as well as structural analysis of the pipe to determine the cause of failure.

The survey included measurements of the depths to each deformation and comparison to record survey data. Original construction reports, photographs and surveys were reviewed and collated with measured data and original design calculations. In most cases historic survey data indicated that the horizontal location of the riser pipe at the current waste surface varied from the location of the initial dike penetration significantly, in some cases as much as 50 horizontal feet through the entire 120 foot depth. Observations of the inspection videotapes confirmed sweeps and angular offsets from vertical throughout the lengths of pipes, particularly at welded joints.

Historical photographs from the project records revealed a construction sequence that included fabrication of the HDPE riser pipe from the sloped section of the intra-cell berm, through the elbow, and the first vertical section of pipe, before the compacted clay layer had been constructed. This sequence made adequate backfilling and structural support below the elbow very difficult, if not impossible. Figure 5 shows this sequence at the tie-in between subcell A-North and subcell D. Figure 6 shows the loose condition of the bedding soil at the reducer and elbow sections.

Figure5. – A-North riser at subcell D tie-in

Other data evaluated included soil boring information and shallow test pits excavated around one of the risers. The original design specified the placement of sand bedding around the vertical portions of the outer protective sleeves extending around the pipe diameter for 5 feet. The sand bedding was intended to assist in limiting downdrag forces as well as to provide a buffer against waste placement in direct contact with the outer protective pipe. Boring information revealed that the sand bedding was not in

Figure 6. - Condition of bedding soil around elbow

place throughout the entire length of the riser. In addition, test pits excavations revealed buried drums in direct contact with the protective sleeve.

Structural analysis of the HDPE riser pipes and outer sleeves to resist the forces imposed by lateral and vertical pressures was also completed based on ASTM F1759 – Design of HDPE Manholes for Subsurface Applications. Historic settlement data from settlement plates installed in late 2000 and estimates of settlement based on standard penetration test of the waste were used in the analysis. Settlement analysis suggest that the total waste settlement could be highly differential, varying significantly from place to place and could be as high as 10 to 15 percent under fills of 100 feet or more.

Results of the analysis indicated that the long-term strain in the waste fill could exceed the critical buckling strain of the outer protective sleeves and that the outer sleeves were likely to buckle from waste down drag forces alone at depths of about 50 feet or greater. Load transfer to the inner pipe can occur because the outer sleeve deforms downward from down drag forces and come in contact with the inner pipe at points of angular offset from vertical. Axial buckling of the inner riser pipe was evaluated for the condition where down drag forces are transferred to the inner pipe installed at an angular offset of 30 degrees from vertical. This condition imposed both an axial load component and a bending load component on the inner pipe. The axial load component was evaluated using ASTM F1759 and the bending component was analyzed using a solution for a beam on an elastic foundation (Hetenyi, 1946). The results showed that the inner riser pipes could fail in axial buckling if down drag forces were transferred to them.

In general, the results of the evaluation concluded that possible poor backfilling below the elbow, lack of sand bedding around the pipe and angular offsets from vertical likely contributed to transfer of down drag forces to the inner pipe causing buckling failure. Other installation defects, such as equipment impact damage or poor waste placement techniques, may have caused or contributed to the observed distress in the pipe.

Evaluation of Repair Methods

How do you repair a distressed pipe that terminates in a sump over 120 feet below the surface of a hazardous waste landfill? To answer this, numerous methods were thoroughly evaluated, including:

- Open excavation;
- Sliplining with a smaller pipe and use of micro pumps;
- Pipe bursting technologies;
- Braced excavations;
- Micro-tunneling;
- Directional drilling;
- Conventional tunneling below the landfill
- Internal re-rounding of the pipe followed by structural polymer reinforcing; and
- Drilled access shafts to the sump with complete riser replacement.

The evaluation of the repair methods considered several design challenges. Regulatory oversight by the Michigan Department of Environmental Quality (MDEQ) and the US Environmental Protection Agency (EPA) would require regulatory approval. Environmental impacts also needed to be considered. Any repair undertaken must be environmentally protective and could not impact the environmentally sensitive secondary leak detection system. Any contamination of the secondary leak detection system could potentially be construed as a leak in the landfill liner system.

The most straightforward solution, and potentially the least costly, was open excavation to the distressed elbows and a direct repair of the riser pipes. This method was not considered desirable because it would require relocation of over 500,000 cubic yards of hazardous waste. Not only was there insufficient permitted area to relocate the waste to, but also would potentially expose the environment to airborne contaminants and undesirable odors. Further, this method would nearly completely disrupt regular site operations.

Cost and risk analysis was completed for most of the options that were considered viable. Internal re-rounding coupled with structural reinforcement of the risers was considered to be the least intrusive, had the least risk to the environment, and had the least operational impact. However, the technology had never been applied in this application. Therefore, the owner elected to proceed with the design of drilled access shafts to occur concurrently with the development of the re-rounding option.

Internal pipe re-rounding is a technology typically applied to horizontal PVC pipe with two-way access. That is, the equipment is inserted into a manhole and is pulled through a length of pipe from a second manhole. To apply this technology to rerounding of the HDPE riser pipes in a vertical orientation with only one-way access limited use of this technology. A contractor was located who had successfully applied rerounding techniques to vertically oriented HDPE pipes. Working with the contractor, modifications to the equipment were made to incorporate site conditions. Some of the constraints to the modifications included the requirement to get through the deformed elbows without damaging the pipe or contamination of the secondary liner system. In addition, all work must be conducted within the exclusion zone of the landfill, and therefore required work in Level C personal protective equipment (PPE) per Federal OSHA requirments.

Trials of the modified rerounder equipment were conducted at the contractor's facility (Figure 7) to determine optimum configurations and operating pressures to prevent damage to the pipe. After several months of research, bench scale trials and equipment modifications, field trials were begun in early 2004 (Figure 8).

Field trials successfully rerounded one of the upper deformations of the riser pipe, but failed to make the turn through the deformed elbow section. Serveral modifications to the equipment were made, and after numerous unsuccessful attempts at rerounding, the technology was ultimately abandoned.

The shaft accessed repair method became the focus of the repair design effort.

Figure 7. - Bench scale trials of rerounder

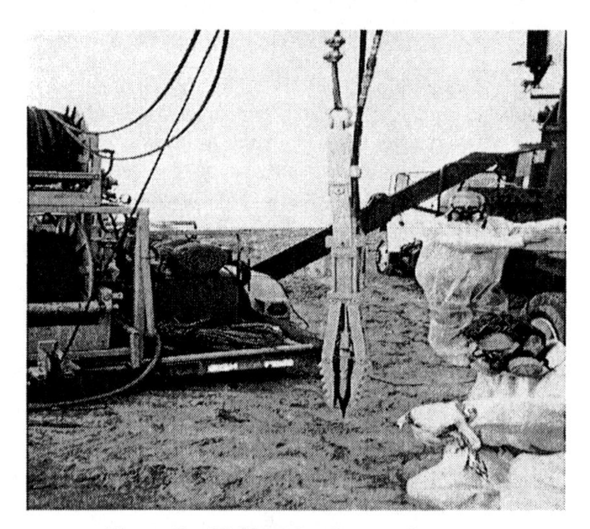

Figure 8. - Field trials of rerounder

Shaft Accessed Repair Design

A shaft-accessed repair involved numerous design challenges. The concept would requires:

- Drilling through over 120 feet of hazardous waste;
- Put personnel in the shaft to hand excavate through the last 5-feet of hazardous waste;
- Breach the primary 80 mil HDPE geomembrane liner;
- Hand excavate through 5-feet of compacted clay liner;
- Complete the above work at the lowest point in the cell;
- Control leachate and secondary consolidation water;
- Prevent contamination of the secondary leak detection system;
- Accurately excavating to the end of the secondary riser pipes within a 3-foot by 5-foot sump.

Additional challenges with logistics also needed to be addressed. Regulatory approval was needed to cut a 7-foot diameter hole in the primary containment liner near the low point of the subcells. Drilling a shaft would also require a crane with over 120 feet of boom to be in place on top of the landfill. Because of the proximity to an adjacent major airport, the crane would infringe on FAA and airport management airspace, thereby requiring FAA approvals. And finally, we needed to identify an appropriate contractor to implement the work.

After several meetings with both the EPA and MDEQ, the draft work plan and conceptual design was approved. With regulatory approval in place, the project team moved forward with addressing the remaining challenges.

Initially, the FAA for one of the repair locations granted a "Notice of no Hazard to Air Navigation" determination. However, the remaining three repair locations within the landfill were denied. After several months of negotiation by the owner, the airport management agreed to close the impacted runway to air traffic during construction. This action allowed planning to proceed without the need for FAA evaluation of the proposed hazard.

Due to the construction complexities and specialty sub-contractor needs of the conceptual repairs, the contractor was solicited early in the final design phase of the project. This allowed the contractor and major subcontractors to provide input as the final design was developed. Working as a single team, the Owner, designer and contractor provided valuable input in preparation of the final design. Of primary concern to all parties was the safe execution of work. To this end, the contractor developed a comprehensive Health and Safety Plan to guide each step of the work.

The final repair design is packaged in such a way that a drawing, detailed construction sequence, Quality Assurance requirements, and Health and Safety requirements for each of the anticipated 17 major tasks is presented on it's own sheet relating specifically to that task. This system will allow the team to have easy reference to all pertinent information based upon the current task in progress.

The final design includes extending a 10-foot diameter steel cased shaft to a depth of approximately 60 feet below waste surface. The shafts will be excavated using conventional drilled shaft equipment. The shaft then steps down to a 9-foot diameter steel cased shaft extended to a total depth of approximately 120 feet and to within 8-feet of the primary geoemebrane liner. The final 8 feet of excavation will be by hand from within sectional liner plates, each 18-inches in height, to the geomembrane. The gasketed steel liner plates are to be fitted with grout injection ports to facilitate the control of liquid into the excavation, if needed. The final section of liner plate will be cut to conform to the shape of the sump and will be fitted with gaskets to protect the geomembrane against damage.

Because massive objects had been disposed at the facility, at least two borings were made at each shaft location. Borings were extended to within 10-feet of the geomembrane to determine the nature of waste to be anticipated during excavation. Although all locations are believed to be "drillable" without slurry techniques, problem areas were identified early. For example, at the A-North shaft location, a 2-foot thick layer of "rubber" waste was encountered 95 feet below waste surface. This layer is immediately underlain by a large metal obstruction that was eventually drilled past with the borings.

Once the geomembrane liner is exposed and leachate seepage is controlled, a 7 ½-foot diameter, 3-foot tall cylinder of geomembrane will be extrusion welded to the existing liner. This cylinder will be the primary feature that separates the "contaminated" waste and liquids from the exposed primary clay and secondary sump. Following welding, the entire interior of the shaft and the surface of the liner will be pressure washed to remove any contaminants. Cleaned, existing geoemembrane will be cut out in a 7-foot diameter circle, and the underlying 5-foot thick primary clay liner will be hand excavated from within sectional liner plates (Fig. 9). Because atmospheric conditions within the waste are unknown, all of the work within the shaft will be completed with personnel using Level B personal protection equipment. That is, a full-face respirator with supplied air, and disposable waterproof outer clothing.

Restoration of sampling and pumping capabilities includes complete replacement of the secondary riser pipe. A new liquid storage pipe and associated riser will be constructed from within the shaft extending to ground surface. The new riser pipe will consist of 10-inch SDR 7.0 HDPE pipe with a wall thickness of 1.5 inches. An outer 24-inch diameter SDR 7.0 HDPE pipe will be installed to serve as a protective sleeve to minimize any future stress to the inner pipe due to down drag forces.

After installation of the first 10-foot long section of riser pipe, the sump drainage materials will be restored and the 5-foot thick layer primary liner will be backfilled using cement-bentonite grout. The cement-bentonite grout must duplicate the low permeability characteristic (less than $1X10^{-7}$ cm/sec) of the compacted clay, but have enough compressive strength to support the weight of the outer protective sleeve and subsequent backfill without significant deformation.

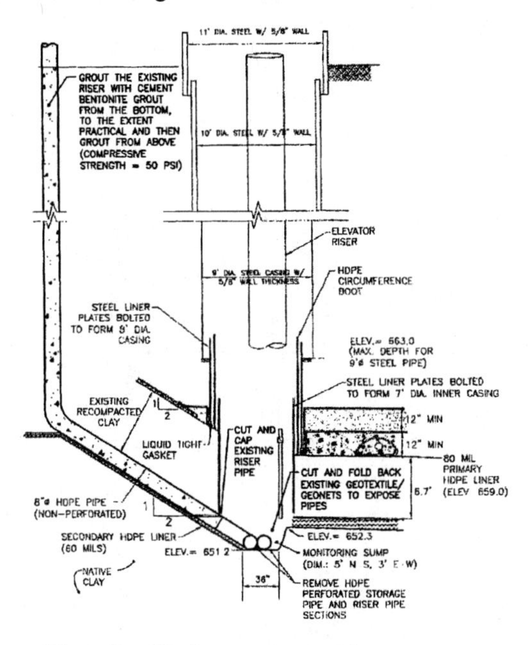

Figure 9. – Shaft access to secondary sump

After sufficient set of the cement-bentonite backfill, the primary geomembrane liner must be restored. This step in the process is the most critical to the success of the project. The primary geomembrane repair must be made leak-proof under less than ideal welding conditions. Therefore, multiple redundant measures were incorporated. For the repair of the liner itself, the extrusion weld will be capped with an additional layer of geomembrane. In the area where the pipe penetrates the liner, an extrusion-welded boot will be installed with a second boot completely covering the first (Fig 10.) All liner welds will then be covered with a layer of bentonite paste. The entire area within the HDPE cylinder, trimmed to leave approximately 12 to 18-inches in place, will be filled with low shrinkage cement based grout. In this manner, the HDPE cylinder filled with grout will serve to direct leachate away from the repair welds. Each of the primary repair welds are also backed up with a second redundant welded repair.

Construction quality assurance will be provided via remote, real-time video monitoring. A special pan-tilt-zoom camera will be remotely operated from the surface.

After cement based grout backfill has sufficiently set, the shaft will be backfilled with rounded 1½-inch nominal size aggregate incrementally as the riser pipe and protective sleeve are extended. Depending upon actual conditions encountered, the steel shaft casing will attempt to be extracted. If the casing requires grouting in place due to voids, it is likely that the casing will be left in place.

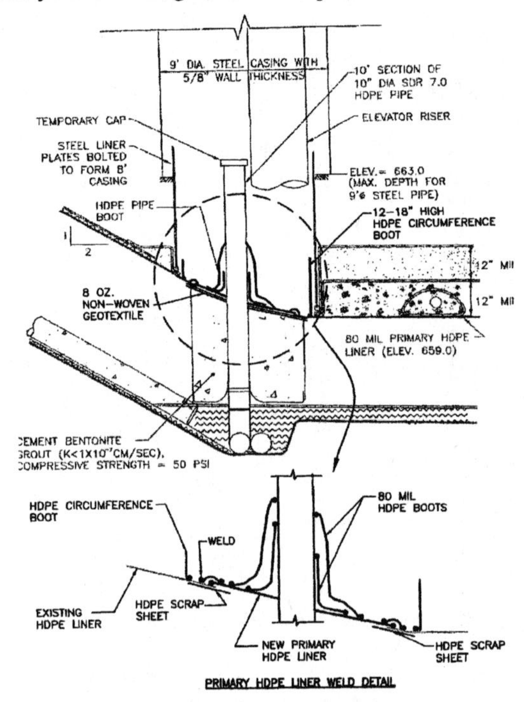

Figure 10. – Restoration of liner system

The critical nature of this project, both to the owner and potentially to the environment, has required that every aspect of the design be scrutinized with regard to constructability, contingencies and safety. One major effort completed to assure constructability, involves construction of a full-scale mock-up. There was concern that personnel in full Level B PPE and with all required tools and support equipment may not be physically able to complete the required welds within the confines of the 7-foot diameter shaft with the new 10-inch diameter riser pipe in place. A full scale mock-up would demonstrate the ability to make high quality welds within the confined space anticipated (Fig 11.)

The owner constructed a test pad area with similar slope and configuration anticipated within the landfill. The surface was lined with 80-mil HDPE geomembrane that intentionally included a wrinkle.

A section of 7-foot diameter steel casing was placed over the geomembrane liner and a 10-inch diameter pipe was installed near the center. Anticipated equipment such as air supply lines, lighting, air monitors, video cameras, simulated air exchange equipment

Figure 11. – Full scale mockup

and electrical lines were included within the confines of the casing. The contractor then completed all welding described on the drawings and work plan, including repair of the intentionally placed wrinkles, while fully dressed in Level B PPE. Although several problems needed to be addressed, the mock-up successfully demonstrated that quality welds could be made within the confines of the shaft. The contractor also learned that the maximum amount of time a person could work under the anticipated conditions was slightly more than one hour.

Conclusion

Sideslope riser pipes, fully supported on competent compacted clay, performed as designed during a vertical expansion of a hazardous waste landfill. Riser pipes extending vertically through significant waste depth during a vertical expansion experienced significant deformations. Protective sleeves intended to reduce down drag forces on the riser pipes, lacked sufficient factors of safety against buckling strain. This condition was exacerbated by construction defects (pipe offsets from vertical, lack of designed bedding, and possible construction/waste placement damage) causing the outer protective sleeve to transfer loads to the inner pipe.

Where riser pipes are constructed or extended vertically through waste, particularly in the case of vertical expansions, close attention to additional stress and strain imposed on the pipes is necessary. In addition, a well-implemented quality control program should be developed to assure that landfill infrastructure is constructed strictly in accordance with the design.

Various methods for repair of vertical riser pipes at depth, with one-way access, have been thoroughly evaluated. Although re-rounding and reinforcement may be an option for straight sections of pipe, elbows and fittings preclude this method of repair. A shaft accessed repair method has been developed that addresses anticipated issues such as leachate control, compacted clay liner repair and restoration of the geosynthetic components of the liner system. Field mock-up of the repair methodology (primarily for geosynthetic repair) has successfully demonstrated that high quality repairs can be made under the spatial constraints of a vertical shaft. Implementation of such a repair will be conducted a hazardous waste landfill during the spring of 2006.

INVESTIGATION OF DAMAGE TO A MASONRY CONDOMINIUM BUILDING FROM THE 1994 NORTHRIDGE EARTHQUAKE

A. Jain[1], Ph.D., M.ASCE, C. C. Simsir[2], Ph.D., A.M.ASCE, A. P. Dumortier[2], and G. C. Hart[3], Ph.D., M.ASCE

ABSTRACT

This paper presents the findings of a structural engineering evaluation of earthquake damage to an 8-story reinforced masonry residential building in Los Angeles. The load in some of the upper level masonry walls of the building is transitioned at the 2-story subterranean parking garage levels through reinforced concrete transfer beams and columns. On-site damage investigation revealed severe cracking of the concrete transfer beams. The cracks were most significant in the mid-span of the beams as opposed to the termination point of masonry wall above closer to the beam ends where shear and moment transfer and consequent damage would be expected to occur. A staged finite element analysis of the building subjected to gravity loads, to replicate the sequence of construction, revealed that the walls had separated from the transfer beams during construction when the temporary shoring was removed. This is consistent with field observations of separations along the beam-wall interface. This separation was incorporated into a detailed computer model of the building for evaluation of the building's response to the 1994 Northridge Earthquake. The locations and sizes of observed beam cracks correspond well with the location of predicted large moment demands in the beam elements. Predicted regions of overstressed masonry walls also correlated well with locations of observed wall cracks.

---

[1]Associate Principal, Weidlinger Associates Inc, 2525 Michigan Avenue, D2, Santa Monica, CA 90404; PH (310) 998-9154; FAX (310) 998-9254; email: jain@wai.com

[2]Senior Engineer, Weidlinger Associates Inc, 2525 Michigan Avenue, D2, Santa Monica, CA 90404; PH (310) 998-9154; FAX (310) 998-9254; email: simsir@wai.com , dumortier@wai.com

[3]Principal and Head of Division, Weidlinger Associates Inc, 2525 Michigan Avenue, D2, Santa Monica, CA 90404; PH (310) 998-9154; FAX (310) 998-9254; email: hart@wai.com

Introduction

This paper contains the results of a structural engineering investigation to determine the effects of the 1994 Northridge Earthquake on a reinforced masonry building located in Los Angeles. The purpose of the evaluation was to determine the cause and extent of damage to the structural components and recommend the repairs required to restore structural strength.

A major issue of damage that was observed during the field investigation is the severe cracking of reinforced concrete transfer beams at the transition from residential to garage floors which appear to have been adequately designed per the codes in force at the time of design according to available building plans. The paper focuses on the reasons for the damage to the transfer beams; however, evaluation of the earthquake related structural damage in other parts of the building is also discussed.

Building Description

The building, located in Los Angeles, California, was constructed in 1972 according to Building Department records. The building consists of a six-story residential structure over two levels of parking. The upper parking level is on grade and the lower one is subterranean. The building is constructed of reinforced masonry bearing and shear walls composed of single-wythe 203 mm (8") nominal concrete masonry unit (CMU) blocks, and the flooring consists of 203 mm (8") thick precast, prestressed hollow-core concrete planks that span 9.1 m (30 feet). The planks have a specified weight of 2.15 kPa (45 psf) and bear directly on the CMU block walls. The planks are 1.2 m (4 feet) wide separated by grouted joints through which they are connected to the CMU walls via the vertical reinforcing bars in the walls (Ø16@1200mm (#5@48") with Ø29 (#9) or Ø32 (#10) bars at wall ends). The lowest two levels of the plank floors are topped with a 64 mm (2.5") thick layer of reinforced concrete for added fire separation from parking levels. Reinforced concrete transfer beams and columns are provided at the parking garage levels to accommodate the wide driving bays and transition from the residential areas to parking levels. These beams also support the weight of the CMU walls in the residential area directly above the beams. The larger of these beams are 914 mm by 813 mm (36" by 32") and they are located in the upper parking (second floor) level spanning 9.4 m (31 feet) over the interior driving bays supported by 406 mm (16") diameter circular columns. The smaller beams over the exterior 9.4 m (31 feet) bays are 556 mm by 762 mm (22" by 30") and supported by 406 mm (16") wide square columns. The exterior walls, interior partition walls between units, and the South walls of the hallway are constructed of CMU. Interior non-bearing walls of the residential units and the North walls of the hallway are constructed of metal studs sheathed with drywall. The roof is flat and covered with a built-up roofing system.

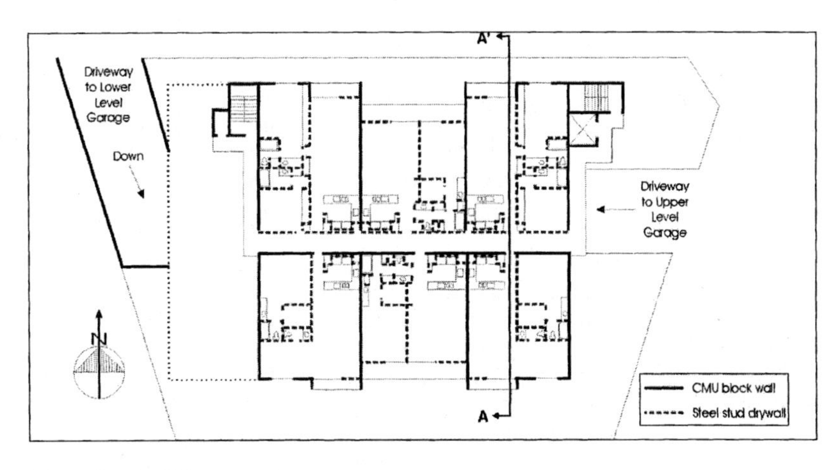

Figure 1. Building plot plan.

A plot plan of the building with the layout of units on a typical floor is presented in Figure 1 and a cross-section through the building is shown in Figure 2. The building has plan dimensions of approximately 25 m by 27.4 m (82 ft by 90 ft), and a height of approximately 19 m (62 ft) from the ground surface (first floor level in Figure 2). Vertical loads are transferred from the roof, 7th through 1st floors to the interior and exterior concrete masonry block walls, the transfer beams and columns in the parking levels, and then down to the foundation. The roof and the floor diaphragms, the interior and exterior concrete masonry block walls resist lateral loads. The building's properties were determined from on site observations and available building plans.

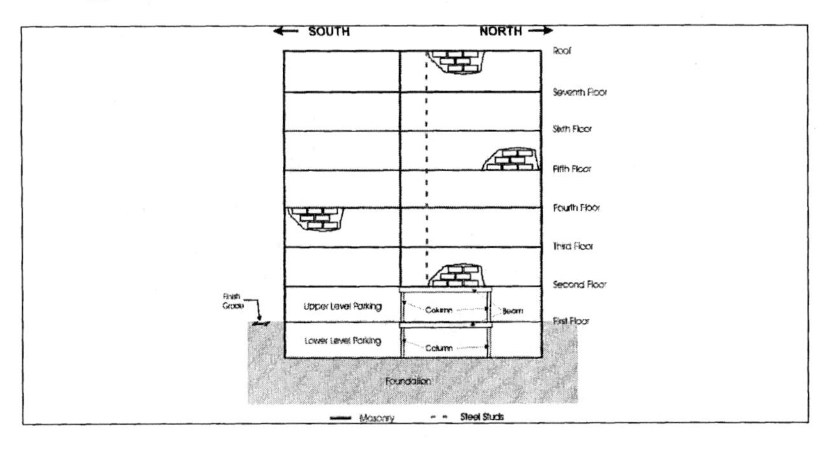

Figure 2. Building floor levels (Section A-A').

Earthquake Ground Motion at the Site

The building site is located approximately 19.3 km (12.0 miles) southeast of the epicenter of the main shock of the 1994 Northridge Earthquake. The magnitude 6.7 event occurred on January 17, at 4:31 am PST. The hypocenter was about 32.2 km (20 miles) west-northwest of Los Angeles in the San Fernando Valley at a depth of approximately 19 km (12 miles) (EERI 1995).

The ground motion at the building site was estimated using a spatially weighted averaging technique (King et al. 2004) for the recorded ground motion response spectra at the five closest stations on similar soil conditions. The ground motion recording stations are all located within approximately 2.4 km (1.5 miles) of the building site. The corrected data obtained from these stations were rotated if necessary to align with the North-South and East-West directions.

The effective peak acceleration (EPA) values of the estimated earthquake ground motion at the site are 0.296 g in the North-South direction, 0.125 g in the vertical direction, and 0.221 g in the East-West direction. Figure 3 shows a comparison of the 5% damped horizontal acceleration response spectra from the estimate of the ground motion experienced at the site during the 1994 Northridge Earthquake with the code-specified values from 1970 UBC (ICBO 1970) when the building was designed and the current 2003 IBC (ICC 2003). The 1970 UBC did not define a response spectra; the constant value of acceleration for 1970 UBC in Figure 3 is equivalent to the design base shear divided by the weight of the building. The 2003 IBC spectrum is for the design level earthquake.

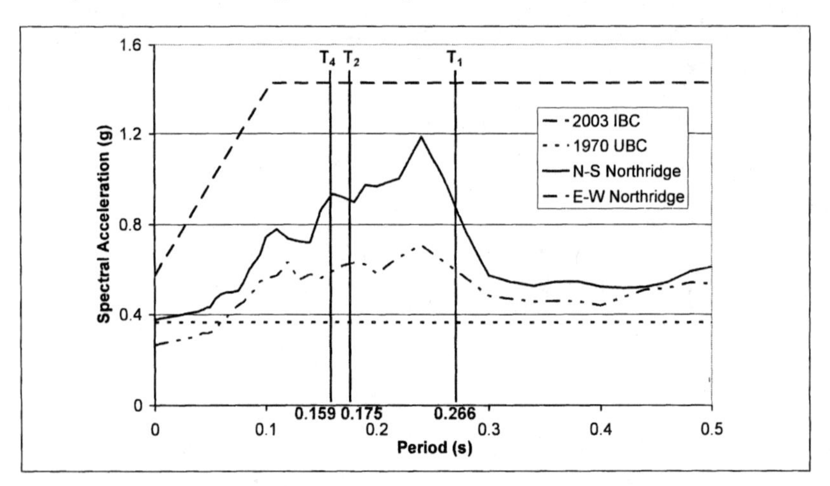

Figure 3. Comparison of estimated site response spectra with design response spectra and the building's 1st, 2nd, and 4th mode periods of vibration.

On-Site Damage Investigation

Visual and non-destructive investigation of the building was performed to identify damages to the structural components. Numerous cracks ranging in width from hairline to 4.8 mm (3/16 inches) were observed in the reinforced concrete transfer beams at the second floor level. The configuration of the cast-in-place concrete beams in second floor level is illustrated in Figure 4. As shown in Figure 5, the cracking of the cast-in-place concrete beams is most significant in the mid-span of the beams at gridlines 3 and 4 as opposed to the termination point of CMU walls near the beam ends where damage was expected to occur. Most of the beam cracks were observed to penetrate the entire section of the beams and they also showed signs of water penetration through the cracks.

In the residential units directly above the cracked beams, CMU shear walls were separated from the floor concrete topping along the edge of the garage beams below that support the walls above. Diagonal cracks were also observed on these CMU walls in the second floor level where the wall support changes from CMU wall to the beam above the garage driving bay.

Other structural issues that were discovered in the two garage levels are cracks in the reinforced concrete columns, significant amount of cracks ranging in width from hairline to 2.4 mm (3/32 inches) on the bottom surface of the second floor planks, numerous cracks ranging in width from hairline to 0.8 mm (1/32 inches) on the top of the first floor level elevated planks and slab-on-grade, and separation at the CMU block wall and elevated plank interface.

There were various bed joint and head joint cracks in the CMU walls on the building exterior, separations along orthogonal CMU wall-to-wall interfaces, and broken CMU face shells along these interfaces. Structural damage commonly observed on the building's interior included cracks in the precast concrete floor planks and in the grouted joint along the interface of adjacent planks, concrete floor topping cracks with vertical offsets of up to 6.4 mm (¼ inches).

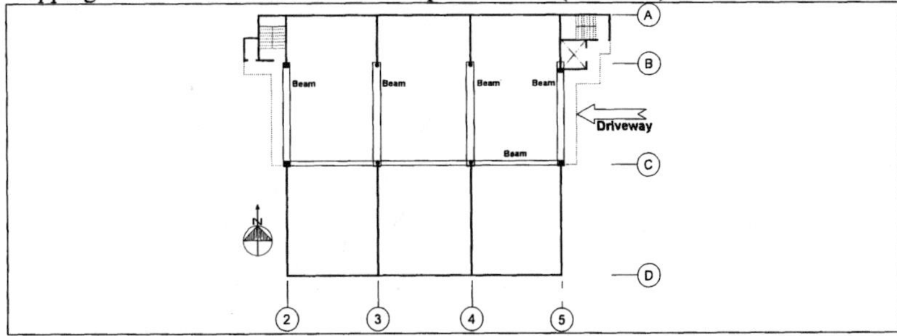

Figure 4. Second floor level plan with gridlines showing transfer beam and column locations.

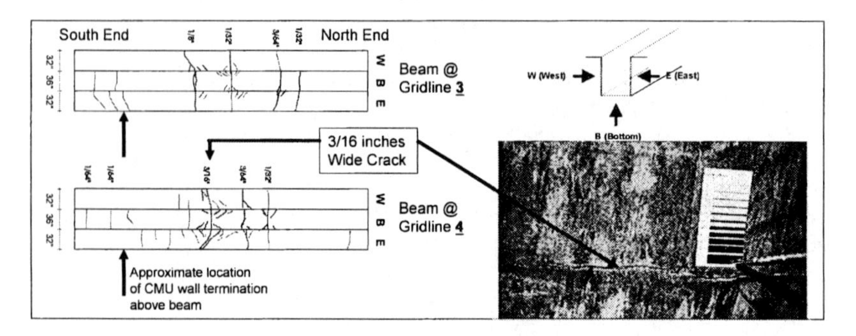

Figure 5. Observed cracks in beams on gridlines 3 and 4.

Structural Engineering Analysis

The Structural Model

A detailed three-dimensional computer model of the entire building was developed using the structural engineering software SAP2000 (CSI 2001) as shown in Figure 6. Columns, beams and lintels were modeled with frame elements, floor slabs and walls were modeled with shell elements. Fixed boundary conditions were assigned to the foundation and the walls of the lower parking level in contact with the ground. The material strength values used in the model were as specified in the structural drawings used for the construction of the building. The model utilized cracked section properties as recommended by the FEMA 356 (2000) document for the effective stiffness of reinforced concrete beams and columns and reinforced masonry shear walls. Thus, the shear stiffness was reduced by 60% for these members; the flexural stiffness was reduced by 50% for beams and columns and by 20% for walls. A uniform thickness was used for the uncracked hollow-core pre-cast concrete slabs which were modeled as orthotropic shell elements. The equivalent thickness was obtained by performing an analysis to calculate plank properties in bending and shear in each orthogonal direction.

The computer model considers the masses and dead loads due to the self weight of the structural components of the building plus a uniformly distributed load of 0.36 kPa (7.5 psf) due to the calculated weight of the partition walls in the residential units. An estimated live load of 1.2 kPa (25 psf) is realistically assumed for the parking floor slab instead of the code prescribed 4.8 kPa (100 psf). Similarly, the live load for all other floor slabs is taken as 0.48 kPa (10 psf). This value is one quarter of the design live load and represents a realistic estimate of the imposed live load at the time of the 1994 Northridge Earthquake. In addition to gravity loads, seismic loads were applied by subjecting the model to the ground acceleration response spectra estimated for the site during the earthquake in all three orthogonal directions (two horizontal and one vertical). The model was also subjected to the

2003 IBC (ICC 2003) design basis earthquake ground motion response spectra in the horizontal and vertical directions.

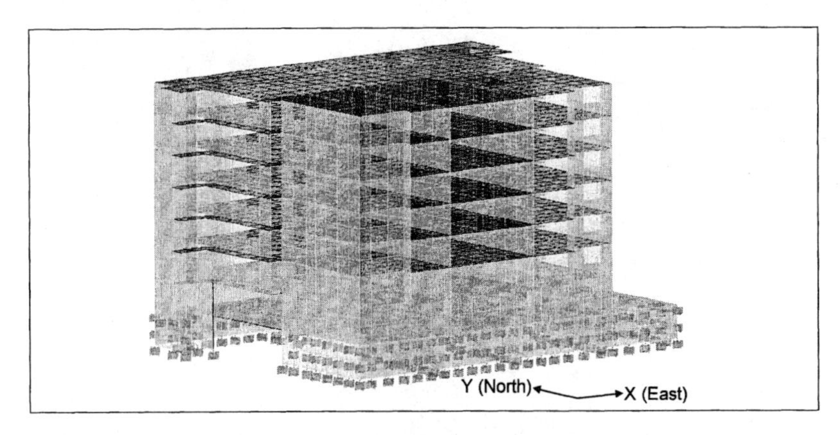

Figure 6. The three-dimensional structural model of the building.

Gravity Load Analysis of CMU Walls in the North-South Direction

The structural design drawings indicated that the second floor transfer beams were to be shored until the construction reached up to the fifth floor. A SAP2000 (CSI 2001) finite element model of the second floor beam along gridline 4 and the wall above it in the North-South direction was created. Only the dead load from the walls and floor slabs was applied to this computer model. As determined from the dead load study of the model, the vertical stresses in portions of the second floor walls had exceeded the yield strength of the vertical reinforcing bars (Ø16@1200mm (#5@48") on center) in the walls, assuming no real capacity in tension for the mortar bedjoint between the wall and planks. Therefore, the transfer beam was separated from the wall in these regions of large stress resulting in even larger stresses in the neighboring nodes of the beam due to load distribution. In the final stage, all wall nodes with stress values exceeding the yield capacity of the Ø16 (#5) reinforcing bar were released from the beam. Figure 7 illustrates the staged gravity load analysis. While releasing the wall from the beam, the vertical loads from the wall were still kept on the beam. Otherwise, the CMU wall would have acted as a very deep cantilever beam carrying its own load without transferring anything to the beam.

The outcome of the gravity load analysis was incorporated into the detailed three-dimensional model in which the transfer beams on gridlines 2, 3, 4 and 5 at the second floor level were released from the second floor CMU walls in the mid-span region of the beams. This was consistent with the observations at the site of a significant separation along the beam-wall interface at the second floor (Figure 8).

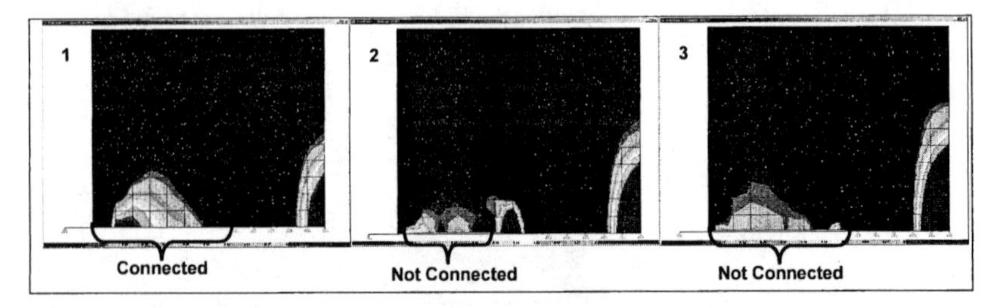

Figure 7. Vertical force diagrams for the CMU shear wall above the transfer beam on gridline 4 in three stages of gravity load analysis.

Figure 8. Separation between CMU shear wall and the concrete topping at the second floor.

Dynamic Analysis

The first 120 modes of vibration were included in the modal analysis, contributing to more than 99% of the mass participation in the East-West, North-South and vertical directions. The fundamental natural period of vibration of the building was determined as 0.266 seconds in the East-West direction, and the second mode period of vibration as 0.175 seconds in the North-South direction. The fourth period of vibration for the torsional mode is 0.159 seconds. Figure 3 illustrates the periods of modes 1, 2 and 4 in relation to the estimate of site response spectra. The periods of vibration for the modes with the highest participation coincide with the period range where the peaks of the response spectra occur. The 2003 IBC spectrum also has the maximum amplitude in this period range. The proximity of the periods of vibration to the peaks of the response spectra results in amplified structural response due to the earthquake.

Evaluation of Second Floor Beams on Gridlines 3 and 4

The bending moment demand, as determined from the computer model, on the North-South beam at gridline 4 at the second floor level is shown in Figure 9 for gravity loads, and gravity and Northridge earthquake loads combined. The maximum moment demand values are compared in Figure 10 with the capacity curve obtained from a moment-curvature analysis of the beam. The beam has 12Ø36 (12#11) longitudinal steel bars at the bottom for the first 3.7 m (12 feet) from its South end to support the termination point of the CMU wall above, but only 2Ø36 (2#11) bars for the rest of its span, and 2Ø19 (2#6) for top reinforcement for the entire span. All longitudinal steel is Grade 420 (60). At 3.7 m (12 feet) from the South end of the beam where the beam is lightly reinforced, the moment demand from the combined gravity and Northridge earthquake loads was computed as 677.0 kN-m (499.3 kip-ft), 12% larger than the beam's yield moment capacity of 603.6 kN-m (445.2 kip-ft). The locations of the observed beam cracks (Figure 5) and their sizes correspond well with the locations of the moment demand in the beams beyond their capacity.

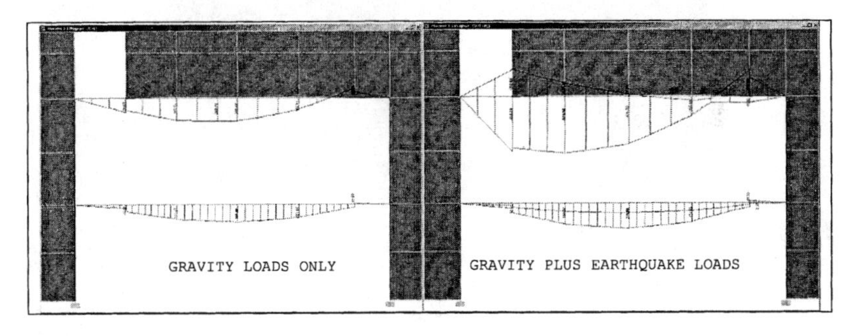

Figure 9. Moment demand for beams on Gridline 4 (Units: Kips, ft).

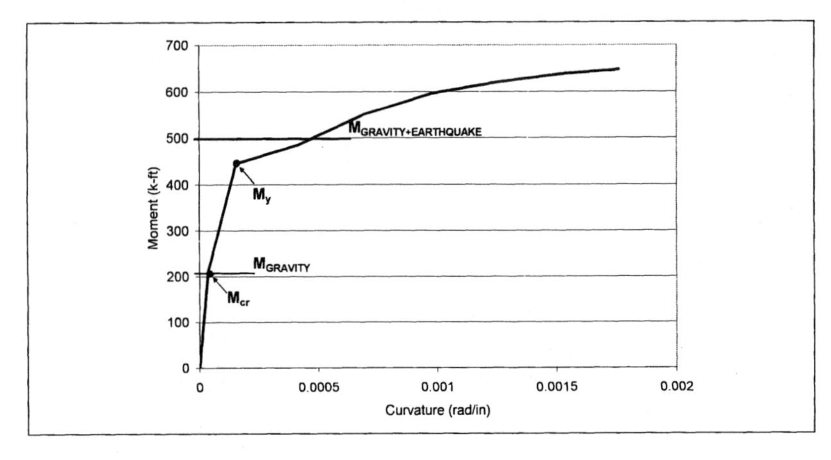

Figure 10. Moment-Curvature diagram for second floor beam section at 3/16" crack location.

Evaluation of Building Walls

Figure 11 illustrates the horizontal stresses in the plane of the reinforced masonry walls on gridlines 4 and 5. The stresses are plotted for the combination of gravity and Northridge earthquake loads. Cracking in the walls is expected for the red (dark) colored areas in Figure 11 where the tensile stress demands exceed the tensile strength capacity of the masonry mortar joints just above the 2nd floor transfer beams. Stress plots for other gridlines are similar to that of Figure 11. The location of overstressed regions above the 2nd floor transfer beams corresponds well with visual observations of cracks in CMU walls.

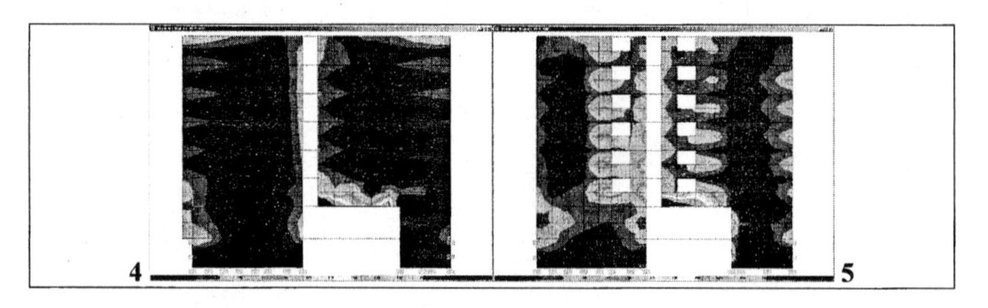

Figure 11. Gridlines 4 and 5 horizontal stress (S11) diagrams (Units: Kips, in).

Evaluation of Building Floors

The calculated stresses in the second floor slab from gravity and Northridge earthquake loads along with an overlay of observed cracks are illustrated in Figure 12. The normal stress demands in the plane of the slab in the East-West and North-South directions are shown in Figure 12. The red (dark) colored regions along the gridlines correspond to stress levels that exceed the tensile capacity of the grout joints between the hollow-core planks. These regions correlate well with the areas in the building where cracking was observed as shown in the crack maps overlaying the stress plots in Figure 12. Crack maps in Figure 12 are located in areas where damage survey was conducted. Figure 12 illustrates the excellent correlation obtained from the field data and computer model results.

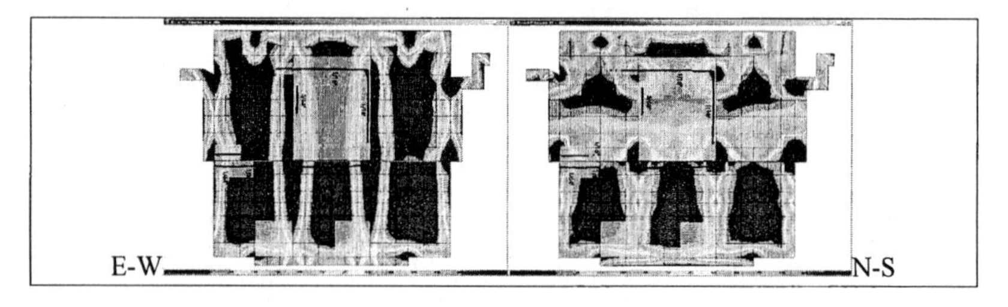

Figure 12. Overlay of crack map and E-W, N-S normal stress plots (Top of 2nd floor slab).

Evaluation of Building Columns

It was determined from the axial load – biaxial moment interaction diagrams for the reinforced concrete columns that the 2003 IBC design earthquake demands exceeded either the tensile or the moment capacity of the square columns on the grid nodes C2, C5 in the lower and upper level parking and B5 in the upper level parking. The columns that had 79.1 kN-m (700 kip-in) or more moment demand from the 2003 IBC earthquake in either direction had their moment capacity exceeded at their axial tension demand levels. However, the 1994 Northridge Earthquake demands did not exceed capacity for any of the columns. The capacity of the circular columns on gridlines 3 and 4 in the interior of the building was sufficient for both earthquake demands.

Conclusions

Field observations of structural damage were validated with structural behavior and analysis. The staged gravity load analysis indicated that the vertical stresses in the second floor CMU wall above the transfer beam exceeded yield strength of the vertical reinforcing bars in the wall. This was consistent with field observations. This load redistribution from gravity loads alone resulted in the unanticipated shifting of peak moment demand to a lightly reinforced section of the garage transfer beams.

For the case of combined gravity and Northridge earthquake loads, the moment demand on the reinforced concrete transfer beams at gridlines 3 and 4 were up to 12% more than the beams' yield moment capacity at their less reinforced mid-span regions. The tensile stress demand on the floor slabs exceeded their tensile capacity at the grout joints between the precast concrete hollow-core planks. The tensile stress demand on portions of the CMU bearing and shear walls exceeded the tensile capacity of the wall mortar joints. The locations of observed cracks in beams, slabs and walls correlate well with the predicted overstressed regions where demand was in excess of section capacity.

Although much higher levels of ground motion were recorded elsewhere in Los Angeles during the Northridge Earthquake, the proximity of the periods of vibration to the peaks of site specific response spectra resulted in amplified structural response to the relatively lower levels of ground motion.

Design engineers should anticipate the large stresses that develop in the transfer beam from the masonry wall immediately above it due to gravity loads. Performing a staged construction analysis may be appropriate to avoid the spurious condition of the wall acting as a deep beam that would not transfer the correct magnitude of moment to the transfer beam resulting in unduly low demands. Sufficient reinforcement should be provided along the transfer beam and in the connection of the beams to the masonry wall to avoid beam-wall separation that could cause failure of the beam in a future earthquake.

References

CSI (2001). SAP 2000 Structural Analysis Program Non Linear Version 8.2.7., Computer and Structures, Inc, Berkeley, California.

EERI (1995). Northridge Earthquake Reconnaissance Report, Volume 1, Earthquake Spectra 11 (C), Earthquake Engineering Research Institute.

FEMA 356 (2000). Prestandard and Commentary for the Seismic Rehabilitation of Buildings, Federal Emergency Management Agency, Washington, D.C.

ICBO (1970). Uniform Building Code, International Conference of Building Officials, Whittier, CA.

ICC (2003). International Building Code, International Code Council, Whittier, CA.

King, S. A., Hortacsu A., and Hart, G. C. (2004). "Post-Earthquake Estimation of Site-Specific Strong Ground Motion." 13th World Conference on Earthquake Engineering, Vancouver, Canada.

Guidelines for Determination of Wind versus Water Causation of Hurricane Damage

James. M. Hinckley, P.E.[1]

[1]PADCO [Planning and Development Collaborative International], 1025 Thomas Jefferson Street NW Suite 170, Washington, DC 20007; PH (202) 337-2326; FAX (202) 944-2351; email: jhinckley@padco.aecom.com

ABSTRACT:

In the wake of a hurricane event, the attribution of damage causation to "wind" or to "flooding and storm surge" is essential to the process of cost-recovery and the determination of insurance applicability. In such a catastrophic event, the damage left in the aftermath is often seen as chaotic and inseparable into these two categories. Knowledge of the storm dynamics and careful observations on the ground, however, can provide clues leading to an equitable attribution. In contrast to complete forensic analysis, for which there is usually insufficient time, a concentration on applied forces, observed damage, and debris geometry is often sufficient to arrive at a segregation of damage between the two or a determination of simultaneous action. Drawn from experience in Louisiana following Katrina, this paper seeks to set forth a series of guidelines for the conduct of such an examination and contains descriptions of commonly-encountered patterns of damage with the attributions suggested by each.

INTRODUCTION:

While most are aware of the issues surrounding the question of "Wind vs. Water" damage assessment and the need for determining, as accurately as possible, which damages were caused by which, there may be some lingering doubt as to the best means of addressing that task. While the issue is, indeed, a thorny one, there *are* some considerations which, if judiciously applied, may assist measurably in coming to an equitable determination.

Many of these considerations are common to any good forensic examination, but those pertaining to the determination of the causative forces, as contrasted with the "collapse mechanism," may help to grossly segregate the damage defensibly into the two categories of "wind" or "water" individually, or into a third, consisting of the simultaneous application of both.

For the purposes of this paper, "water" damage is considered to be anything caused by the action of salt water (not spray) either as a dynamic (rapidly-moving as in "surge") or static (flood) entity. When this is identified, the remaining damage is either "wind" (hurricane force wind plus fresh-water rain) or pre-existing damage present before the recent hurricane events (so-called "pre-incident" damage).

TIME-DEPENDENT APPLICATION OF FORCES:

Forensic investigation involves a process wherein one attempts to reconstruct events leading up to the production of the observed damage, closing the gap between what the building *was* and what it *is now*. This gap can be closed from both sides: by reasoning backwards from the debris to the events that produced it and by reasoning forward from what is known or suspected concerning the forces that were applied. Both of these efforts involve the reconstruction of some sort of time-line, however inexact, that accounts for all of the observed evidence. A time-line of forces is useful, especially when attempting to identify which of these forces, wind or water, was causative on a particular site.

For wind-water determinations in the southern parishes, it was helpful to consider the sequence of events of the hurricane in question. Archived reports from the National Hurricane Center were used to plot the storm track on NOAA Marine Chart #411 of the Gulf of Mexico. From each "storm center" location, winds of defined speeds are reported by the NHC as distances in nautical miles in each of the sub-cardinal directions (NE, SE, SW, and NW), as well as eye-wall sustained winds and gusts. These were plotted on Chart 411 for each center starting with the onset of 34 kt winds on land. Smoothly connecting the plotted winds gave a history of these winds for any location in southeastern Louisiana and western Mississippi. Care was exercised in the interpretation of this data, as speeds and, especially, locations of those speeds are highly variable during the event.

From data collected from official sources, notably the Louisiana State University Hurricane Center, and reported for FEMA by Mark Pierepiekarz, P.E., the history of the flood events was approximated as well. For example, from "Hurricane Katrina Time-Dependent Storm Surge and Wind Data," February 28, 2006, by Mark Pierepiekarz, P.E., and S.E. (after the ADCRIC model of LSU Hurricane Center) the data for Slidell, Louisiana indicates a surge of 10 feet at 9:00 AM (CDT) August 29th. This surge height was selected as representing levee overtopping in New Orleans. The actual onset of flooding was some time earlier. From data, the maximum flood is given as approximately 15 feet at 11:00 AM. Variations of height with time in such events are rarely linear, but rather reflect the direction of the surge (from the east), the resistance to flood flows (shoaling ground and marshes) and the speed of advance of the hurricane (slow).

Other factors such as "fetch," or the open water distance across which the surge-building winds blow, and the direction of winds relative to the direction of advancing flood surge complicate the determination of flood onset. If, however, one *approximates* the flooding as linear, an onset time of approximately 6:00 to 7:00 AM results which is sufficiently accurate.

From the data, it is apparent that elevated winds associated with the outer bands began to reach Plaquemines and St. Bernard Parishes (on the post-landfall storm track) as early as noon of August 28th, some 18 hours prior to landfall and apparently quite ahead of storm surge effects. Thus the winds, rising markedly from about 34 kts (40 mph) to gusts of over 160 kts (184 mph) at landfall were apparently acting on the structures of Plaquemines Parish for most of a day prior to landfall.

With the approach of the eye and notably on the leading, easterly, and trailing edges of the storm, rainfall of significantly high intensity was associated with the hurricane winds. As the eye made landfall, forensic meteorologists with whom we have consulted see an increase in vertical wind shear as the "drag" of the land caused the top of the storm to progress more rapidly along the storm track than the bottom of it. While the landmass of Plaquemines is not large, they believe that it was likely sufficient to cause the organization of the storm to deteriorate enough to spawn small tornados randomly about the eye. With and slightly preceding landfall, the storm surge made up of wind-driven waves on the easterly and northeasterly sides of the storm and a "dome" of raised water in the extremely low-pressure eye came ashore, precipitating levee breaches and overtopping all along the central and southern portions of Plaquemines and areas of St. Bernard Parishes.

With the passage of the eye, and in those areas within about 20 to 30 miles of the storm track, horizontal winds abated somewhat or ceased altogether, since most winds were of an upwardly spiraling convective nature. Following passage of the eye, however, the entire sequence of events for these areas was repeated in reverse order, with wind directions generally switching about 180° to come from the opposite direction.

The specific nature of the events will vary somewhat from one area to another depending upon their location relative to the storm track, proximity to sea or riverine shore or canal structures, topographical features which may either "funnel" or mitigate wind, and the effect of surrounding structures providing a lee. The *detailed* wind, rain, and surge data for the southern parishes may be assembled from the data. But if one envisions the following general sequence, it may place the events in some chronological perspective:

1. Elevated winds as much as a day prior to landfall, depending upon location, building to maximum force for the particular area, gusting up to 180 mph in marine areas,

2. Rainfall in intensities of over 1" per hour with total accumulations of 8" to 10",
3. Tornados with tightly-wrapped cyclonic winds possibly ranging from 110 to 200 mph in limited areas near the storm track,
4. Storm surge landfall before, after, or during peak winds and tornadic activity,
5. Extremely low barometric pressure in areas on storm track as the 20- to 30-mile diameter eye passes,
6. Increase in rainfall from zero in the eye to 1" per hour or more,
7. Rapid increase in wind speed to near earlier values, but from the opposite direction,
8. Possible continued inundation from damaged levees.

From the sequence of loadings noted above, expect that there may be much more of the observed damage attributable to hurricane winds than might be the case if one thought of the storm as an "instantaneous" event. In addition to the high distributed loadings of wind, the dynamic variations in wind speed will produce fatiguing of thin metal siding and lighter cold formed members. In previous papers on hurricane damage, the contribution of wind-borne debris has been widely documented as more destructive than the wind itself. Where tornadic activity can be documented, spectacularly violent damage can be inflicted by this source alone. When confronted by winds rising from perturbation to full hurricane strength over an 18-hour period, substantial damage can result even *prior to the onset* of flooding.

Examination of surge data produced by Louisiana State University discloses that the flood inundation occurred rather gradually. For example, a rate on the order of five feet in two hours (1/2" per minute) in the Slidell area is both reported in the computer model output and verified by competent eyewitness reports. Much faster rates, with associated higher velocities, can be expected in the vicinity of levee breaches. The water-borne transport of *non-buoyant* materials alone over a long distance requires high velocities with turbulent up-currents in the floodway. Absent a proximity to a levee break with its associated velocities and turbulence, consider wind transport as a working hypothesis in such situations.

SEQUENCES OF DAMAGE:

The *sequence* of damage to structures will depend upon not only the sequence and severity of the environmental forces to which they are subjected (as discussed above), but also to the very localized conditions in the area of the structure which may either amplify or attenuate those forces, as well as the very nature of the structure itself. These factors would include:

1. *Height of structure relative to surrounding structures and topographic features:*

Higher structures are obviously exposed to higher wind speeds and the effect of those winds on the structure will be greater. Lower structures or those in the "lee" of other features around them will be more sheltered. Adjacent features may also act to "funnel" winds and accelerate their speed, especially in urban or semi-urbanized areas. Keeping in mind that many locations experienced a reversal of wind direction, any significant sheltering on one side or the other could suggest which of the two wind-events (before or after eye passage) caused the damage.

2. *Proximity to shore or riverine formations and canal structures:*
   Shore areas are, of course, generally more exposed to early elevated winds from an approaching storm on the Gulf. But they are also more susceptible to the effects of storm surge. Expect that since wind moves faster than water, there will be a delay in the arrival of storm surge on a river system compared with storm surge effects on even nearby shore areas and both effects may be delayed relative to the arrival of winds. Conversely, expect that structures near levee breaks will be inundated sooner than those near intact flood-control structures and the speed of the advancing water may be significantly higher.

3. *Elevation of the structure relative to known flood depths in the area:*
   We tend to think of the ground as "flat", but obviously it is not. Expect significant variations in flood depths from one area to another in fairly close proximity. Variations in flood depth may not always indicate dynamic (wave-action) flooding, but such cannot be automatically ruled out. Water also flows around adjacent structures in a similar manner as wind, so envision the same "funneling" and "lee" effects with water damage that you would for wind. Be alert, also, for the effects of waterborne debris, which can cause impact-type penetrations of exterior walls and windows.

4. *Presence of nearby utilities and stormwater structures funneling water to the site in reverse flow:*
   Originally intended to drain water *away* from structures and landscaped areas, inundated stormwater culverts or perforated drain lines can and will act in the reverse direction if the soils are saturated or the outfalls of these structures are flooded. The effect is to accelerate the arrival of significant hydrostatic forces at or below the foundations of structures well ahead of the arrival of surface flooding in the area. Ample evidence of this exists in structures inspected in St. Bernard Parish.

5. *Nature of soils, gravels, or voids below the structure:*
   Many of the structures inspected to date show signs of gradual subsidence or settling over time owing to inadequate bearing capacity of the site soils, either to footings (rare) or piles (common). When fine alluvial soils are saturated, they can become "fluid" (thixotrophic) and the effects of vertical and horizontal loads on and of the structure may have a greater effect than

they would under non-saturated conditions. Also, gravel (rare) or crawl spaces (common) below structures can cause flood water, rising only a few feet above the floor, to exert significant hydrostatic uplift on the floor, which deformation can precipitate or accelerate the subsequent failure of interior and exterior walls resting on it.

6. *Permeability of the structure to wind:*
   In a structure such as a pavilion composed of little more than steel framing supporting a competent roofing system, wind effects would be expected to be very limited. On the other hand, a "tight" unvented structure can support significant differences in air pressure from one side of a barrier to the other. It is not unheard of for the very low pressure in the eye of hurricanes and tornadoes to cause an explosion of glass due to the relatively high atmospheric pressure trapped inside a tight structure exerting pressure against the low pressure of the passing eye. Most structures are vented sufficiently that this is not common, but it is believed to have occurred in at least one structure inspected to date.

7. *Permeability of the structure to water:*
   Equal and oppositely-directed water pressure on both sides of a barrier will have little or no effect on the barrier. If water builds up higher on one side of a barrier than the other, however, even seemingly insignificant differences in height can have dramatic effects. Each foot of such difference (if it is sea water) will exert about 64 pounds per square foot of pressure at the base of the barrier. A 4 or 5-foot of difference, then, is a very significant finding. This has been seen not only on exterior walls, but on interior walls enclosing rooms which are nearly water-tight. Keep in mind also that the volume of water creating the elevation difference has no bearing on the exerted pressure. It is only the *height* of this difference that determines the pressure. If one floods the cores of a 20-foot concrete block wall with salt water, the pressure outward on the faces of the bottom course of block is still 1,280 psf, regardless of the small volume of the water.

8. *Materials and nature of construction:*
   The manner in which any structure reacts to the imposition of loads is determined by the materials of which it is built and the manner in which those materials are joined together. We may subconsciously assume that a building is competently constructed and, further, that it is assembled in a manner familiar to us from our own experience. Many of the structures we are inspecting may differ from others of our experience in small but very significant ways. Older structures may employ materials and methods no longer in use today but common many years ago. Locally-available aggregate, in many structures examined to date, is not of a crushed granitic material, but of more commonly-available and less angular stone. This can significantly impact the expected strength of the concrete of which it is a part. Keep in mind, for those of us in northern climates, that roofs here are

more commonly constructed against wind and its associated horizontal and uplift vertical forces than gravity loads, since snow is rare. This will significantly alter what we might "assume," from our experience, is the nature of roof structural systems.

9. *Observed damage to adjacent structures:*
   Damage to adjacent, weaker structures is a valuable clue in at least two ways. Being "weaker" they will display a more obvious reaction to the forces exerted in the area and, as such, give insight into those forces. If a weaker building has been blown flat but left dry, this alone will indicate that the area was subject to wind and not water. Conversely, if the adjacent, weaker structure is standing, has an intact roof, but is significantly damaged by saturation and mold, the absence of significant winds, but the presence of water to the height of the scum line is strongly suggested. Secondly, large-scale damage to adjacent structures produced debris from those structures. Expect that the debris became either wind- or water-borne and look for evidence of damage to the inspected structure as a result of the impact of that debris.

10. *Observable pre-incident damage to the structure:*
    As stated above, many structures in the area have undergone damage pre-incident to the hurricanes, due to subsidence or prior storm events. Gauging what is pre-incident and what is current can be tricky, but the effect of such damage is important in determining how the structure reacted. Cracks or other structural damage and deterioration existing prior to the current hurricanes abridged the strength of the structure, effectively creating either gaps or "hinges" which permit the structure to move laterally or rotate at the location of this damage. As an example, cracking at the bases of concrete columns which were originally poured monolithically with a grade beam will produce a "hinge" at the base of the column. This will predispose the structure to either wind or water-induced sidesway. This may manifest in additional loading on masonry infill or in new cracking at the tops and in the body of the hinged columns.

HOW TO LOOK:

1. Know as much as possible about the loads of the subject hurricane on the site as possible. Consult published accounts from knowledgeable sources and develop track and sector wind-speed plots for the study area.
2. Examine from the "outside in" working from the entire region inward to the structure of interest in a complete circle with the structure at the center. This will provide a comprehensive examination and helps to assure that no relevant observations are overlooked.
3. Note any and all effects of the subject hurricane, however insignificant they may seem at the time. The smallest details, when assembled, may provide a clearer picture than the more spectacular gross damage or deformations.

4. Note the pattern of damage to adjacent structures in the area and the severity of that damage. Weaker structures are expected to have greater damage and both the nature and severity of that damage can give evidence of the load magnitudes to which the structure-of-interest was subjected.
5. Note the pattern of debris dispersal in the area, directional, bi-directional, radial, and the proximity of storm-attributable debris to its original location in the structure to which it was attached. This may suggest the direction of the causative forces.
6. Note the nature of storm-attributable debris. Could it float? Could it be wind-driven? Was it wet? Does it have a significant salt residue? Does it show evidence of pre-incident deterioration?
7. Note the nature of changing debris with depth in the debris pile. Items on top were deposited last and the debris pile generally reflects a reverse-chronology of deposition with increasing depth.
8. Note the nature of structural damage. Were walls displaced inward or outward? Did they "crumple" with debris falling to *both* sides? Inward or outward gross displacement may indicate the result of a distributed load and the direction of that load. Debris to both sides may suggest a concentrated load with the direction of the load suggested by the displacement of the lower portion of the wall. Think of it as striking a person at the back of the knees. The lower leg goes forward, but the upper body comes straight down or backward. For concentrated forces above the floor level, the bottom of the wall is driven in the direction of the force. The top portion then guillotines downward, striking the sloped face of the lower portion, which drives it in the opposite direction.
9. Look, obviously, for scum lines and the location of floatable debris. Scum lines may be multiple, showing incremental drops in flood levels as water was evacuated. Floatable debris *significantly* higher than scum lines can suggest a dynamic wave action which "sloshed" the incoming debris higher than the static flood level.
10. Look for "patterns" or strange replications that may be subtle, but significant when fitted into an overall "picture" of the event. There is no single conclusion to be drawn from such patterns, but their presence is rarely accidental. An attempt to discover the reason for such patterns in damage may pay huge dividends in discovering the sequence of events which produced them.
11. By all means "paint" the entire area with high-resolution photography with at least 25% overlap between photos. For observations that seem significant, take a series of orientation photos which start well back and gradually zoom in on the significant item. In this way the location of the item can later be determined by anyone examining the photos in order. If site plans or evacuation plans (for schools, public buildings, or hotels) are available, use these plans and a ball point pen to show the orientation of the subsequent photo or photo series and photograph the plan with the pen properly positioned.

12. In ALL cases, know WHAT you are looking at and be able to ascribe its significance to the chronology of events leading up to the final damage.

CONCLUDING REMARKS:

1. The primary differences between water and wind are weight, speed, and viscosity.
2. Water, being heavier, can exert tremendous hydrostatic forces at little or no dynamic speed and with very little elevational difference across the barrier. With speed, however, these forces are greatly magnified on the order of the square of the speed. The more viscous nature of water means that it cannot penetrate small cracks or openings in structures and is more inclined to act on the entire structure as a crack-free surface than a less viscous, more "fluid" agent. Additionally, and also due to its viscosity, water will flow around and behind structures more readily than the less viscous agent. Water-borne debris is, generally speaking, floatable debris and can be carried hundreds of feet from its original location. If speeds are involved, this debris can inflict serious penetrating barrier damage at or near the flood elevation at the time of impact.
3. Wind is, of course, nearly weightless with respect to water. The speed of wind, however, is what produces the significant forces that it exerts on windward surfaces and the speeds of hurricane winds are several magnitudes greater than that of water. Unlike water, these speeds are rapidly fluctuating in magnitude creating a "buffeting" effect that will cause fatigue failure of metal siding at girts and midpoints, shredding of thin materials in the direction of wind travel, and an aerodynamic "rippling" effect on metal roofing that will pull it over the heads of fasteners. Further, tornadic wind effects are seen as a very *sudden* and *violent* damage to structural elements, such as the ripping of shingles which leaves the broad-headed shingle nails in place. Such tornadic damage may also be suggested by rotational damage to large trees which are observed to be rotated and punched down locally in the same direction. Wind can permeate cracks that are non- or slightly permeable to water, thus more readily equalizing aerostatic loads on opposite sides of the barrier. Owing to its lower viscosity, however, it exerts a *negative* pressure on sides of the structure away from the originating direction. Wind borne debris need not necessarily be light, but wholly or partially floating debris can be propelled by the wind over distances that are greater than expected from the action of water alone.
4. There are very few instances in the areas I have inspected where the damage seen is attributable solely to either water or wind. In almost every case, there is damage attributable to either force acting in sequence, in concert, or both. Assembling all of the available evidence and formulating a defensible hypothesis as to the chain of events which incorporates the significant elements of that evidence is necessary in order to assure that an equitable attribution of "wind vs. water" damage and causation is made.

5. In cases of significant economic impact, a greater stress may be placed upon attribution. In such cases, assistance with evidentiary interpretation may be desirable. To facilitate such assistance, care must be exercised to collect as much evidentiary data as possible, especially overall and detailed photography.

CHECKLIST OF TYPICAL FINDINGS:

Although there is absolutely no substitute for a careful and insightful application of the guidelines and methods outlined above, there is, perhaps, value in examining certain manifestations which have been discovered to date in the southern parishes and which appear to be associated with either wind or water (as defined above) in the majority of cases examined. The following represents a partial list of those manifestations and the causative force or forces from which they have been found largely derivative:

1. *Concrete slab uplift and cracking associated with full or partial exterior wall collapse:*
   This condition will be seen as frank and incident-related cracking of the perimeter areas of concrete slabs, with or without residual "bulging" displacement of the slab upward. This has so far also been associated with full or partial destruction of adjacent masonry walls or infill where wall debris has been noted both inside and outside of the building perimeter. This has been, to date, attributed rather conclusively to water. Flooding fills the crawl space below the slab and, as waters rise on the outside of intact walls and doors, a significant uplift force is exerted on the slab. This causes an upward buckling of the slab, associated rotation of the edge in the direction of center-slab bulging, and subsequent failure of the exterior masonry as a result of both mounting exterior hydrostatic force and the loss of base support created by the rotation of the slab edge.

2. *Event-derived cracking of concrete columns at tops, bottoms, or through column body:*
   Cracking of columns must first be identified as incident- or preincident-related. Many column bases will be found with cracks that are seen to have been "repaired" or parged in the past. If done absent epoxy injection, such repairs still provide a hinge at the column base which is then exploited by incident-related forces to cause damage elsewhere. Note flood water height indicated by scum lines or the height of floatable debris. Note also if there is apparent shredding or removal of roofing materials. If the flood depth is at or below the 1/3 point of the column height and roof shredding is noted, consider a horizontal "racking" of the structure in response to wind, utilizing the crack hinges at the column bases as the base of racking deformation. Such racking will almost always cause incident-related cracking at the tops of columns or through the column body as inflection points are established during this "sidesway" deformation. In the absence of the shredding of

roofing material and when flood levels are to a significant height on exterior walls, consider racking of the building frame in the same manner, but due to hydrostatic or hydrodynamic loading, rather than wind.

3. *"Sardine can" rollback of metal roofing:*
   This is manifest as, literally, a "rolling up" of light-gauge metal roofing in a direction suggestive by debris dispersal as the causative wind direction, although the rolled roofing debris may not be seen on the roof, but detached and found some distance downwind. This is nearly always caused by wind, as water is rarely discovered at roofing heights in the areas where this type of damage is seen.

4. *Shredding of roofing, ceiling insulation/vapor barriers:*
   Shredding, or the reduction of large sheets into smaller strips, can be seen quite commonly in light horizontally-installed materials, such as light-gauge roofing, plastic vapor barriers, and insulating materials. Shredding is nearly always the result of forces which cause a flagellation or longitudinal flapping of the thin materials as a result of minor variations in speed, direction, or both. Consider wind as the only one capable of such variations and at the same time facilitating the "flapping" movement necessary to produce shredding of this type. Consider it conclusive if shredded materials are above the documented flood elevation.

5. *Straight "cutting" of metal siding at girts and mid-points:*
   Close examination of the "cut" edges of the material may disclose residual evidence of bending both to the interior and exterior of the structure. Commonly, such straight line "cutting" may be attributable to bending fatigue of the material. Such fatigue, by definition, must be the result of repeated, often rapid, deformation of the material in both directions cyclically. Such alternating deformation is judged to be caused only by wind, as water forces tend to be steadily from one direction and then, in some cases, steadily from the other. This would provide, at the most, a few cycles of alternating bending, probably below what is needed to fatigue the material.

6. *Kinking of metal siding at girts and midpoints:*
   Again, close examination of the ridges of light-gauge metal siding may disclose a series of kinks at girts and at midpoints. This kinking may show both a compression buckling of the ridge portion of the siding and a localized tear at the kink, indicative of tension cracking of the material. This is also a sign of alternating forces as discussed above, but may only require a few alternations of direction to produce. If there is associated shredding of the siding, consider wind as the causative force. If there is an associated large-scale bending deformation of the siding between girts, consider flooding, with localized swirling, as capable of producing both the kinks and the large-scale bending of the panels.

7. *Complete/partial destruction/removal of roofing/deck:*
   If large sections of the roof and deck have been removed by the storm, the key to a determination of causation is the flood elevation. If the roof is low and evidence exists of floatable debris in the roof joists, consider flood as having created a buoyant force, facilitated by floating debris, to pop the roof from the joists. If the flood elevation is significantly below the roof and no floatable debris is seen anywhere on the roof joists, then consider wind as the cause. A word of caution is offered here, though. In one instance, so-called "floatable debris" was seen lodged above the roof joists in one corner of a structure whose roof had been completely removed in the storm, but evidence elsewhere on the site could only document a flood elevation of 2 ½ to 3 feet. Close examination of the "floatable debris" by high-resolution digital photography disclosed that the material in question was a pre-incident bird's nest, originally built between the roof deck and the ceiling and accessed via an opening under the eave.

8. *Buckling or translation of vertical and horizontal girt system:*
   Girt systems may be seen commonly as either set atop a masonry wall, or as extending from the roof to the ground. Such girts are usually sized only to reinforce and support light-gauge metal siding, and are not considered as "stand alone" portions of the structure. Deformations to such systems in the form of a local buckling of cold-formed members or of a translation of an entire wall section have been noted in both Plaquemines and St. Bernard Parishes. Again, the height of the flood can be used to index a likely causation. If situated atop a masonry wall, the cause is nearly conclusively wind. If extending from roof to ground and associated with other flood-induced damage, the conclusion suggested is as strongly in favor of water as the cause.

9. *Isolated penetration damage to exterior walls:*
   Isolated penetration damage to exterior walls used to be clearly attributable, as the object inflicting the damage could be seen still lying there. With debris removal operations far advanced in some areas, the object(s) may have been carted away. As a suggestion, consider the size and location of the penetration. If large and located at or below the documented flood elevation, it was likely caused by floating debris, and the amount of energy needed to cause the penetration is a clue as to the rough speed of the advancing water. If small, or if located above the flood elevation, consider wind-borne debris as the causative agent.

10. *Failure of interior partitions:*
    Unless there is a direct route to the failed partition from outside the building afforded by other wall and partition damage, failed interior partitions can nearly always be plausibly attributed to flood. The mechanism may be a softening of the wall materials and subsequent gravity-load failure, a buildup

of hydrostatic water to one side and not the other (as discussed above), hydrodynamic (surge) forces, water-borne floating debris, or a collateral manifestation of hydrostatic uplift on the slab below the partition. To date, there have been no structures inspected by this writer where a failed interior partition, as an isolated manifestation, was attributable to wind.

11. *Translation of unanchored exterior accessory equipment:*
In one or two interesting cases, there have been significant translations of outside accessory equipment (notably a trash compactor) as a result of the storm. In the case of the compactor, evidence of the "skidding" of the steel support legs in the dirt told a story of flooding, buoyancy of the compactor, wind on the high compactor hopper catching the wind like a sail, breaching of the compactor throat to flood the ram and box, and re-settlement of the compactor at some distance from its original location. Other examples of this can be anticipated with other accessory equipment whose buoyancy is either later abridged (as in this case) or whose buoyancy is sufficient to just clear the ground with later re-settlement caused by abating flood water elevations.

12. *Failure of all or most of exterior masonry infill:*
If evidence of significant wind-induced deformation of the building frame is found and if the masonry infill is not competently anchored to this frame, consider that wind-induced racking may have "cracked out" the rigid infill as a result of distortion of the infilled portal. If there are other evidences of flood-induced damage and if the masonry is well-anchored to the building frame, consider a hydrostatic load of flood water as a defensible cause.

13. *Pattern breakage of exterior glass windows:*
Normally, pattern breakage of glass in a well-ventilated building is considered to have been caused by hydrostatic or extensive water-borne debris forces. Causation in the absence of evidence of hydrostatic load (i.e. low flood elevation) was discussed above. Any discontinuities of pattern should be checked to determine if unbroken glass is similar in thickness and age to the broken panes. Check, also, to verify that they are actually glass. In one instance, unbroken "glass," after inspection, was discovered to be plastic.

14. *Pattern dispersal of debris including interior contents:*
If debris dispersion is seen to be strongly directional, consider flood as a major factor in the overall damage to the structure. The speed and direction of water tends to be less variable than wind, thus wind-borne debris is more widely scattered and les directional, whereas debris transported by flowing water, unless acted upon by wind forces as discussed above, tends to be strongly dispersed downstream of the flow. Keep in mind that the direction of this flow, in small areas, may be impacted by the presence of barriers to

the water, so deviations from a strict "one-direction" translation are not only possible, but are likely.

15. *Helical distortion of site vegetation, especially large trees:*
   In at least one instance so far, evidence of an axial twisting of major trees on a site all in the same direction was noted in the form of a helical twisting of the fibers of the trunk. In addition, other trees of similar size and identical species on the adjacent property were unaffected. This has been judged as evidence of tornadic activity, a condition which has been more completely discussed above.

16. *Uplift and rippling of interior wood flooring, especially gymnasia and civic centers:*
   In several institutions, hardwood flooring has been observed to have been lifted and/or buckled in a direction at right angles to the direction of the grain. This is conclusively caused by water, as the buoyant properties of the wood made it susceptible to hydrostatic uplift and the absorbent qualities of the lignin between the grain fibers predispose it to a much greater swelling in the cross-grain direction than the longitudinal.

17. *Translation of exterior steel or aluminum fenestration:*
   In all cases so far inspected, translation of doors and windows has been determined to be the effect of flood. The direction of translation is an indication of flooding direction and the timing of the flooding of the interior. Keep in mind that equal elevations of water on both sides of a barrier will, theoretically, produce no net force in either direction. There are undoubtedly exceptions to this, such as the failure of a full-story window system as a result of wind racking of the structure, but this has not been observed to date.

18. *Base rotation and/or mast distortions of tall and slender structures such as light poles:*
   If there is no rotation of the base, the height of the mast deformation must be the clue as to cause. If above the flood elevation, then wind is the likely cause. If below, then flood OR wind may be at fault, but likely flood, as wind forces increase with height, and a higher bend location would be expected. If a base rotation is present, mast bend must be carefully documented, as the base rotation may give the appearance of bend where none exists. Base rotations attributable to flood can occur if saturated alluvial soils permit a rotation of a deep footing or shallow piles. For this reason, examine the base for signs of cracking and the base of the mast for signs of impact from water-borne debris. If the base is cracked, and no impact point is seen, wind is the likely agent. If the mast is uncracked and/or an impact point can be identified, consider flood as having softened the soils and/or conveyed heavy debris to the mast to cause the rotation.

19. *Shredding of thin-segment overhead roller doors:*
    Several sites have been inspected where thin-segment overhead roller doors have been shredded by stripping segments from one another and fatiguing the material at a batten or splice point. This has been attributed, so far, exclusively to wind, as the fatigued metal would have to undergo cyclic flagellation as discussed above. Examine the ends of such segments for evidence of cyclical bending in both directions.

20. *Translation of entire portions of wooden structures remaining structurally intact:*
    Sometimes, entire roof structures have been seen, intact, several feet from the original site. If the nature of the roofing material is such that wind removal would be expected, examine the underside of the roof perimeter to determine the means by which it was affixed to the top of the wall. If the connection appears to have been competent, consider flood as having failed the walls and translated the roof structure by floatation to its present location. If the connection appears weak, consider that wind acting on the gable end may have acted to dislodge the roof without exerting sufficient force to remove roofing. If the roof were easily removed by the wind, there may well have been insufficient resistance to the wind by the connection to cause the shingles to be stripped from this direction.

References:

Grateful acknowledgement is made for the contributions of the following to the assembly of a picture of the Katrina time-dependent dynamics:

Pierepiekarz, Mark P.E. and S.E., *(Memorandum)* "Hurricane Katrina Time-dependent Storm Surge and Wind Data", Feb. 28, 2006

NOAA, (NWS-National Hurricane Center) (2005). "Hurricane Katrina Advisory Archive"
http://www.nhc.noaa.gov/archive/2005/refresh/KATRINA%2Bshtml/171011.shtml

SOUTH CLEAR WELL ROOF COLLAPSE:
Hydraulic Uplift or Excessive Construction Loading ?

C. Roarty, Jr., P.E. [1], J. Sivak, P.E. [2], P. Vogel, P.E. [3], and K.V. Ramachandran, P.E. [4]

Abstract: This paper presents a failure investigation case history of a partial roof collapse of a below-grade water storage facility during rehabilitation. The evaluation considers the original design and maintenance of the facility, rehabilitation design, and construction sequencing. The cause of the roof collapse was the failure of selected columns under approved construction equipment loading. The columns that failed were initially damaged by hydrostatic uplift of the base slab during a severe rainfall event and subsequently loaded repetitively by construction operations. Hydraulic and structural models developed during the evaluation accurately predicted the response of hydraulic systems, general crack patterns and specific crack locations; thereby confirming the failure mechanism.

Introduction

This paper presents the results of our investigation of the collapse of a portion of the roof of the South Clear Well at the Lake Huron Water Treatment Plant in Fort Gratiot Township, Michigan. The South Clear Well is a 330–foot long by 370–foot wide, approximately 17-foot deep, reinforced concrete structure designed to retain approximately 15 million gallons of filtered water. The clear well contains an 18-foot wide influent channel along its east wall and a beam and post support system for future transmission piping in the northernmost 45 feet. The balance of the clear well construction consists of a 12-inch thick floor slab, 16-inch diameter columns with top capitals and bottom pedestals at a 22-foot center-to-center spacing, and a 10-inch thick roof slab. It is located south of the high lift pump station and west of the filter building.

---

[1]M. ASCE, Vice Pres., NTH Consultants, 480 Ford Field, 2000 Brush, Detroit, MI 48226; PH (313) 237-3900; FAX (313) 237-3909; email: croarty@nthconsultants.com
[2] M. ASCE, Principal, Nehil-Sivak, 414 S. Burdick St., Suite 300, Kalamazoo, MI 49007; PH (269) 383-3111; FAX (269) 383-3112; email: jsivak@nehilsivak.com
[3] M. ASCE, Principal, Greeley and Hansen, 211 W. Fort St. Suite 710, Detroit, MI 48226; PH (313) 628-0730; FAX (313) 967-0365; email: pvogel@greeley-hansen.com
[4] Head Engineer – Field Engineering, Detroit Water and Sewerage Department, 3501 Chene, Detroit, MI 48207; PH (313) 833-8443; FAX (313) 833-8420; email: ramachandran@dwsd.org

Improvements to the clear wells and the high lift suction well were designed in 1995. The construction of the improvements began in 1997. The collapse of a portion of the South Clear Well roof occurred at approximately 2:00 p.m. (EDT) on June 22, 1999 during final topsoil placement operations on the roof of the clear well. No persons were injured or construction equipment damaged by the failure.

Figure 1. View of collapsed roof section

The evening of the collapse, the plant suffered a power outage and dewatering pumps within the gate wells in the clear well were rendered inoperable. As a result, approximately 10 inches of water from gate leakage covered the floor of the clear well during the initial post-collapse inspections by Detroit Water and Sewerage Department (DWSD), contractor, and consultant personnel. The initial inspection teams stayed primarily under the beam and post section of the clear well out of concern for safety. The two primary observations during the initial inspections were the water appeared to be shallower in the center of the clear well and the most severe roof and column damage had generally occurred beneath the areas of top soil placement on the roof of the clear well.

DWSD requested that NTH Consultants, Ltd. lead a team of local consultants to perform an independent evaluation to determine the cause of the collapse and develop schedule and budget for replacement options. At the request of DWSD, the firms of Nehil-Sivak and Greeley and Hansen were retained to provide structural engineering services and an operational assessment of the plant, respectively. Greeley and Hansen also performed an evaluation of the storm water and under drain systems.

Investigation Methodology

In order to develop cause and effect relationships between the various factors involved in the collapse, a chronological approach was used to create an understanding of the project. We reviewed and evaluated available information on the construction including: the original construction plans; current construction basis of design reports, plans, and specifications; historical maintenance and repair records for the clear wells; available subsurface data; daily field reports; interviews of resident engineering and plant personnel; and measurements and observations made during visits to the site after the failure.

We developed simplified assessments of short term and long term loading conditions for comparison with the original design and evaluated the effect of construction procedures and adjacent construction activities on the basin. We then reconstructed the sequence of events during the construction period and ascertained potential causes of the failure. Based on various load combinations, we then evaluated the structural adequacy of as-designed and as-built conditions to resist the forces associated with the existing conditions at the time of failure.

Original Clear Well Construction

The south clear well was constructed in the 1970's in an open cut excavation with the high lift building and the identical north clear well. The base and roof slabs have thickened sections at the perimeter walls that transition to the typical thickness over a distance of 4 to 5 feet from the face of the exterior walls. The beam and post supported section of the clear well with thickened roof and base slabs also transitions to the typical thickness in the general clear well area over a distance of 4 feet from the face of the beam and post support.

The top of base slab elevation for the clear wells and high lift suction well are 587.5 and 579.5, respectively. The original under drain system, installed to prevent groundwater accumulation beneath the base slab, combines relief wells adjacent to north and south sides of the high lift building under the clear well base slabs, a 3-inch thick sand drainage blanket under most of the clear well, collector piping under the clear well at invert elevation 584.5, and perimeter collector pipes backfilled with pea gravel around the clear wells at invert elevations ranging from El. 584.5 for the north clear well to 587.5 at the southeast corner of the south clear well. All under drain piping was originally connected to a dedicated under drain pump station located at the northwest corner of the north clear well. A review of the original construction drawings indicates the high lift suction well, the clear well areas beneath the beam and post system, and the exterior clear well walls do not have a sand drainage blanket and are not serviced by the under drain system.

Planned Rehabilitation

The purpose of Project No. 4, Water Storage Reservoir Improvements, was to eliminate the problem of surface water ponding on top of the reservoirs. The ponded water potentially could leak through the soil cover and into the reservoirs. The basis of design report recommends the regrading of the surface of the clear wells with lightweight concrete fill to promote drainage; installation of a polyethylene liner over the top of the lightweight fill; construction of storm sewers and french drains to collect surface runoff and infiltration; construction of new north and south storm water detention basins; construction of new north and south pump stations to handle combined surface and under drain flow; abandonment of the existing under drain pump station; repair of cracks, construction joints, and control joints in the interior roof slabs, walls and base slabs; gate repair; installation of new air vents; and repair of clear well roof hatches.

The calculations for the new storm water collection system, detention basin, and pump station indicate the north and south systems are sized for a 10-year storm at maximum operating conditions. The pumps in each pump station were sized to provide a total 800-gpm capacity and are controlled to maintain a 10-minute pumping cycle time at minimum operating conditions of 250 gpm.

Rehabilitation Plans and Specifications

The contract plans and specifications were consistent with the basis of design report. To address potential hydrostatic uplift during construction, the specifications required phased construction of the north pump station, demolition of the dedicated under drain pump station, dewatering and rehabilitation of the north clear well, construction of the south pump station, and dewatering and rehabilitation of the south clear well. The contract documents required dewatering of the clear wells for a period of 6 months over two consecutive winter seasons to facilitate the repair work in the basins. The additional requirement to remove all surface backfill to place the new drainage system reduces the dead weight available to resist hydrostatic uplift forces.

To address the potential for construction equipment overloading, the plans contain a note prohibiting the contractor from placing more than 225 psf of load on the roof of the reservoir including the proposed lightweight concrete fill, soil, and equipment.

Shop Drawing and Construction Change Documentation

The contractor performed independent structural calculations and a Cat 936 end-loader with a three cubic yard bucket and 42,000 pound truck was approved for soil removal/filling operations. The loaded trucks would only travel on the beam and post supported section of the clear well and empty trucks could be driven on the remainder of the slab only if the soil was removed.

Construction Chronology

At the completion of activities at the north clear well, DWSD took possession of the north clear well and turned the south clear well over to the contractor. At the time of the collapse, the north pumping station was pumping the surface drainage from the north and the under drainage from the north and south clear wells. The south pump station was not operational and the temporary surface drainage was accomplished by submersible pumps placed in the manholes of the new storm water collection system.

The soil removal operation was conducted without incident using the same methods used on the north clear well and interior and exterior crack repairs were completed in late March, 1999. DWSD plant and field engineering personnel conducted an inspection of the reservoir for cleanliness at the completion of the work within the clear well. During the crack sealing work and periodic inspections, while the clear well was well lit, no signs of structural distress were noted in the floors, walls, roof or columns of the south clear well. After March 29, 1999, entry into the clear well was limited to maintenance of dewatering pumps at the northwest and southeast corners.

The lightweight concrete placement on the roof commenced on April 28, 1999. The contractor pumped and shaped the lightweight concrete with no equipment on the roof. Membrane placement across the roof of the clear well was accomplished using a forklift. Sand placement took place working from the southwest corner to the north and east. A protective board was placed over the membrane prior to placing sand with the Cat 936 loader. A D-5 bulldozer with low-pressure tracks was used to grade the sand. Topsoil placement took place working from the middle of the beam and post supported sections at the north end of the clear well. Loaded trucks backed down the ramp from the east and dumped the topsoil on the thickened roof slab section. The Cat 936 loader retrieved the topsoil and spread it over the basin.

On June 22, 1999, topsoil had been placed over approximately the east half of the clear well. The loader operator noted a small hole had developed in the roof and soil was running into the clear well. The operator moved the loader off of the roof of the clear well prior to the collapse at approximately 2:00 p.m. (EDT).

Post Failure Field Observations

Initial Inspection - The team noted that six columns had collapsed under the failed roof section. All six collapsed columns appeared to have broken off at the connection between the column and the capitals/pedestals. The column capitals and pedestals were still connected to the roof and floor slabs, respectively.

Numerous columns outside of the collapsed roof area were also damaged. Concrete spalling was noted at the top and bottom interface between the column and the capital/pedestal on opposite sides of the columns. Some severely damaged columns had been displaced on shear cracks running diagonally through the column at the top. Severely damaged columns were concentrated in areas with similar crack orientation.

No new cracks were noted in the roof or the walls. Previously repaired cracks appeared to be intact. Some displaced floor cracks were noted parallel to the face of the wall, but the depth of water and sediment on the clear well floor made it impossible to map.

Subsequent Detailed Inspections - Inspection teams confirmed the observations made during the initial inspections and also discovered a pattern of damage consisting of:

- The collapsed portion of the roof had separated from the clear well on the east, north, and west sides and was still connected on the south edge north of column line 6f (See Fig. 2). The remaining roof slab on the north edge of the collapsed area lines up with a family of previously grouted roof cracks approximately four feet from the face of the beam and post supported section (See Fig. 3). The remaining roof slab on the east and west edges of the collapsed area extends from the edge of column capital up to four feet from the capitals. The lower mat of roof reinforcing bars were stripped from the underside of the remaining slab in the east, north and west sides (See Fig. 4). It appears the northern edge of the collapsed roof section broke free and encountered the clear well floor first.

Fig 2. Hinge formed at south edge of collapsed roof section at column line 6f

Fig 3. Collapse along existing cracks

Fig 4. Roof slab steel stripped

- All columns on the three outside column lines of the clear well were damaged. Damage consists of spalled or delaminating concrete above the bottom pedestal oriented towards the center of the clear well and below the top capital oriented towards the exterior walls. The inspection confirmed the presence of horizontal cracks at the joint between the column section and the column capitals. In addition, a hairline diagonal shear crack in the column section immediately beneath the capital was confirmed for columns in the outer two rows at the midpoint of the exterior walls.

- A group of four failed columns were noted along the mid point of the east wall of the clear well. The failed columns had a similar pattern of damage, more pronounced spalling and had been displaced along the upper shear cracks. Reinforcing steel was visible and had yielded. See Figure 5 for a view of a failed column in this group.

- A single failed column (6f-Qq) was noted southeast of the access hatch located in the northwest corner of the clear well with similar damage to the group of four columns described in the previous bullet.

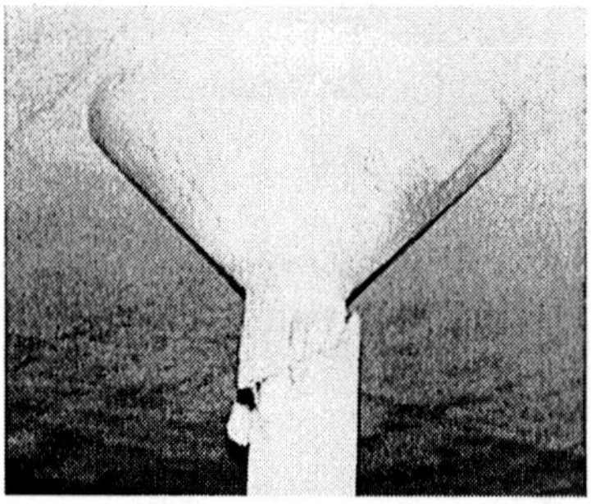

- Floor cracks parallel to and three to four feet from the face of each perimeter wall and the beam and post-supported section of the clear well were noted. A family of floor cracks parallel to the west wall of the basin, ranging from 2 ½ to 14 ½ inches from the face of the wall, indicated a vertical separation ranging from 0 to 3/8 inches. Soil had been carried into the clear well through some of the floor cracks. Diagonal floor cracks in the interior of the clear well were noted.

Figure 5. Failed Column

For ease of presentation, we have classified the column damage into four categories:

- "Undamaged" columns have no structural defects that can be detected by the naked eye. These 54 columns are located within the center of the clear well indicated by the dashed line on Figure 6.

- "Damaged" columns have spalled concrete at base column/pedestal interface oriented towards the center of the clear well, spalled concrete at roof column/capital interface oriented towards the exterior of the clear well. At the midpoint of the exterior walls away from the corners, horizontal hairline cracks at both the top and bottom column/capital/pedestal interfaces and diagonal shear cracks immediately below the roof column/capital interface were noted. These 110 columns are located at the perimeter of the clear well on Figure 6.

- "Failed" columns have undergone lateral displacement along the upper shear cracks in addition to the same characteristics as the damaged columns. At these columns, the roof does not appear to be damaged. These ten columns are indicated by a hexagonal symbol on Figure 6.

- "Collapsed" columns are located in the area of the roof collapse. These six columns are indicated by a triangular symbol on Figure 6.

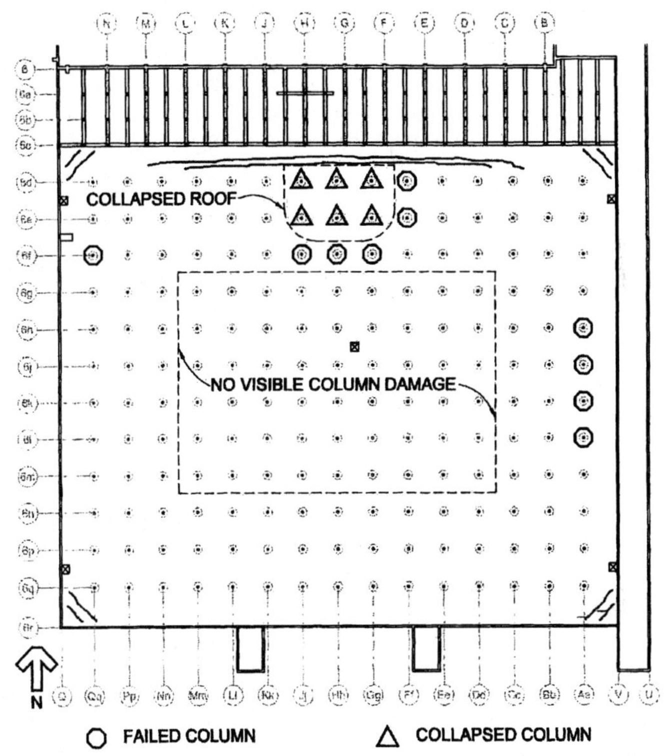

Figure 6. Plan view of clear well roof slab with damaged column notation

<u>North Pump Station Observations</u> - On July 12, 1999, an inspection of the north pump station was conducted to measure on/off float elevations and ascertain if a high water mark existed. The high water mark on the side walls of the pump station was indicated by a line of dark scum and grass clippings above the storm water inlet pipe at approximate Elevation 595. The float elevations were determined by measuring from the underside of the pump station top slab (Elevation 609.0) with a weighted tape. The inlet piping from the under drain system was located 4 feet below the

design inlet elevation. The "pump off" float was 1 ft above the inlet piping, instead of 3 feet below by design, rendering the inlet continuously submerged. The visit was made under dry weather conditions and the pumps cycled at an approximate 45-minute interval to handle the flow from the under drain inlet.

Greeley and Hansen later obtained more precise measurements of pump station features from a survey benchmark and testing of the controls system to verify the operating characteristics of the pump station. The high level alarm in the pump station is powered from the same power source as the pumps through a 480V-120/240 transformer, primary circuit breaker, panel circuit breaker, and single-pole 15-amp GFI circuit breaker. The high level alarm is annunciated at the plant control panel as general alarm with no battery backup. A loss of power to the pump station or a nuisance trip of the control circuit GFI renders the pumps inoperable and no alarms will be generated.

Construction Records Review

A review of the construction chronology and discussion of the means and methods used indicates the contractor generally complied with the provisions of the documents. The contractor's methods and equipment used on top of the reservoir were the subject of detailed analysis and discussion at the beginning of the project. Load restrictions were agreed with the contractor prior to commencement of work on the north clear well.

The plant took beneficial occupancy of the north clear well and the north pump station after the work on the north clear well was complete. The contractor apparently made no provisions to monitor water levels or dewater the under drain system after the south clear well was dewatered. A review of the contract documents indicates there were no specific contractual requirements to do so.

DWSD site representatives have indicated the contractor complied with the approved means and methods during the soil removal and filling operation on the south clear well and did not report any deficient work by the contractor. However, a review of photographs taken after the collapse indicates there was deficient work.

Photographs indicate stockpiles of topsoil were not located exclusively on the beam and post supported section of the roof and were on the flat slab portion of the roof at the time of the collapse. The topsoil piles were reportedly dumped from trucks that had backed down the northwest ramp to the edge of the clear well. Therefore, the contractor may have had trucks on the two-way slab section of the clear well in violation of the contract documents.

Post-collapse observations include tire tracks of a truck that had passed over the sand fill to exit at the southwest of the clear well in violation of the contract documents. Whether this truck was full or if this was an isolated incident is not known.

Assessment of Data and Chronology

The damage to the south clear well has several characteristics:

- The pattern of "damaged" columns and cracks in the floor slab is symmetrical, indicating a uniform loading condition caused the defects. The damage pattern suggests hydrostatic uplift as a possible cause.

- The location of "failed" and "collapsed" columns are located at the midpoint of the flat slab perimeter and coincide with areas of topsoil placement.

- The column capitals/pedestals in the area of the collapsed roof are still attached to the roof and the floor slabs, indicating the roof slab did not fail in shear at the columns.

- The roof failure formed a hinge in the roof at column line 6f, indicating the roof failed first at the north side in the vicinity of the previously grouted shrinkage cracks.

- The roof failure occurred across three bays of the clear well, indicating the underlying columns failed first and the roof slab followed.

Based on our review of the data presented in the preceding paragraphs, we evaluated four possible causes for the damage to the south clear well. These four causes are:

1. Existing material or workmanship defects in the original construction,
2. Stresses in the roof slab due to the volumetric changes from shrinkage,
3. Noncompliant work by the contractor and overloading of the slab, and
4. Hydrostatic uplift of the base slab of the tank.

<u>Existing material or workmanship defects</u> - A comparison of the original contract drawings with the exposed concrete and reinforcing steel at the perimeter of the collapsed slab indicates the bar sizing and spacing was correct. A review of the collapsed roof section against the construction joint pattern contained in original contract drawings indicates the collapsed area encompassed portions of four separate concrete placements. Compressive test results on concrete core samples from the south clear well indicate the concrete meets the requirements for 4500-psi minimum compressive strength. Tensile tests results on reinforcing bar samples from the collapsed roof section indicate the reinforcing steel meets the requirements for Grade 40 steel. Field observations and tests on selected samples indicate the existing materials and workmanship for the south clear well meet the requirements of the original contract. Therefore, we do not consider material or workmanship defects from the original construction to be a cause of the collapse.

<u>Volumetric changes from shrinkage</u> - The clear well roof may have undergone some temperature-related volume change when the tank was empty and the soil cover

removed. However, the family of existing shrinkage cracks in the flat slab roof located between column lines 6c and 6d (south of the beam and post section) were grouted prior to the current rehabilitation effort. Because the concrete cracked and was repaired, no residual tensile stresses remained in the north end of the roof slab at the commencement of the current rehabilitation work. An analysis indicates additional shrinkage would cause the maximum stresses to occur at the south end of the clear well. In addition, the noted shrinkage cracks exist east of the collapsed area, underwent similar loading conditions, and did not collapse. Therefore, cracking due to shrinkage or temperature related volume changes is unlikely to be a primary cause of the failure.

Deficient work by the contractor - The structural analysis done before the work commenced and additional analyses since the failure indicate the loading generated from the approved equipment and materials used to make the surface improvements would not have exceeded the allowable design load of the roof slab or the columns. The analysis performed for this investigation addresses a variety of loading conditions, including dead, live, unbalanced, and dynamic loading in various combinations. All load cases generate moments, thrusts, and shears under the allowable values for the roof slab and the columns.

Hydrostatic uplift - The general pattern of structural damage in the south clear well away from the collapsed roof portion is similar. This suggests that a hydrostatic uplift event had occurred after the final inspection on March 29, 1999 and prior to the collapse.

All reports indicate the north pump station was operational and no alarms were noted during the period from March 29, 1999 until the collapse. A review of the critical uplift elevations that correspond to various stages of surface filling operations indicates that critical uplift elevations could have been reached prior to placement of the sand fill without exceeding the alarm setting in the pump station. Therefore, an uplift event could have occurred without the knowledge of the plant or the contractor.

A review of local rainfall records indicate several significant rainfall events occurred between the last inspection on March 29, 1999 and the collapse on June 22, 1999. Using radar images to reconstruct the actual rainfall pattern at the site, Nexrain Corporation provided gage-adjusted radar rainfall estimates for April 22nd, May 18th, and June 12th, 1999. The most significant event occurred when approximately 4 inches of rain fell in approximately 75 minutes at the LHWTP site on the afternoon of June 12, 1999. The characteristics of this event relate to a return period frequency in excess of 50 years. The input of a site and time specific hyetograph into the hydraulic model developed by Greeley and Hansen indicates the volume of water associated with the June 12th event caused flooding of the north pump station and pressurized the under drain system. The hydraulic model predicts a high alarm condition at 4:54 p.m. that is consistent with an entry in the Lake Huron Water Plant Operator's log book at 5:00 p.m. indicating a severe rain storm with thunder and lightning. The hydraulic model also predicts a high water elevation of 595.3 that is consistent with the high water mark at an approximate elevation of 595 measured in the north pump station on July 12, 1999. Because this rainfall event took place over the weekend between the completion of the sand backfill placement and the commencement of the topsoil placement, the contractor was not aware of any impact this rainfall event would have on the work.

The elevation of the under drain system outlet pipe at the north pump station, the constant recharge of the system by the service water from the high lift building, the flat slopes of the under drain system, and the control sequence for the pumps result in an under drain piping system that is partially full and may be saturated at all times. The under drain piping system and sand blanket beneath the clear wells is confined by relatively impermeable clay soils or concrete structures that force the system to maintain a constant volume. Therefore, hydrostatic pressure within the under drain system would develop in the under drain system relatively quickly once the partially filled system became saturated. The hydraulic model indicates the high water levels in the north pump station that occurred as a result of the June 12, 1999 rainfall event caused an increase in pressure beneath the base slabs of the north and south clear wells. Since the north clear well was in operation and full of water, the net pressure acting on the base slab was incapable of lifting the floor slab. Since the south clear well was empty, the net pressure acting on the base slab was sufficient to lift the base slab.

Structural Evaluation

If one considers the base slab of the clear well a flat plate fixed at the perimeter, some useful analogies are available to explain the noted behaviors. A flat plate fixed at the edges under uniform load will have its maximum induced moment and shear at the center of the edge spans and lesser moments and shears at the center and in the corners of the slab. A structural analysis of the empty basin under an uplift pressure indicates that heave of the base slab would induce a level of stress in the floor slab and columns that varies with the position within the clear well.

A finite element program was used to model elastic behavior of the south half of the south clear well under an incremental application of uplift pressure to a maximum of 8.8 feet (corresponding to Elevation 595.3 predicted by the hydraulic model). The weight of the structure and the required cover except for the topsoil would equal the uplift at approximately 70 percent of the uplift load (Elevation 592.7) and loss of contact between the base slab and the soil would result. Figure 7 is a graphical representation of stress levels within the base slab at an uplift head (Elevation 592.8) sufficient to exceed the cracking stress of the concrete at the extreme fiber in tension. The stress patterns are consistent with the noted pattern and location of column damage and cracks in the floor slabs at the perimeter walls after the failure.

The structural model prepared by Nehil-Sivak to analyze the performance of a column bay indicates the ultimate capacity of the perimeter three rows of columns (156 of the 180 columns) would be exceeded at an uplift pressure corresponding to a water level within the north pump station at Elevation 594.4 (approximately 90 percent of the total uplift predicted by the hydraulic model). The structural model predicts the formation of plastic hinges at the column/capital interface at each end of the column, crushing of the concrete on the interior column face at the bottom, crushing of the concrete on the exterior column face at the top, and the formation of shear cracks within the upper and lower thirds of the column. See Figure 8. The post-failure conditions documented in previous sections of this paper are consistent with the conditions predicted by the structural model except for the formation of the lower shear crack. The floor cracks at the perimeter of the clear well result from an improper lap of reinforcing steel in the floor, reducing the ability of the floor to resist the moments generated by the uplift.

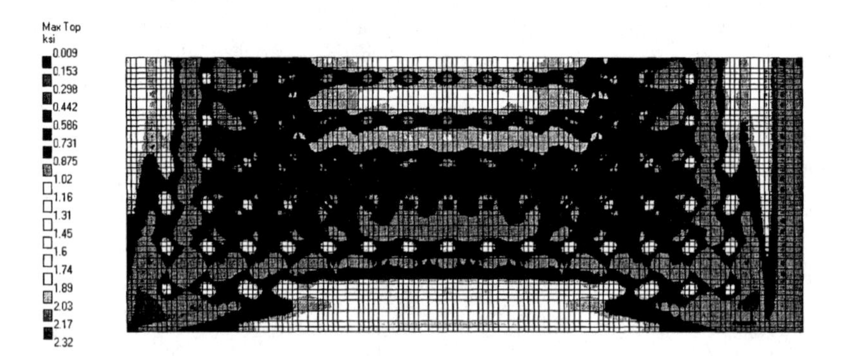

Figure 7. Floor Slab Principal Stress – Uniform Uplift vs. Dead Wt. Plus Topsoil

Structural analysis of the base slab and columns indicate that uplift pressures less than the maximum levels predicted by the hydraulic model and measured in the field are sufficient to cause the generally symmetrical damage noted in the clear well. The actual severity of floor and column damage varies with the general position in the clear well, the specific end conditions (i.e., the stiffness of the walls, adjacent beam and post system, or the filtered water conduit), the specific material properties of the concrete and steel at each column or floor section, and the ability of the under drain system to deliver the uplift pressure to an area of the floor slab. The pattern of failed columns, however, is not consistent with the general pattern of damage. Therefore, a source of additional loading was required to bring specific areas to failure. A review of the construction sequence indicates the collapsed columns and the scattered failed columns have two factors in common. First, the columns are located adjacent to the perimeter of the clear well where the moments in the floor and columns are largest. Second, the failed columns are all in areas where topsoil fill had been placed. The area of the roof collapse was the offloading point for the topsoil and underwent the highest traffic loading.

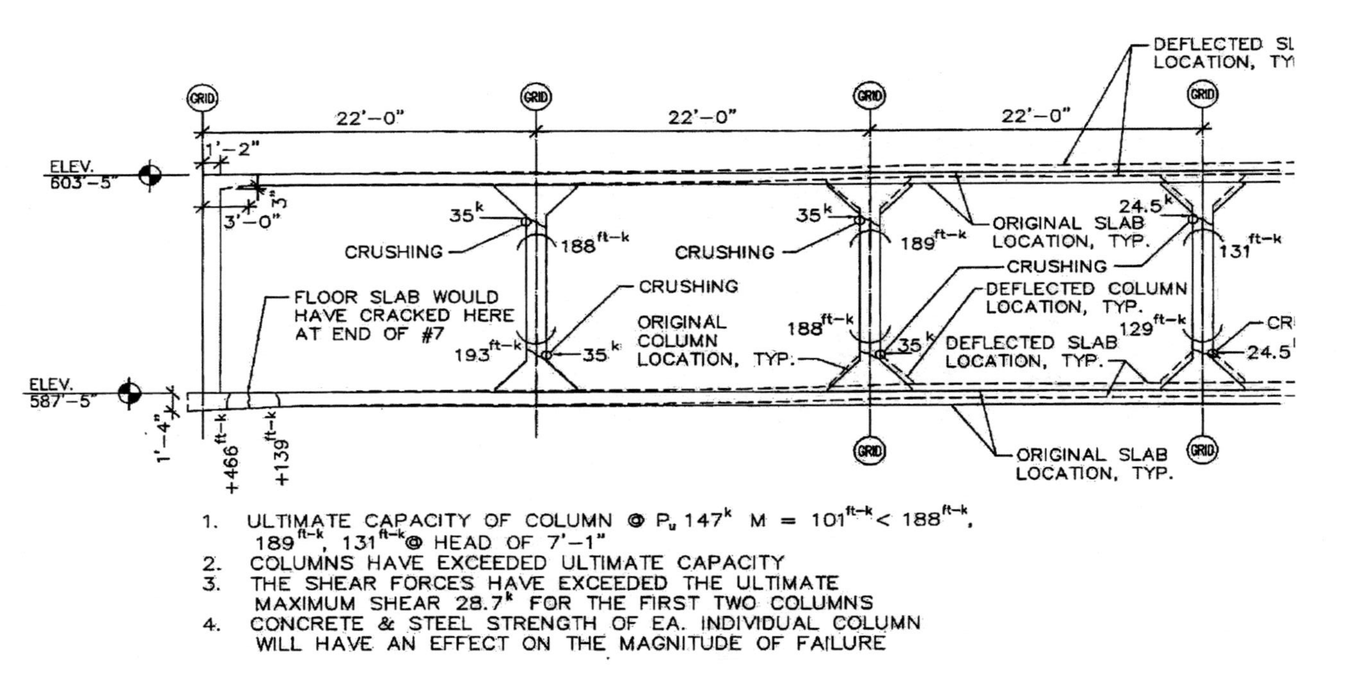

Figure 8. Deflected Slab Shape and Resultant Shear Forces and Moments

Conclusion

Based on our review of the data, we have concluded the roof collapse was caused by a failure of the underlying columns. Our analysis of the available data indicates hydrostatic uplift of the base slab occurred during a storm event on June 12, 1999. The characteristics of this event relate to a return period frequency in excess of 50 years. The volume of water associated with the event caused flooding of the north pump station and pressurized the under drain system. Heave of the base slab induced a level of stress in the columns that varied with their position within the clear well. The perimeter three rows of columns (126 of the 180 columns) yielded, crushing the concrete as plastic hinges formed at each end of the column, and forming shear cracks immediately below the column capitals. The subsequent construction activity of placing topsoil on the clear well roof loaded the recently formed hinges and shear cracks. In high traffic areas, the yield point of the columns supporting the roof of the clear well was reached, the longitudinal bars buckled between the points of restraint provided by the lateral ties, and the failure occurred suddenly. Without the support of the columns, a section of the roof collapsed into the clear well.

The area of the roof collapse corresponds to the area of heaviest traffic. The severe structural distress noted in selected additional columns resulted from the application of similar, albeit reduced, traffic loads.

References

1. ACI 318-99, Building Code Requirements for Structural Concrete, American Concrete Institute, Farmington Hills, MI, June 1999
2. ADOSS V7 Equivalent Frame Software, Portland Cement Association
3. National Weather Service Technical Paper 40 Type II Rainfall Distribution, 2nd Ed., June 1986
4. STAAD-PRO Finite Element Software, Research Engineers International

Prowers Bridge Study: Experimental and Analytical Techniques for Wind Loading Analysis at an Historic Truss Bridge

Veronica R. Jacobson[1], Frederick R. Rutz[2], Kevin L. Rens[3]

[1]Research Assistant, Department of Civil Engineering, Graduate Student, University of Colorado at Denver, Campus Box 113, PO Box 173364, Denver, CO 80217-3364
[2]PhD, PE, Senior Project Manager, J.R. Harris and Co., 1776 Lincoln Street, Suite 1100, Denver, CO 80203-1080
[3]PhD, PE, Associate Professor, Department of Civil Engineering, University of Colorado at Denver, Campus Box 113, PO Box 173364, Denver, CO 80217-3364

Abstract

The University of Colorado at Denver has been studying the relationship between wind loading and structural response in historic truss bridges adapted to pedestrian use. Currently, many historic truss bridges with traditional timber decks would be inadequate for pedestrian conversion using the traditional "skeleton" method of modeling and the current American Association of State Highway and Transportation Officials (AASHTO) *Guide Specifications for Design of Pedestrian Bridges* for lateral (wind) design loads (AASHTO 1997) on the windward bottom chord members (eyebars). An experimental and analytical study was completed on the Prowers Bridge over the Arkansas River, constructed in 1909, which is located near Lamar, Colorado. The experimental study utilized data from anemometers and clamp-on modular strain transducers to provide verification of an analytical deck model of the current Prowers Bridge. This paper presents the equipment, results with methodology, and engineering applications based on the experimental and analytical response to the lateral (wind) loads at the Prowers Bridge. The overall results indicate that increasing the dead load of the deck and accounting for the stiffening effect of the deck in the analytical model allows the windward bottom chord eyebars to satisfy AASHTO lateral (wind) loading requirements. In summary, this research provides useful applications to aid rehabilitation and restoration of historic vehicular truss bridges for pedestrian use.

Introduction

Developments in the engineering design and construction of truss bridges illustrate a history of engineering innovations. While historic bridges present these developments in a forum readily accessible for public viewing, only a small fraction of this engineering heritage remains. Historic truss bridges from the late 19th and early 20th centuries are vanishing rapidly. 60% of Colorado through-truss bridges in existence 20 years ago have been removed (Rutz 2004) and it is estimated that 50% of the nation's truss bridges have been removed over the same time period (DeLony 2005). At this rate of attrition, the American engineering legacy of the truss bridge may soon be relegated to the history books.

One avenue for preservation is rehabilitation of such bridges for pedestrian use. Conversion to pedestrian use permits ready public access to historic structures and has the added advantage of providing incentives for continued maintenance. Unfortunately, the engineer for today's historic bridge preservation project often finds that the windward bottom chord eyebars has insufficient lateral strength to satisfy modern requirements (Rutz and Rens 2004). This is due to two circumstances that combine together, hampering preservation projects: (1) today's design wind load is significantly higher than that used for the original design a century ago and (2) the use of traditional structural analysis can lead to an incorrect conclusion that wind load overstress the windward bottom chord eyebars.

While design wind load is not addressed here, it is a goal of this study to improve upon traditional analytical methodology and verify the method by field tests.

Summary of Prowers Bridge

To aid preservation efforts, real bridges were examined with non-traditional structural analysis and on-site field wind data. This paper focuses on the Prowers Bridge shown in Figures 1 and 2. The bridge is a pin connected Camelback Pratt through-truss. It has moment-resisting end post member at the portals and horizontal trusses consisting of rod X-bracing intended to resist lateral loads.

Figure 1. Prowers Bridge over the Arkansas River, near Lamar, Colorado. This bridge consists of one five panel Pratt pony truss built in 1921, three nine panel Camelback Pratt through trusses built in 1909, two six panel Pratt through truss built in 1902 and 1906. The 49-meter (160-feet) Camelback Pratt through truss span, seen in this photo, was instrumented because it received the greatest wind exposure. It has steel floor beams with steel stringers and asphalt filled corrugated steel deck. It has steel bottom chord eyebars and diagonals with steel rod counterbracing. The railing is a steel lattice with single angle top and bottom rails. Virtually all paint has weathered away. It survived a major flood of the Arkansas River in 1921 and served as a highway bridge until its abandonment in 1994 when a replacement bridge was constructed nearby.

Figure 2. Prowers Bridge Deck. The steel stringers are riveted to the floor beams. Asphalt pavement, now significantly weathered, was installed over a corrugated bridge deck.

Prowers Bridge Modeling

Figures 3 through 6 for Prowers Bridge illustrate the modeling technique used for Prowers Bridge with the RISA-3D software (RISA 2002).

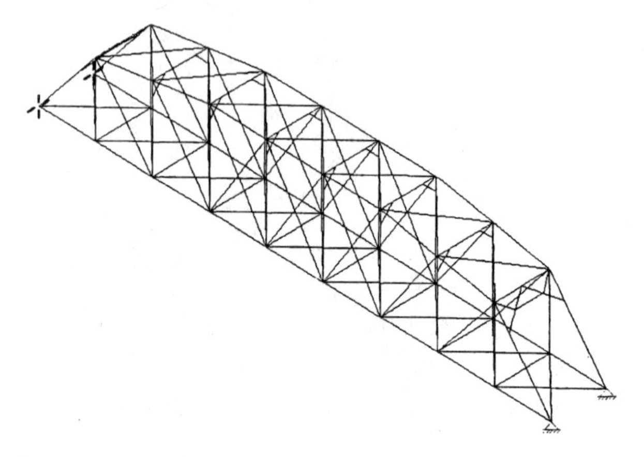

Figure 3. Prowers Bridge. Illustration of the traditional skeleton method of modeling based on the steel members only.

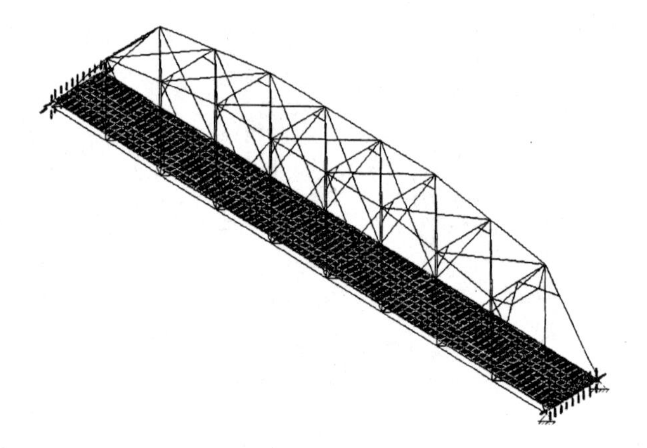

Figure 4. Prowers Bridge. Skeleton model with stringers and deck.

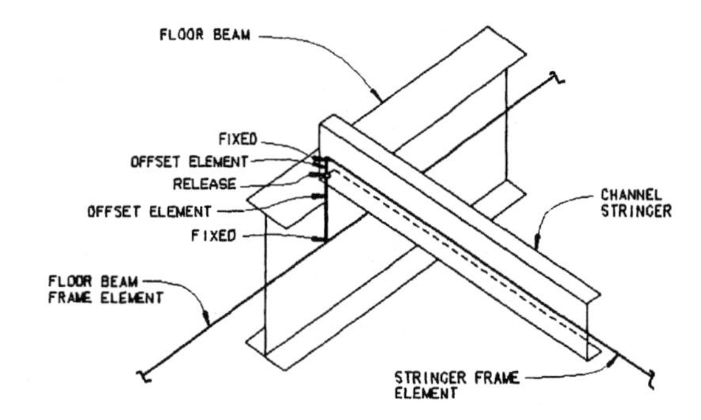

Figure 5. Prowers Bridge. Offset members and release locations

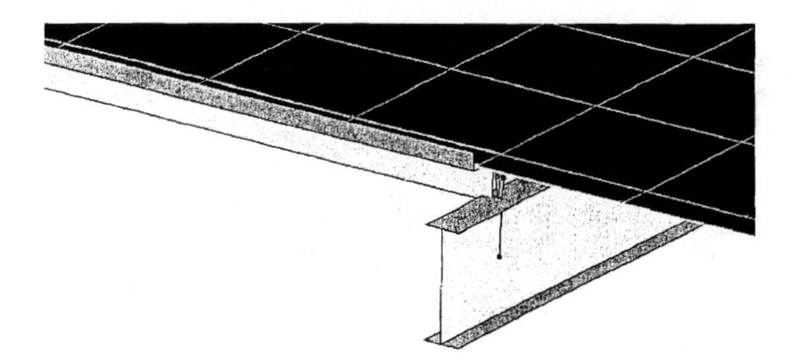

Figure 6. Prowers Bridge. Rendering of equivalent corrugated steel deck with deck edge angle on steel stringers on steel floor beam. The deck was modeled as an equivalent solid.

The Prowers deck model was based on a skeleton model plus 810 plate elements with steel stringers, as shown in Figure 4 through Figure 6. The corrugated deck was represented as a thin plate. The corrugated steel deck has one stiffness in the corrugated direction and a different stiffness in the non-corrugated direction. The RISA 3-D model is not capable of defining stiffnesses of orthogonal plates/shells. The method used here was to model a range of deck stiffnesses for the corrugated steel deck, treating the deck as if it were isotropic. Table 1 shows axial forces of the windward bottom chord eyebars from the original timber deck skeleton model, the existing deck skeleton model, two models with different deck stiffnesses for the corrugated steel deck models, and the deck modeled as a rigid diaphragm. The wind pressure criteria was based on the AASHTO wind pressure of 3.59 kPa (75 psf), perpendicular to the bridge.

Table 1. Prowers Bridge: Summary of Maximum Axial Forces in Windward Bottom Chord Eyebars. Forces are expressed in kN (kips), followed by percent of reduction of compression (or increase in tension) compared to the "traditional" existing deck skeleton model. (Positive = tension; negative = compression).

Model	Axial force due to dead load only	Axial compression due to wind load only	Net axial force due to wind plus dead load
Case 1:			
Skeleton Model	160.1	-226.8	-59.60
(Timber Deck Dead Load =	(36.0)	(-51.0)	(-13.4)
0.72 kPa (15 psf) - Figure 3)			
Case 2:			
Skeleton Model	284.2	-226.8	69.8
(Existing Deck Dead Load =	(63.90)	(-51.0)	(15.7)
2.3 kPa (47 psf) - Figure 3)			
Case 3:			
Deck Model	278.9	-152.1	129.8
(E = 5.94 MPa (861 ksi) -	(62.74)	(-34.2)	(29.2)
Figure 4)	1.8%	32.9%	86.0%
Case 4:			
Deck	276.7	-118.3	161.0
(E = 60 MPa (8700 ksi) -	(62.20)	(-26.6)	(36.2)
Figure 4)	2.6%	47.8%	130.0%
Case 5:			
Diaphragm	279.0	-27.1	260.7
(Rigid Deck)	(62.71)	(-6.1)	(58.6)
	1.9%	88.0%	272.0%

* Note values are not necessarily identical to (D+W) because tension–only members may not be the same for the (D+W) case as for the individual D or W cases.

The percent change from the skeleton case (existing deck) was determined for the deck model from:

$$\%\ \text{change} = 100 \times \left| \frac{F_{skeleton} - F_{deck}}{F_{skeleton}} \right| \tag{1}$$

and for the rigid diaphragm model from:

$$\%\ \text{change} = 100 \times \left| \frac{F_{skeleton} - F_{diaphragm}}{F_{skeleton}} \right| \tag{2}$$

where:

$F_{skeleton}$ = calculated force in windward bottom chord from the skeleton model
F_{deck} = calculated force in windward bottom chord from the deck model
$F_{diaphragm}$ = calculated force in windward bottom chord from the rigid diaphragm model

The skeleton model with the original timber deck (Case 1) indicates that the windward eyebars would be in compression, thus it would not pass the current AASHTO Pedestrian criteria (AASHTO, 1997). The skeleton model with the existing deck (Case 2) shows that the windward eyebars would be in tension; thus, this model conforms to the AASHTO Pedestrian criteria. This shows that an increase in deck dead allows the windward bottom chord eyebars to conform to current wind load criteria.

The deck model with Young's Modulus of 5,936 MPa (861 ksi) (Case 3) for the deck plates shows the windward eyebars to be in tension and thus conform to AASHTO wind criteria. This Young's Modulus used was approximately 3 percent of Young's Modulus for steel. The reason for the small percentage was to take into account the corrugation effects and unknown welds of the deck to the steel stringers. The deck model with Young's Modulus of 59,984 MPa (8,700 ksi) (Case 4) shows a higher tensile force in the windward eyebars so it also conforms to AASHTO wind criteria. The Young's Modulus used was approximately 30 percent of the Young's Modulus of steel, to examine the effect of a different overall deck stiffness. The deck model treated as a rigid diaphragm (Case 5), which was the stiffest theoretically possible for the deck. This case would provide an upper bound for deck stiffness to examine this effect. It is recognized that a rigid deck would not be reasonable to use for design.

After field testing, the deck model with Young's Modulus of 5,936 MPa (861 ksi) (Case 3) was selected and used for the remainder of the modeling studies because of the close comparison to the field testing data. It was also believed that this stiffness was based on better representation of the probable deck to stringer welds.

Prowers Bridge Field Equipment and Results

A Campbell Scientific CR5000 Data Logger (Campbell Scientific 2001) with remote data downloading capability collected all Prowers Bridge field data. Figure 7 shows the field installation of the strain transducer and wind sensors cables to the CR5000 Data Logger. Remote downloading capability was provided through the use of a program downloaded to the data logger, a digital cellular modem, and an antenna. Figure 8 shows the digital cellular modem and antenna used during the Prowers test.

Over five weeks of field-testing was performed to obtain a high wind event with a wind direction perpendicular to the bridge; therefore, the wind and the strain data were downloaded from the data logger remotely every day and was reviewed for the highest wind event and closest wind direction perpendicular to the Prowers Bridge. On the April 4 2005, at 7:13 pm, a wind event occurred with max wind speed of 43 mph at approximately at 290 degrees (20 degrees from "broadside") and was used for verification of the model.

Figure 7. Strain transducer and wind sensors cables installation to the data logger prior to placing into service.

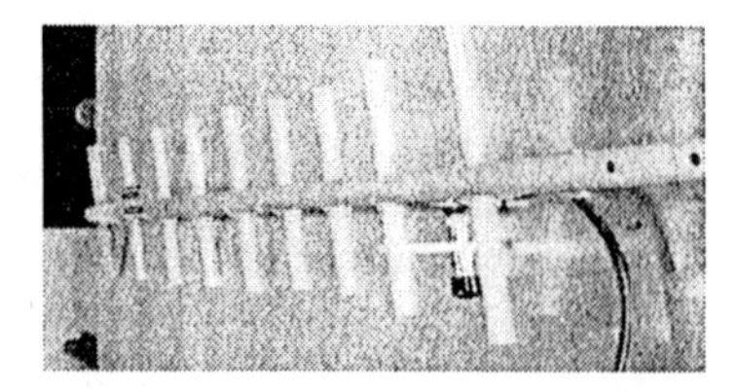

Figure 8. Redwing Cellular Digital Modem from AirLink Communication and Yagi Antenna.

Figure 9 shows one of the seven anemometers installed on the bridge. Figure 10 shows the location of the anemometers. The bridge is oriented approximately north to south, and the wind directions sensor was oriented parallel to the bridge's longitudinal axis, making the local north direction the equivalent of global north. Wind from the west would be directed broadside to the bridge.

Figure 9. Anemometer (WS1) and Wind Direction sensor (WD) installed at Prowers Bridge. They are located on the upwind side approximately at the centriod of the bridge.

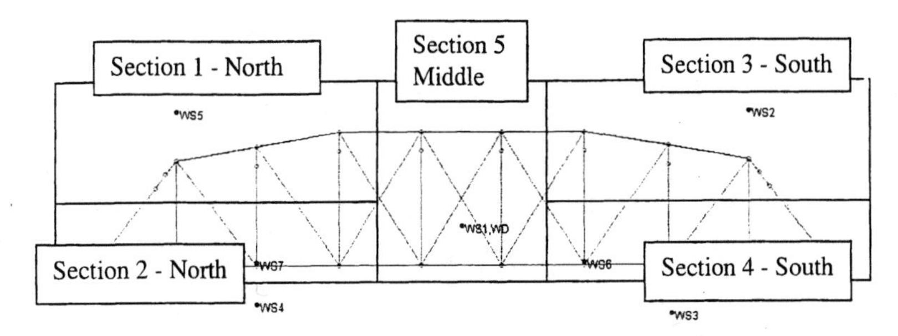

Figure 10. Diagram of Prowers Bridge illustrating the locations of the anemometers (WS1-WS7) and wind direction sensor (WD) and sections subjected to different uniformly distributed wind pressures in the verification analysis. North is to the left. WS1 was positioned directly upwind of the centroid of the wind intercept area. WS2 and WS5 were located 3 meters (approximately 10 feet) above the top of the end post members in the portals. WS3 was positioned 3 meters (approximately 10 feet) below the bridge deck, at an elevations mid-height between the bridge deck and the water surface below. WS4 was positioned 2.5 meters (approximately 8 feet) below the bridge deck, at an elevations mid-height between the bridge deck and the existing ground below. WS6 and WS7 were positioned approximately at the deck elevation. Wind pressure on Section 1 was determined from a weighted average from the velocities measured at WS5 and WS1. Wind pressures at the other sections were similarly determined except for Section 5, which was determined from WS1 only.

The wind pressure is really the variable of interest and wind pressure varies with the square of velocity. Therefore, the wind velocities from the anemometers were converted into wind pressures at each location based on the Bernoullli's theorem (Weidner and Sells 1966) incorporating drag coefficients. The field velocity data from the seven anemometers are shown in Figure 11 and field wind direction data is shown in Figure 12.

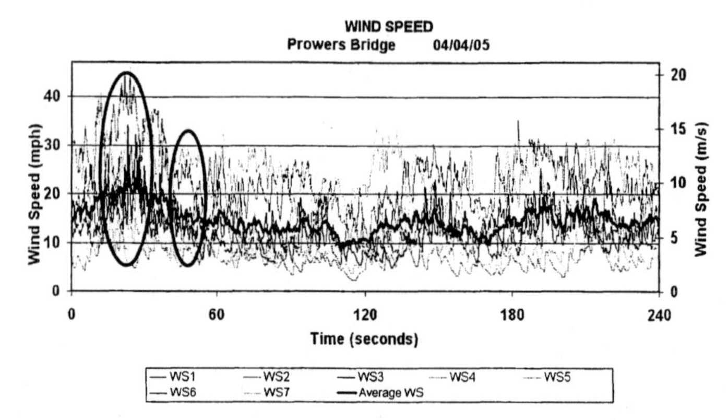

Figure 11. Prowers Bridge. Illustrating the wind speed as measured by the seven anemometers. The bold line shows the average of all seven anemometers. The circle areas were the "high" spike event and "low" spike event.

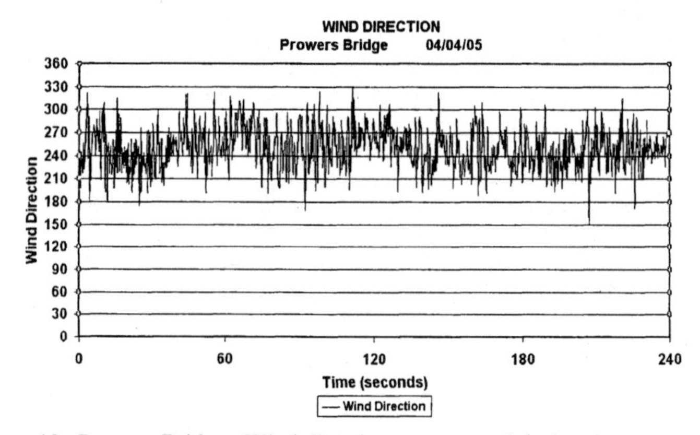

Figure 12. Prowers Bridge. Wind direction as measured during the test.

The data from the seven anemometer revealed that the instantaneous wind speeds are not the same at different parts of the bridge. Therefore, the wind pressures are not uniform along the bridge. The bridge model was divided up into five sections to account for the different pressure at different areas of the structure as shown in Figures 10 and 13. Each set of the values were averaged to remove some of the local fluctuation out of the results. Table 2 shows the measured field pressures used on the Prowers Bridge model.

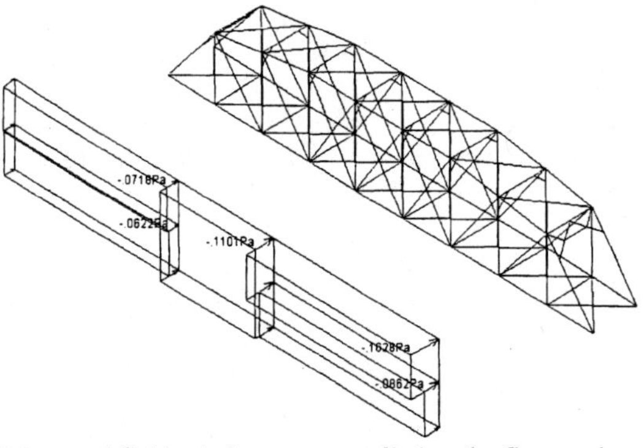

Figure 13. Measured field wind pressure applied to the five sections for analysis. North is to the left.

Table 2. Prowers Bridge. Measured Field Wind Velocities, Section Average Velocities, and Section Average Pressures.

<table>
<tr><th>Section</th><th>Anemometer</th><th>Velocity from anemometer m/s (mph)</th><th>Section - average velocity m/s (mph)</th><th>Section - average pressure Pa (psf)</th></tr>
<tr><td>Section 5-Central</td><td>WS1</td><td>10.0 (22.3)</td><td>10.0 (22.3)</td><td>110.1 (2.3)</td></tr>
<tr><td rowspan="2">Section 3 -South Upper</td><td>WS2</td><td>14.1 (31.6)</td><td rowspan="2">12.1 (27.0)</td><td rowspan="2">162.8 (3.4)</td></tr>
<tr><td>WS1</td><td>10.0 (22.3)</td></tr>
<tr><td>Not in any Section</td><td>WS3</td><td>7.1 (15.8)</td><td>Not in any section</td><td>Not in any section</td></tr>
<tr><td>Not in any Section</td><td>WS4</td><td>1.0 (2.2)</td><td>Not used in any section</td><td>Not used in any section</td></tr>
<tr><td rowspan="2">Section 1 -North Upper</td><td>WS5</td><td>5.5 (12.2)</td><td rowspan="2">7.7 (17.3)</td><td rowspan="2">71.8 (1.5)</td></tr>
<tr><td>WS1</td><td>10.0 (22.3)</td></tr>
<tr><td rowspan="2">Section 4 -South Lower</td><td>WS6</td><td>7.5 (16.7)</td><td rowspan="2">8.7 (19.5)</td><td rowspan="2">86.2 (1.8)</td></tr>
<tr><td>WS1</td><td>10.0 (22.3)</td></tr>
<tr><td rowspan="2">Section 2- North Lower</td><td>WS7</td><td>3.2 (7.2)</td><td rowspan="2">3.5 (7.9)</td><td rowspan="2">62.2 (1.3)</td></tr>
<tr><td>WS1</td><td>10.0 (22.3)</td></tr>
</table>

Figure 14 shows the location of the clamp-on modular strain transducers on the Prowers Bridge. Each bottom chord member consists of two pin-connected eyebars. At the mid-span of the bridge, a location where wind-induced strains were expected to be the maximum, a strain transducer was clamped to each side of each eyebar. Figure 15 shows the clamp-on strain transducers installed on the eyebars on the Prowers Bridge.

The clamp-on modular strain transducers were mounted at the location of the knee brace near the top of the end post and above the pin location at the bottom of the end post of the portal frame. These locations were selected due to anticipated high moments from lateral (wind) loading. Figure 16 shows one of clamp-on strain transducers on the lower end post member.

Data collected from the strain transducers required some processing. The raw data was processed as follows:

- Make a correction to account for the resistance of the wire or cable from the strain transducer to the data logger.
- Apply pre-determined transducer factors to each strain gage reading to account for the difference between measured strain at the transducer and true strain in the member.
- Filter out electrical noise by a rolling average.
- Obtain delta strains from 3-second time interval of a "high" spike event and 3-second time interval of a "low" spike event.

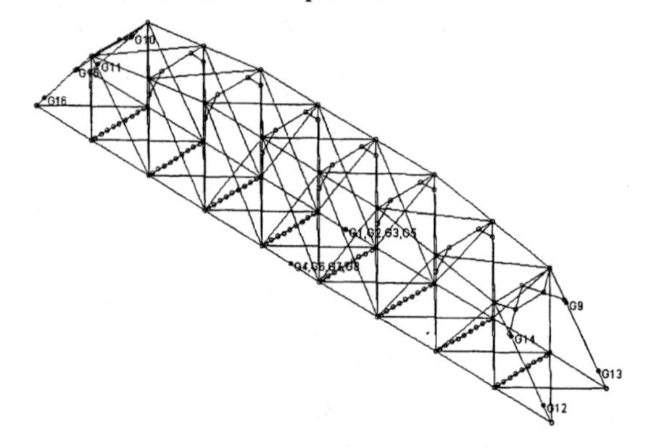

Figure 14. Diagram of Prowers Bridge. Illustrating the locations of the strain transducers. North is to the left. The wind direction was from the west, orthogonal to the bridge. Strain transducer numbers, G1-G3, G5 were clamped to the leeward bottom chord eyebars. G4, G6-G8 were clamped to the windward bottom chord eyebars. G9, G12-G14 were clamped to the end post at the south portal. G10-G11, G15-G16 were clamped to the end post at the north portal. See Figure 15 for actual field location of transducers on the bottom chord eyebars.

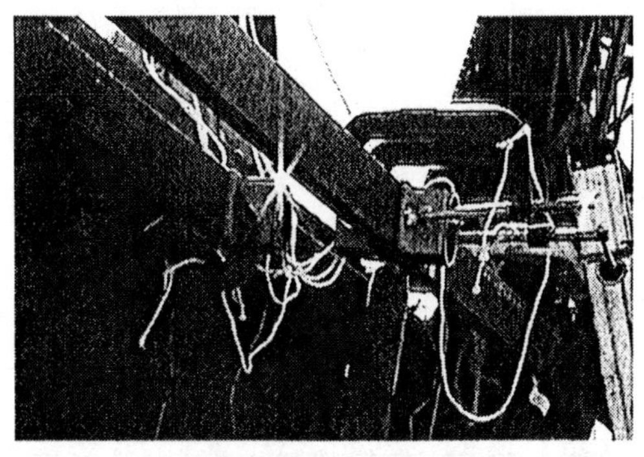

Figure 15. Strain transducers (G4 & G6 – G8) installed on the bottom-chord eyebars on the west side at Prowers Bridge. Each device has a safety string attached to the bridge as a precaution against falling off the bridge during installation.

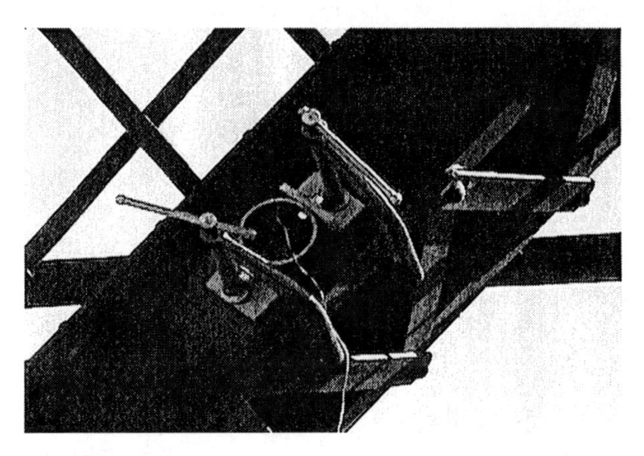

Figure 16. Strain Transducer installed on the outside face of the north end post (portal) at Prowers Bridge. It is located 1.5 meters (2 feet) above the lower pin.

A local "low" spike of a 3-second interval was averaged at the bottom of the event and a local "high" spike of a 3-second was averaged at the top of the event to obtain a strain difference (delta strain) to eliminate any residual strain from the strain gage. A 2-second rolling average was performed to filter out electrical noise from the raw data on the bottom chords as shown in Figure 17. It also was performed on raw data for the end posts members (portal) as shown in Figure 18.

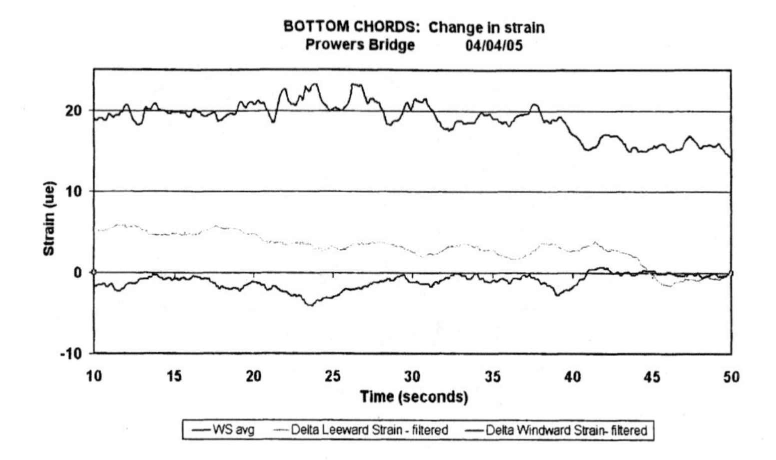

Figure 17. Prowers Bridge. Enlargement of the traces for windward (bottom line) and leeward (middle line) bottom chord eyebars measured strains. Both are traces of the filtered data. Thus, they represent the change in measured strain starting from the same point in time as the corresponding change in wind velocity. The wind velocity (top line) is shown for reference to an arbitrary scale.

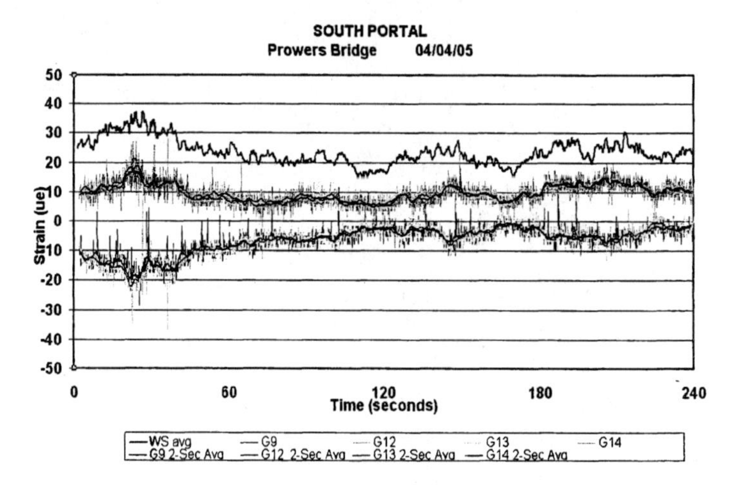

Figure 18. Prowers Bridge. Illustrating the strains measured at the south end post. Note how the strain response correlates with changes in the wind velocity, shown at the top to an arbitrary scale.

Table 3. Verification Summary. Comparison of total calculated forces to total measures forces expressed in kN (kips) and m-kN (foot-kips) per bridge member.

Bridge Member	Total Calculated Force	Total Measured Force	Correlation: % Difference
Windward Bottom Chord	-2.71 kN (-0.61 kips)	-3.11 kN (-0.70 kips)	14%
Leeward Bottom Chord	2.74 kN (0.62 kips)	3.10 kN (0.70 kips)	13%
North End Post (Upper Portal)	1.52 m-kN (1.12 ft-k)	0.67 m-kN (0.50 ft-k)	56%
North End Post (Lower Portal)	1.76 m-kN (1.30 ft-k)	0.47 m-kN (0.36 ft-k)	73%
South End Post (Upper Portal)	2.44 m-kN (1.80 ft-k)	1.33 m-kN (0.98 ft-k)	45%
South End Post (Lower Portal)	2.54 m-kN (1.87 ft-k)	1.48 m-kN (1.09 ft-k)	42%

Prowers Bridge Verification Test Discussions

There are good correlations with the measured data and calculated data for the bottom chord members. A fair correlation was observed for south end post members but not the north end post members. A possible reason for a weak correlation for the north end post may be because that the actual roller bearing in the field had shifted off the roller, which could possibly allow rotation in any direction similar to a ball joint. In the RISA-3D model, the roller is released about the pin and restrained in other two directions as if the roller was not shifted. Lower end post of the north portal was also shielded wind due an existing embankment of a newer bridge about 15 meters to the west. The wind pressure in RISA-3D model could possibly be higher than the bridge really experienced in the field due to shielding effects of the embankment on the end posts.

Conclusions

It was shown that the significant increase of dead load from the asphalt filled corrugated steel deck when compared to the original timber deck results in an increased tension in the windward bottom-chord eyebars so the bridge modeled as a skeleton conformed with the AASHTO wind design loading requirements. The Prowers Bridge was also modeled with the current deck to show the stiffening effects of deck to the design wind loads. The deck model shows the structure is laterally stiffer. The windward bottom-chord eyebars remain in tension; therefore, the members conform to the AASHTO wind loading requirements.

This work stems from earlier work completed by Carrol (2003), Herrero (2003) Rutz (2004), and Rutz and Rens (2004). For additional information about the Prowers Bridge, refer to Rutz et al (2005) and Jacobson (2006).

Acknowledgements

The University of Colorado at Denver has completed this work funded in part by Grant # MT-2210-04-NC-12 from the National Center for Preservation Technology and Training and Grant # 2004-M1-019 from the State Historical Fund of the Colorado Historical Society. Further, the cooperation of the Bent County of Colorado the bridge owner and the town of Lamar are acknowledged. The assistance Val Moser and Don Brown of Campbell Scientific, Inc. is gratefully acknowledged. The following University of Colorado at Denver students and others assisted in laboratory and field investigations throughout the project: Sam Brown, Nick Clough, Kazwan Elias, Aaron Erfman, Helen Frey, Shohreh Hamedian, Paul Jacob, and Clint Krajnik.

References

AASHTO (1997). *Guide specifications for design of pedestrian bridges*, American Association of State Highway and Transportation Officials, Washington, D.C.

Campbell Scientific (2001). *CR5000 measurement and control system operator's manual*, Campbell Scientific, Inc., Logan, Utah.

Carroll, D. (2003). *Analysis of historic pin connected through truss bridges for conversion into pedestrian use*, Dept. of Civil Engineering, Univ. of Colorado at Denver, Denver, CO.

DeLony, E (2005). "Rehabilitation of historic bridges", *Journal of Professional Issues in Engineering Education and Practice*, American Society of "Civil Engineers, Reston, VA, July, 131, 3, 178.

Herrero, T. (2003). *Development of strain transducer prototype for use in field determination of bridge truss member forces*, Dept. of Civil Engineering, Univ. of Colorado at Denver, Denver.

Jacobson, S.H. (2006) *Analysis of Prowers Bridge,* M.S thesis, University of Colorado at Denver *in press*

RISA (2002). *RISA-3D*, ver. 4.5, Risa Technologies, Foothill Ranch, CA.

Rutz, F.R. (2004). *Lateral load paths in historic truss bridges*, PhD Thesis, Civil Engineering, University of Colorado at Denver, 31-32.

Rutz, F.R., and Rens, K.L. (2004). "Alternate load paths in historic truss bridges: new approaches for preservation," *Proceedings of the 2004 Structures Congress*, Ed. by G.E. Blandford, Structural Engineering Institute of the American Society of Civil Engineers, May 22-26, Nashville, TN, ASCE, Reston VA.

Rutz, F.R., K.L. Rens, V.R. Jacobson, S Hamedian, K.E. Elias, W.B. Swigert, (2005) *Load Paths in Historic Truss Bridges*, MT-2210-04-NC-12, prepared by Dept. of Civil Engineering, University of Colorado at Denver for National Center for Preservation Technology Transfer, Natchitoches, LA, 2005.

Weidner, R.T., and Sells, R.L. (1966). *Elementary classical physics*, Allyn and Bacon, Boston, 456.

Analysis and Testing of the Historic Blue River Bridge Subjected to Wind

By Shohreh Hamedian[1], Frederick R. Rutz[2], Kevin L. Rens[3]

[1] Structural Engineer, J.R Engineering, 6020 Greenwood Plaza Blvd., Greenwood Village, CO 80111 (Master of Science graduate student University of Colorado)
[2] Senior Project Manager, PhD, PE, J.R. Harris & Company, 1776 Lincoln Street, Suite 1100, Denver, CO 80203
[3] Associate Professor, PhD, PE, Department of Civil Engineering, University of Colorado at Denver, Campus Box 113, PO Box 173364, Denver, CO 80217-3364

ABSTRACT

The overall purpose of this research is to analyze an historic truss bridge called the Blue River Bridge near Dillon, Colorado under wind load to investigate the stiffening effect of the deck. The bridge, located in the Rocky Mountains has a timber deck with relatively high stiffness in the lateral direction. The traditional method of analysis is based on a skeleton frame with no deck and alternative load paths are neglected. Analytical modeling was completed using finite element software. The American Association of State Highway Transportation Officials (AASHTO) wind load of 75 psf was applied to the models to demonstrate the stiffening effect of the deck. The deck analytical model was verified by a field test under real wind conditions. In summary, Blue River Bridge was analyzed under AASHTO wind load for two different systems, skeleton frame and skeleton with stringers and deck and again under wind pressure determined experimentally for the skeleton with stringers and deck. The results were compared for critical members. Despite existing distress in the truss and abutments, it was found that the lateral stiffness of the deck was near its theoretical maximum.

Historic and Modern Analysis

Until the middle 19th century, there were no accurate or accepted methods to calculate the force for each member in a truss bridge. Since 1847, engineers have designed truss bridges using the method of joints. This historic method is based on a "skeleton" frame analysis where alternative load paths are neglected. Today's engineers use finite element software to analyze and calculate forces in each member of a truss bridge. Computer analysis is also based on a "skeleton" frame. Therefore, traditional and modern methods follow the same fundamental basis, just with different techniques.

Bridge History

In 1895, Colorado House Bill No, 37 provided funds for the construction of a vehicular bridge over the Blue River, north of the town of Breckenridge. The contract to construct the bridge was awarded to the Kansas City Bridge Company of Kansas City, Missouri. The bridge design consisted of a Pratt through truss configuration with five approximately equal-length panels. The bridge spanned 80 feet with a nominal clear road width of 16 feet. The bridge was constructed using steel and consisted entirely of pinned connections. At a later date, this bridge was relocated to its current location near the tailrace of the Dillon Dam. This structure no longer carries vehicular traffic, but carries regular pedestrian traffic.

Bridge Geometry

The Blue River Bridge is a steel Pratt truss bridge that has an 80 feet span with five 16 feet bays as shown in Figure 1. The bridge road width is 16 feet and the truss height is also 16 feet. The bridge truss consists of double channel lattice posts, top chords, and portals. The bottom chords are double eye bars and the diagonal members are either rods or double plates. The top and bottom chord x-braces are also rods. The deck consists of longitudinal and transverse timbers on steel stringers that bear on the floor beams. The bridge bearings are buried.

Figure 1. Blue River Bridge over Blue River.

Test Set Up

The university research team utilized several instruments to obtain measurements of wind speed, wind direction, and strains from selected members during windy conditions. A total of five anemometers were used to obtain wind speed data. Wind direction was determined by a wind direction sensor with respect to the bridge orientation. Wind direction was expected to be consistent over the bridge span. A total of sixteen strain transducers were installed at different members to obtain the strain data. These members were selected because of relatively high axial forces or moments due to lateral load. For the Blue River Bridge, eight strain transducers were installed back to back on each eyebar at mid-span. Four transducers were installed on end stringers at mid-span and four on top and bottom of the south end posts. Figures 2 and 3 show the test set-up for the Blue River Bridge.

Figure 2. Blue River Bridge. Strain Transducer Installation.

Figure 3. Blue River Test set- up in progress. The deck consists of longitudinal "running boards" on transverse timbers on steel stringers. The orthogonal criss-crossing of running boards and deck timbers creates a much more continuous deck than that at Fruita Bridge. The steel stringers bear on and are mechanically attached to the steel floor beams.

Modeling and Analysis

The Blue River Bridge was first modeled as a skeleton structure as shown in Figures 4 through 8 and then modeled with a deck shown in Figures 9 through 11. The choice for boundary conditions for both configurations was pinned joints at one end and roller joints at the other. The roller joints were restrained from lateral translation but free to translate in the longitudinal direction the bridge. All internal member connections were assumed to be pinned, that is released to rotate.

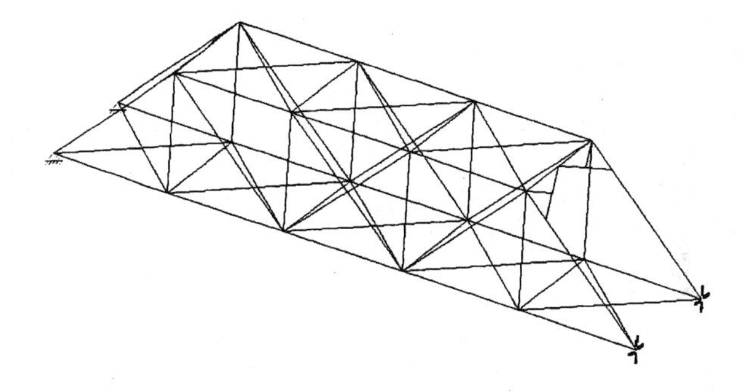

Figure 4. Blue River Bridge: Illustrating the traditional skeleton based on the primary members only.

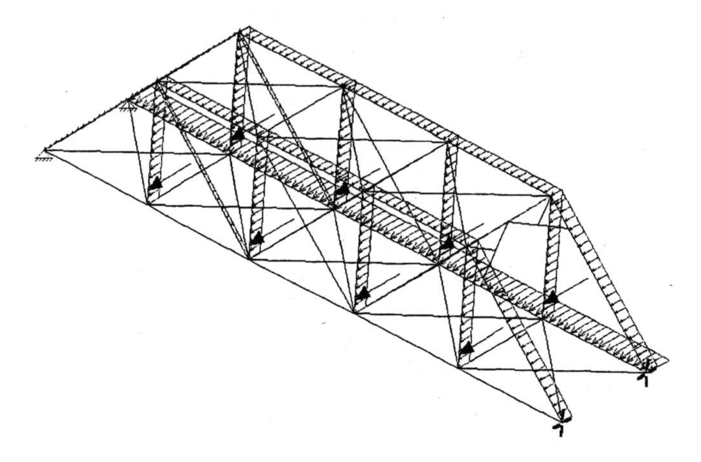

Figure 5. Blue River Bridge: Lateral wind pressure on the bridge.

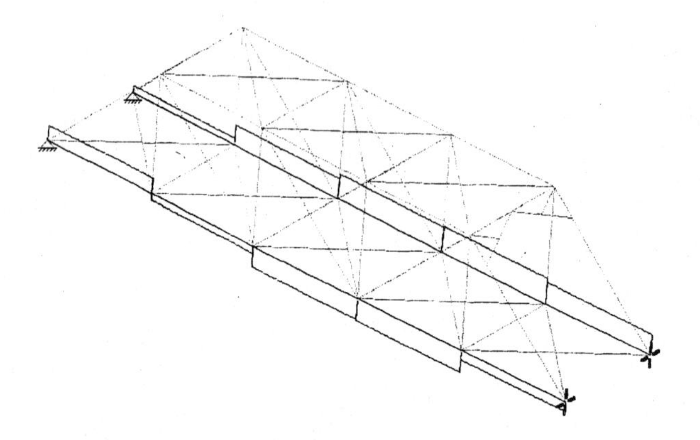

Figure 6. Blue River Bridge: Graphical representation of axial forces in the bottom chord eyebars due to wind on the skeleton structure.

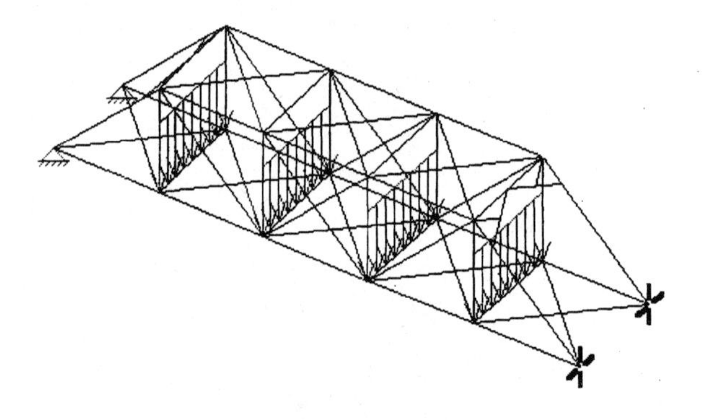

Figure7. Blue River Bridge: Representation of superimposed gravity loads.

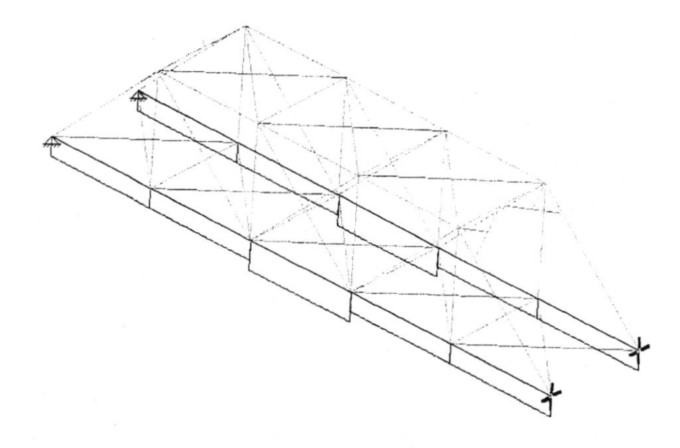

Figure 8. Blue River Bridge: Relative axial forces in the bottom chord eyebars due to gravity loads on the skeleton structure.

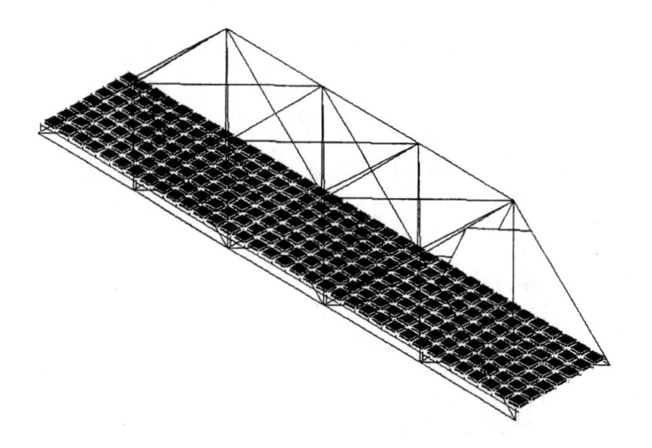

Figure 9. Blue River Bridge: Skeleton model with stringers and deck.

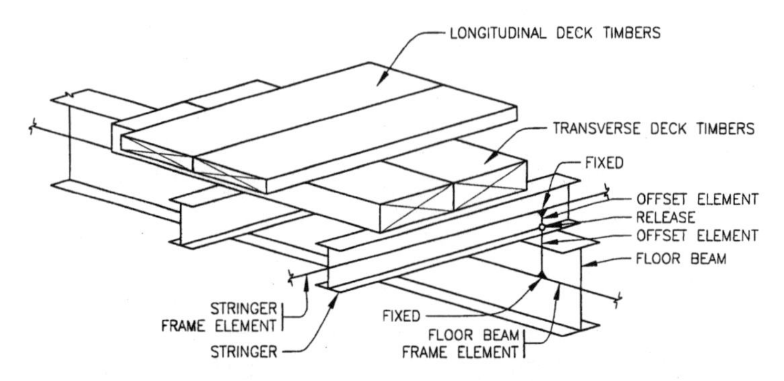

Figure 10. Blue River Bridge. Offset members and release locations.

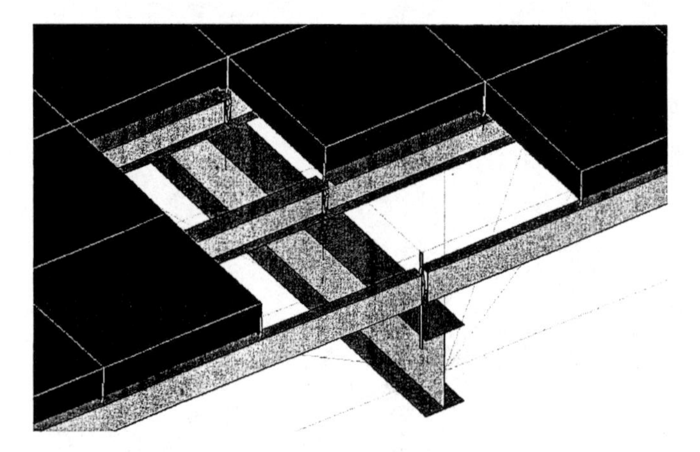

Figure 11. Blue River Bridge: Rendering of deck timber deck on steel stringers on steel floor beam. The mutually orthogonally deck timbers have been treated as a single monolithic solid.

As it is shown in Figures 9-11 the transverse deck boards and the longitudinal running boards oriented perpendicular to them were treated as a solid instead of as individual boards. This was because the two mutually perpendicular layers, well spiked together, were believed to act similar to a single solid material. As such, this timber deck had considerably stiffness in the lateral direction. This can be validated by comparing the deck model to the diaphragm model, discussed below. Table 1 shows the results for bottom chord eyebars in three different cases.

Table 1. Blue River Bridge: Summary of Maximum Axial Compressive Forces in Bottom Chord Eyebars. Forces are for windward side and are expressed are in kN (kips), followed by percent reduction in compression (or increase in tension) compared to the traditional skeleton value. (Positive = tension; negative = compression). Note the deck and diaphragm values are virtually identical, suggesting the deck, as modeled, is about as stiff as possible.

Model	Axial force due to dead load only	Axial compression due to wind load only	Net axial force due to wind plus dead load*
Case 1:			
Skeleton	57	-60	-1.4
	(12.6)	(-13.3)	(-0.3)
Case 2:			
Deck	56	-50	9.4
	(12.5)	(-11.2)	(2.1)
	1%	16%	600%
Case 3:			
Diaphragm	56	-46	13.2
	(12.5)	(-10.4)	(3.0)
	1%	22%	900%

* Note values are not necessarily identical to (D+W) because tension–only members may not be the same for the (D+W) case as for the individual D or W cases.

Comparison

As shown in Table 1 the deck and diaphragm values for dead load are identical. For wind load the results are within 6%. Since a diaphragm model is a theoretically rigid model the deck model is concluded to be near to its maximum theoretical rigidity in the lateral direction.

Eye bar Condition

As it is shown in Figure 12 the eye bars were buckled significantly under the applied loads. This condition was assumed to be a result of the abutment failure. Because the abutment wall shown in Figure 13 had failed, soil pressure was apparently causing a net compressive force down the longitudinal axis of the bridge which caused buckling in the bottom chord eye bars. Despite the eye bar buckling, the bridge was still functional for pedestrian use. The stiff deck can carry some tensile force caused by gravity and lateral loads. In addition arching can occur and compressive forces can go to the end posts. Gravity load tests compared very well with analytical results suggesting the buckled condition of the eyebars was not significant to the members respons (Hamedian, 2006)

Figure 12. Buckled eyebars in the Blue River Bridge.

Figure 13. Abutment failure in the Blue River Bridge.

Verification Results

Figures 14 through 19 and Table 2 show the location of instruments such as strain transducers, wind sensors, and general results for the Blue River Bridge.

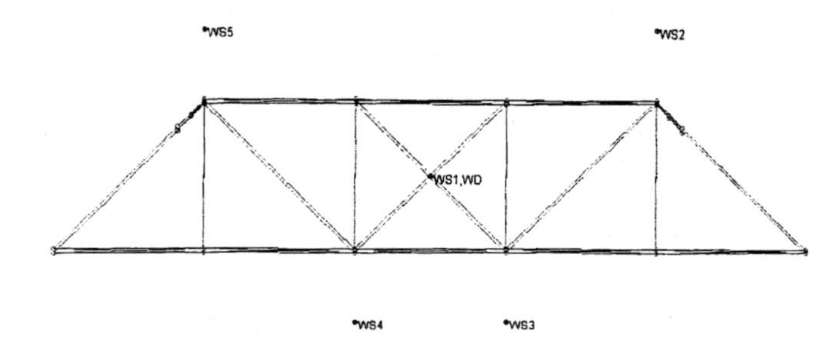

Figure 14. Diagram of Blue River Bridge, illustrating the locations of anemometers (WS1-WS5) and wind direction sensor (WD). North is to the left. WS1 was positioned at the approximate center of the wind intercept area. WS2 and WS5 were located above the end diagonal in the portals. WS3 and WS5 were located below the bridge an the elevation mid-height between the bridge deck and the water surface below.

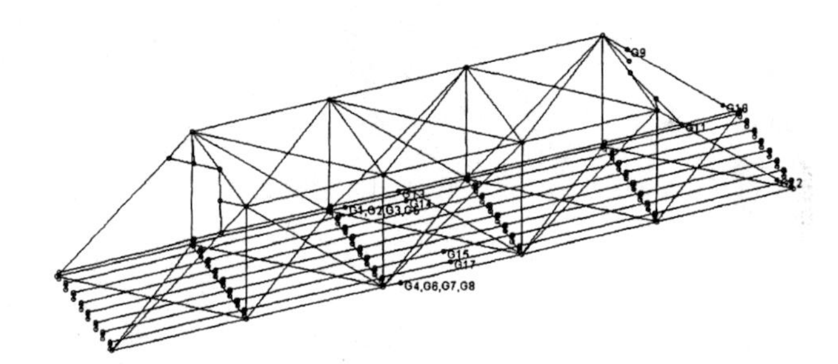

Figure 15. Diagram of Blue River Bridge, illustrating the locations of the strain transducers. North is to the left. The wind direction was from east to west, orthogonal to the bridge. Strain transducers G1, G2, G3 and G5 were clamped to the windward bottom chord eyebars. G4, G6, G7 and G8 were clamped to the leeward bottom chord eyebars. G9 and G11 were clamped to the top of the south portals. G12 and G18 were clamped to the bottom of the south portals. G13 and G14 were clamped to the east stringers in the middle bay and G15 and G17 were clamped to the west stringers.

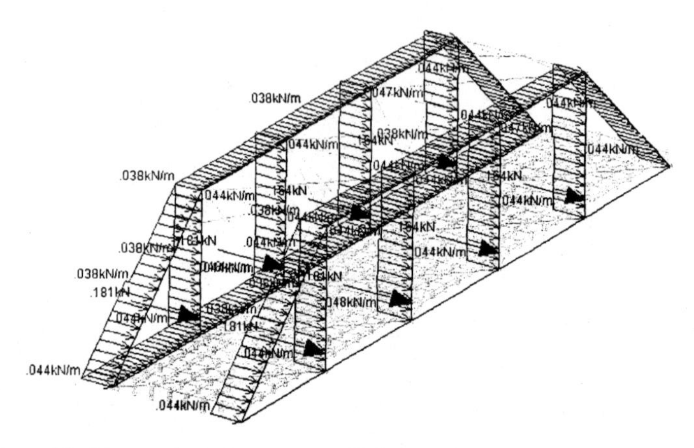

Figure 16. Measured wind pressure applied to the four quadrants for analysis. North is to the right.

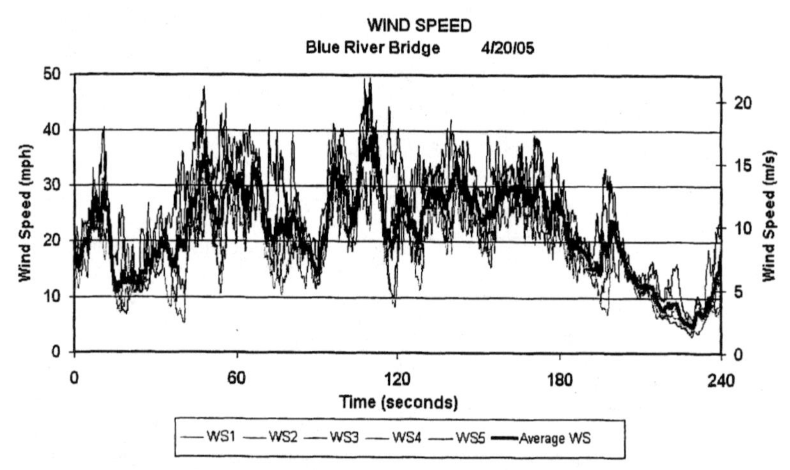

Figure 17. Blue River Bridge: Wind Speed as measured by the five anemometers. The average of all five anemometers is shown by the bold line.

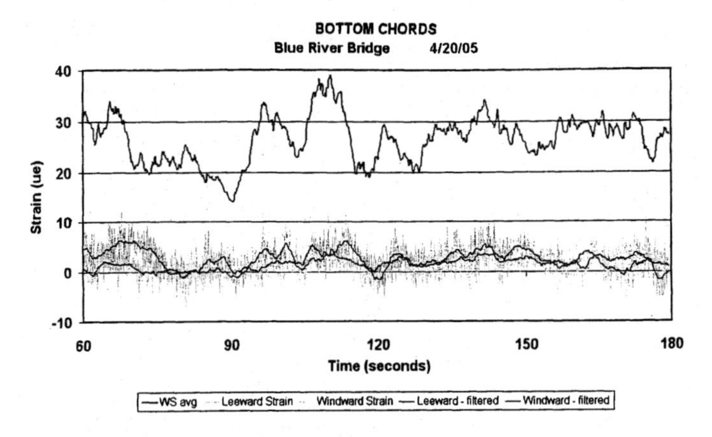

Figure 18. Blue River Bridge. Strain measurements for the windward and leeward bottom chord eyebars. The average wind speed, to an arbitrary scale, is also shown as the bold line at the top. Strain in the leeward eyebar is shown above strain in the windward eyebar. Both the raw data and a filtered line that removes much of the signal noise are shown.

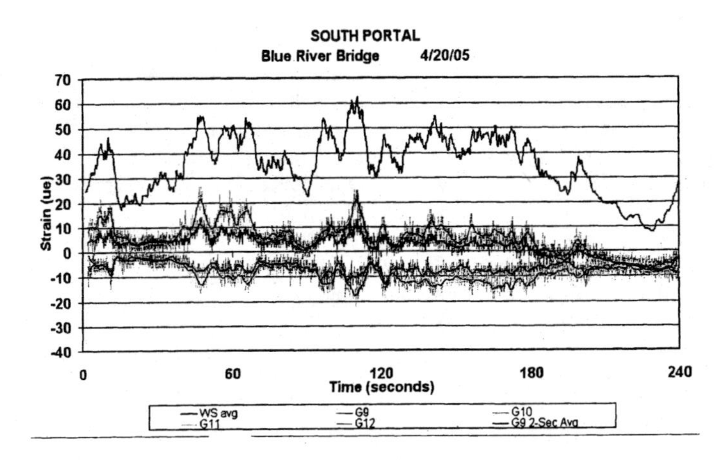

Figure 19. Blue River Bridge. Measured strains at the south end posts are shown.

Table 2. Blue River Bridge Verification Summary. Comparison of calculated forces to measured forces expressed in kN (kips) and kN-m (ft-kips). The north portal was not instrumented.

Member	Calculated Force	Measured Force	Correlation: % difference
Windward bottom chord	-0.44 kN (-0.1 kips)	-0.54 kN (-0.122 kips)	23%
Leeward bottom chord	1.44 kN (0.32 kips)	0.78 kN (0.18 kips)	45%
South portal upper	0.77 kN-m (0.56 ft-k)	0.70 kN-m (0.51 ft-k)	15%
South portal lower	1.25 kN-m (0.91 ft-k)	0.55 kN-m (0.40 ft-k)	56%

* E = 10.34 x 10^6 KPa (1.5 x 10^3 ksi) and E = 13.79 x 10^6 KPa (2.0 x 10^3 ksi) were both used in RISA-3D modeling and both gave the same results, therefore the deck stiffness had little or no effect on member forces. E = 13.79 x 10^6 KPa (2.0 x 10^3 ksi) is considered an upper bound on the stiffness of the wood in the deck, so higher stiffnesses were not considered.

Blue River Conclusion

Good correlations were achieved for the windward bottom chord suggesting that the deck model is a reasonable approximation. The actual bearings could not be observed because they were buried in soil. The bearings were presumed to be rusted to a "frozen" condition and were treated as fixed in the calculations. However, they may have permitted some small rotation, that is they may have exhibited some small degree of partial fixity, which would have altered the calculated results.

This work stems from earlier work completed by Carrol (2003), Herrero (2003) Rutz (2004), and Rutz and Rens (2004). For additional information about the Blue River Bridge, refer to Rutz et al (2005) and Hamidian (2006).

ACKNOWLEDGEMENTS

The University of Colorado at Denver has completed this work funded in part by Grant # MT-2210-04-NC-12 from the National Center for Preservation Technology and Training Grant # 2004-M1-019 from the State Historical Fund of the Colorado Historical Society. Further, the cooperation of the bridge owner, Summit County, CO, is acknowledged. In addition to Principal Project Contact Kevin L. Rens and Principal Investigator Frederick R. Rutz, the following University of Colorado at Denver students and others assisted in laboratory and field investigations throughout the project:
Mohammad Abu-Hassan, Paul Bountry, Sam Brown, Jennifer Davis, Kazwan Elias, Aaron Erfman, Helen Frey, Veronica Jacobson, Chris Kline and Peter Marxhausen.

REFERENCES

Carroll, D. (2003). *Analysis of historic pin connected through truss bridges for conversion into pedestrian use*, Dept. of Civil Engineering, Univ. of Colorado at Denver, Denver, CO.

Hamidian, S.H. (2006) *Wind Analysis of the Historic Blue River Bridge,* M.S thesis, University of Colorado at Denver

Herrero, T. (2003). *Development of strain transducer prototype for use in field determination of bridge truss member forces*, Dept. of Civil Engineering, Univ. of Colorado at Denver, Denver.

Rutz, F.R. (2004). *Lateral load paths in historic truss bridges*, PhD Thesis, Civil Engineering, University of Colorado at Denver.

Rutz, F.R., and Rens, K.L. (2004). "Alternate load paths in historic truss bridges: new approaches for preservation," *Proceedings of the 2004 Structures Congress*, Ed. by G.E. Blandford, Structural Engineering Institute of the American Society of Civil Engineers, May 22-26, Nashville, TN, ASCE, Reston VA.

Rutz, F.R., K.L. Rens, V.R. Jacobson, S Hamedian, K.E. Elias, W.B. Swigert, (2005) *Load Paths in Historic Truss Bridges*, MT-2210-04-NC-12, prepared by Dept. of Civil Engineering, University of Colorado at Denver for National Center for Preservation Technology Transfer, Natchitoches, LA, 2005.

Wind Load Analysis of a Truss Bridge at Rifle Colorado

William B. Swigert[1], PE SE; Frederick R. Rutz[2], PhD, PE; Kevin L. Rens[3], Phd, PE

Abstract: Current AASHTO requirements for pedestrian bridges may prove some historic truss bridges to be under-strength when applying wind loads to simple skeleton models. The Rifle Bridge over the Colorado River at Rifle, Colorado, is a historic steel truss structure that was one of five in a study to analyze actual wind loads on existing structures. This paper discusses the effects of including stiffening elements in 3D models by comparing actual and calculated wind loads. During the six week wind study period, maximum wind loads measured were in excess of 60 mph, which resulted in easily measured strains. Analytical models include the metal deck with asphalt as a stiffening element, which is treated as plate elements with a modulus representative of the composite section. Recommendations for modeling the deck are provided.

1.1 Introduction/Goals

The focus of research at the University of Colorado at Denver is on the identification of alternate load paths in historic truss bridges. Modern methods are being employed to aid in the preservation of this technological heritage. There is strong evidence that alternative load paths do exist – they have just been overlooked for the past century or so. While the overall purpose of the project is to aid in preservation efforts for historic iron and steel truss bridges, the specific goal of this project is to demonstrate a new methodology to account for increased strength from non-traditional (but real) load paths. While some bridges may possess several different types stiffening elements, there are generally one or two that not only have the greatest effect, but are found on most all historic bridges.

[1] Senior Structural Engineer, Schmueser Gordon Meyer, Inc., 118 W. 6th St, Glenwood Springs, Colorado 81601 (Graduate student, UCD)
[2] Senior Project Manager, J.R. Harris & Company, 1776 Lincoln St., Suite 1100, Denver, Colorado 80203-1080
[3] Associate Professor, Department of Civil Engineering, University of Colorado at Denver, Denver, Colorado 80217-3364

1.2 Truss Bridge at Rifle Colorado

It is believed that to be of the greatest practical value in aiding preservation efforts that real bridges be examined. This paper focuses on the Rifle bridge shown in Figures 1 and 2.

The structure is a 240 ft span Pennsylvania truss, built in 1909 and in service until the late 1960's. The Rifle bridge is unique in that it has the longest span of those studied, which results in a span to width ratio of 12. Additionally, it was retrofitted with a deck covering heavier than the original construction. This had the effect of maintaining tensile forces in the slender bottom chords that would have otherwise reversed stress direction resulting in a compressive stress condition for higher wind load events.

Figure 1. Rifle Bridge over the Colorado River at Rifle, Colorado. This 73 meter (240-foot) span Pennsylvania truss comprises the longer of two different spans at that location. It has steel floor beams with steel stringers, covered by a corrugated metal deck with asphalt pavement. It has steel eyebar bottom chords and diagonals and steel rod counterbracing. The railing is a steel lattice with double angle top and bottom rails. It has been abandoned since the late 1960's, when a replacement bridge was constructed upstream.

Figure 2. Rifle Bridge deck. The deck of (weathered) asphalt pavement on a corrugated metal bridge deck on of steel stringers was modeled.

1.3 Loads

Superimposed dead load and superimposed live load are still computed manually, the same way as the 19th century designer. Self-weight may be computed manually, or may be determined by software. The *AASHTO Guide Specifications for the Design of Pedestrian Bridges* (AASHTO, 1997) prescribes the live load value. It may vary between 3.11 kPa to 4.07 kPa (65 psf to 85 psf), depending on the area of the walkway. The late 19th century and early 20th century designer selected design wind loads on a case-by-case basis, which varied typically from 1.44 kPa to 2.39 kPa (30 psf to 50 psf) applied to the projected area of the components (Smith 1881, Waddell 1898, and Cooper 1905). Today's AASHTO *Guide Specifications for the Design of Pedestrian Bridges* mandates 3.59 kPa (75 psf) applied to the same area.

1.4 Customary Analysis

Today's typical analysis for truss bridges used for pedestrian crossings would be based on:

- AASHTO wind load determined from a pressure of 3.59 kPa (75 psf).
- "Pin" boundary conditions for both bearings at one end and "roller" boundary conditions for both bearings at the other end.
- Internal member-to-member connections treated as pinned
- Probably a 3D skeleton analysis, although some engineers still use 2D analysis of the vertical trusses and for the top and bottom horizontal trusses and combine the results. For a 3-D model, a true "pin" boundary condition is a joint that is restrained from translation in all three degrees of freedom (DOF), but the three rotational DOF's are released. The "pin" support acts like a ball joint. A true "roller" boundary condition is the same as a "pin" with an additional release for translation in the bridge longitudinal direction.

It may be noted that the term "pin", meaning "free to rotate" as used in modern structural analysis, derives from 19th century analyses of trusses that had true physical pins, as does the Rifle bridge.

2.1 Bridge Modeling

The Rifle Bridge was modeled as shown in Figures 3 through 8. Three models were constructed: Skeleton, Deck, and Diaphragm. The skeleton model was constructed using traditional truss elements composed of chords and web members longitudinally, and floor beams together with x-bracing and lateral trusses in the transverse direction.

The deck model included the stringers acting along with the asphalt topped corrugated metal deck to act as a more rigid floor system.

The diaphragm model used a rigid plate member at the floor elevation. The diaphragm model was used as an upper-bound for stiffness that is thought could be achieved at the floor level.

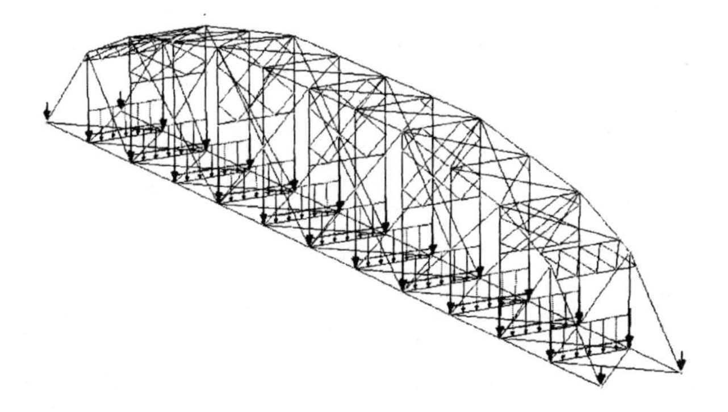

Figure 3. 3D model of bridge, illustrating the traditional skeleton based on the steel members only, using the software RAM Advanse. Representation of superimposed gravity loads (D + L) is shown.

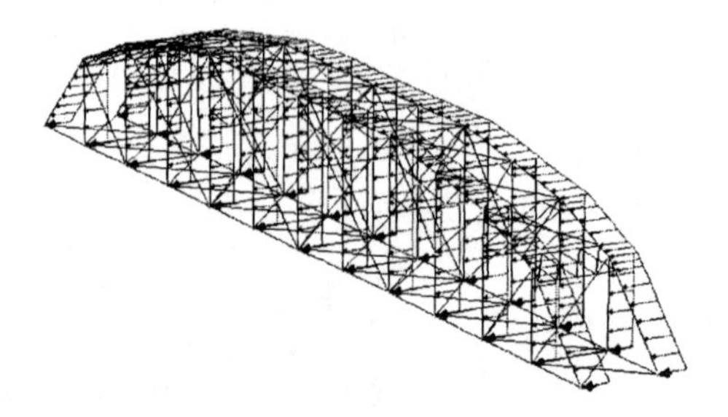

Figure 4. Representation of transverse wind pressure on the bridge. The wind load event studied was skewed, and therefore included a longitudinal component (not shown).

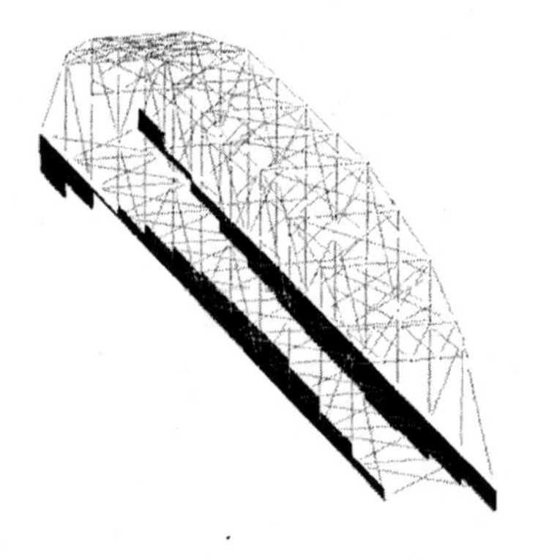

Figure 5. Graphical representation of axial forces in the bottom chord eyebars due to wind for the skeleton structure.

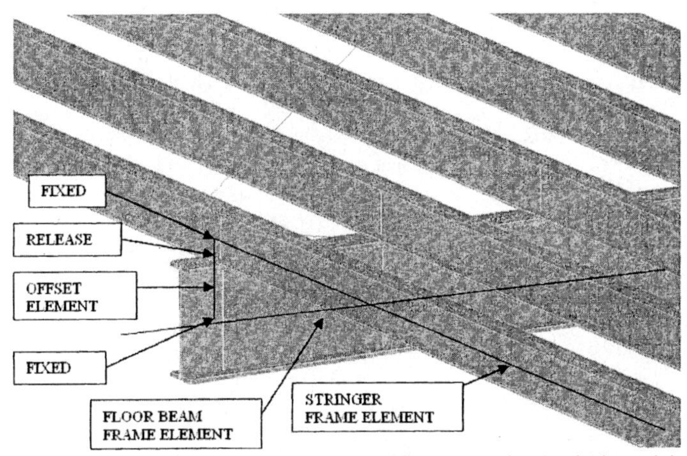

Figure 6. Rendering of stringers on a steel floor beam for the deck model. The rendering was produced by RAM Advanse 3D. The dummy offset elements used to connect the floor beam and stringer have a rotational release at the beam-to-stringer interface.

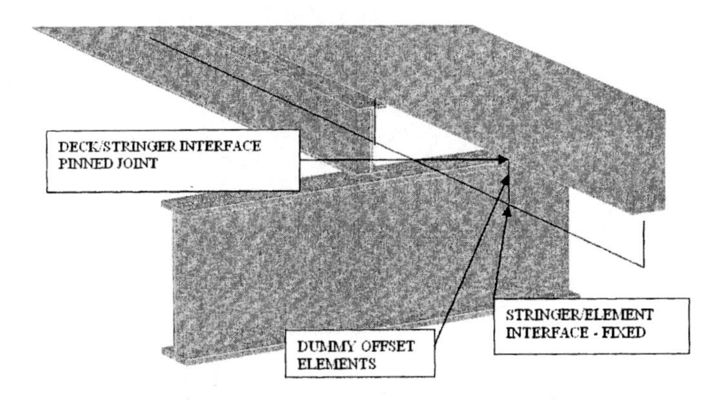

Figure 7. Rendering of the 1" asphalt paving over 2.25" steel deck planks on steel floor beam. For clarity, near stringers are not shown. Offset elements, from the beam centerline to the beam/stringer intersection to the stringer centerline to the deck/stringer interface were used.

The modeled representation of the deck shown in Figure 7 represents the actual deck shown in Figure 2. The 2.25" x 0.125" corrugated metal bridge deck is topped with an average of 1" of asphalt pavement, all of which was modeled as approximately square plate elements. This analysis included the asphalt as a contributing factor to the stiffness. This was accomplished by calculating a composite stiffness by using a transformed modulus composed of both steel deck and asphalt

This study analyzed several cases, of which the assumptions are believed to bound the results presented in 3.1 Model Verification. The cases are as follows:

Case 1 – Steel Plate Deck. This model uses the steel deck as a flat steel plate with a thickness of 0.125", and E = 29,000x10^3 psi.

Case 2A, 2B, 2C – Steel Deck and Asphalt. These models use a transformed modulus composed of the 2.25" x 0.125" corrugated steel deck and 1" asphalt topping. The stiffness of the steel deck was calculated by a finite element analysis, accounting for the warping effect of the corrugations. The asphalt modulus was determined from Park, S.W. and Lytton, R.L. (2004), and Huang, Y.H., (2004). A composite value of E = 6,231 x10^3 psi is based on studies evaluating pavements subjected to stress; a value of E = 4,412 x10^3 psi and E = 1,212 x10^3 psi are recommended for typical pavements at 70°F and 100°F respectively.

Case 3 – SDI Steel Deck. This model used a modulus of E = 2,309 x10^3 psi, which is calculated using Steel Deck Institute guidelines for stiffness of corrugated metal deck. This stiffness takes into account deflections due to shear, flexure, and warping of a light-guage corrugated metal deck section. It should be noted that these values are developed assuming low span to depth ratios, and that shear deflection is the major contributor.

Case 4 – No Deck. This model used a modulus of E = 1 psi, ie. virtually zero, to arrive at a lower bound, assuming the deck adds no stiffness to the model.

Table1. Summary of Maximum Axial Compressive Forces in Bottom Chord Eyebars. Forces are for windward side due to AASHTO loads, and are expressed in kN (kips). (Positive = tension; negative = compression).

Model	Axial force due to dead load only	Axial force on windward side due to wind load only	Net axial force due to wind plus dead load*
Skeleton (Figure 3)	572 (129)	-565 (-127)	23 (5.2)
Deck (Figures 6 & 7)	551 (124)	-342 (-77)	203 (46)
Diaphragm (Not shown)	541 (122)	-151 (-34)	388 (87)

* Note values are not necessarily identical to (D-W) because tension–only members may not be the same for the (D+W) case as for the individual D or W cases. The Deck model used Case 2, with $E = 2{,}392 \times 10^3$ psi.

3.1 Model Verification

Figures 10 through 14 show the location of the instruments and instrument recordings. Tables 2 and 3 provide a comparison of assumed and measured results for the Rifle Bridge.

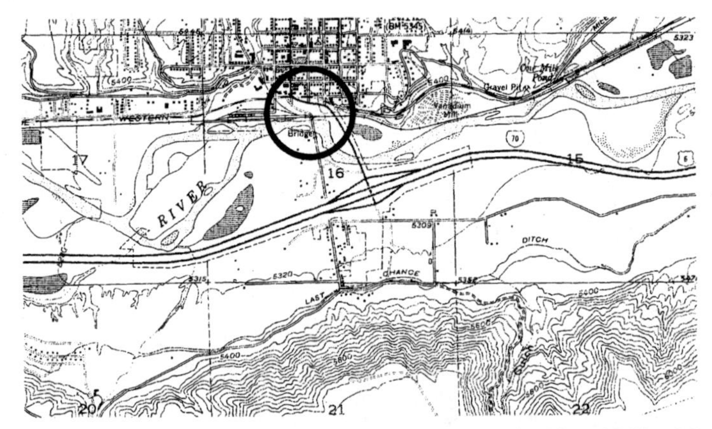

Figure 8. Location of Rifle Bridge. It is located between the City of Rifle, CO on the north, and Interstate 70 on the south (USGS 1982). The winds are predominately from the west, and the Colorado River Valley in this area is very open and flat both east and west.

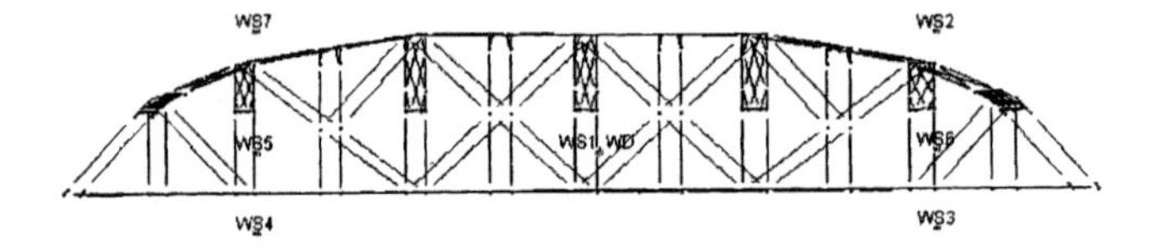

Figure 9. Diagram of Rifle Bridge, illustrating the locations of anemometers and wind direction sensor. WS1, WS5, and WS6 were positioned directly upwind of the centroid of the wind intercept area. WS2 and WS7 were located above the second diagonal frame, near the portals. WS3 and WS4 were positioned 2.5 meters below the bridge deck, at an elevation mid-height between the bridge deck and the water surface below.

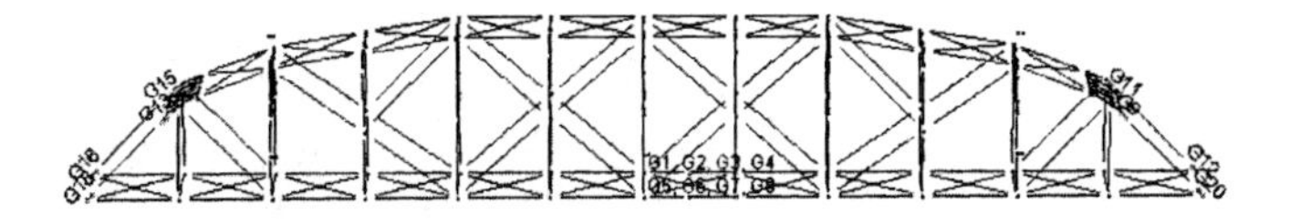

Figure 10. Diagram illustrating the locations of the strain transducers. The wind direction was generally from the west, orthogonal to the bridge. Strain transducer numbers G1-G4 were clamped to the leeward bottom chord eyebars. G5-G8 were clamped to the windward bottom chord eyebars. G9, G20, G11, and G12 were clamped to the end posts at the south portal. G13, G18, G15, and G16 were clamped to the end posts at the north portal.

Transducer measurements indicated each region was subjected to a different wind pressure. Pressures used in analysis were averaged from adjacent regions. Wind velocities varied from 1.5 m/s to 14.9 m/s at the different anemometer locations; although it is the difference in the maximum and minimum velocity at a location for the period of time studied that was used to establish pressure on the model.

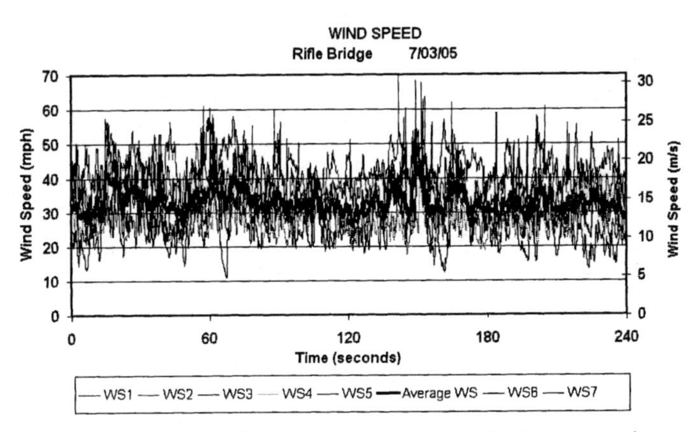

Figure 11. Wind speed data from seven anemometers. The heavy trace is the average of all seven, and will be used as a reference in the subsequent graph.

Figure 12 portrays the effects of the wind event used for the Verification study. Note at Time 150 seconds the spike in the wind velocity and the subsequent spike in the measured strains at the same time. The strains were applied to the properties of the bottom chord eyebars to develop measured forces. Tables 2 and 3 show a comparison of calculated and measured forces for the Rifle Bridge.

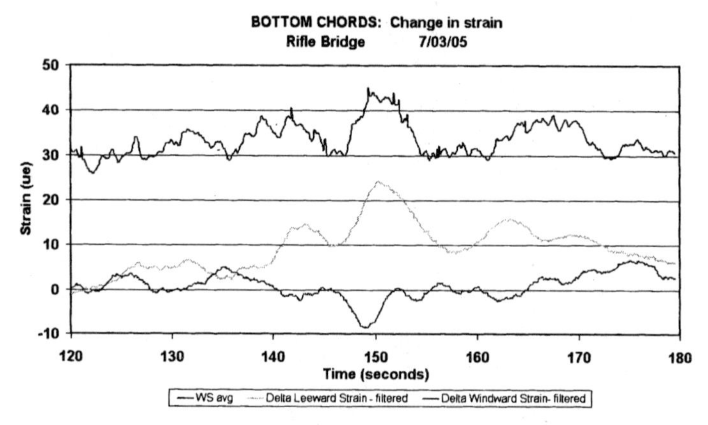

Figure 12. Enlargement of the traces for windward and leeward bottom chord eyebar measured strains. Both are baseline traces of the filtered data. Thus, they represent the change in measured strain starting from the same point in time as the corresponding change in wind velocity. The wind velocity is shown for reference at the top. The spike in the strain at 150 seconds correlates to the spike in the wind velocity used for this study.

Table 2. Rifle Bridge Verification Summary.

Comparison of calculated forces to measured forces in the windward bottom chord expressed in kN (kips).

Deck Model	Calculated Force	Measured Force	Correlation: % difference
Case 1 E = 29,000 x10³ psi Steel Plate	-6.14 (-1.38)	-8.41 (-1.89)	37%
Case 2A E = 6,231 x10³ psi Steel Deck & Asphalt	-7.57 (-1.69)	-8.41 (-1.89)	12%
Case 2B E = 4,412 x10³ psi Steel Deck & Asphalt	-7.83 (-1.76)	-8.41 (-1.89)	7%
Case 2C E = 1,212 x10³ psi Steel Deck & Asphalt	-8.94 (-2.01)	-8.41 (-1.89)	6%
Case 3 E = 2,309 x10³ psi SDI Steel Deck	-8.36 (-1.88)	-8.41 (-1.89)	1%
Case 4 E = 1 psi No Deck	-11.61 (-2.61)	-8.41 (-1.89)	38%

Table 3. Rifle Bridge Verification Summary.

Comparison of calculated forces to measured forces in the leeward bottom chord expressed in kN (kips).

Deck Model	Calculated Force	Measured Force	Correlation: % difference
Case 1 E = 29,000 x10³ psi Steel Plate	7.78 (1.75)	16.81 (3.78)	54%
Case 2A E = 6,231 x10³ psi Steel Deck & Asphalt	9.07 (2.04)	16.81 (3.78)	46%
Case 2B E = 4,412 x10³ psi Steel Deck & Asphalt	9.30 (2.09)	16.81 (3.78)	45%
Case 2C E = 1,212 x10³ psi Steel Deck & Asphalt	10.27 (2.31)	16.81 (3.78)	39%
Case 3 E = 2,309 x10³ psi SDI Steel Deck	9.79 (2.20)	16.81 (3.78)	42%
Case 4 E = 1 psi No Deck	12.50 (2.81)	16.81 (3.78)	26%

3.2 Findings

While the windward bottom chord measured and calculated forces correlated well within the range of deck models used, forces did not correlate well for the leeward bottom chord. Measured forces in the leeward bottom chord eyebar differed from calculated forces by up to 54%, suggesting a non-symmetry in the structure – possibly due to an outboard sidewalk on the east (leeward) side that was not accounted for in the model stiffness. Additionally, a 1" square x-brace in the floor system near the center of the bridge was noted to be disconnected and hanging into the river below. The Skeleton model assumed it to be connected; the verification deck model had it removed. In viewing the graphic deformation results available in the RAM Advanse software, there is an obvious discontinuity at this location. That coupled with the fact that the absolute value of the windward and leeward strain guage readings were significantly different would lead one to believe that this anomaly could be due to either or both of these effects.

Verification of the boundary conditions at the bearings could not be field verified within the scope of this study. Boundary conditions at the north end were modeled as pinned, and at the south end as a roller as discussed in Section 1.4. The roller provides an additional translation release in the longitudinal direction.

The important point to note is that within a range of magnitude of $E = 6{,}231 \times 10^3$ psi and $E = 1{,}212 \times 10^3$ psi, that the difference in calculated bottom chord force is 1.37 kN (0.31 kips), a difference of 16% of the total force. This would suggest that any reasonable assumption of deck stiffness, whether it is based on the steel deck alone, or acting compositely with an asphalt topping, will not have major effects on the final force. This analysis indicates the forces in the bottom chord are not sensitive to changes in deck stiffness within the ranges discussed here, but are sensitive to deck vs. no deck (Case 4).

3.3 Conclusion

Results support the inclusion of the bridge deck in modeling. This analysis would indicate the deck contributes 40% to the overall stiffness of the system, when comparing a reasonable deck model to that of no deck at all. Inclusion of the asphalt topping, while arguably helps, is not an important contributor to stiffness. Skeleton models alone do not account for the stiffnesses measured in actual models. The type of deck assumed, connections to supporting members, and connections between deck members all play a part in the stiffness of the model to transfer forces to supports, and are left to the designer to estimate; however provided the deck is included, the model is not overly sensitive to the range of deck stiffness values that are typically assumed.

ACKNOWLEDGEMENTS

The University of Colorado at Denver has completed this work funded in part by Grant # MT-2210-04-NC-12 from the National Center for Preservation Technology and Training and Grant # 2004-M1-019 from the State Historical Fund of the Colorado Historical Society. Further, the cooperation of the bridge owner is acknowledged: Garfield County, CO.

The following University of Colorado at Denver students and others assisted in laboratory and field investigations throughout the project: Paul Bountry, Sam Brown, Nick Clough, Helen Frey, Paul Jacob, Chris Kline, and Andy Pultorak.

REFERENCES

Park, S.W. and Lytton, R.L. (2004). *Effect of Stress-Dependent Modulus and Poisson's Ratio on Structural Responses in Thin Asphalt Pavements,* Journal of Transportation Engineering, Vol. 130, No. 3, May 1, 2004

Huang, Y.H., (2004). *Pavement Analysis and Design, Second Edition,* Pearson Education, Inc., Upper Saddle River, NJ 07458

Vulcraft Steel Floor & Roof Deck 1993, A Divison of Nucor Corporation, PO Box 59, Norfolk, Nebraska 68702

ICBO Report No. 2078, November 1981, *Verco Steel Decks, PO Box 14667, Phoenix, Arizona, 85603*

Luttrell, L.D., (1983). *Diaphragm Design Lecture and Workshop Notes,* Denver, Colorado, April 26, 1983. West Virginia University

United States Geological Survey (USGS) (1982)

Analysis & Verification Testing of the Historic San Miguel Bridge

By Kazwan M. Elias, E.I,.[1], Fred Rutz,
Ph.D., P.E.[2] and Kevin L. Rens, Ph.D., P.E .[3]

[1] Structural Project Engineer, Area Design and Engineering Corporation, 810 Quail Street, Unit D, Lakewood, Colorado 80215 (Graduate student UCD); PH (720) 987-7068; FAX (303) 237-4258; email: k1elias@hotmail.com
[2] Senior Project Manager, J.R. Harris & Company Structural Engineers, 1776 Lincoln Street, Suite 1100, Denver, Colorado 80203-1080; PH (303) 860-9021; FAX (303) 860-9537; email: fred.rutz@jrharrisandco.com
[3]Associate Professor, Department of Civil Engineering, University of Colorado at Denver, Campus Box 113, Denver, Colorado 80217-3364; PH (303) 517-8527; FAX (303) 556-2368; email: kevin@rens-engineering.com

Abstract

The Civil Engineering Department at the University of Colorado at Denver has been involved in a bridge research project funded by the National Center for Preservation Technology Transfer and by the State Historical Fund of the Colorado Historical Society. The subject topic of this paper is the San Miguel Bridge, located in Montrose County. The lateral stiffness of the existing bridge was compared to a finite element analysis. The analysis accounted for the extra stiffing offered by the deck. The results are presented and issues with the data are illustrated and discussed.

Rehabilitation for Preservation

Rehabilitation for pedestrian use is a practical and popular way to preserve these historic structures. But the American Association of State Highway and Transportation Officials (AASHTO) *Guide Specifications for the Design of Pedestrian Bridges* (AASHTO 1997) throws an obstacle in the way. It mandates modern wind load design criteria. Structural engineers attempting to rehabilitate historic bridges from former highway to modern pedestrian use often discover that the old structures lack the strength to resist the AASHTO wind load criteria. This can contribute to either a "heavy-handed" design approach, which is both expensive and detrimental to the historic character to be preserved in the first place, or condemnation of the bridge.

A dichotomy presents itself: On the one hand, historic truss bridges often do not possess "code-compliant" lateral resistance for current wind load requirements. Yet in case after case observations reveal *no* physical evidence to suggest that wind has caused damage or distress, even after a century of exposure. At this age, bridges have indeed weathered many severe windstorms.

Summary of San Miguel Bridge

It was believed that to be of the greatest practical value in aiding preservation efforts that real bridges be examined. This paper focuses on the San Miguel Bridge, shown in Figure 1.

Figure 1. San Miguel Bridge over San Miguel River near Uravan, Colorado.

The 43-meter (142–foot) Pratt truss was built by the Phoenix Bridge Company in 1886 and used wrought iron as the primary structural material. The bridge was part of a five-span Fifth Street Bridge over the Grand (now Colorado) River at Grand Junction, Colorado, shown in Figure 2. One span was relocated to the San Miguel River location in the 1930s. It has a roadway of gravel on an unusual system of semi-circular lengths of corrugated metal pipe set between steel stringers. It has wrought iron eyebar bottom chords and diagonals and wrought iron rod counterbracing. A steel vehicular rail has replaced the original railing. It served the mining industry in western Colorado until the 1980s. Abandoned since 1990, it remains the oldest bridge originally built in Colorado.

Figure 2. Fifth Street Bridge over the Colorado River at Grand Junction, Colorado in 1886 (courtesy of the museum of western Colorado, Lloyd File research library).

The deck shown in Figure 3, believed to have been installed in 1964, consists of semi-circular corrugated metal pipe segments which bear on the bottom flanges of steel stringers. The metal pipe has been replaced with timber blocks in a few places where the corrugated metal pipe has rusted away. The semi-circular corrugated metal pipe segments are topped with a gravel roadbase.

This deck is much heavier than its original timber deck. The gravel roadbase produces a heavy dead load of approximately 3.54 kPa (74 psf), much higher than the original timber deck, and estimated at approximately 0.62 kPa (13 psf).

Figure 3. San Miguel Bridge deck as it appears today near Uravan, Colorado.

The pin connected skeleton frames, shown in Figures 4 through 6, were analyzed under AASHTO loads for pedestrian bridges. This analysis was compared to the deck model, shown in Figure 7, subject to the same loads. As-built dimensions and section properties from San Miguel Bridge were used. Because it was desired to investigate the problem using software tools that were readily available to practicing engineers, RISA-3D software (RISA 2002) was used for both analyses. RISA-3D is exemplary of readily available software that includes both frame elements and plate/shell elements.

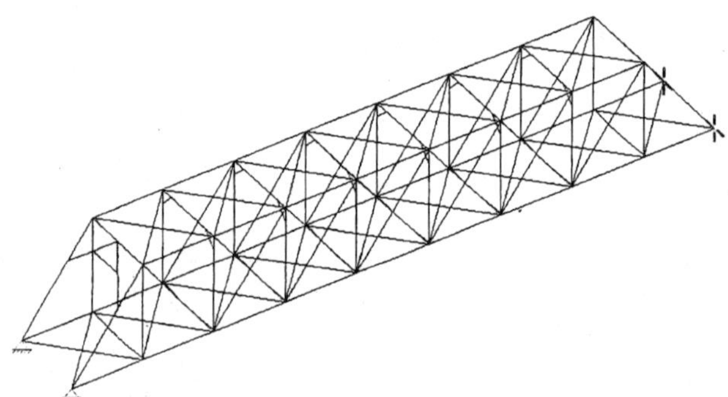

Figure 4. Illustration of the traditional of skeleton structure based on steel members only.

The San Miguel Bridge boundary conditions were pinned at one end and rollers at the other end as indicated in Figure 5. The rollers are restrained from translation in the lateral direction.

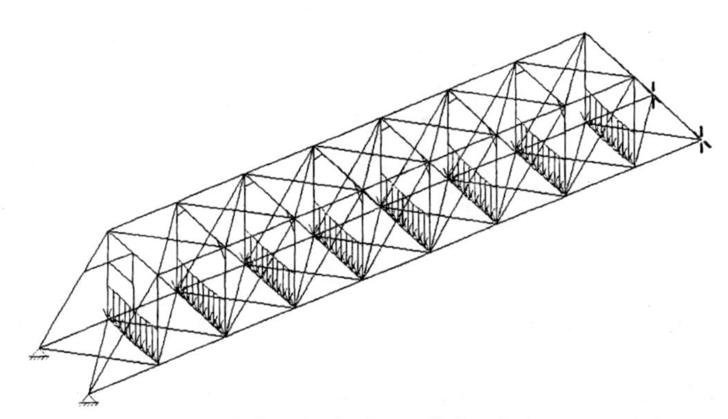

Figure 5. Gravity loads applied to skeleton structure.

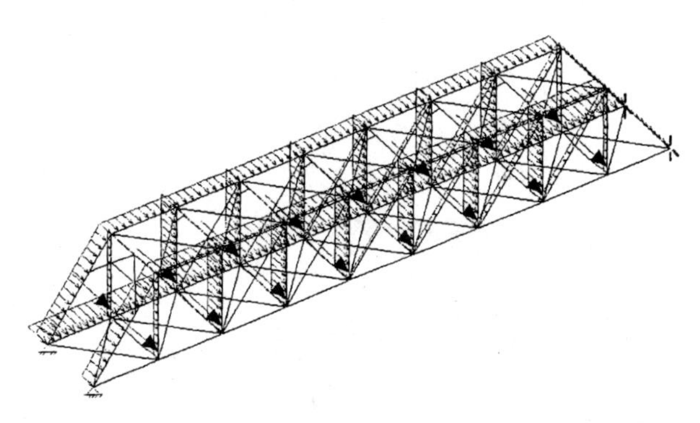

Figure 6. Wind load applied to skeleton structure. The wind load shown is based on the AASHTO criteria of 3.59 kPa (75psf).

The interior stringers shown in Figures 7-9, are 12 inch deep wide flange beams while the outer stringer is a 15 inch deep channel, all of which is intended to model the floor system that was installed in 1964.

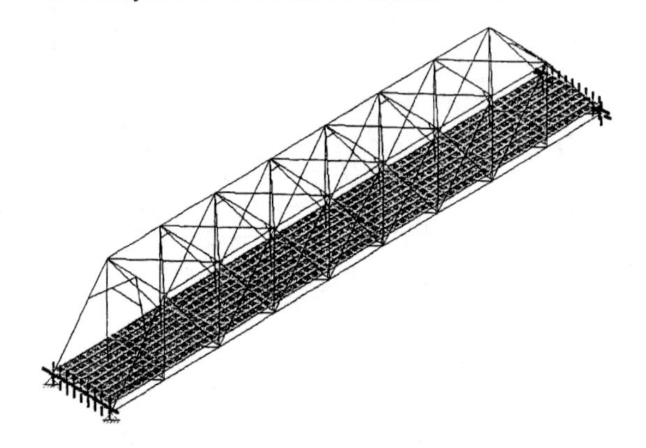

Figure 7. San Miguel Bridge: Skeleton model with stringers and deck.

The deck was modeled using RISA 3D and interconnected plate elements were used to represent the gravel roadbase. Offset elements, shown in Figure 8, were modeled such that the deck elements were at the same elevation as the stringer top flanges, but otherwise similar to Figure 8. This was done for modeling simplicity.

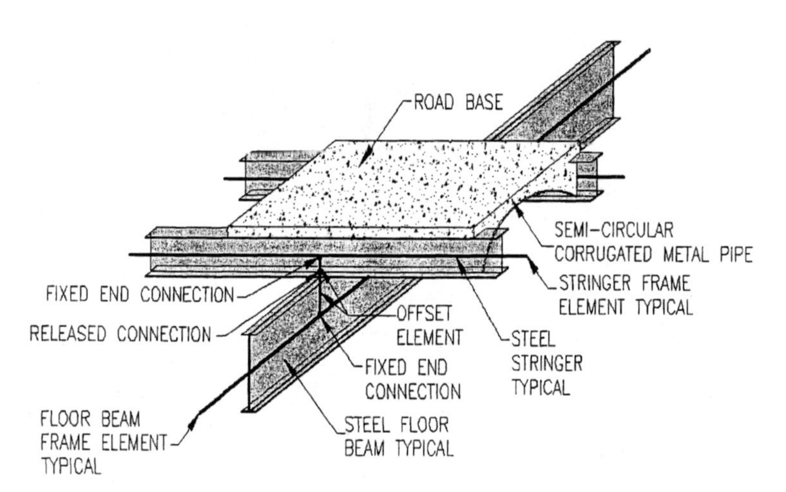

Figure 8. Offset members and release locations.

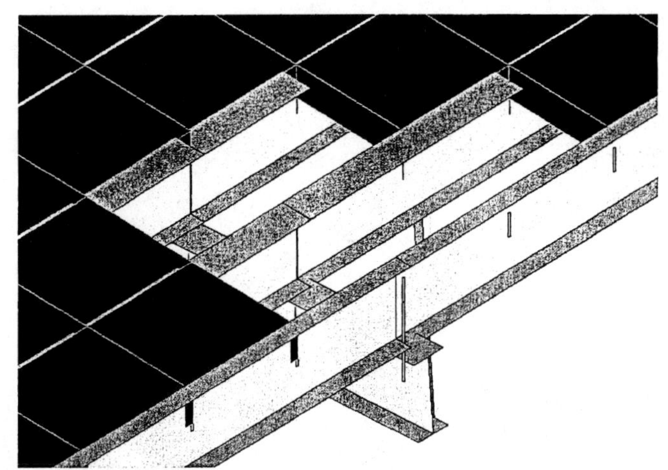

Figure 9. Rendering of floor beam, stringers, and deck. Note that the deck is represented as centered on the top flange of the stringers, intended to approximate the actual roadbase material, which envelops the stringer top flanges.

Studying the bridge and deck in Figures 1-3 and utilizing RISA 3D software in Figures 4 through 9 was used to construct the skeleton model with the applied gravity and wind loading. The deck was modeled as an elastic solid that enveloped the top flange of the stringers, as shown in Figure 9. The San Miguel Bridge deck, with its large amount of gravel roadbase, has a relatively heavy dead load. Table 1 shows axial forces of the windward bottom chord eyebars.

Table 1. Summary of Maximum Axial Compressive Forces in Bottom Chord Eyebars.

Model	Axial force due to dead load only, kPa (psf)	Axial compression due to wind load only, kPa (psf)	Net axial force due to wind plus dead load* kPa (psf)
Case 1:			
Skeleton	399	-222	180
	(89.7)	(-49.9)	(40.5)
Case 2:			
Deck	317	-105	209
	(71)	(-23.7)	(46.9)
	21%	53%	16%

* Note values are not necessarily identical to (D+W) because tension–only members may not be the same for the (D+W) case as for the individual D or W cases.

Forces are for windward side bottom chord and are expressed in kN (kips), followed by percent reduction (or increase in tension) compared to the traditional skeleton value. (Positive = tension; negative = compression).

The percent change from the skeleton case was determined for the deck model from:

$$\% \text{ change} = 100 \times \left| \frac{F_{skeleton} - F_{deck}}{F_{skeleton}} \right| \tag{2.1}$$

And for the diaphragm model from:

$$\% \text{ change} = 100 \times \left| \frac{F_{skeleton} - F_{diaphragm}}{F_{skeleton}} \right| \tag{2.2}$$

Where:

$F_{skeleton}$ = calculated force in windward bottom chord from the skeleton model
F_{deck} = calculated force in windward bottom chord from the deck model

Figure 10 is a topographic map of the vicinity of San Miguel Bridge. Figures 11 through 18 and Table 2 show the location of instruments and results for the San Miguel Bridge.

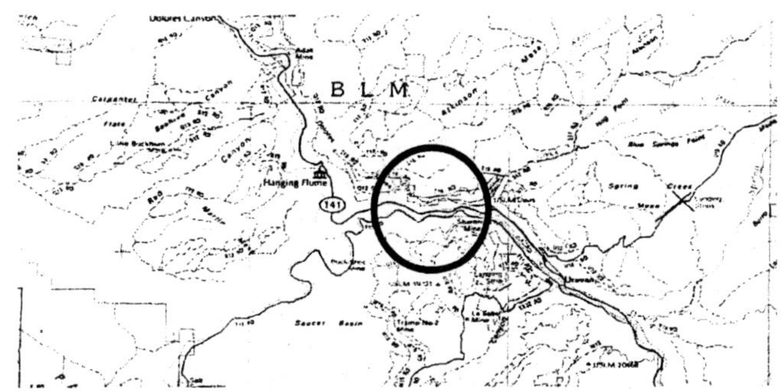

Figure 10. Location Map. Note that wind from the west could become channelized in the river canyon. One of the five original spans was relocated from Grand Junction, CO to a now abandoned County Road northwest of Uravan in Montrose County, CO spanning the San Miguel River (DeLorme 1997). The bridge still resides in this location, but was closed to vehicular traffic in 1988 (Fraser, 2000).

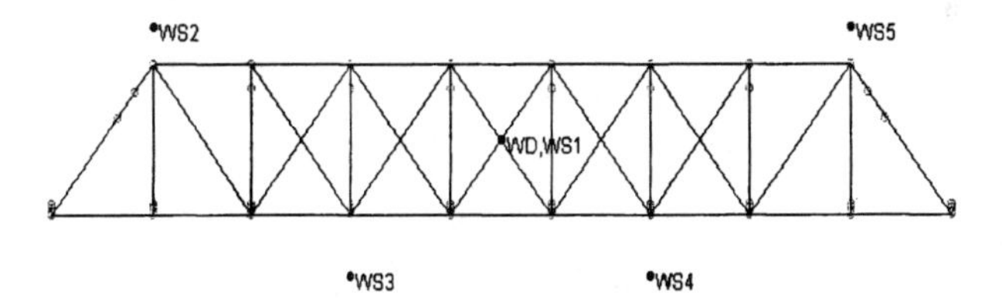

Figure 11. Diagram illustrating the locations of anemometers (WS1 – WS5) and wind direction sensor (WD). North is to the left. WS1 was positioned directly upwind of the approximate center of the wind intercept area. WS2 and WS5 were located 1.5 meters (approximately 5 feet) above the top of the end diagonal members in the end post. WS3 and WS4 were positioned 2 meters (approximately 7 feet) below the bridge deck, at an elevation mid-height between the bridge deck and the water surface below.

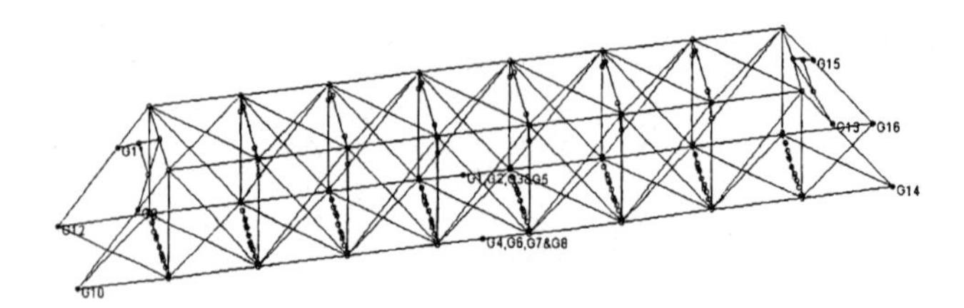

Figure 12. Diagram illustrating the locations of the strain transducers. North is to the left. The wind direction was from the west, orthogonal to the bridge. Strain transducer numbers G1, G2, G3 & G5 were clamped to the leeward bottom chord eyebars. G4, G6, G7 & G8 were clamped to the windward bottom chord eyebars. G9 – G12 were clamped to the end diagonals at the north end post. G13 – G16 were clamped to the end diagonals at the south end post.

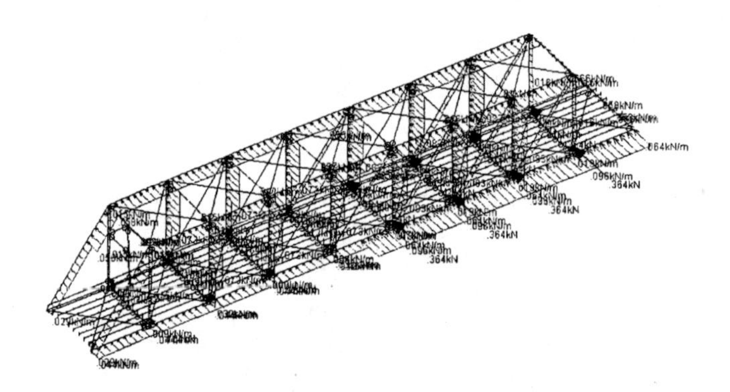

Figure 13. Wind Pressure applied to the four quadrants for analysis. North is to the left. Adding another component of wind pressure from the south (not shown here) to simulate the measured wind direction from the southwest was also investigated but the result was inconclusive.

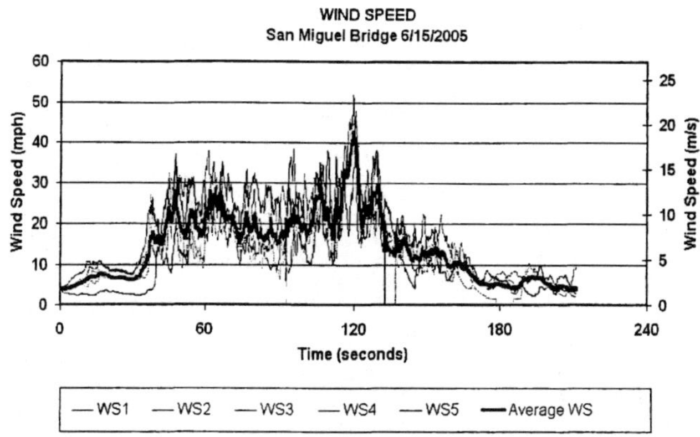

Figure 14. Wind Speed as measured by the five anemometers. The bold line shows the average of all five anemometers.

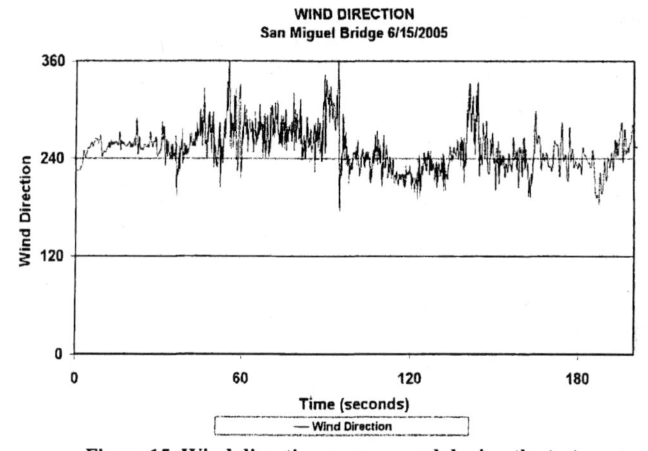

Figure 15. Wind direction as measured during the test.

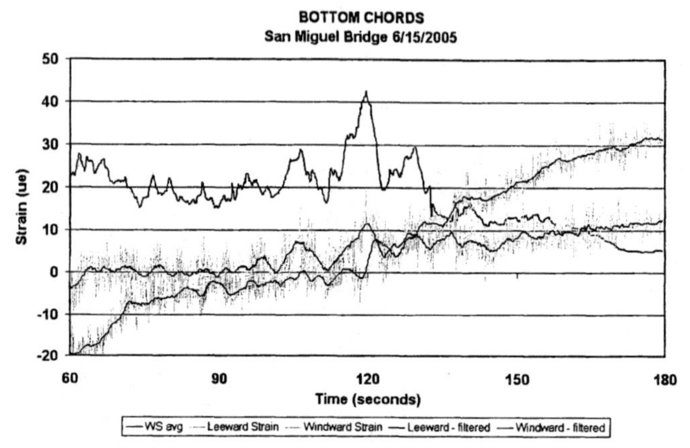

Figure 16. Strain measurements for the windward and leeward bottom chord eyebars.

The bold trace at the top of Figure 16 is the average wind speed, to an arbitrary scale. Measured strain for both leeward and windward bottom chords is shown. Both the raw data and a filtered line that removes much of the signal noise are shown. (Note: The increase in the strain value beyond 120 second time is most likely due to some signal drift – it does not appear to be due to mechanically-induced strain because the wind pressure is diminishing in this range)

Both are baseline traces of the filtered data. Thus, they represent the change in measured strain starting from the same point in time as the corresponding change in wind velocity. The wind velocity is shown for reference at the top of the graph to an arbitrary scale. The verification study data is summarized in Table 2.

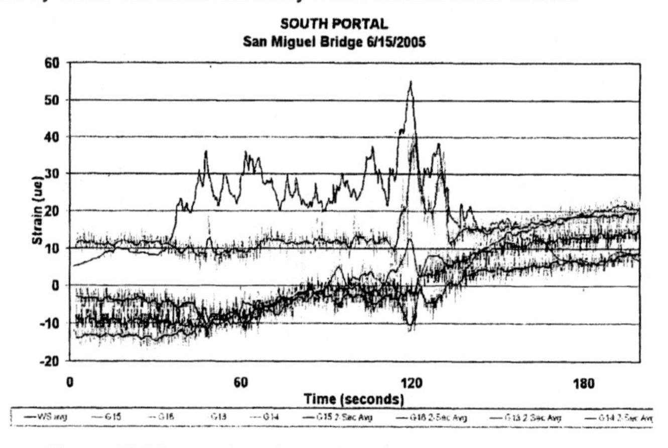

Figure 17. Measured strains at the south end post are shown.

Table 2. Verification Summary.
Comparison of calculated forces to measured forces expressed in kN (kips) and kN-m (foot-kips)

Member	Calculated Force	Measured Force	Correlation: % difference
Windward bottom chord	-1.5 kN (-0.3 kips)	0.22 kN * (0.05 kips)	114% *
Leeward bottom chord	4.3 kN (1 kips)	9.61 kN (2.16 kips)	55%
North end post upper	Broken Member	n/a	n/a
North end post lower	3.9 kN-m (2.9 ft-k)	0.99 kN-m (0.73 ft-k)	75%
South end post upper	3.1 kN-m (2.3 ft-k)	0.92 kN-m (0.68 ft-k)	70%
South end post lower	-4.6 kN-m (-3.4 ft-k)	-2.58 kN-m (-1.9 ft-k)	39%

The weak correlation for the bottom chords may be attributable to a probable anomaly in the field data, particularly for the windward bottom chord.

Weak Correlation:

By comparing the results of the calculated forces to the measured forces in Table 2 reveal the weak correlation of the bottom chords, especially the windward bottom chord data results. Listed below are some of these issues that might be helpful in clarifying some of these results:

- The existing boundary conditions of the bridge might have changed from its original conditions, due to rust, dirt deposits and debris and weather changes. The changes in boundary conditions might switch from being pinned to being fixed or vise versa. These changes impact the measured strains.

- Hidden elements like deck attachments and their connections, whether fixed, pined, or hinged should be verified and studied. These attachments might have different impacts on evaluating the moments and forces along the deck.

- The amount of gravel to sand mixture and their assumed E and G modulus values undoubtedly vary from what was used in the analysis.

- Non-perpendicular wind events impacting the bridge might lead to the poor correlations at the end posts, which indicate the difficulty in modeling the longitudinal component of wind pressure.

Drift:

The drift affecting the reading at time of collecting data as shown in Figure 16 and Figure 18 for the strain measurements at bottom chords and south end post could be explained by the following reasons:

- Possible slipping or movement of the C- clamps at the transducers following the wind velocity spike.
- The fast temperature changes at dusk might have affected the readings. Engineers and analysts should be aware of the time of day that data is being collected.
- The movement of the wires connected between the transducers and the data logger under high wind pressure might be one of the drift reasons. The transducer strain result output is in units of millivolts per volt of excitation. The wire resistance to the high wind might have created some electrical noises that have affected the small voltage.

ACKNOWLEDGEMENTS

The investigation of six historic bridges was completed and analyzed by the University of Colorado at Denver. The funding agencies include Grant # MT-2210-04-NC-12 from the National Center for Preservation Technology and Training and Grant # 2004-M1-019 from the State Historical Fund of the Colorado Historical Society. In addition, acknowledgement is given to Montrose County, CO for allowing access to the bridge. The assistance of Campbell Scientific, Inc. was utilized in programming the data logger.

The assistance of the following University of Colorado at Denver students is acknowledged: Shohreh Hamedian, Veronica Jacobson, William Swigert, Sam Brown and Chris Kline. These students assisted in laboratory and field investigations throughout the project.

This work stems from earlier work completed by Carrol (2003), Herrero (2003) Rutz (2004), and Rutz and Rens (2004). For additional information about the San Miguel Bridge, refer to Rutz et al (2005) and Elias (2006).

REFERENCES

AASHTO (1997). *Guide Specifications for Design of Pedestrian Bridges*, American Association of State Highway and Transportation Officials, Washington, D.C.

Carroll, D. (2003). *Analysis of historic pin connected through truss bridges for conversion into pedestrian use*, Dept. of Civil Engineering, Univ. of Colorado at Denver, Denver, CO.

DeLorme, *Colorado Atlas & Gazetteer*, Yarmouth , Maine, 1997.

Elias, K.E. (2006) *Wind Analysis of the Historic San Miguel Bridge,* M.S thesis, University of Colorado at Denver *In Press*

Fraser, C.B (2000c), "Historic Bridge Inventory, Site No. 5MN4984, San Miguel River Bridge", Colorado Historical Society, Denver, Colorado, 2000.

Herrero, T. (2003). *Development of strain transducer prototype for use in field determination of bridge truss member forces*, Dept. of Civil Engineering, Univ. of Colorado at Denver, Denver.

RISA (2002). RISA-3D, version 4.5, RISA Technologies, Foothill Ranch, CA, 2001.

Rutz, F.R. (2004). *Lateral load paths in historic truss bridges,* PhD Thesis, Civil Engineering, University of Colorado at Denver.

Rutz, F.R., and Rens, K.L. (2004). "Alternate load paths in historic truss bridges: new approaches for preservation," *Proceedings of the 2004 Structures Congress*, Ed. by G.E. Blandford, Structural Engineering Institute of the American Society of Civil Engineers, May 22-26, Nashville, TN, ASCE, Reston VA.

Rutz, F.R., K.L. Rens, V.R. Jacobson, S Hamedian, K.E. Elias, W.B. Swigert, (2005) *Load Paths in Historic Truss Bridges*, MT-2210-04-NC-12, prepared by Dept. of Civil Engineering, University of Colorado at Denver for National Center for Preservation Technology Transfer, Natchitoches, LA, 2005.

FORENSIC INVESTIGATION OF SECTIONS 390103, 390108, 390109, AND 390110 OF THE OHIO SHRP U.S. 23 TEST PAVEMENT

S.M. Sargand, I.S. Khoury, J.L. Figueroa

Abstract

The Ohio Strategic Highway Research Project (SHRP) Test Road, constructed on U.S. 23 about 25 miles (40 km) north of Columbus, Ohio, consists of forty test sections in the SHRP SPS-1, SPS-2, SPS-8, and SPS-9 experiments. During the summer of 2002, after the appearance of localized distress in Section 390103, a forensic study of this and Sections 390108, 390109, and 390110 in the SPS-1 (Asphalt Concrete) experiment was completed through a series of destructive and non-destructive tests.

Distress surveys were conducted on these four sections in accordance with SHRP-P-338 "Distress Identification Manual for the Long-Term Pavement Performance Project." Non-destructive testing included Falling Weight Deflectometer (FWD), transverse profiling, and Dynamic Cone Penetration tests (DCP). Trenches were excavated at locations with various levels of distress to measure transverse layer profiles, and to obtain material samples for laboratory testing. Collected data was analyzed to determine the causes of localized distresses.

The investigation revealed substantial variability in stiffness and high levels of moisture in the subgrade soil at all four pavement sections. These trends indicate that the severity of distress in Sections 390108, 390109, and 390110 would soon be similar to those in Section 390103 if the sections were left open to traffic.

Despite the use of various base materials and the presence of edge drains in Sections 390108, 390109 and 390110, higher than anticipated levels of subgrade soil moisture reaching saturation were present in all four pavement sections. Excessive moisture was determined to be the underlying cause of rutting and cracking. While edge drains probably removed some moisture infiltrating down from the pavement surface, they provided little relief from moisture migrating up through the subgrade.

1 INTRODUCTION

In 1987, Congress established the Strategic Highway Research Program (SHRP), a five-year, $150 million research effort to improve the performance of highway pavements and bridges. Eight broad study areas were set up within the program to meet the prescribed goals. One of these areas was defined as the Long Term Pavement Performance (LTPP), which was aimed at extending the life of asphalt concrete and Portland cement concrete pavements. A series of experiments known as the Specific Pavement Studies (SPS) was developed within LTPP to assess the effect of various structural parameters on pavement performance.

The Ohio SHRP Test Road, located about 40 km north of Columbus Ohio on US23 in Delaware County, encompasses forty test sections in the SHRP SPS-1, SPS-2, SPS-8 and SPS-9 experiments. The SPS-1 experiment was a strategic study of the effectiveness of various structural factors on the performance of flexible pavements.

As part of its pavement research program, the Ohio Research Institute for Transportation and the Environment (ORITE) was charged with the task of monitoring these test sections. Severe localized distresses were observed in Section 390103 of the SPS-1 experiment during the spring of 2002, resulting in emergency lane closure and repairs. Deflection analyses indicated distress was imminent in Sections 390108, 390109, and 390110. A forensic investigation consisting of destructive and non-destructive testing was scheduled to determine the possible causes of rutting and other distress observed in these sections, designed and built according to the detail outlined in Table 1. Environmental instrumentation was installed in Section 390110 at the time of construction to monitor temperature, moisture, and frost penetration within the pavement structure. Although the forensic investigation was completed for all four sections, this paper reviews data collected from section 390103, with the findings from this section applying to the remaining three.

Table 1 Pavement Structure

Section	Thickness mm		Base Type	Drainage
	AC	Base		
390103	101	203	Asphalt Treated Base	No
390108	178	304	101 mm Permeable Asphalt Treated Base 203 mm Dense Graded Aggregate Base	Yes
390109	178	406	101 mm Permeable Asphalt Treated Base 304 mm Dense Graded Aggregate Base	Yes
390110	178	203	101 mm Asphalt Treated Base 101 mm Permeable Asphalt Treated Base	Yes

Forensic Procedure

To complete the forensic investigation of the four sections under study, several tests were conducted in each of them. The testing was subdivided into non-destructive and destructive procedures. Non-destructive testing included photographs of selected areas referenced by station, distress surveys conducted according to SHRP-P-338 Distress Identification Manual for Long-Term Pavement Performance Project[1], Falling Weight Deflectometer (FWD) tests, and transverse profiles taken to measure rutting.

Destructive testing included dynamic cone penetration (DCP) tests, trenching, and recovering cores and sections of AC for laboratory testing. During the excavation of trenches material sampling, moisture, and stiffness testing were also completed.

Load History

The SPS-1 sections were opened to traffic on August 14, 1996, however the highway was intermittently closed, between September 1996 and November 1997, because of rutting in some of the AC sections. Section 390103 developed localized distress starting on March 8, 2002. After a review of the remaining SPS-1 sections, it was determined that distresses were imminent in Sections 390108, 390109 and 390110. On April 24, 2002 the SPS-1 sections were again closed to traffic and a forensic study was conducted on these sections. A load cell based weigh-in-motion (WIM) system was used to continuously monitor traffic flow on the Ohio SHRP Test Road. Accounting for all the closures, the actual service life of these sections was 4.5 years.

Water Table

Water table elevation measurements were initiated on December 1996, and continued through the life of these sections. The wells are located at SHRP station 3+73 and 2+75 for Sections 390103 and 390108 respectively. The data shows a significant seasonal variation in water elevation. The peak water elevations occur mainly in the spring (April and May) of each year, while the low water levels are recorded in late fall.

The highest water table elevation recorded at SHRP station 3+73 was in March of 1997 indicating a position 1.64 m below the subgrade surface. Similarly, the highest water table elevation at SHRP station 2+75 was just 1.04 m below the subgrade surface on July 2nd 2001. With the water table being so close to the surface of the subgrade in this area, good drainage systems are crucial to the endurance of the pavement.

2 NON-DESTRUCTIVE TESTING

Pavement Distress

Distresses were recorded in accordance with SHRP-P-338 Distress Identification Manual for the Long-Term Pavement Performance Project[1] and are summarized in Table 2. In all four sections; fatigue cracking, wheel path longitudinal cracking and non-wheel path longitudinal cracking were present with severity levels ranging from low to high.

Section 390103 showed the worst overall distress with high levels of rutting and longitudinal cracking. The longitudinal cracking was present in the center of the lane for the entire length of this section, with a mean crack width of about 12.7 mm. Also, longitudinal cracking of the same magnitude was noted along the inside edge of both wheel paths. Cold patch had been used to repair this distress (Figure 1) but had deteriorated to high severity. Fatigue cracking was present in approximately 25% of the right and left wheel paths and could be classified as moderate in severity. Some alligator cracking was also noticed in the wheel path and on the shoulder.

Table 2 Distress Summary of Sections. (Percentages Refer to Section Length)

Section	Average Distress Classification	Maximum Rut Depth (mm)	Longitudinal Cracking in Center of Lane	Longitudinal Cracking in Wheel Path	Average Crack Width (mm)
390103	High	69.85	100%	90%	12.7
390108	Moderate	20.32	50%	50%	9.525
390109	Low	19.05	10%	10%	6.35
390110	Low	12.7	15%	10%	6.35

Figure 1 Severe Distress, Section 390103

Longitudinal Cracking

Longitudinal cracking observed in all four sections during trench excavation, initiated on the pavement surface and was considered to be top-down with low to moderate severity as shown in Figure 2. The longitudinal crack consistently occurred in the center of the lane in all four sections and did not stray from a straight line path. The longitudinal wheel path cracking occurred on the edge of the wheel path and also followed a straight line. The wheel path cracks were shallower than the center line cracks and none of the longitudinal cracks extended all the way to the base, and most were within the top 10 cm.

Similar observations were made during a recent Colorado top-down crack study[7]. Of twenty-five longitudinal crack sites in Colorado, 72% were top-down cracking and 67% of the top-down cracking associated with visual segregation at the bottom of the surface layer. Figure 2 shows the segregation, where a relatively large portion of coarse aggregates is distributed in the bottom half of the surface layer. The Colorado study further identified the source of the segregation and it was attributed to the edges of the slat conveyors located at the center point of certain models of pavers, thus explaining the straight line longitudinal cracks.

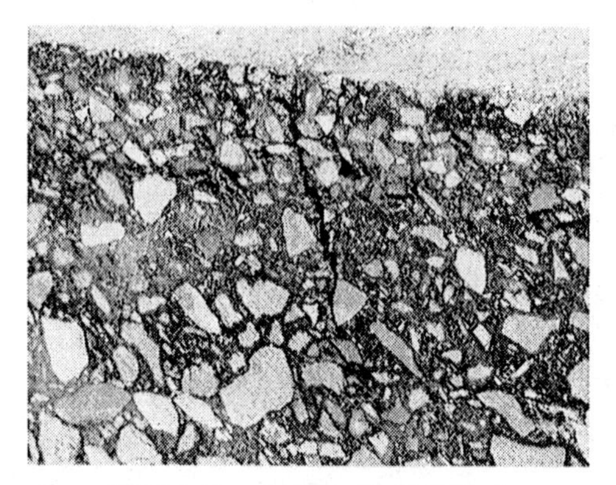

Figure 2 Close-up of Longitudinal Crack

Rutting

Two methods were used to determine the transverse pavement profiles; the ORITE automatic profilometer and a dipstick. Thirteen profiles were taken, plotted and compared for each section approximately 12.2 m (40 ft) apart.

Transverse profiles from Section 390103 for the profilometer and dipstick, shown in Figures 3, illustrate significant variations from the proposed original design. This section exhibited the greatest rutting of all. It should be noted that the profiles measured with the dipstick and the profilometer were very similar with the differences between the two attributed to each instrument starting at a slightly different reference point on the pavement.

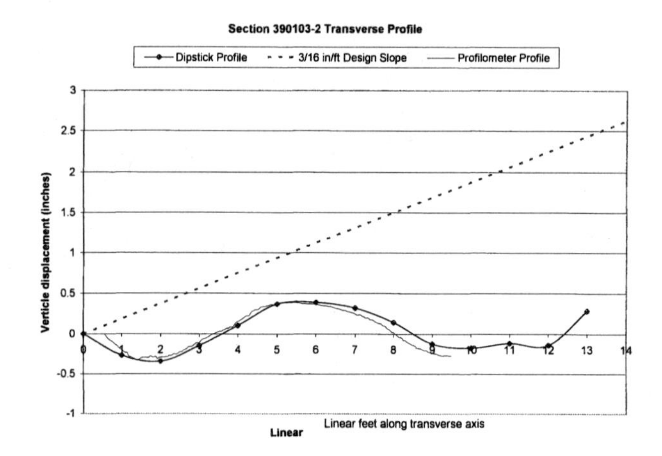

Figure 3 Transverse Profile 2 of Section 390103 (1 in =2.54 cm, 1 ft =0.3048 m)

Falling Weight Deflectometer Testing

Staggered FWD tests were conducted every fifty feet along the centerline and right wheel path on the surface of the AC pavement during the first week of August 2002. Data collected indicate that the stiffness varied along the section length in the center of the lane and wheel path.

FWD tests were also conducted during the initial construction of these sections. In order to compare the two sets of data, the FWD data taken at the time of the forensic study was normalized to a load of 4.45 kN (1 kip). Examining the normalized FWD data at the time of the forensic study for Section 390103 (Figure 4) a large deflection spike is observed, while FWD deflection plots obtained at the time of construction do not show spikes in the AC or ATB layers. Therefore the distress occurring in the pavement at that location can not be attributed to any flaw during construction in the pavement base layers including the subgrade.

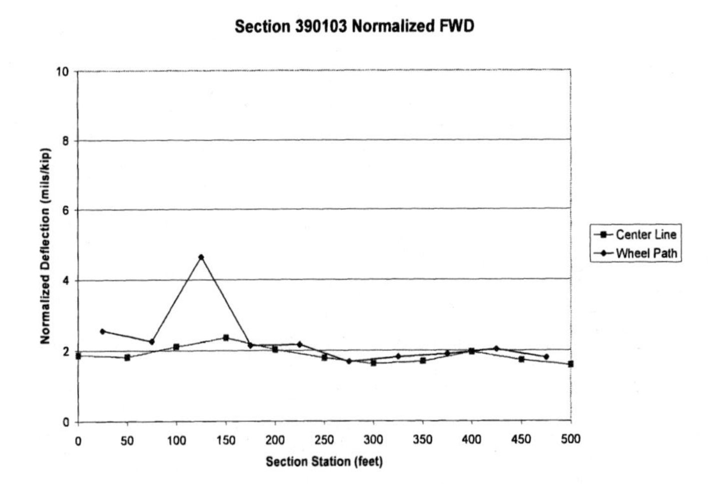

Figure 4 Normalized FWD Plot of Section 390103 (1 ft=0.305m; 1mil/kip=5.71 nm/N)

3 DESTRUCTIVE TESTING

Trenches

Lateral trenches approximately 1.5 m wide and 1.2 m deep were excavated (Figure 5) at locations within each section exhibiting various levels of distress. Three trenches were excavated in Sections 390103 and 390110, and two in Sections 390108 and 390109. Approximate section stations and levels of distress noted at the trench locations are shown in Table 3.

Table 3 Trench Distress Classification and Locations

Section	Trench	Section Station	Indicated Distress Classification
390103	#1	1+27	High
	#2	3+00	Low
	#3	4+05	Moderate
390108	#1	3+32	Moderate
	#2	5+00	Low
390109	#1	2+00	Moderate
	#2	3+27	Moderate
390110	#1	1+10	Low
	#2	2+52	Moderate
	#3	5+63	Environmental Trench

Figure 5 Trench Excavation

Layer and Lift Thickness

The thickness of individual pavement and base lifts were measured at one meter intervals with a standard tape measure along the length of each trench and estimated to the nearest 50 mm. The results of measurements for section 390103 are presented in Table 4. While the extent of rutting was very apparent in some sections, as shown in the transverse profiles, pavement layers remained consistent with their design thickness throughout the transverse length of all trenches, indicating minimal, if any rutting in the AC. However some consolidation of the base layers was noted.

Table 4 Layer /Lift Thickness for Section 390103

Section 390103 layer thicknesses in inches											
	Trench 2			Trench 3						At time of construction	
Layer	0+00	0+03	0+06	0+09	0+00	0+03	0+06	0+09	0+12	Design	Actual
AC Surface	1.75	1.75	1.75	1.75	1.75	2	1.75	1.5	2.25	1.75	1.71
AC intermediate	2	2	2	2	2.5	1.75	2.5	2	2	2.25	2.16
ATB	3	3.75	3.5	3.75	3.5	2.5	2	2.5	2.75	8	8.04
	5	5	4.75	5	4.5	4.75	4.25	4.75	4		

Section 390103 layer thicknesses in cm											
	Trench 2			Trench 3						At time of construction	
Layer	0+00	0+03	0+06	0+09	0+00	0+03	0+06	0+09	0+12	Design	Actual
AC Surface	4.45	4.45	4.45	4.45	4.45	5.08	4.45	3.81	5.72	4.45	4.34
AC intermediate	5.08	5.08	5.08	5.08	6.35	4.45	6.35	5.08	5.08	5.72	5.49
ATB	7.62	9.53	8.89	9.53	8.89	6.35	5.08	6.35	6.99	20.32	20.42
	12.70	12.70	12.07	12.70	11.43	12.07	10.80	12.07	10.16		

Stiffness Variability

In-situ subgrade stiffness was measured at different depths within the trenches using a Humboldt Stiffness Gauge with typical data shown in Figure 6. Measurements were taken at 15 and 30 centimeter intervals at the same transverse location where the trenches were excavated. As shown in this figure, the data showed considerable variability in stiffness with depth in each section. Section 390103 revealed a significant increase in stiffness at a depth of 60 centimeters below the surface. This trend was present in all sections in at least one trench.

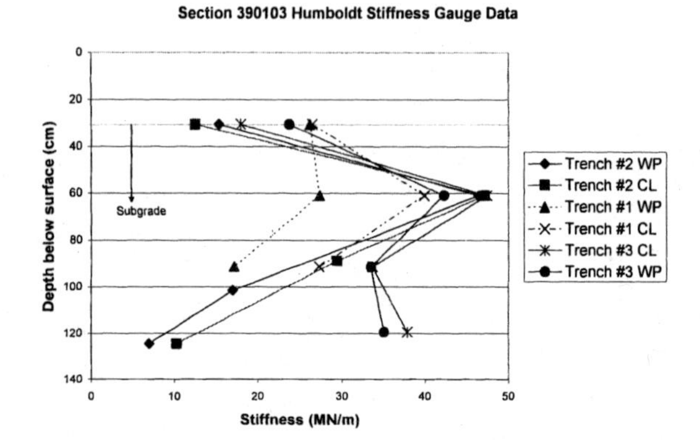

Figure 6 Humboldt Stiffness Gauge Plot of Section 390103

Moisture Content

Subgrade moisture content was determined by three methods; traditional laboratory methods, a nuclear gauge, and Time Domain Reflectometry probes installed during construction. Soil samples and nuclear gauge readings were taken at depth of approximately 15 and 30 cm during trench excavation. Soil samples were then transferred to ORITE's laboratory in sealed containers, and tested for moisture.

Data from soil samples generally showed moisture content increases to depths of 64 to 102 cm, followed by a decrease at deeper points in all four sections. The highest and lowest moisture content readings are shown in Table 5 for each section. As expected, an inverse relationship between stiffness and moisture content was noted. Further examination of the stiffness moisture relationship suggested that moisture was seeping up through the subgrade, and also down through the pavement structure. The increase in stiffness indicated by the Humboldt Gauge readings and lower moisture content at a depth of 60 cm may indicate that moisture had not reached this point from either direction.

From the nuclear gauge data, it was determined that the subgrade soil was very close saturation in all four sections assuming a specific gravity of 2.7. Nuclear gauge readings were taken (when possible) and showed reasonable agreement with the in-situ moisture samples values.

Free water was observed between layers in the subgrade, between the asphalt concrete pavement and the base layers, and also water could be seen "seeping" from the subgrade (Figure 7). This indicated a lack of adequate drainage in the four pavement structures, even though Sections 39108, 390109, and 390110 were constructed with edge drains.

Table 5 Gravimetric Moisture Content Range

Section	Highest Moisture Content	Lowest Moisture Content
390103	30.2%	14.7%
390108	25.6%	5.8%

390109	23.3%	5.6%
390110	27.0%	12.9%

Figure7 Water Seeping from Interface Layer

Dynamic Cone Penetrometer

Dynamic Cone Penetration (DCP) tests also showed non-uniform subgrade stiffness with depth in each test section. Tests were conducted every 31 m in both the centerline and wheel path, and also at trench locations. These data were used to calculate the resilient modulus of the subgrade as described in the following section.

4 TEST RESULTS

Subgrade Resilient Modulus

The resilient modulus M_R of the subgrade was calculated and plotted for each section from the DCP tests. The M_R was determined from DCP data using the following equation:

$M_R = 1200*CBR$ (units are in kip per square inch (ksi))

Where $\log(CBR) = 2.20 - 0.71 \log(PI)1.5 \pm 0.075$
and PI is the DCP penetration index (mm/blow). (1 ksi = 6.89 MPa)

The average resilient modulus of the upper level of subgrade in Sections 390103, 390108, 390109, and 390110 was calculated to be 86.74 MPa (12.58 ksi), 168.4 MPa (24.42 ksi), 145.4 MPa (21.09 ksi), and 155 MPa (22.50 ksi), respectively. According to the report Characterization of Materials & Data Management for Ohio SHRP Projects, U.S. 23 [3] the AASHTO subgrade soil classifications are A-6 for section 390103 and A-4 for sections 390109 and 390110, while section 390108 did not have a soil class identified in the report.

The resilient modulus for each structural layer was also estimated by a back calculation technique using FWD data. The back calculation program selected for this analysis was MODULUS 4.2. [5] This program, developed by the Texas Transportation Institute, is based on the multilayer linear elasto-static theory.

The input parameters for this analysis include sensor locations, layer thickness, initial modulus ranges, and Poisson's ratio. The initial values used for the analysis based on the FHWA/OH-2002/031 report [4] are shown in Table 6, where the subgrade layer was assumed to be semi-infinite. The back calculated modulus values are summarized in Table 7 for each section. The deflection matching errors per sensor for all of the back calculation runs were within acceptable limits.

Table 6 Initial Input Parameters

	Initial modulus range				
Material type	low (ksi)	high (ksi)	low (MPa)	high (MPa)	Poisson's ratio
AC	200	2500	1400	17000	0.35
DGAB	10	150	70	1000	0.35
PATB	50	1000	350	7000	0.25
ATB	150	4000	1000	28000	0.25
Subgrade	3	30	20	200	0.40

Table 7 Summary of Back Calculated Modulus

		Thickness		Backcalculated Modulus (ksi)			Backcalculated Modulus (MPa)			Absol. Error/ Sensor
Section	Layer	(in)	(cm)	CL	WP	Average	CL	WP	Average	
390103	AC	4	10.16	2127	1978	2052.5	14665.1	13637.8	14151.5	1.30%
	ATB	8	20.32	42.1	24.4	33.25	290.3	168.2	229.3	
	Subgrade	-	-	11.2	10.9	11.05	77.2	75.2	76.2	
390108	AC	7	17.78	283	346	314.5	1951.2	2385.6	2168.4	0.60%
	PATB	4	10.16	57.8	56	56.9	398.5	386.1	392.3	
	DGAB	8	20.32	13.7	10.5	12.1	94.5	72.4	83.4	
	Subgrade	-	-	21.9	23.4	22.65	151.0	161.3	156.2	
390109	AC	7	17.78	342	396	369	2358.0	2730.3	2544.2	0.60%
	PATB	4	10.16	140.5	96.5	118.5	968.7	665.3	817.0	
	DGAB	12	30.48	18	16.1	17.05	124.1	111.0	117.6	
	Subgrade	-	-	25	26.5	25.75	172.4	182.7	177.5	
390110	AC	7	17.78	523	524	523.5	3606.0	3612.9	3609.4	0.70%
	ATB	4	10.16	214.4	167	190.7	1478.2	1151.4	1314.8	
	PATB	4	10.16	33	28	30.5	227.5	193.1	210.3	
	Subgrade	-	-	20.9	22.3	21.6	144.1	153.8	148.9	

Note: Absolute Error/Sensor is the difference between deflection actually recorded by FWD at one of the sensors and the backcalculated value of the same deflection, expressed as a percentage.

Since FWD tests were conducted immediately after construction the same technique was used to estimate layer resilient modulus [4]. Results are compared with the calculations from the forensic investigation and shown in Table 8. It is clearly seen that all layers in the four sections experienced a decrease in the resilient modulus.

This indicates that the pavement structure was weakened by increased moisture and various distresses. The back calculated resilient modulus for the subgrade was also compared with the resilient modulus calculated from DCP test data, showing a fair agreement in modulus values between the two methods as seen in Table 9. Resilient modulus values were also determined from measured moisture content and interpolation from curves determined during the laboratory characterization of materials[3], assuming a deviator stress $\sigma_d = 13.8$ kPa. While the average M_R determined from the moisture content is smaller than the other values, this may be due to increased moisture from water used during the asphalt concrete cutting process.

Table 8 Comparison of Back Calculated Modulus

Section	Layer	Thickness (in)	Thickness (cm)	M_R (ksi) After Construction	M_R (ksi) After Distress	M_R (MPa) After Construction	M_R (MPa) After Distress	Change
390103	AC	4	10.16	2762	2053	19043.3	14154.9	-25.7%
	ATB	8	20.32	83	33.3	572.3	229.6	-59.9%
	Subgrade	-	-	16	11.1	110.3	76.5	-30.6%
390108	AC	7	17.78	781	315	5384.8	2171.8	-59.7%
	PATB	4	10.16	83	56.9	572.3	392.3	-31.4%
	DGAB	8	20.32	41	12.1	282.7	83.4	-70.5%
	Subgrade	-	-	24	22.7	165.5	156.5	-5.4%
390109	AC	7	17.78	843	369	5812.3	2544.2	-56.2%
	PATB	4	10.16	124	118.5	854.9	817.0	-4.4%
	DGAB	12	30.48	24	17	165.5	117.2	-29.2%
	Subgrade	-	-	23	25.8	158.6	177.9	12.2%
390110	AC	7	17.78	708	524	4881.5	3612.9	-26.0%
	ATB	4	10.16	163	190.7	1123.8	1314.8	17.0%
	PATB	4	10.16	51	30.5	351.6	210.3	-40.2%
	Subgrade	-	-	19	21.6	131.0	148.9	13.7%

Table 9 Comparison of Subgrade Modulus

Section	Average M_R (MPa) Original Lab Value	Moisture Based	DCP based	Backcalculated
390103	108.2	69.6	86.9	76.5
390108	131.0	88.9	168.2	156.5
390109	79.3	82.7	146.2	177.9
390110	89.6	93.1	155.1	148.9

Core Density and Air Voids

One of the common detection methods for aggregate segregation is the comparison of mix volumetrics between segregated and non-segregated areas. When the air void difference between segregated and non-segregated areas is larger than 2 to 6%, it is considered medium to high level segregation[6]. The segregation of the surface course in the vertical direction observed in this study would also cause high air void levels. Table 10 shows the air voids of cores taken from the test pavements. As expected, cores obtained from the wheel-path were further compacted by traffic and had generally lower air voids than cores taken from the center-lane. Overall, air voids of most cores were significantly higher than the normal range of new pavements, which might have been caused by the vertical segregation previously discussed. Mix properties typically affected by segregation are lower modulus, lower strength, and higher moisture sensitivity. Top-down cracks observed on the test pavements might be explained by the segregation.

Density determination of cores taken from the sites showed much higher air voids than usual. This is believed to be caused by the segregation of asphalt mixes which also contributed to the top-down cracking.

It should be noted that the average percentage of air voids in the surface layers ranges from 9.45% for section 390108 to 11.25% for section 390110, while the overall average for all four sections is 10.1%. These values are much higher than the 7.1% recorded at the time of construction and shown in Table 2. For intermediate layers, a similar trend is observed: the averages range from 10.9% in section 390108 to 12.95% in section 390110, for an overall average of 11.9% compared to 6.6% recorded at the time of construction.

Table 10 Density of cores taken from test pavement

Pavement Layer	Location	AC %	Bulk Sp. Gravity	Max Sp. Gravity	% AIR VOIDS
Section 390103					
Surface	CL	6.7	2.193	2.450	10.5
	WP	6.8	2.238	2.453	8.8
Intermediate	CL	6.0	2.192	2.468	11.2
	WP	6.0	2.208	2.476	10.8
Base	CL	4.6	2.242	2.486	9.8
	WP	4.6	2.28	2.521	9.6
Section 390108					
Surface	CL	6.7	2.225	2.461	9.6
	WP	6.4	2.241	2.470	9.3
Intermediate	CL	6.1	2.172	2.465	11.9
	WP	6.1	2.222	2.467	9.9
Section 390109					
Surface	WP	6.5	2.222	2.460	9.7
	WP	7.0	2.202	2.454	10.3
Intermediate	CL	5.9	2.143	2.469	13.2
	WP	5.7	2.165	2.465	12.2
Section 390110					
Surface	CL	6.2	2.178	2.465	11.6
	WP	6.6	2.186	2.453	10.9
Intermediate	CL	5.6	2.153	2.469	12.8
	CL	5.7	2.147	2.472	13.1
Base	CL	5.3	2.24	2.461	9.0
	WP	5.3	2.285	2.454	6.9

5 SUMMARY AND CONCLUSIONS

An in-depth forensic study of Sections 390103, 390108, 390109, and 390110 on the Ohio SHRP Test Road was performed to determine the causes of localized distresses occurring in these sections. Various degrees of distress were present in the four sections. Section 390103 exhibited the highest distress with severe rutting, fatigue cracking in the wheel paths, and longitudinal cracking extending the length of the section. Low to moderate distresses of a similar nature were present in the other three sections.

The service lives of 1.3, 1.8, 4.3, and 2.8 years as predicted by the AASHTO equations for sections 390103, 390108, 390109, and 390110 respectively, agreed with the performance observed in the field, assuming actual ESALs equal the design ESALs. Sections were opened to traffic on August 14, 1996 and closed for replacement on April 24, 2002. During this time span, the sections were closed at various times totaling about one and a half years, resulting in an actual service life of about 4½ years.

The investigation revealed substantial variability in stiffness and high levels of moisture in the subgrade of all four pavement sections. According to the stiffness test results back calculated from the FWD deflections, distresses observed were not due to any flaws in the pavement at the time of construction and correlation between stiffness and moisture was evident. These trends indicate that the severity of distress in Sections 390108, 390109, and 390110 would soon be similar to those in Section 390103 if the sections were left open to traffic.

Despite the use of various base materials and the presence of edge drains in Sections 390108, 390109 and 390110, higher than anticipated levels of subgrade moisture were present in all four pavement sections. The subgrade in these sections was saturated; thus moisture was the underlying cause of rutting and cracking. While edge drains probably removed some moisture infiltrating down from the pavement surface, they provided little benefit for moisture migrating up through the subgrade.

Laboratory measurements and field tests showed the subgrade soil in Sections 390108, 390109, and 390110 to have similar values of resilient modulus, stiffness, and moisture content regardless of the type of base used. Section 390103 also had similar subgrade soil characteristics with the exception of the average resilient modulus being significantly less than at the other three sections.

The thickness of the AC pavement layers remained constant along the transverse axis, and consistent with their design specifications in all test sections, while some consolidation of the base material was noted. This would indicate that rutting observed in the pavement surface had accumulated from the base and the subgrade.

The longitudinal cracking observed in all test sections was the result of the segregation of aggregates in the asphalt mix at the time of construction, with the augers or spreaders in the paver being the most likely cause of segregation. Segregation is also reflected in the much higher percentage of air voids seen in the core samples taken for the forensic study, compared to those recorded at the time of construction.

In comparing the resilient modulus, there was good agreement between the back calculation technique and the DCP test results. The decrease of the layer modulus between the time of construction and the forensic investigation is attributed to the distress developed along the pavement.

6 REFERENCES

1. AASHTO *Guide for Design of Pavement Structures*, © 1986 American Association of State Highway and Transportation Officials, Washington, D.C.

2. Harmelink D., Aschenbrener T., "Extent of Top-Down Cracking in Colorado", Rpt CDOT-DTD-R-200-7, 2003.

3. Masada T. et al, FHWA/OH-2001/07 "Laboratory Characterization of Materials & Data Management for Ohio-SHRP Projects U.S. 23", © 2002 Ohio Department of Transportation, Columbus, Ohio.

4. Michalak, C.H. and Scullion, T., "Modulus 5.0: User's Manual", Texas Transportation Institute, College Station, Texas, 1995

5. Sargand S. at al, FHWA/OH-2002/031 "Determination of Pavement Layer Stiffness on the Ohio SHRP Test Road Using Non-Destructive Testing Techniques", © 2002 Ohio Department of Transportation, Columbus, Ohio.

6. *SHRP-P-338 Distress Identification Manual for the Long-Term Pavement Performance Project*, © 1993 National Academy of Sciences, Washington, D.C.

7. Stroup-Gardiner, M., "Influence of Segregation on Pavement Performance" AAPT vol 69, 2000, pp 424-454.

Repair of Harbor Facilities

Richard A. Mirth, P.E.[1]

Abstract

This paper describes work to provide civil engineering based arguments for the maintenance and repair of a harbor lakewall and breakwall system. A condition survey for three important harbor facilities was completed in an attempt to introduce some factual data into a discussion that seemed to be undefined and perhaps endless. Three concrete structures were inspected, their condition evaluated, and recommendations made for work that was needed and realistic. The results were introduced into the debate.

Introduction

This paper describes work to provide civil engineering based arguments for the maintenance and repair of a harbor lake wall and breakwall system in Dunkirk, a Lake Erie city of about 13,000 in Western New York. The harbor has been a focal point of city activities since the mid nineteenth century when the Erie Railroad, then the longest railroad in the world, reached the great lakes at Dunkirk (Chard, 1971). Commercial shipping activity has declined to where it now involves only occasional coal boats supplying the power plant coal dock just inside the outer breakwall. The harbor is still home to a large number of pleasure boats, and the sport fishing draws people from the surrounding states and Canada. Therefore, the harbor continues to be a focal point of the city's economy. This study began in late 2003 in response to public remarks by private citizens, harbor commission members, and elected officials expressing concern about the condition of the harbor system, but few remarks were fact based and several sought to fix blame rather than solve any problems that might be present. The city operates on a tight budget and funds are not available to conduct engineering studies. The city staff does not include an engineer and the technicians on the staff are very busy keeping up with the street maintenance and the water and sewer maintenance programs. This study attempted to interject engineering based

[1] Associate Professor Emeritus, Northern Arizona University, Life Member ASCE, Resident Dunkirk on Lake Erie, 119 E. Green Street, Dunkirk, New York 14048, PH (716) 366 7754; email: dickandsally@netsync.net

arguments into the discussion of what should be done to preserve the harbor structures.

The study broke the system into three geographically separate segments, analyzed the condition of each segment, and recommended courses of action.

Overview

The walls are all constructed of portland cement concrete. The three segments include an inner harbor segment that runs west to east along the inner harbor, a segment to the east of the harbor that protects a residential neighborhood, and an outer breakwall that protects the harbor from winds from the north (Figure 1). The two lake walls were built about 1927 to 1930. The outer breakwall also includes large stones on the outside and some replacement concrete on the top at the west end near the ship channel. Its construction date is not known. Common to all of the old concrete was evidence of alkali-aggregate reaction in the form of white laitance patterns on the surface of the concrete. The laitance is formed by the reaction of the alkali in the cement with the silica in the aggregate. The alkali-aggregate reaction causes expansion and resulting cracking in the hardened concrete over a period of years (Kosmatka and Panarese, 1988). The expansion and cracking reduce strength and allow water intrusion which can result in spalling of concrete from the surface during the freeze thaw cycles. Since the walls were built, a better understanding of the reaction has led to a countermeasure. This consists of adding fine silica fly ash to the wet concrete mix. The fine fly ash has a large surface area which allows all of the cement alkali to react with the silica prior to the hardening of the concrete. No alkali is left to cause expansion. This cannot help the lake walls, but it does suggest that the reaction might have ended after more than 70 years. Monitoring of the white laitance patterns on the wall surfaces may verify this assumption. No change may indicate the reaction has ceased. While the alkali-aggregate reaction was common to all of the walls, other conditions were peculiar to individual sections. Each section of the three will be analyzed separately, and conclusions and recommendations noted. The present state of action will also be discussed.

Inner Harbor Lake Wall

The inner harbor lake wall consists of 34 mass concrete sections each 65 feet long. The cross section of the wall is a trapezoid with a 6'-6" foot tall vertical front face, a two foot wide top, a tapered back, and a 3 foot base (Figure 2). The sections have a keyway between them. This is set on a 1'-6" foot thick deck slab which extends about one inch beyond the toe of the wall and itself is founded on a sheet pile about 1'-3" back of the front face and wood piling about 7 feet back of the front face. The deck slab is a total of 9 feet wide. Drawings show a counter weight 35 feet behind the slab, but the presence of the counter weight and the wood piling could not be verified in the field. The drawings also show a keyway between the wall and the slab. The wall was anchored to the slab by 5/8 inch bars 18 inches on center at the toe and heel of the wall. The earth back fill comes to the top of the wall. There are no structures

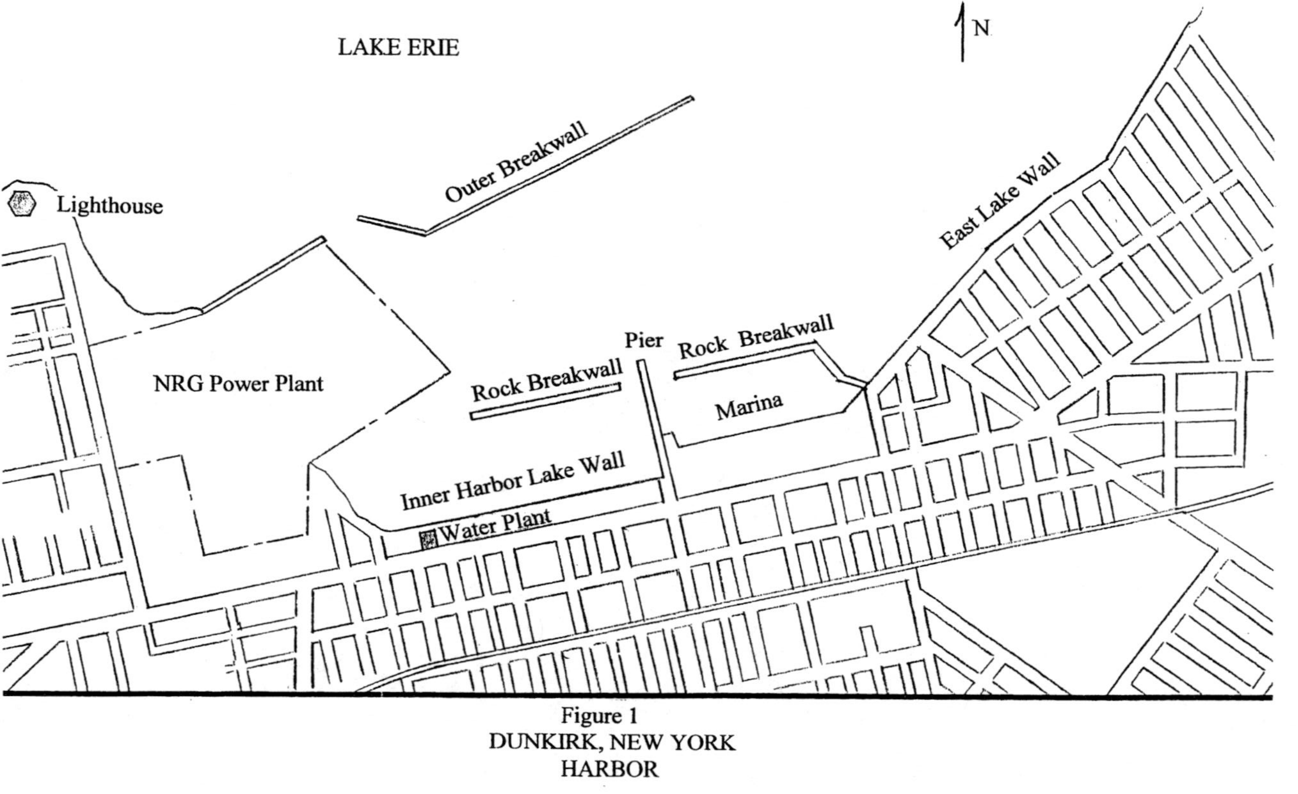

Figure 1
DUNKIRK, NEW YORK
HARBOR

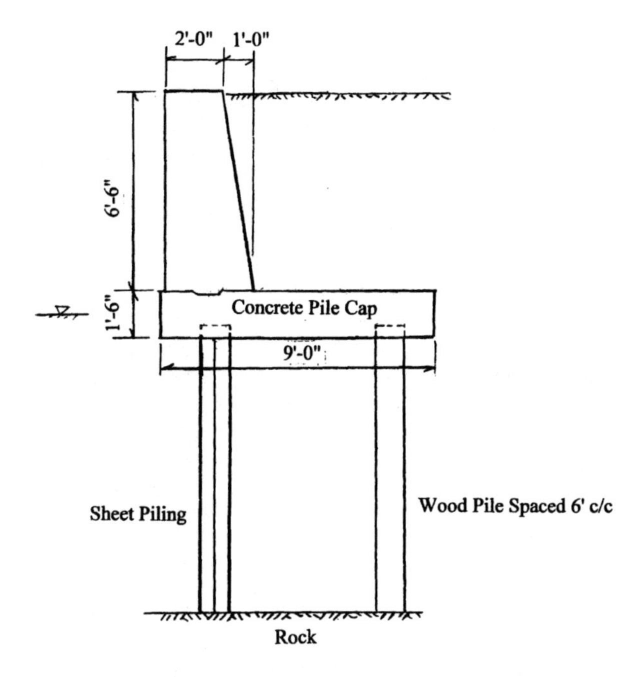

Figure 2
WALL CROSS-SECTION

close behind the wall except behind the seventh through ninth sections where the addition to the water treatment plant extends to 15 feet from the back top of the wall. A road/bike trail runs along the back of the wall through most of the length of the wall. In back of the water treatment plant the asphalt pavement runs up to the back of the wall.

Damage to the section of the wall behind the water treatment plant consists of tipping of the top of the wall outward, shearing of the keyway between sections from relative displacement, spall particularly at corners and along the top edge of the wall, and the ever present white laitance from the alkali-aggregate reaction.

The remainder of the wall was less than 3 inches out of plumb from top to bottom in most places. Two sections near the west end have tipping with an overhang of 7 inches top to bottom while two sections in back of the water treatment plant have an overhang of more than 8 inches. Major keyway failure at the ends of this three section segment provides the most visual evidence that caused concern about the possible failure of the wall.

Conclusions

The wall rotation about the toe of the wall, particularly behind the water treatment plant, and the resulting overhang are a sign of progressive failure of the system. If not checked it will eventually lead to collapse of the wall system. The rotation is probably caused by water freezing in the ground behind the wall and causing a small increment of deflection each year. Soils fall into the void when the ice melts and the cycle repeats. The deflections accumulate. The rotation is more pronounced behind the water treatment plant because the foundation is more rigid than the surrounding soil. This causes all of the deflection to occur in the direction of the lake where the lake wall is less rigid than the building foundation. A portion of the damage may have been caused by heavy vehicles moving between the water treatment plant and the wall. Heavy loads applied close to the wall will cause lateral forces, the resultant of which will be close to the top of the wall. These forces will cause large overturning moments. These loads may have been present during the construction of the water treatment plant addition.

The pavement that is immediately adjacent to the top of the wall may also contribute a significant outward force on the top of the wall. When the pavement is heated on warm days it expands and presses out at the edges. The pavement is restrained on the side opposite the wall by a berm and the building foundation.

The vertical joint keyway failures cause a loss of continuity, but there is probably no realistic economical fix. The loose concrete pieces might be dropped into the lake so they do not fall when someone climbs on them.

The spall is unsightly but is not a threat to the wall at this time.

Recommendations

The lake wall is an expensive and important structure. Its life can be extended indefinitely if collapse can be prevented.

The position of the wall must be documented to determine if it is still moving .The tip at each end of each concrete section can be recorded and rechecked against this baseline data every couple of years. Sections making unacceptable movement can be examined and a fix designed.

The section behind the water treatment plant should be treated on a more urgent basis. The distance between the wall and the building should be measured as accurately as possible. These dimensions should be measured twice each year to see if they are changing. If the dimension was increasing a potential fix using a foam backpacking was recommended. This would consist of a wall of closed cell foam in a trench about 4 feet deep and about 18 inches behind the wall. The thought is that the foam would dissipate the force caused by the ice formation and keep the wall from moving out. The road/bike trail would be repaved leaving an expansion joint between the top of the wall and the pavement. The measurements should be continued immediately after the construction and semiannually until it is clear the fix has worked.

Results

In response to the report on this wall the city set up permanent markers to measure the distance between the wall and the water treatment plant foundation and has been measuring this distance periodically. The first measurement was taken in March 2004. The measurement was checked four times through January 2005 with no changes. In April 2005 an increase of about 0.06 foot was recorded. The dimension has again stayed about the same through March 2006.

The author measured the overhang of each wall section and also made a photo record of the entire length of the wall. These data were distributed to the city , harbor commission and others in hope that they would be available if someone wished to check for any changes that might have taken place.

Lake Wall East of the Harbor

This section of wall extends about 0.8 mile along a residential neighborhood east of the inner harbor. A sidewalk, bike path and city street parallel the top of the wall. Sand beaches lie in front of the ends of the wall while shallow water lies off the center portion. An off shore reef breaks down the large waves, but the wall is still hit by numerous smaller waves. In high winds drivers frequently park along the top of the wall and let the spray plumes fall over the car. Fifty mile an hour winds off the lake are fairly common.

The wall is mass concrete cast in sections 64 and 48 feet long with a vertical keyway between the sections. No drawings are available, but it appears similar to the wall in the inner harbor shown in Figure 2. It is 6 feet high plus a one foot high cap that was added later. It is two feet wide at the top and the back may taper. The wall sits on a concrete leveling course which in turn rests on the slate of the lake bottom.

Conclusions

The wall is showing its age, but is generally in satisfactory condition. The face of the wall is vertical, indicating tipping has not occurred. Five problems occur to various degrees along the length of the wall. These are: Spall of the concrete from the face of the wall, spall of the concrete at the joints, deterioration of the cap of the wall, alkali-aggregate reaction and subsidence of the sidewalk behind the wall. These problems are discussed in more detail below.

The spall from the wall face and at the joints is serious in a few places. A length of about 440 feet near the center of the wall is badly eroded. The one foot cap is intact but the old concrete from the original wall has been severely eroded. In a few places the sidewalk slab is undermined by the combination of the wall erosion and the flow of water over the top of the wall. In addition, several joints are badly eroded on the lake side where water enters a small joint at the back of the wall and then causes extensive spall and erosion on the front side of the wall. This damage is believed to be caused by a combination of the alkali-aggregate reaction opening the concrete up to water penetration and runoff water coursing over the face of the wall and through the joints. Freezing of the water within the concrete spalls off the surface layer. During the winter the water contains road salt which makes the problem worse as the salt in the cracks attracts additional water which in turn freezes and increases the spall.

The cap of the wall varies from excellent condition to an almost gravel state. About 290 feet of the cap is badly deteriorated and needs to be replaced. This part of the wall is important because it anchors the safety railing which runs the length of the wall. The cap concrete in the section that is badly deteriorated was probably cast with too high of a water cement ratio which makes it susceptible to damage from freeze thaw cycles.

The sidewalk subsidence is caused by water running under the side walk and down the back of the wall. The water carries soil particles with and the resulting void causes the sidewalk to crack and settle.

Recommendations

Periodic maintenance should be performed on the cap and railing. Cap sections can be cut out and replaced as needed. The joints with the most serious spall could be patched. Chipping back to sound concrete may be a problem because of the alkali-aggregate reaction. The patches will not bond to the wall and will not match the wall in color. Most of the wall is serviceable as is. Patching is probably not worth doing except in the worst cases. The back of the joints can be sealed to reduce the flow through the joints.

The 440 feet of wall that has the worst erosion should be covered full height with a minimum 6 inch thick wall of temperature reinforced concrete. The facing should be tied to the existing concrete and cast from a high slump, low water cement ratio, air-entrained concrete.

The 290 feet of deteriorated cap section should be removed and replaced. The replacement concrete should have a low water cement ratio and be air entrained. The city should consider not using salt for ice control on the streets that drain over the wall. If necessary portions of the road might be closed when conditions warrant rather than salting.

Results

The above study for the lake wall east of the harbor will be given to the engineer selected to design the repair. The city is awaiting a grant to pay for the design. A second grant will be needed to do the work. Salting of the road behind the wall has been reduced.

Outer Breakwall

The concerns about the outer breakwall were focused on the erosion of the concrete and the vegetation that grows on the wall. The vegetation was thought to be unsightly and causing damage to the wall. The wall is about 3,200 feet long and sets as an island across the north side of the harbor. It receives the full fury of storms from the northwest and northeast. It is exciting to watch the waves burst over the wall in a big blow. The breakwall consists of mass concrete sections on the east and west ends with a center section of large cut stone blocks. The breakwall is built on a relatively shallow section of the lake bottom. An 1855 drawing shows the depth along the present course of the wall to be about eight feet. Presently the water depth south of the stone block section is only one to three feet in many places. The main mass of the wall is underwater where the main force on the wall probably comes from ice jamming against the wall. The wall is subjected to weathering from temperature changes, freeze-thaw cycles, wave action and plant roots. There is evidence of alkali-aggregate reactions in what appears to be the older concrete on the wall. It appears patches have been added to the original material.

The concrete sections of the breakwall show some deterioration from the combined forces acting on the wall. On the west end the concrete erosion is probably due to the combination of wave action and freeze -thaw cycles. The east end of the wall also shows some loss of concrete probably also due to wave action and freeze-thaw cycles, but here the weak point is not alkali-aggregate reaction but several cold joints between the various lifts of concrete. The stone sections appear to be in good condition. Protected from the direct pounding of the waves a number of bushes and small trees have a toehold on the harbor side of the wall. There is at least one plot of milkweed which is critical to Monarch Butterfly habitat. A variety of birds including terns, gulls, herons, cormorants, swallows, and small wading birds may be observed. We have also seen a pair of rock ptarmigan on the wall, but the local bird experts question our observation.

Conclusions

The outer breakwall is also showing it age and the effects of its environment. However, the small amount of mass that has been lost is insignificant compared to the mass that remains. The degradation caused by the vegetation is on a geologic time scale and is not significant. Indeed the weathering and vegetation have created a character that makes the breakwall an interesting and beautiful place. The bird population further enhances the setting, and argues against removing the vegetation.

Recommendation

The concrete of the outer breakwall does not need work at this time.

Results

The above report on the outer breakwall was reviewed by the harbor commission and reported in the local newspaper. We believe it will impact thinking when future harbor projects are prioritized. It has also drawn the attention of the local bird watchers to the outer breakwall.

General Conclusion

This report has been well received by city officials and the harbor commission. It will help to interject engineering rational into further discussions on work to be performed in the harbor area. It has also defined specific items that can be accomplished within a realistic economic scope. Providing the report to the city decision makers was an attempt to demonstrate the value of rational fact based discussion to the decision making process. In this regard, it was successful for the projects involving the harbor facilities. How the idea of fact based arguments as contrasted to emotion based statements will affect future discussions and decisions remains to be seen.

References

Chard, Leslie F., 1971, The City of Dunkirk, Dunkirk, New York

Kosmatka and Panarese, 1988, Design and Control of Concrete Mixtures, Portland Cement Association, Skokie, Illinois.

BLAST CAPACITY AND PROTECTION OF AASHTO GIRDER BRIDGES

A.K.M. Anwarul Islam, Ph.D., P.E., M.ASCE[1], Nur Yazdani, Ph.D., P.E., F.ASCE[2]

[1]Assistant Professor, Department of Civil & Environmental Engineering, Youngstown State University, One University Plaza, 2460 Moser Hall, Youngstown, OH 44555. Tel: (330) 941-1740, Fax: (330) 941-3265, Email: aaislam@ysu.edu.

[2]Professor and Chairman, Department of Civil & Environmental Engineering, University of Texas at Arlington, Box 19308, 425 Nedderman Hall, Arlington, TX 76019. Tel: (817) 272-5055, Fax: (817) 272- 2630, Email: yazdani@uta.edu.

Abstract

AASHTO has specified probability-based design methodology and load factors for designing bridge piers against ship impact and vehicular collision. Currently, no specific AASHTO design guideline exists for bridges against blast loading. Structural engineering methods to protect infrastructure systems from terrorist attacks are required. This study investigated the most common types of concrete bridges on the interstate highways and assessed the capacities of the critical elements. A 2-span 2-lane bridge with Type III AASHTO girders was used for modeling. AASHTO Load and Resistance Factor Design methods were used for bridge design. The girders, pier caps and columns were analyzed under blast loading to determine their capacities. This study determined the blast capacities of the AASHTO girders, pier caps and the columns, and the required standoff distance of explosion from the columns that may possibly protect the bridge from failure.

Keywords: Blast, explosion, bridge design, terrorism, AASHTO girders.

INTRODUCTION

Terrorist attacks around the world have created concern over the safety and protection of the nation's infrastructure. Bridges are the most common among the infrastructures in the nation's highway system. As a result of the terrorist threats and attacks, engineers and transportations officials are becoming more active in physically protecting the bridges from potential blast attacks. It is likely that these bridges were not designed to resist blast loads.

Since 9/11 terrorist attacks, numerous research and demonstration initiatives have been undertaken to find cost-effective and efficient retrofit, security and rapid reconstruction techniques for important buildings. The bridge and highway infrastructure engineers face new and largely unexpected challenges relating to the security of critical structures against terrorists attacks. In response to this need, the American Association of State Highway and Transportation Officials (AASHTO) Transportation Security Task Force sponsored the preparation of a guide to assist transportation professionals in identifying critical highway structures and to take appropriate action to reduce their blast vulnerability. In order to provide guidance to bridge owners and operators, the Federal Highway Administration (FHWA) formed the Blue Ribbon Panel (BRP 2003) on bridge and tunnel security. The panel's report includes recommendations on actions that can be taken by bridge and tunnel owners and operators or by FHWA, and other state and federal agencies, that will result in improved security and reduced vulnerabilities for critical bridges and tunnels. Additionally, to develop and transfer knowledge rapidly within the bridge community, a series of workshops were conducted in early 2003 under the National Cooperative Highway Research Program (NCHRP) Project 20-59(2).

AASHTO has probability-based design methodology for designing bridges against various dynamic loads such as seismic, ship impact and vehicular collision. However, it has no specific guidelines for blast resistant bridge design. In response to this vital need and growing concern about the safety of highway bridges, NCHRP has sponsored a research project to develop design and detailing guidelines for blast-resistant highway bridges that can be adopted in the AASHTO Bridge Design Specifications (NCHRP 2005).

The American Society of Civil Engineers (ASCE) has guidelines entitled "Design of Blast Resistant Buildings in Petrochemical Facilities" (ASCE 1997). This report provides general guidelines for structural design of blast resistant petrochemical facilities.

The National Center for Explosion Resistant Design (NCERD 2005) at the Department of Civil and Environmental Engineering at the University of Missouri-Columbia has been conducting workshops on "Explosion Effects and Structural Design for Blast." The workshop focuses on the fundamentals of explosion effects, determining blast loads on structures, computing structural response to blast loads, and the design and retrofit of structures to resist blast effects.

The results of ongoing research to develop performance-based blast design standards were summarized and discussed in an article for the incorporation of physical security and site layout principles into the design process (Winget 2005). The paper discussed the potential effects of blast loads on bridges and the probable solutions to counter these effects.

Bridges are less protected as compared to other structures such as high-rise buildings, federal and state offices, and other important structures. As traffic flow continues over the interstate and state highways 24 hours a day, it is a common perception that the bridges are protected to some extent by the moving traffic. The government has adopted utmost security measures for protecting important bridges in the United States, such as the Golden Gate Suspension Bridge in San Francisco, Sunshine Skyway Cable-Stayed Bridge in Tampa, and the Brooklyn Suspension Bridge in New York City. On the other hand, typical interstate and highway bridges are largely unprotected and vulnerable to terrorist attack.

AASHTO girder, suspension, cable-stayed and box girder bridges are the most commonly used concrete bridge types. Because of the complex nature of suspension, cable-stayed or box girder bridges, the blast performance of an AASHTO girder bridge was investigated herein. The intent of this study was to investigate the capacity of a typical AASHTO girder bridge under blast loading. The results are expected to help determine necessary structural design criteria or retrofit techniques to reduce the probability of catastrophic structural failure, which in turn may lessen human casualties, economic losses and socio-political impact due to explosion.

The passive protection of public buildings to provide life safety in the event of explosions is receiving renewed attention. This highly effective approach of blast resistant design is only feasible where a standoff zone is available and affordable. For many urban settings, the proximity to unregulated traffic brings the terrorist threat to or within the perimeter of the structure. For these structures, blast protection has more modest goal of containing damage in the immediate vicinity of the explosion and the prevention of progressive collapse. In suburban and government facilities, the minimum standoff zone can be easily established. In the urban area, it is relatively harder to attain these standoff distances surrounding the facilities.

MODEL BRIDGE

A typical Type III AASHTO girder simply supported bridge of 24.38 m span length, common for Type III AASHTO girders, was selected for this study. The model bridge was assumed to contain two 3.66 m lanes, 3.05 m and 1.83 m shoulders, and two 457 mm wide barriers, producing an overall bridge width of 13.11 m. Seven Type III AASHTO girders with center-to-center spacing of 1.83 m were used in the model bridge.

Clear roadway width of 12.19 m and overhang width of 1.07 m were used in the design. The cross-section view of the model bridge is shown in Fig. 1. One intermediate pier and two end bents supported the girders, which were connected with 203 mm thick cast-in-place concrete deck slab. Seven 457 mm square prestressed precast concrete piles, spaced at 1.83 m, supported each end bent. Two identical columns of 1.07 m diameter on two separate footings supported the pier cap. Four 610 mm square prestressed concrete piles supported each footing of dimension 3.05 m by 3.05 m with the column located at the center of the footing. The end bent cap and the pier have dimensions of 914 mm by 914 mm and 1.22 m by 1.22 m, respectively.

FAILURE CRITERIA

Failure criteria can be used in combination with information about stresses in a structure to predict the load levels a structure can withstand before failure. A widely accepted general failure criterion was followed in this study – if the applied load parameter exceeds the capacity of the section, the component is assumed failed.

Bridge columns and the pier caps are more vulnerable to blast attack because of easy access. Therefore, they are considered the most critical elements of a bridge structure in terms of failure and stability. Columns are more accessible and susceptible to damage in case of a ground-based or water-borne blast attack underneath the bridge. Although one column failure may or may not initiate bridge collapse, failure of both columns will definitely collapse the entire model bridge. Pier caps may be attacked over or under the bridge. Under-the-bridge explosion will directly attack the pier caps, while over-the-bridge explosion loads will be transmitted to the pier caps through the deck-girder system.

The girders play an important role in securing the superstructure and substructure by creating redundancy against sudden collapse. Girders are typically designed to withstand moments and shears caused by vertically downward loads. On the other hand, a vertically upward load, caused by blast load from underneath the bridge, produces negative moments and shears. The composite action of the deck-girder system increases the negative moment capacity of the section to a minor extent. The deck slab contributes to the most of the resistance. The girders exhibit minimal resistance because they are primarily designed for positive moments.

MODEL BRIDGE DESIGN AND ELEMENT STRENGTH

All major components of the model bridge were designed following the AASHTO LRFD Bridge Design Specifications, 2nd Edition, 1999, with Interims through 2003 (AASHTO 2003). The strength of the critical elements were determined from the design and later compared with the applied stress on the respective element to determine their blast capacities. The respective strength of each individual element was determined using simple support condition for the girders, and fixed support condition for connection between the pier cap and the columns.

Deck Slab

Following the Empirical Method of the AASHTO LRFD Specifications, the deck slab was designed with #5 reinforcement spaced at 305 mm on center at top and bottom in both directions with clear cover of 50 mm at top and bottom. In addition to this typical reinforcement, additional steel on top of the deck perpendicular to the direction of traffic was also used to secure connection between the barrier and the deck. To resist the negative flexure over the pier support, the slab was reinforced with additional reinforcement along the direction of traffic.

AASHTO Girder

Type III AASHTO girders were designed using the FDOT LRFD Prestressed Beam Design Program (FDOT 2001). The girders were prestressed with 13 mm diameter 1862 MPa low relaxation 30 straight strands of which 4 strands were debonded for 3.96 m at each end. The jacking force per strand was limited to 138 KN.

The ratio of the ultimate moment capacity to the applied moment due to LRFD Strength I loading was 1.17. Therefore, the girder design was acceptable under normal traffic loading.

Prestressed girders composite with the deck slab react with their fullest capacity against vertically downward loads resulted from over-the-bridge explosion. From the beam design output, the maximum positive moment capacity of the prestressed girder was 6.10 MN-m, while the maximum shear force capacity was 1.24 MN.

When the bridge deck experiences vertically upward loads generated through an explosion underneath the bridge, the girder-slab composite is subjected to negative moments. In this situation, the prestressing steel located near the bottom flange is ineffective in providing the needed tensile strength. Only the non-prestressed longitudinal reinforcement in the slab provides limited tensile strength. Manually analyzing the section properties of an inverted T-beam, the maximum negative moment capacity of the girder-slab composite was determined as 1.09 MN-m. The shear capacity of the girders remained unchanged at 1.24 MN.

Pier Cap

The pier cap and the end bent cap were designed using commercially available software RC-Pier (LEAP 2001). Multiple piles support the end bent caps with short spans between the piles, which decreases the applied moment and increases the cap stiffness. Moreover, the bottom of the end bent caps stays in touch with the ground. This makes them less critical and more protected compared to the pier cap in assessing the bridge performance. Therefore, end bent caps were excluded from this study. The pier cap was reinforced with 13 #10 bars at top and bottom with 75 mm typical clear cover at each side. Two-leg #5 double stirrups spaced at 150 mm on center were used to satisfy vertical shear and torsion. The ratio of the resisting moment to the applied moment caused by regular dead and live load was 1.9, which exceeds the minimum acceptable limit of 1.

Since the pier cap was reinforced with identical reinforcement at top and bottom, the negative or positive flexural strength remains the same. From RC-Pier output, the maximum resisting positive or negative moment of the pier cap was 4.00 MN-m. The maximum shear strength was 3.43 MN.

Column

The columns and the footings were also designed using the RC-Pier. Minimum reinforcement requirement governed in the column steel selection. Because columns are possibly the most critical component of the bridge, they were reinforced with the maximum permissible reinforcement of 20 #10 bars with #4 ties spaced at 305 mm, so that the maximum efficiency of the columns could be achieved. The ratio of the ultimate moment capacity to the applied moment due to normal traffic loading was 1.29, which is acceptable.

Using the FDOT Biaxial Column Program (FDOT 2001), the maximum resisting moment and axial capacity of the column were 3.28 MN-m and 8.90 MN, respectively. Footings and piles were not considered as critical components of failure since they were the buried components of the model bridge.

BLAST LOAD

Blasts can create very powerful and extreme loads. Even a small amount of explosive can inflict sizeable amount of damage to a structure if they are set at critical locations. Blast pressure can create loads on structures that are many times greater than normal design loads. If explosion occurs on top of the bridge, deck slab experiences the downward thrust of the overpressure, which is transmitted to the supporting girders, pier caps and columns. If blast load is applied at the bottom of the bridge, pier caps, prestressed girders and deck slab are subjected to vertically upward pressure, and the columns experience horizontal or inclined pressures.

Load Application Hypothesis

Similarities and differences between seismic and blast loading are noticeable. Both of these loads are dynamic loads and they produce dynamic structural response. The structural behavior in response to these loads is inelastic as well. Blast load damages structures through propagating spherical pressure waves, while earthquake damages structures through lateral ground shaking. The focus of structural design against these loads is on life safety as opposed to preventing structural damage. Therefore, the designs are normally performance-based that include life safety issues, progressive collapse mechanisms, ductility of certain critical components, and redundancy of the whole structure. There are growing interests in determining the effects of seismic design on blast resistance of structures. Although blast load is a dynamic load and it impacts the structure for a very short duration, equivalent static loads due to explosion were used in assessing the structural performance. There may be some minor variation in the results between equivalent static and dynamic analysis of the model bridge because of impact and sustained loading, but the overall performance of a structure would be fairly close in each of these two types of analyses. Although earthquake produces dynamic load, similar analogy of converting dynamic loads into static loads has been used with acceptable accuracy for structural designs. Therefore, performance of bridge elements under equivalent static loads can be considered as reasonably similar to that under the original dynamic blast loads.

The method of determining equivalent static load due to an explosion is a complex phenomenon. The blast pressure diminishes with distance from the point of explosion. In the TM 5-1300 Manual (DoD 1990), an empirical formula, as shown in Eq. 1, was used to find the scaled distance. The amount of blast pressure generated due to an explosion is inversely proportional to the scaled distance, which is presented in a chart form in the Manual. The empirical formula for the scaled distance, Z, is:

$$Z = R/W^{(1/3)} \tag{1}$$

Where, R = Distance of target from point of explosion, and W = Equivalent TNT weight of explosive.

Using this formula and the chart in TM 5-1300, Applied Research Associates, Inc. (ARA) developed a software named ATBlast (ARA 2003) to calculate the blast loads for known values of charge weight and standoff distance. The ATBlast was developed for the US General Services Administration to convert dynamic blast load into equivalent static load. In fact, ATBlast software is widely used and recommended by professionals to determine the equivalent blast pressure due to an explosion.

ATBlast estimates the developed blast loads during an open-air explosion. The software allows the user to input the minimum and maximum range, increment, explosive charge weight, and the angle of incidence. From this information, ATBlast calculates the following values: Range Distance (m), Shock Front Velocity (m/msec), Time of Arrival (msec), Pressure (MPa), Impulse (MPa-msec), and Duration (msec). The results are displayed in a tabular format and can be printed. In addition, the resulting pressure and impulse curves can be displayed graphically. ATBlast is proprietary software developed by ARA, and is provided at no cost to the users.

For an explosion underneath the bridge, it is reasonable to assume that a regular truck or any other vehicle commonly used to carry explosive charges cannot go closer than 1.22 m to a bridge column, and this minimum standoff distance from the point of explosion to the column surface was used herein. The typical minimum vertical clearance of 4.88 m between the bottom of the girder and the top of the roadway underneath was considered in the analysis. The bottom of the girder and the deck slab were determined to be 3.96 m and 5.18 m away, respectively, from the point of explosion, considering the charge was placed on the truck bed at 914 mm above the ground. On the other hand, when explosion occurs on top of the bridge deck, the truck bed, where the explosive is placed, also acts as a barrier between the explosion and the deck surface. Considering this barrier effect and 914 mm height of the truck bed from the deck surface, it was conservatively assumed that the minimum distance between the point of explosion and the deck surface was 1.83 m.

Following trial and error procedure, amount of TNT was determined for each critical case with minimum and maximum ranges as explained above, and was converted into equivalent static loads using ATBlast at every 305 mm increment. The resulting static loads, as presented in Table 1 for various amounts of TNT explosion, were applied on the model bridge at different critical locations following load cases presented in Table 2. Figure 2 illustrates variation of pressure with respect to distance from the point of explosion. Closer proximity of explosion to the structure produces more severe resulting pressure and the likelihood of increased structural damage.

Figure 3 shows the plan view of a typical blast pressure distribution on the bridge surface. If the explosion occurs 1.83 m above the deck surface, the spherical distribution of the pressure extends 3.05 m in each direction assuming a 30-degree angle of projection of the pressure wave.

Blast load on structures has no definite direction of application. It can affect the structure from any direction at any angle. For the sake of simplicity, only the governing vertical or horizontal components of the inclined loads were applied at critical locations on the members.

For simplifying the method of blast distribution, it was assumed that the blast pressure beyond the 30-degree projected region has negligible impact on the structure. Weighted average of the vertical components of these inclined pressures on each girder was calculated according to the distribution shown in Fig. 3. The greatest pressure, generated due to an explosion of 226.8 kg of TNT at a height of 1.83 m above the deck, is 10.2 MPa from Table 1. Girder B, directly under the point of explosion, experiences the highest average pressure of 5.1 MPa for a length of 6.10

m, which is approximately 50% of the peak pressure of 10.2 MPa. The adjacent girders A and C are subjected to a pressure of 3.06 MPa from Table 1 along a length of 6.1 m, which is approximately 30% of the peak pressure. These assumptions were verified for three different intensity of explosions (226.8 kg, 45.4 kg and 22.7 kg of TNT) found to be acceptable, and were used in applying average blast pressure for different load cases on the bridge components.

FEA BRIDGE MODEL

The bridge was modeled in the STAAD.Pro (REI 2004) software for analysis, as shown in Fig. 4. The centerlines of the elements were coded in and connected as beam elements. The deck slab was modeled as a 203 mm thick plate composed of plate elements. A total of 58 beam elements and 128 plate elements were used in the model.

Beam elements 1 to 8 and 17 to 24 formed two end bent caps, and 9 to 16 made up the pier cap. The barriers consisted of beam elements 25, 33, 34 and 42. The girders were represented by beam elements 26 to 32 in the first span and 35 to 41 in the second span. The piles supporting the end bent caps consisted of beam elements 43 to 49 and 52 to 58. Beam elements 50 and 51 represented the two columns.

Cross-sections of all the beam elements were defined as per geometry of the respective members. The equivalent areas of the girders and barriers were coded in the software because of their irregular shapes to account for the dead loads. The deck slab was made integral with the girders to represent the composite behavior.

The girders, end bents and pier cap including the columns, and the deck slab were assumed to be made of concrete Material 1, Material 2 and Material 3, respectively, with properties defined in Table 3. Since displacement and camber of girders are normally considered for traffic riding comfort, rather than predicting member efficiency, they were not used in assessing the capacity of the model bridge elements.

The deck slab is mostly affected due to possible explosion on top of the bridge. It is a highly redundant member due to the presence of alternate load paths and integral connection with the girders through shear keys. Any possible slab damage due to blast is likely to be localized. The loads are distributed on the deck slab and ultimately applied as uniformly distributed loads along the centerline of the girders. Thus, the deck slab performance was excluded from this study in order to focus on the more critical elements of failure, such as girders, pier cap and columns.

BRIDGE CAPACITY

Using STAAD.Pro bridge model analysis, it may be worthwhile to determine the actual capacity of the bridge components for the chosen design. The blast capacities of the components were calculated and presented in the following sections.

AASHTO Girder Capacity

Only load Cases 1 and 2 locations were considered herein for the blast capacity analysis of the Type III AASHTO girders. In case the explosion occurred underneath the bridge for Case I loads, as shown in Fig. 5(a), the blast load produced due to an explosion of 6.58 kg of TNT underneath the bridge at 914 mm above the ground was found to be the maximum load the Type III AASHTO girder could resist before failure. The girders and the bottom of the slab were assumed to be at the height of 3.96 and 5.18 m, respectively, above the explosion. The average pressure due to this explosion at the height of 3.96 m was 125 KPa (Table 1, using 50% Distribution) on girder 29, and 75 KPa (Table 1, using 30% Distribution) on girders 28 and 30. The equivalent uniformly distributed loads due to these pressures were calculated as 69 KN/m and 42 KN/m, respectively. The bottom of the slab at the height of 5.18 m above the explosion was subjected to an average pressure of 41 KPa (Table 1, using 30% Distribution) for a width of 1.22 m in between two adjacent girders, and a length of 6.1 m along the length of the bridge. The uniformly distributed load produced by this pressure was calculated as 50 KN/m equally distributed along the centerline of each girder. The total load applied in the middle 6.1 m along the centerline of girder 29 was 119 KN/m, and on girders 28 and 30 was 92 KN/m.

The maximum negative moment produced by these upward pressures on girder 29 is 1.04 MN-m, which is less than its capacity of 1.09 MN-m. The maximum shear force produced by these loads is 373 KN, which is less than its shear capacity of 1.24 MN. Therefore, the model bridge Type III AASHTO girders of the chosen design can resist pressures produced by 6.58 kg of TNT explosion underneath the bridge.

When the explosion occurred on top of the bridge for Case 2 loads, as shown in Fig. 5(b), the blast load produced due to an explosion of only 0.77 kg of TNT at 1.83 m above the bridge deck was found to be the maximum the Type III AASHTO girder could resist before failure. The average pressures due to this explosion are 145 KPa (Table 1, using 50% Distribution) on girder 29, and 87 KPa (Table 1, using 30% Distribution) on girders 28 and 30. The equivalent uniformly distributed loads are 265 KN/m and 159 KN/m, respectively, applied along the centerline in the middle 6.10 m of the girders.

The maximum bending moment produced on girder 29 by these loads is 5.65 MN-m, which is less than its capacity of 6.1 MN-m. The maximum shear force on the girder due to these loads is 1.24 MN, which is equal to its capacity. For over the bridge explosion, shear capacity controlled the amount of explosive the Type III AASHTO girder could resist before failure. In general, the prestressed girders are not designed with much excess capacity beyond the minimum requirement. In this model bridge, the girders had only 17 percent more flexural capacity than required for the AASHTO Strength I loading combination. Therefore, it appears reasonable that the girders can resist pressures caused by only 0.77 kg of TNT explosion over the bridge.

Pier Cap Capacity

The maximum amount of blast load that the pier cap can resist before failure was produced by an explosion of 0.77 kg of TNT at 1.83 m above the bridge deck. For pier cap capacity, explosion was assumed to occur right on top of the pier cap, similar to load Case 3, as shown in Fig. 6(a). The pressures produced were converted to equivalent uniformly distributed loads, and applied on the middle three girders on each side of the pier cap for a length of 3.05 m. The pressure applied on the middle girder and the adjacent two girders were 145 KPa (Table 1, using 50% Distribution) and 87 KPa (Table 1, using 30% Distribution), respectively. The equivalent uniformly distributed load for 145 KPa pressure was calculated as 265 KN/m, and applied on girder 29 and 38 along their centerline for a length of 3.05 m each way from the centerline of the pier cap. The equivalent uniformly distributed load for 87 KPa pressure was calculated as 159 KN/m, and was applied on girder 28, 30, 37 and 39 along their respective centerline for a length of 3.05 m each way from the centerline of the pier cap.

The maximum bending moment on the pier cap produced by these loads is 3.89 MN-m, which is less than 4.0 MN-m, the capacity of the pier cap. The maximum amount of shear force on the pier cap as a result of this explosion is 2.72 MN, which is below the shear capacity of 3.43 MN. Therefore, flexural strength controlled the failure of the pier cap, which can resist blast pressures produced by 0.77 kg of TNT explosion.

Column Capacity

The column blast capacity was investigated herein for two different scenarios of explosion, on top of the bridge and under the bridge. Applying loads on top of the bridge followed the same method and analogy of Case 3 loading pattern, as shown in Fig. 7(a). In the first scenario, pressure produced by the explosion of 2.59 kg of TNT 1.83 m above the deck was applied on the middle girders of both spans for a length of 3.05 m each way from the centerline of the pier cap. Pressure produced by the above explosion was 361 KPa (Table 1, using 50% Distribution) on girders 29 and 38. This load was converted to an equivalent distributed load of 660 KN/m and applied along the centerline of each girder. The other girders 28, 30, 37 and 39 experienced a pressure of 216 KPa (Table 1, using 30% Distribution), which was converted to 395 KN/m and applied along the centerline of each girder.

The applied moment and axial force on the column produced by these loads are 3.17 MN-m kip-ft and 6.93 MN, respectively, which are less than its moment capacity of 3.19 MN-m and the corresponding axial capacity of 7.12 MN.

The second scenario, explosion underneath the bridge, followed Case 5 blast loading location, as shown in Fig. 7(b). In this scenario, pressures generated by the explosion of 5.9 kg of TNT at a standoff distance of 1.22 m from the column surface were applied on the column and the bottom of the pier cap. Column 51 at a distance of 1.22 m from the point of explosion experienced a pressure of 1.47 MPa (Table 1, using 50% Distribution), and the pier cap elements 13, 14 and 15 at a distance of 2.74 m from the explosion were subjected to a pressure of 270 KPa (Table 1, using 50% Distribution). The converted uniformly distributed loads due to these pressures were 1.34 MN/m applied on the column 1.22 m above the ground for a length of 2.44 m, and 329 KN/m applied on the pier cap for a length of 5.49 m. All these loads were applied along the centerline of the elements.

From this loading pattern, the applied moment and corresponding axial force on column 51 are 2.52 MN-m and 1.13 MN, respectively, which are less than its moment capacity of 2.54 MN-m with the corresponding axial capacity of 1.16 MN.

It is evident from these comparisons that the column can resist the blast pressure generated by 5.9 kg of TNT if it is applied underneath the bridge. When the blast generates over the bridge, it can withstand the pressure produced by 2.59 kg of TNT explosion without failure. The reason for this type of performance is the moment arms of the applied loads. Over the bridge explosion produces larger moment arm for the columns than under the bridge explosion.

The capacity of each of the typically designed critical members under consideration is summarized in Table 4.

Minimum Standoff Distance for Column

Physical security of bridge components against blast loading is of great importance to highway departments and bridge engineers. One major component of this concept is placing passive standoff barriers around bridge components, which will prevent easy access to these components for blast attack purposes. Such barriers are a very economic and convenient alternative to bridge retrofit for existing bridges and design enhancement for new bridges. Standoff barriers are only applicable around bridge columns; it is not possible to provide such barriers for above-the-bridge explosion without closing one or more traffic lanes. The minimum standoff distance for a 226.8 kg of TNT explosion near the model bridge pier was determined as follows.

A separate model was developed and analyzed herein with the blast location similar to load Case 5. By using trial and error, it was found that the model bridge column could resist blast pressures generated due to 226.8 kg of TNT, which was adopted as a typically expected amount of blast load from Table 3 of Blue Ribbon Panel's Report, if the explosion occurred at a distance of at least 4.88 m from the face of the column. The full exposed height of the column was subjected to a horizontal blast pressure of 1.07 MPa (Table 1, using 50% Distribution) due to an explosion of 226.8 kg of TNT at a distance of 4.88 m from the column surface. The equivalent uniformly distributed load of 978 KN/m, calculated for this blast pressure, was applied along the centerline of column 51 for its exposed height of 3.66 m. Simultaneously, the pier cap, at a distance of 5.49 m, experiences a horizontal blast pressure of 507 KPa (Table 1, using 30% Distribution). This blast pressure was converted to an equivalent uniformly distributed load of 618 KN/m, and applied on pier cap elements 13, 14 and 15. The moment and axial force on column 51 produced by these loads are 2.67 MN-m and 2.7 MN, respectively. These loads are almost equal to and slightly less than the column moment capacity of 2.69 MN-m at axial loads of 2.71 MN. Therefore, if the standoff distance of at least 4.88 m from the face of the columns is maintained, the model bridge columns are capable of resisting the blast pressure generated due to 226.8 kg of TNT explosion.

Logistics may or may not allow the enforcement of this minimum standoff distance for actual bridges. It may be possible to place physical barriers around the columns of bridges over waterways, if sufficient clearance can be maintained for navigational channels (if present). For bridge overpasses over land, such high standoff distances may not be feasible because they may encroach on adjacent traffic lanes or other property.

CONCLUSION

The following conclusions may be made based on the results of this study:

1. Typical Type III AASHTO girder bridges are very susceptible to extensive damage and possible failure from blast loading that may occur over or under the bridge.
2. If the explosion is set over the bridge, Type III AASHTO girders as well as the pier cap may resist only 0.77 kg of TNT explosion. For an explosion underneath the bridge, the girder may resist as much as 6.58 kg of TNT explosion. The larger explosion capacity underneath the bridge is due to the greater standoff distance, as compared to over the bridge explosion.
3. For over the bridge explosion, the bridge column may resist explosions produced by up to 2.59 kg of TNT. The columns can withstand more than twice this amount for under the bridge explosion, about 5.9 kg of TNT explosion.
4. Type III bridge girders may fail due to the blast loads as specified above in either flexural or shear mode, depending on the location of the blast set off. The pier caps will most likely fail in a flexural mode. The columns will likely fail in a combined flexural-axial mode.
5. Typical Type III AASHTO Girder type bridge components have very low blast load capacity, certainly much lower than the typical truck-carried 226.8 kg of TNT load as published by the AASHTO Blue Ribbon Panel.
6. Passive protection of bridges through provision of limited access to bridge piers may be a very economic and convenient deterrent to potential blast attack. This may be provided through ensuring minimum stand off distance around pier caps provided through physical barriers. For the typical Type III AASHTO girder bridge studied herein, a minimum stand off distance of 4.88 m may be sufficient for the substructure to resist the resulting blast loads without failure.

Appendix I: References

Recommendation for Bridge and Tunnel Security (2003), AASHTO Blue Ribbon Panel on Bridge and Tunnel Security.

National Cooperative Highway Research Program (2005), Project No. 12-72.

Design of Blast Resistant Buildings in Petrochemical Facilities (1997), ASCE Task Committee on Blast Resistant Design.

Explosion Effects and Structural Design for Blast (2005). A Two-Day Training Course, National Center for Explosion Resistant Design, Department of Civil and Environmental Engineering, University of Missouri-Columbia.

Winget, D.G., Marchand, K. A., Williamson, E.B. (2005). "Analysis and Design of Critical Bridges Subjected to Blast Load." ASCE Journal of Structural Engineering, Vol. 131, No. 8, pp. 1243-1255.

Load and Resistant Factor Design (2003), Bridge Design Specifications, American Association of State Highway and Officials.

Florida Department of Transportation (2001), LRFD Prestressed Beam Program, English v1.85.

LEAP Software, Inc., RC-Pier 2001, Tampa, Florida.

Florida Department of Transportation (2001), Biaxial Column Program, v2.3.

TM 5-1300 (1990), Structures to Resist the Effects of Accidental Explosions, US Department of Defense.

ATBlast Software (2003), Developed by Applied Research Associates (ARA), Inc.

Research Engineers International (2004), STAAD.Pro 2004, USA.

Table 1 – Converted Pressure for Various Amounts of TNT Explosion

Range (m)	Pressure (Mpa) due to TNT Explosion				
	226.8 kg	6.58 kg	5.9 kg	2.59 kg	0.77 kg
(1)	(2)	(3)	(4)	(5)	(6)
1.22	17.31	3.14	2.94	1.73	0.73
1.52	12.99	2.07	1.92	1.08	0.43
1.83	10.20	1.43	1.32	0.72	0.29
2.13	8.26	1.03	0.94	0.51	0.20
2.44	6.83	0.76	0.70	0.37	0.15
2.74	5.74	0.58	0.54	0.28	0.12
3.05	4.88	0.46	0.42	0.22	0.09
3.35	4.18	0.37	0.34	0.18	0.08
3.66	3.61	0.30	0.28	0.15	0.06
3.96	3.14	0.25	0.23	0.12	0.06
4.27	2.75	0.21	0.19	0.11	0.05
4.57	2.42	0.18	0.17	0.10	0.06
4.88	2.14	0.16	0.14	0.08	0.05
5.18	1.90	0.14	0.12	0.08	0.04
5.49	1.69	0.12	0.11	0.06	0.04

Table 2– Critical Load Cases for Model Bridge Analysis

Load Cases (1)	Location (2)	Members Affected (3)	Blast Set-backs (4)
1	Under the bridge, at mid-span.	Deck slab, girders	914 mm above ground.
2	Over the bridge, at mid-span.	Deck slab, girders	1.83 m above deck.
3	Over the bridge, over pier cap.	deck slab, girders, pier cap	1.83 m above deck.
4	Over the bridge, at span end.	Deck slab, girders	1.83 m above deck.
5	Under the bridge, at 1.22 m away from column.	Column, pier cap.	914 mm above ground.

Table 3 – Concrete Material Properties used in Model Bridge

Concrete Designation (1)	Element (2)	f_c (MPa) (3)	Poisson's Ratio (4)	W_c (kg/m^3) (5)	E_c (MPa) (6)
Material 1	Girders Bent Cap	44.82	0.2	2,400	33,700
Material 2	Pier Cap Columns	37.92	0.2	2,400	31,000
Material 3	Deck Slab	31.03	0.2	2,400	28,040

Note: E_c = modulus of elasticity of concrete;
f_c = 28-day compressive strength of concrete;
W_c = unit weight of concrete;

Table 4 – Blast Load Capacity of Critical Components of Model Bridge

Typically Designed Member (1)	Maximum Weight of TNT Explosion Resisted: Over-the-Bridge Explosion (2)	Under-the-Bridge-Explosion (3)
Type III AASHTO Girder	0.77	6.58
1.22 m X 1.22 m Pier Cap	0.77	--
1.07 m Diameter Column	2.59	5.90

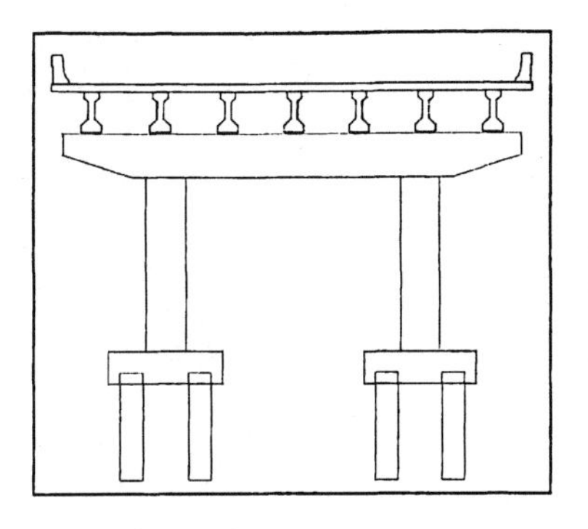

Figure 1: Model Bridge Cross-Section

Figure 2: Variation of Blast Pressure with Distance.

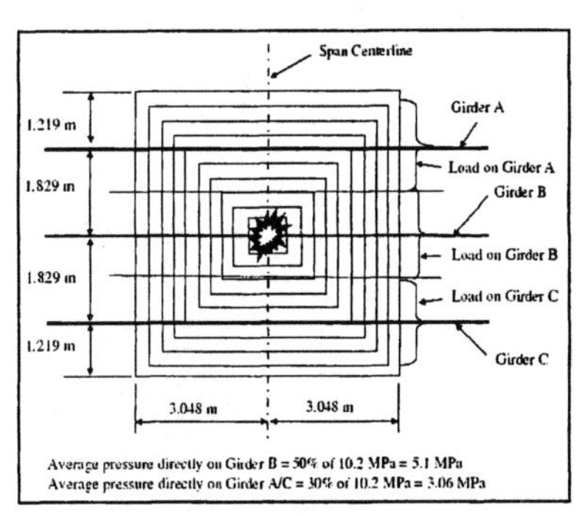

Figure 3: Blast Pressure Distribution on Bridge Deck.

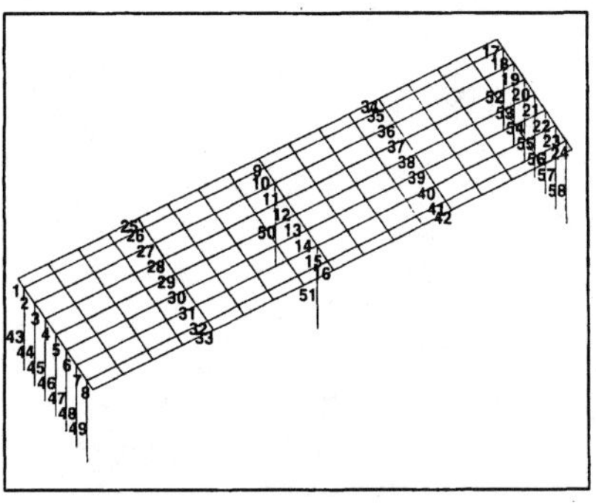

Figure 4: Model Bridge in STAAD.Pro

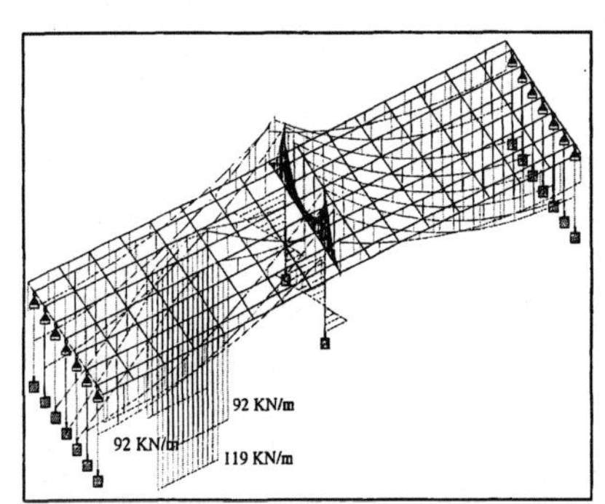

(a) Load Case 1

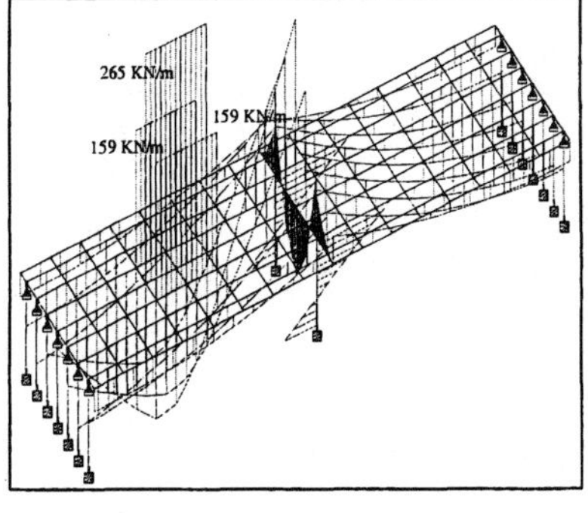

(b) Load Case 2

Figure 5: Load and Moment Diagrams for Girder Capacity.

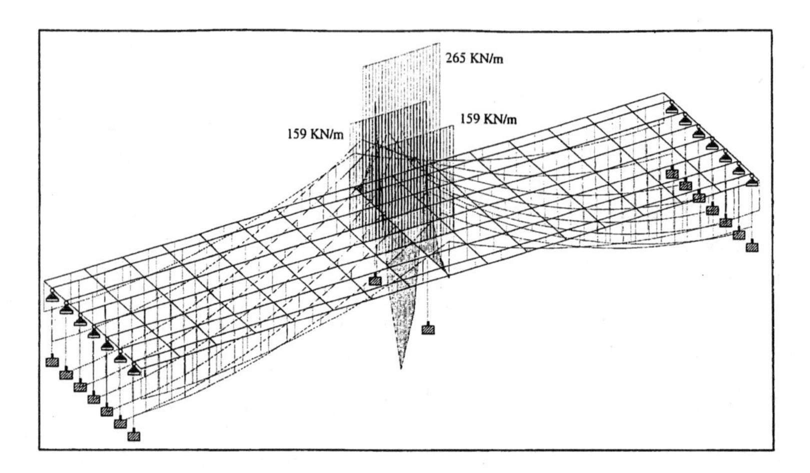

Load Case 3
Figure 6: Load and Moment Diagrams for Pier Cap Capacity.

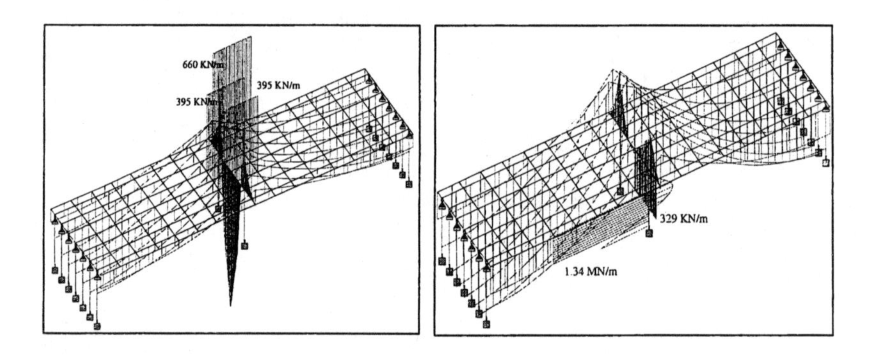

(a) Load Case 3 (b) Load Case 5

Figure 7: Load and Moment Diagrams for Column Capacity.

Environmental Tragedy of Love Canal

Yung-Tse Hung[1], F-ASCE, Paul A. Bosela[2], M-ASCE, and Alicia Saunté Phillips[3]

[1]Professor, Department of Civil and Environmental Engineering, Cleveland State University, Cleveland, Ohio 44115-2214; PH (216) 687-2596; FAX (216) 687-5395; email: y.hung@csuohio.edu
[2]Professor and Chair, Department of Civil and Environmental Engineering, Cleveland State University, Cleveland, Ohio 44115-2214; PH (216) 687-2597; FAX (216) 687-5395; email: p.bosela@csuohio.edu
[3]Graduate Student, Department of Civil and Environmental Engineering, Cleveland State University, Cleveland, Ohio 44115-2214; PH (216) 687-2400; FAX (216) 687-5395

Abstract

The purpose of the paper is to provide the historical background of Love Canal tragedy and the lessons learned from the environmental tragedy. In 1948, it was discovered that hazardous waste had contaminated homes and schools in the Love Canal area. During the summer of 1978, the Love Canal first came to international attention. On August 7, 1978, the United States President Jimmy Carter declared a federal emergency at the Love Canal, a former chemical landfill which became a 15-acre neighborhood of the City of Niagara Falls, New York. The Love Canal became the first man-made disaster to receive such a designation based on a variety of environmental and health related studies. As a result of grass roots and interest and media attention, the Love Canal provided an impetus for dramatic interest in changes to environmental concerns worldwide. The tragedy shows the importance of identification of hazardous waste and the proper disposal of hazardous waste for protection of public health.

Keywords
Superfund cleanup, Landfill failure

Introduction

The Love Canal is a neighborhood in Niagara Falls, New York. It covers 36 square blocks in the far southeastern corner of the city, along what is now known as 99th Street. Two bodies of water define the northern and southern boundaries of the neighborhood: Bergholtz Creek to the north and the Niagara River one-quarter mile (400 m) to the south.

The name Love Canal came from the last name of William T. Love, who in the early 1890s envisioned a canal connecting Lake Ontario and Lake Erie. He believed it would serve the area's burgeoning industries with much needed hydroelectricity. After 1892 Love's plan changed to incorporate a shipping lane that would bypass Niagara Falls. Due to economic depression, Love's plan failed. Only one mile (1.6 km) of the canal, stretching northward from Niagara River, was ever dug (http://en.wilkipedia.org/wiki/Love_Canal; September 21, 2005).

The project was abandoned when a section of the canal about 1000 m (3200 ft) long and 24 m (80 ft) wide had been excavated to a depth of the order of 6 m (20 ft). In 1942, the Hooker Chemical and Plastic Company (Hooker) purchased the abandoned excavation site from the Niagara Power and Development Company and began using the canal excavation as a dumpsite for industrial wastes that included pesticide residues, process slurries and waste solvents. In total, approximately 22,000 tons of waste contained in metal drums was placed in the excavation during an eleven year period. Later studies would show that more than 200 different chemical compounds including at least 12 known carcinogens were present. Once filled, the excavation was capped with a loose soil cover (Brown, M. H., 1979)

History and Background

The former Love Canal landfill is a rectangular, 16-acre tract of land located in the southeast end of the City of Niagara Falls (est. pop. 77,050) in Niagara County (est. pop. 242,200) on the western frontier of New York State.

Aerial photography from 1938 depicts the canal as being about 3,000 feet long and almost 100 feet wide, extending in a north-south axis, with the southern end approximately 1,500 feet from the Niagara River. Much of the canal bed contained impounded water and there was no visible evidence of waste disposal in 1938. The excavation was reportedly used as a swimming hole for local residents for several decades into the twentieth century.

Manufacturing of chemical and allied products was and is a major industrial enterprise of Niagara County. According to 1970 data from the New York State Department of Commerce, there were in the county nine major chemical-producing companies employing a total of 5,267 people. Recent surveys by the State Department of Environmental Conservation point to the presence of approximately 100 chemical dump sites in the county (New York State Department of Health, 1981)..

One of these is the Love Canal landfill, in which the Hooker Electrochemical Company, now the Hooker Chemical and Plastics Corporation (Hooker), admits to the deposition, between 1942 and 1953, of 21, 8000 tons of chemical wastes form its plants in Niagara Falls. At was common at the time; the company did not install a liner to prevent leaching (Colten, C., Skinner P., 1996). The wastes included various chlorinated hydrocarbon residues, processed sludge, fly ash and other materials, including municipal garbage that the City of Niagara Falls had disposed there for a number of years, concluding in 1953 (New York State Task Force on Toxic Substances Files, 1981). Approximately 200 chemicals and chemical compounds have been identified there, originally disposed as liquids and solids in metal drums and other types of containers, according to a November 1978 memo from the New York Commissioner of Health (New York State Department of Transportation Files, 1989). Occidental later explained that the site had been chosen because it was sparsely populated at the time, even though six homes were already constructed adjacent to the canal (Silverman, G., 1989). Another factor was that the local geology provided some degree of natural containment due to deposits of soft clay underneath the canal that provided low permeability, thus limiting the potential for groundwater contamination within the layer of glacial till below.

In April 1953, Hooker sold the Love Canal property, to which it then held title, to the City of Niagara Falls Board of Education. Homebuilding directly adjacent to the landfill was accelerated in the mid-1950s and in 1954 a public elementary school was built on the middle third of the Love Canal property.

Aerial photography from 1956 shows continuing residential development and soil banks, some of them as high as 15 feet, surrounding parts of the canal bed. By 1966 these hills were no longer apparent, and two streets crossed the landfill north and south of the public elementary school. By 1972, virtually all houses with backyards directly abutting the landfill were completed (New York State Task Force on Toxic Substances Files, 1981)..

History of the Problem

Although the disposal of hazardous waste at Love Canal dates back to the early 1940s, the contamination of homes located near the site did not become evident until the mid-1960s when residents complained of fumes and minor explosions. During the construction of the LaSalle Expressway, noxious fumes, corrosive waters and oily materials were encountered, according to State personnel and local residents. When Read Avenue was installed some 13 years ago, drums were exposed during excavation work, which allowed the release of noxious fumes and oily liquids, causing several work stoppages. Noxious fumes and hazardous liquid chemicals were detected in various storm sewers, mostly to the west of the site, and at the outfall which collected flow from both the 97th and 99th Street sewer lines.

In addition to these problems, land subsidence in the grammar school playground occurs regularly, and the holes are periodically filled with soil. School personnel reported to the County Health Department that school children handled waste phosphorous and received burns. In 1976, the New York Department of Environmental Conservation (NYDEC) conducted its first investigations of suspected leaching into nearby sewers and basement sumps. Based on these and subsequent testing the following year, NYDEC hired an environmental consulting firm, Calspan Corporation, to conduct its own studies and later enlisted the help of the New York

Department of Health (NYDOH). In February 1978, NYDOH reported finding "quantitatively significant" levels of chemicals such as toluene and several benzene compounds in sump samples from eight homes located directly adjacent to the site. Still nothing was done to rectify the problems. It was not until the summer of 1978 that a widespread contamination of the entire neighborhood became evident (Fletcher, T, 2001; and New York State Task Force on Toxic Substances Files, 1981).

The Niagara Falls and nearby Buffalo communities are known for having harsh winter conditions associated with heavy lake-effect snowfall, due to their proximity to Lake Erie. Added to that, the record-breaking Blizzard of 1978 and several other storms that season resulted in even more winter and spring precipitation than is usual for the area (DeLaney, K., 2000). In the following summer there was a widespread leaching of chemicals at Love Canal, due to what has been called the "bathtub effect" whereby water percolated through the clay cap, mixed with the chemicals and seeped laterally through sand and silt as the trench overflowed . The chemicals previously contained in the canal thus emerged at the ground surface and migrated into the basements of homes. The homes adjacent to the canal were affected most, but the contamination also spread further (Fletcher, T, 2001).

On August 2, 1978, Dr. Whalen declared a medical State of Emergency at Love Canal and ordered the immediate closure of the 99th Street School. The second health order spoke directly to the health of children with recommendation that families with children under the age of 2 years and pregnant women living nearest the canal should relocate temporarily. The designated area included homes adjacent to the canal and those across the street, later regarded as Rings I and II. People living within this boundary were also urged to avoid using their basements and to stop consuming food from their gardens. The order stated that there was "growing evidence that there is a higher risk of subacute and chronic health hazards as well as spontaneous abortions and congenital malformations", presumably limited geographically to the area affected by the health order (Fletcher, T, 2001).

The parameters of the health orders continued to expand over the following days and weeks. Five days after the second health order was issued, President Jimmy Carter declared a federal state of emergency in the area, the first ever for a technological hazard (Silverman, G., 1989).

Toxicological Investigation

Since March 1978, the State Health Department's Division of Laboratories and Research has carried out more than 6,000 analysis of environmental and biological samples associated with the Love Canal.

The U.S. Environmental Protection Agency (USEPA) also conducted extensive air, water and soil sampling in homes and yards throughout the Love Canal neighborhood, following a federal emergency declaration in May 1980.
The primary goals for the environmental and toxicological studies were to:

- identify the chemical compounds present in the Love Canal environment
- to establish whether the kind or degree of chemical exposure bears a relationship to observed health effects
- to determine the extent and means of chemical migration outward from the landfill
- to validate the efficacy of remedial construction work undertaken at the site
- to develop improved methodologies for analyzing toxics in environmental samples and biological specimens.

At the request of the State Interagency Task Force on Hazardous Wastes, the Hooker Chemical Corp. submitted a declaration estimating that 21,800 tons of chemicals wastes had been buried in the Love Canal over a 10 year period, including significant quantities of trichlorophenols (TCP). Laboratory analysis of soil and sediment samples form the Love Canal indicates the presence of more than 200 distinct organic chemical compounds; approximately 100 of these have been identified to date.

Dioxin (2,3,7,8 tetrachlorodibenzoparadioxin (TCDD)), considered one of the most toxic man-made compounds based on animal experimental studies, is one of the chemicals found in the landfill. Since dioxin (TCDD) is a contaminant by-product formed during the manufacture of trichlorophenols (TCPs), its presence in the Love Canal was suspected when 200 tons of TCPs appeared on the list of chemicals buried at the site; its presence was confirmed in April 1979 using sophisticated analytical equipment at the university of Nebraska's Midwest Center for Mass Spectrometry. The Department of Health has since acquired the same type of mass spectrometry and formed its own dioxin analysis capability.

The highest level of dioxin quantified to date at the Love Canal is approximately 300 parts per billion (ppb) in a storm sewer adjoining the canal. Lesser concentrations also have been found in leachate collected from remedial holding tanks, soil samples from the canal and backyards of nearby homes and sediment and marine life of two creeks bordering the Love Canal neighborhood..

The departments of Health and Environmental Conservation launched an intensive air, soil and groundwater sampling program in spring 1978, following qualitative identification of a number of organic compounds in the basements of 11 homes adjacent to the Love Canal.

To determine the extent of chemical migration into private residences, 800 basement air samples from 400 homes within a four block radius of the landfill were analyzed for seven chemical compounds: chloroform, benzene, trichloroethene, toluene, tetrachloroethene, chlorobenzene and chlorotoluene. The mapping of benzene air concentrations revealed no clear patterns of contamination. On the other hand, compounds not present in common household products, such as chlorobenzene and chlorotoluene showed definite clusters of contamination in homes immediately adjacent to the canal, with significantly less evidence of contamination further out (New York State Department of Health, 1981).

Evacuation

Evacuation from Love Canal was a disputed issue that evolved over the course of the crisis. For most residents, the obvious preference was for permanent relocation and the purchase of their homes at fair market value. For those living beyond the boundary of Ring II, the ultimate goal of most residents was to force the government to purchase their homes as well. Though they would eventually achieve success in that regard, there was no way of knowing it when remedial construction work began in October 1978. Remediation began just 1 month after the decision to purchase the homes in Rings I and II, yet even those residents had not yet been moved.

The residents were apprehensive that the digging would increase their exposure to the chemicals underneath the surface if fumes and contaminated dust were picked up by the wind and blown through the neighborhood, and particularly if an explosion of fire occurred. In response to these concerns, the Love Canal Task Force run by the New York Department of Transportation agreed to evacuate people in school buses in the event of an explosion or similar problem. On October 10, 1978, the plan also failed its first test of credibility when the buses did not arrive for a trial run emergency evacuation (Gibbs, L. M.,1981; and Gibbs, L. M., 1998). Nonetheless, the state offered nothing more until February 1979, when the third public health order, issued by Commissioner Axelrod, offered financial assistance for temporary relocation. Even then, however, the benefit was only for families with pregnant women and children under the age of 2 years living in Rings I-III (Fletcher, T, 2001)..

It was not until June 1979 that the evacuation policy was expanded further. The New York Supreme Court ordered temporary relocation for any residents in the area who furnished certificates from physicians attesting that illness or breathing difficulties were associated with the remediation work at Love Canal (Levine, A.G., 1982, and Silverman, G., 1989). They were particularly frustrated that many doctors were reluctant to write certificates that might be interpreted as an assignment to blame to Hooker. This controversy became especially thorny on August 25, 1979, when chemical fumes from the site combined with summer heat and humidity made several residents violently ill. In early September, the New York Supreme Court ruled that the Task Force should relocate any Love Canal resident who complained of poor health effects without medical certification. The number of families living in hotels grew to 120, a total of 425 individuals (New York State Department of Health, 1981, and Silverman, G., 1989). The state government paid $7500 per day for these expenses (New York State Department of Health, 1981).

The residents of Love Canal were allowed to stay in their motel rooms until November 5, 1979, when the deep excavation work was completed (New York State Department of Health, 1981).The residents of Ring III returned to their homes, but it would take another 6 months before they were assured of permanent relocation. On May 21, 1980 Governor Carey made a formal request to President Carter to declare a second state of emergency in the area and to provide aid for the relocation of over 700 families in Rings I, II, and III. This request was prompted by a long series of events, the most recent of which had occurred the day before when angry Love Canal residents held two US EPA representatives hostages for 5 hours, before releasing them unharmed (Silverman, G., 1989). On May 22, during his unsuccessful bid for reelection, President Carter made a series of announcements about Love Canal, one of which was to grant Carey's request for the state of emergency and the extension of permanent relocation to Ring III. The action provided for the purchase of all privately owned properties, including businesses and rental housing (Fletcher, T, 2001). .

Remedial Actions

This site has been addressed in seven stages: initial actions and six major long-term remedial action phases, focusing on:

- landfill containment with leachate collection, treatment and disposal;
- excavation and interim storage of the sewer and creek sediments;
- final treatment and disposal of the sewer and creek sediments;
- remediation of the 93rd Street School soils;
- Emergency Declaration Area (EDA) home maintenance and technical assistance by the Love Canal Area Revitalization Agency (LCARA), the agency implementing the Love Canal Land Use Master Plan;
- buyout of homes and other properties in the EDA by LCARA.

Three other short-term remedial actions:

- the Frontier Avenue Sewer remediation,
- the EDA soil removal,
- the repair of a portion of the Love Canal cap, were completed in 1993 and are discussed below.

(A) Initial Actions: in 1978, New York State Department of Environmental Conservation (NYSDEC) installed a system to collect leachate from the site. The landfill area was covered and fenced and a leachate treatment plant was constructed. In 1981, EPA erected a fence around Black Creek and conducted environmental studies.

(B) Landfill Contaminant: in 1982, EPA selected a remedy to contain the landfill by constructing a barrier drain and a leachate collection system; covering the temporary clay cap with a synthetic material to prevent rain from coming into contact with the buried wastes: demolishing the contaminated houses adjacent to the landfill and nearby school; conducting studies to determine the best way to proceed with further site cleanup; and monitoring to ensure the cleanup activities are effective. In 1985, NYDEC installed the 40-acre cap and improved the leachate collection and treatment system, including the construction of a new leachate treatment facility.

(C) Sewers, Creeks, and Berms: In May 1985, as identified in a Record of Decision (ROD), EPA implemented a remedy to remediate the sewers and the creeks which included:

- hydraulically cleaning the sewers;

- removal and disposal of the contaminated sediments
- inspecting the sewers for defects that could allow contaminants to migrate;
- limiting access, dredging and hydraulically cleaning the Black Creek culverts;
- removing and storing Black and Bergholtz creeks' contaminated sediments.

The remediation of the 102nd Street outfall area, as originally proposed in the 1985 ROD, has been addressed under the completed remedial action for the 102nd Street Landfill Superfund Site. The State cleaned 62,000 linear feet of storm and sanitary sewers in 1986. An additional 6,000 feet were cleaned in 1987. In 1989, Black and Bergholtz creeks were dredged of approximately 14,000 cubic yards of sediments. Clean riprap was placed in the creek beds, and the banks were replanted with grass. Prior to final disposal, the sewer and creek sediments and other wastes (33,500 cubic yards) were stored at Occidental Chemical Corporation's Niagara Falls RCRA-permitted facilities.

(D) Thermal Treatment of Sewers and Creeks Sediments: In October 1987, as identified ins a second ROD, EPA selected a remedy to address the destruction and disposal of the dioxin-contaminated sediments from the sewers and creeks:

- construction of an on-site facility to dewater and contain the sediments;
- construction of a separate facility to treat the dewatered contaminants through high temperature thermal destruction;
- thermal treatment of the residuals stored at the Site from the leachate treatment facility and other associated Love Canal waste materials;
- on-site disposal of any non-hazardous residuals from the thermal treatment or incineration process. In 1989, OCC, the United States and the State of New York, entered into a partial consent decree (PCD) to address some of the required remedial actions.

Also, in 1989, EPA published an Explanation of Significant Differences (ESD), which provided for these sediments and other remedial waste to be thermally treated at OCC's facilities rather than at the site. In November 1996, a second ESD was issued to address a further modification of the 1987 ROD to include off-site EOA-approved thermal treatment and/or land disposal of the stored Love Canal waste materials. In December 1998, a third ESD was issued to announce a 10 ppb treatability variance for dioxin for the stored Love Canal waste materials. The sewer and creek sediments and other waste materials were subsequently shipped off-site for final disposal; this remedial action was deemed complete in March 2000.

(E) 93rd Street School: the 1998 ROD selected remedy for the 93rd Street School property included the excavation of approximately 7500 cubic yards of contaminated soil adjacent to the school followed by on-site solidification and stabilization. This remedy was re-evaluated as a result of concerns raised by the Niagara Falls Board of Education, regarding the future reuse of the property. An amendment to the original 1988 ROD was issued in May 1991; the subsequent selected remedy was excavation and off-site disposal of the contaminated soils. This remedial action was completed in

September 1992. Subsequently, LCARA purchased the 93rd Street School property from the NFBE and demolished the building in order to return the resulting vacant land to its best use.

(F) Home Maintenance: As a result of the contamination at the site, the Federal government and the State of New York purchased the affected properties in the EDA. LCARA is the coordinating New York State agency in charge of maintaining, rehabilitating and selling the affected properties. Pursuant to Section 312 of CERCLA, as amended, EPA has been providing funds to LCARA for the maintenance of those properties in the EDA and for the technical assistance during the rehabilitation of the EDA. EPA awards these funds to LCARA directly through the EPA cooperative agreement for home maintenance and technical assistance. The rehabilitation and sale of these homes have been completed.

(G) Property Acquisition: Section 312 of CERCLA, as amended, also provided $2.5M in EPA funds for the purchase of properties (businesses, rental properties, vacant lots, etc.) which were not eligible to be purchased under the earlier Federal Emergency Management Agency (FEMA) loan/grant. EPA awarded these funds to LCARA through a second EPA cooperative agreement.

(H) Short-Term Remedial Actions:

- The Frontier Avenue Sewer Project required excavation and disposal of contaminated pipe bedding and replacement with new pipe and bedding.
- The EDA 4 Project required the excavation and disposal of a hot spot of pesticide contaminated soils in the EDA and backfill with clean soils; excavated materials were disposed of off-site.
- The Love Canal Cap Repair required the liner replacement and regarding of a portion of the cap. These short-term remedial actions were completed in September 1993 http://www.epa.gov/region2/superfund/npl/0201290c.pdf ; September 21, 2005)

Cleanup Progress

In 1988, EPA issued the Love Canal EDA Habitability Study (LCHS), a comprehensive sampling study of the EDA to evaluate the risk posed by the site. Subsequent to the issuance of the final LCHS, NYSDOH issued a Decision of Habitability, based on the LCHS's finding. This Habitability Decision concluded that:

- Areas 1-3 of the EDA are not suitable for habitation without remediation but may be used for commercial and/or industrial purposes
- Areas 4-7 of the EDA may be used for residential purposes.

In 1998, the wastewater discharge permit issued to OCC was modified to include the treatment of the leachate water form the 102nd Street Landfill site. In March 1999, the Love Canal leachate collection and treatment facility (LCTF) began receiving the

102nd Street leachate water for treatment. The following represent the make up of the various Love Canal waste materials:

- Sewer and Creek Sediment Wastes: 38,000 yards3 @ 1.6 tons/yard3 = 62,240 tons
- Collected LCTF DNAP (2003): 6000 pounds
- Collected 102nd Street DNAPL: 14,400 pounds
- Spent Carbon Filter Wastes (2003): 40,380 pounds
- Treated LCTF Leachate: 4.35 MG
- Treated 102nd Street Landfill Treated Leachate (2003): 0.58 MG

OCC is responsible for the continued operation and maintenance of the LCTF and groundwater monitoring. The Site is monitored on a continual basis through the numerous monitoring wells which are installed throughout the area. The yearly monitoring results show that the site containment and the LCTF are operating as designed.

As shown above, numerous cleanup activities, including landfill containment, leachate collection and treatment and the removal and ultimate disposition of the containment sewer and creek sediments and other wastes, have been completed at the site. These completed actions have eliminated the significant contamination exposure pathways at the site, making the site safe for nearby residents and the environment.

The site was deemed construction complete on September 29, 1999. in September 2003, EPA issued a Five-Year Review Report that showed that the remedies implemented at the site adequately control exposures of site contaminants to human and environmental receptors to the extent necessary for the protection of human health and the environment. The next Five-Year review is scheduled for September 2008 (http://www.epa.gov/region2/superfund/npl/0201290c.pdf ; September 21, 2005).

Conclusions

- Politics, public pressure and economic considerations all take precedence over scientific evidence in determining the outcome.
- Characteristic of such events is that the victims, although hostile to Hooker Chemical, directed most of their rage at an indecisive, aloof, often secretive and inconsistent public health establishment.
- Lawsuits against Occidental Petroleum Corporation, which bought Hooker chemical in 1968, were initiated by both the State of New York and the U.S. Justice department to cover costs of the cleanup and the relocation programs and by over 2000 people who claimed to have been personally injured by the buried chemicals. In 1994 Occidental agreed to pay $94 million to New York in an out-of-court settlement and the following year the federal case was settled for $129 million. Individual victims have thus far won in excess of $20 million form the corporation.

- In early 1994 it was announced that the cleanup of the condemned homes in Love Canal had been completed and it was safe to move back to the area. The real estate company offering the inexpensive refurbished homes for sale had chosen to rename the area "Sunrise City" (http://onlineethics.org/edu/precol/classroom/cs6.html ; September 21, 2005)..

Lessons Learned

- Proper disposal of hazardous waste is important in protecting public health
- Site selection and site preparation are important for hazardous waste disposal
- Citizens need to be educated in environmental protection and health effect of hazardous waste
- Industrial plants need to observe environmental ethics when dealing with disposal of hazardous waste in area adjacent to residential area

References

Brown, M. H., (1979). "Love Canal and the Poisoning of America", The Atlantic, December, pp. 33-47.

Colten, C., Skinner P., (1996). The Road to Love Canal: Managing Industrial Waste before EPA. University of Texas Press, Austin, Texas.

DeLaney, K., (2000). The legacy of the Love Canal in New York State's Environmental History. Paper Presented at the Researching New York: Perspectives on Empire State History Annual Conference, November 2000.

Fletcher, T, (2001). Neighborhood Change at Love Canal: Contamination, Evacuation and Resettlement. Land Use Policy Volume: 19, Issue: 4, October, 2002.pp.311-323.

Gibbs, L.M., (1981). Love Canal: My Story. State University of New York Press, Albany, New York.

Gibbs, L.M., (1998). Love Canal: The Story Continues. New Society Publishers, Gabriola Island, British Columbia.

http://onlineethics.org/edu/precol/classroom/cs6.html ; September 21, 2005

http://en.wilkipedia.org/wiki/Love_Canal; September 21, 2005

http://ublib.buffalo.edu/libraries/projects/lovecanal/science_gif/records/hart1.html; September 21, 2005 http://www.epa.gov/region2/superfund/npl/0201290c.pdf ; September 21, 2005

Levine, A.G., (1982). Love Canal: Science, Politics, and People. Lexington Books, Lexington, MA.

New York State Department of Health, (1981). Love Canal: A Special Report to the Governor and Legislature. State Of New York, Albany, New, York.

New York State Department of Transportation Files, (1989). NYSDOT Files of Commissioner William Hennessey, Love Canal Task Force and Office of Legal Affairs. New York State Archives, 13430-89.

New York State Task Force on Toxic Substances Files, (1981). New York State Task Force on Toxic Substances Files of Chairman and Assemblyman Maurice Hinchey. New York State Archives, L0134.

Silverman, G., 1989. Love Canal: A Retrospective, Environmental Reporter 20 (20-2), 835-850.

The Darlington Building Collapse: Modern Engineering and Obsolete Systems

Donald Friedman[1]

[1]Donald Friedman, Consulting Engineer, 225 Broadway, Penthouse, New York, NY 10007; PH 917-494-1586; FAX 917-591-2189; email: dfriedman@oldstructures.com

Abstract

The 1904 collapse of the Darlington Apartments during construction was a sudden and complete failure: eleven stories fell into a pile of rubble less than 15 feet high in a matter of seconds, killing 25 people. The collapse and forensic analysis were prominently reported in the newspapers and engineering press; thirty years later, the publicity was cited as a deterrent to structural use of cast-iron columns. This failure became permanently linked to the shortcomings of cast-iron structure.

If completed, the Darlington would have been typical of an obsolete structural type: the high-rise cage-frame building. Cage frames, first built in the 1870s, had an iron frame supporting the floor gravity loads and surrounded by a self-supporting masonry wall that provided lateral stability to the building. The use of cast-iron columns in commercial buildings with cage frames had effectively ended by the mid-1890s; the structural engineers who were increasingly used as consultants in commercial high-rise design preferred wrought-iron and steel columns. Wrought-iron and steel were known to have lower allowable direct compression stresses than cast iron, but were ductile and could safely withstand accidental tension and moment. The gradual replacement of cast-iron with the ductile metals in the late nineteenth century was encouraged by fears of collapse caused by the brittleness of cast iron.

Cage frames remained popular in high-rise apartment houses for nearly a decade after they were no longer used in commercial construction. Unlike tall commercial buildings, which were built in cities throughout the United States, tall residential buildings were concentrated in a few cities, especially New York. These buildings were typically designed by residential architects working without consulting engineers. Common practice was for the iron sub-contractor to provide "engineering services," often consisting of sizing steel and cast-iron columns from tables based on the span and floor load schedule. In short, lateral load analysis was not part of the design, so the deficiencies of cage framing were not made visible.

This paper will describe the design and construction background to the failure, the forensic analysis performed at the time, a modern review of the failure, and discussion of cage-frame failures within the engineering community.

Building and Collapse Description

The Darlington Apartments was designed to be one of dozens of similar apartment houses erected in New York City before World War I. The rapid growth of the city's population, from 1,500,000 in 1870, to 3,400,000 in 1900 (following the 1898 consolidation of neighboring counties into "Greater New York"), to 4,800,000 in 1910 combined with the even more rapid growth of the downtown business district to create a climate of nearly continuous construction. New York, as the city with the highest residential and commercial population densities in the country, became a center of new construction technology.

The building, located in midtown Manhattan at 59 West 46th Street, was to have twelve typical stories with a smaller, set-back penthouse. The general plan was 55 feet by 90 feet overall, in a broad "U" around a light court. (Figure 1) The full height was to be 148 feet, with 10'-10" typical floor-to-floor spacing. Neville & Bagge, the architects of record, had designed many smaller apartment houses, tenements, and row-houses. The general contracting firm, Pole & Schwandtner, was also the engineer of record.

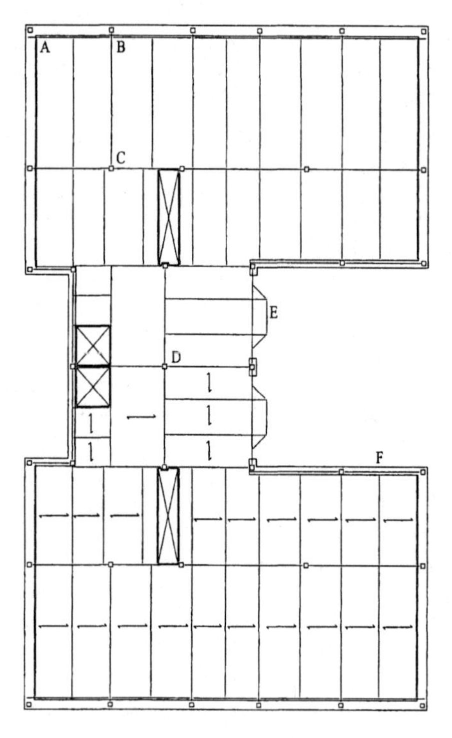

Figure 1: "A" indicates a column eccentrically loaded in both axes, "B" is a column eccentrically loaded around one axis, "C" is an interior column

possibly loaded eccentrically, "D" is a concentrically loaded column, "E" is a metal-clad bay window, and "F" is typical exterior wall.

The structure of the building was relatively simple. The "Roebling flat-slab" floors were cinder-concrete slabs reinforced with rectangular steel bars; the interior partitions were non-structural terra-cotta tile carried on the floor framing; the floor beams were 6-inch, 7-inch, and 9-inch steel channels and I-beams; and the columns were hollow cast-iron sections ranging from 6 inches square with a ¾-inch wall up to 10 inches square with a 1¼-inch wall. Per standard practice, the cast-iron columns were spliced at each floor with pin details (top and bottom flanges cast integrally with the columns received two bolts) and the beams were connected with stiffened seat connections also integrally cast with the column shafts. All connections were loosely bolted. (Figure 2) The design live load was 60 pounds per square foot (psf) on the floors and the design dead load was also 60 psf. The exterior wall was self-supporting – the frame spandrel beams were located entirely within the inboard face of the exterior wall – and increased in thickness from 12 inches at the top to 20 inches at grade in accordance with the New York City Building Code. It should be noted that, per common practice at the time, all masonry other than some front-facade trim was solid, unreinforced brick. ("The Collapse of the Darlington...," 1904)

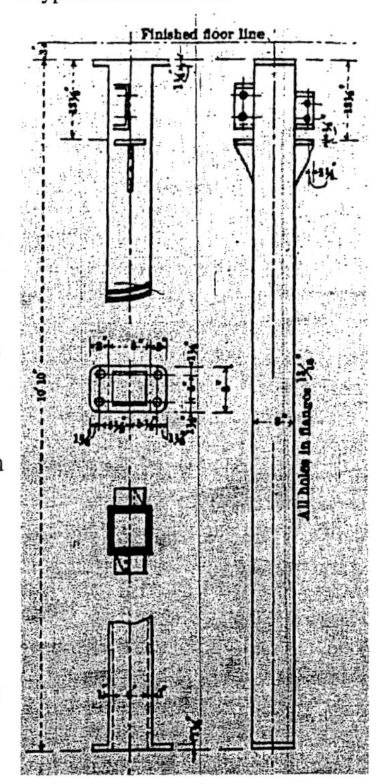

Figure 2: Column connections

Even though the frame did not support the exterior wall, the spandrel columns were completely enclosed within the wall. This was an intentional part of the design: as the wall was intended to resist wind loads, it was necessary for the frame supporting the interior gravity loads and the wall to be structurally linked. Enclosing the columns within solid masonry provided a mechanism for force transfer between the frame and wall. This design meant that all spandrel columns were eccentrically loaded, with offsets between beam and column centerlines as great as 12 inches.

On the afternoon of March 2, 1904, the building collapsed, killing twenty-four laborers and a woman in an adjacent building. At that time, the steel and iron frame had been erected to up to the tenth floor, the concrete floor slabs had been placed to "two or three" floors below the floor beams, and the walls were partially complete to the third floor (rear wall) to sixth floor (side walls) levels. The front wall was not yet complete to the second floor level.

The collapse destroyed the building, leaving portions of the side walls, bent steel beams, and cast-iron columns broken in pieces. (Figure 3) Witnesses observed the upper floors of the frame swaying in various directions but then collapsing inward toward the center of the site. ("The Collapse of the Darlington...," 1904) In the following weeks, the collapse was examined and discussed by various engineers, city officials, and the press.

Figure 3: View from the south after collapse. A portion of the west side wall is visible at left, adjacent to the row house.

Historical Context

Catastrophic failure of buildings, in the form of collapse and fire, was part of urban life in the late nineteenth and early twentieth centuries. The increasing size and complexity of building codes in large cities was a response to both the increasing size and complexity of buildings being constructed and to the perception that new construction technology might alleviate known problems.

Construction of any structure larger than a hut has always been a technological process, requiring tool use, artificially-shaped materials, and organization of groups of people.(Bijker et al, 1987, p. 4) The use of new building technologies after 1850 contributed to the gradual reduction in the rate of death and injury by fire in urban buildings, although the growth in population and increasing urbanization masked this reduction. Public expectations that large-scale fires could be avoided did not come until after the Chicago and Boston fires of 1871 and 1872, despite the existence of some forms of structural fire-proofing before then. The rapid growth of New York City between 1870 and 1910 was coincidentally accompanied by rapid development of new construction technology; other cities have experienced their greatest growth with older technology (e.g., Paris) or with newer technology (e.g., Houston). Late-nineteenth-century New Yorkers could not always clearly distinguish between those changes caused by growth and those caused by technology.

With few exceptions, the general form of building construction in 1870 New York City was traditional wood and masonry, little changed from medieval practice. The industrial revolution had introduced new methods of producing the individual elements used (for example, by introducing powered gang-saws to speed the production of individual lumber "sticks" from felled logs), but had not modified the types of elements used. The floors of buildings were wood plank supported on wood joists; the joists were supported on masonry bearing walls, usually brick or brick with stone veneer. Large open spaces were created through the use of wood girders supporting the joists in lieu of bearing walls. Girders were supported on wood posts, masonry piers or, after 1840, cast-iron columns.

In a city of wood and masonry, little could be done to prevent fires, although there were efforts to limit their spread. The creation of the "fire line" divided the city between the densely built-up portion and the sparser suburban areas. As the city grew north up Manhattan island, the fire line moved northward as well. Buildings south of the fire line were required to have masonry exterior walls, including party walls that marked the boundary between separate buildings on separate lots but were physically enclosed with the buildings; buildings north of the line could be entirely wood. In theory, the use of masonry walls as fire separation could have been taken further by requiring that building interiors be compartmentalized with masonry walls, but this step had not been taken by the time that the introduction of new fireproof materials such as terra-cotta floors made the question moot. (Freyer, 1898, pps. 288-293)

Various fireproof buildings had been erected in New York and elsewhere by constructing floors out of masonry vaults supported on bearing masonry walls, but this method remained in limited use because of its cost. Most "fireproof" buildings of the early and mid-nineteenth century therefore consisted of wood and masonry with some form of protection against external fire spread.

The introduction of new structural materials and new systems in the middle and late nineteenth century changed the nature of the dangers and public and professional understanding of these dangers. Cast iron was introduced into American practice in New York and Boston in the late 1830s, following use in Britain and France. In the United States, this was followed by large-scale wrought iron in the 1850s, steel in the 1870s, and reinforced concrete in the 1890s. The typical pattern of use was introduction as a one-for-one substitute for a well-established building element (such as the first use of cast-iron columns as substitutes for granite piers in storefronts) followed by the introduction of large-scale systems dependent on the properties of the new elements (such as the use of cast-iron columns and wrought-iron beams in interior frames that made possible large commercial buildings with no interior walls).

The substitution of iron for stone or brick provides no obvious improvement in fire-protection, but in some cases the material superseded was wood. Cast iron served the mid-nineteenth century as a symbol of modernity in construction, more specifically of cheap industrial production of buildings. Cast-iron promoters emphasized safety and economy when compared to traditional construction. (Bogardus, 1856, p. 5)

In that era, there was no required testing of the new materials and systems before large-scale use in buildings. New materials were empirically tested by use and the record of their performance under real-life conditions begins in the 1870s, when enough new-technology buildings had been constructed for their performance to be recorded. This period coincides with two of the largest nineteenth-century conflagrations, Chicago in 1871 and Boston in 1872, and therefore began with a public sensitized to fire danger. Reports from those fires emphasized the performance of new-materials buildings, in large part because they had been advertised as "fireproof." (Bogardus, 1856, pps. 12-13) Reports on individual building fires in New York began to focus on whether or not the building burning was "fireproof" and, if so, how it performed compared to traditional construction. Public commentary, including statements from governmental officials concerned with building safety, professionals familiar with the issues, editorial writers from newspapers, and professional journal writers began with the Chicago and Boston fires to examine the effect of new materials in construction and new forms of building control (such as the first comprehensive New York City Building Code, enacted in 1882) on safety. Simply, people were asking whether or not the new forms of building were making New York more safe or less, a question that would continue through the beginning of the twentieth century.

The first new material introduced was cast iron, usually used for columns. Cast iron is a product of the industrial revolution because its production requires intense heat that depended on engine-powered forced air drafts and coal fires. The metal is extremely strong in compression – stronger than most modern steels – but weak and subject to brittle failure in tension. These properties make the metal inappropriate for use in horizontally-spanning beams, where tension and compression exist in equal amounts, but feasible for use in columns, which are substantially in compression. Cast iron columns were first substituted for masonry piers in the late 1830s and became common in building interiors by the 1860s. (Condit, 1982, p. 81)

On a larger scale from material substitution was the use of entirely new structural systems, including the substitution of cage- and skeleton-frames for older masonry-supported structure. Traditional buildings are defined as bearing-wall systems, where masonry walls support the gravity loads of the building itself (the wall weight and the interior floors) and its contents. The walls also serve to resist the lateral push of wind and to provide exterior enclosure. Just as iron framing was substituted for wood floor joists, iron was gradually substituted for the structural functions of the walls, resisting gravity and wind loads.

The main advances in building systems technology were the introduction, circa 1875 and 1890, of cage and skeleton frames. A cage frame is one with a complete metal frame (in which iron or steel floor beams are supported by iron or steel columns) surrounded by a self-supporting masonry wall; a skeleton frame is one in which a wrought-iron or steel frame supports all loads, including the exterior walls. (Freyer, 1898, p. 465) Cage frames are now seen as a transitional form between bearing-wall buildings and skeleton-frame buildings, but were a legitimate method of constructing buildings in the 1880s and 90s. Nearly all cage frames used cast-iron columns.

There are links between fires and non-fire collapses. Obviously, buildings badly damaged by fire often collapse, in part or in whole. Unlike traditional construction, which had hundreds of years of empirical testing through use, new technology was often not well understood when it was put into use. Cast-iron use, for example, went from being considered an improvement in fireproofing to being considered a detriment, and the use of unprotected iron was eliminated by New York's 1882 building code. Iron was no worse in fire than wood, but its more serious flaw was that it could fail without warning if overloaded or improperly manufactured. The superiority of wrought iron and steel, which fail gradually under similar conditions, only became generally recognized among engineers in the 1880s and among other groups in the 1890s or 1900s. The worst collapses were the result of over-loaded or poorly designed cast-iron columns and occurred in cage-frame buildings.

Despite the similarities, we can distinguish between public reaction to ordinary fires and the reaction to failures in modern buildings. The dangers of urban fires were well known before the United States had cities of any significant size and were associated with the combination of wood construction and the domestic use of open flame. In a society where flame was common – in fireplaces, stoves, lamps, and boilers – people were familiar with the dangers of fire. The ordinary person of the late nineteenth century had little familiarity with new building technology, and had to rely on expert opinion about the safety of iron and steel framing, tall buildings, and new frame types. On the other hand, building professionals view catastrophic failures not just as tragedies but as means by which design and construction practice can be studied. As examples of fires in new-technology buildings were discussed in the press, the public understanding of the meaning of fire and fireproofing changed from that based on common knowledge to one influenced by technology as described by professionals.

Building collapses, like fires, have always been part of urban society, but bear a more complicated relationship to the introduction of new technology. The substitution of non-flammable metals for wood posts and fireproof floors for wood joists may not have created the "fireproof building" so often held as an ideal, but they did reduce the gross fire load (the amount of flammable material in a given building) and, when coupled with fire-protection methods, gradually reduced the incidence of severe fires. Collapses, on the other hand, became more frequent in the 1880s and 90s as the average height of buildings in New York increased and as experimental structural systems were empirically tested by use.

There were collapses of other cage-frame buildings before the Darlington, most significantly the August 8, 1895 collapse of the Ireland Building at 3rd Street and West Broadway. The building was planned to be a commercial loft eight stories high and was still in construction, although substantially complete, when its center collapsed, leaving the perimeter walls and minor portions of the interior framing. The collapse killed 16 laborers and buried their bodies under a large pile of debris that impeded rescue of the survivors and recovery of the remains. A coroner's jury performed the formal review of the collapse, including testimony by technical experts and visits to the site to look at the debris. ("Imprisoned in a Wreck," 1895; "Mr.

Constable Inspects," 1895; "Bodies All Taken Out," 1895; "Visited the Ruins," 1895; "To Go Before the Grand Jury," 1895)

At the Ireland Building, at least three contributing causes were identified, including the possibility of inadequate bearing material below the foundations, inadequate design of the interior column footings for the expected loads, and over-loading of at least one interior column with bags of gypsum by the plasterers working on the upper floors. However, the dominant cause was identified as the failure of the interior cast-iron columns under high but not exceptional loads.

An editorial in the *Engineering Record* used the collapse of the Ireland Building to highlight two issues that concerned engineers as a group but were largely unknown to other professionals and the public. ("The Investigation of Building Failures," 1895) The first was the general inadequacy of cast-iron columns. These elements, common twenty years earlier, were not necessarily seen as inadequate by the public and were not treated any differently by building codes than steel and wrought-iron columns, but were described in the editorial as "entirely unfit." For an engineering audience, cast iron was specifically compared to wrought-iron and steel and found wanting, which is not surprising as the state of the art in structural metals had advanced significantly during the 1870s and 80s. Cast iron is brittle and can therefore fail without warning and completely as it did at the Ireland Building, while the other metals are ductile and fail "only after the most extreme or violent distortion, and even then enough will frequently hang together to sustain the existing loading." This engineering distinction between cast iron and the two other metals was not widely understood.

The other issue identified was expertise. Coroner's juries were then a normal mechanism for investigation of untimely death, but jury members were selected in the same random manner as criminal and civil juries. The *Record* editorial states that the Ireland jury, by focusing only on the inadequate footings as the immediate cause of failure missed an opportunity to "confer great benefits upon the city of New York in connection with the construction of modern buildings" by addressing the general inadequacy of cast-iron columns. The editorial bluntly states that coroner's juries were inadequate for investigation of disaster because of lack of expertise: "The questions which come before such juries are largely of a purely engineering character, which can efficiently be treated only by men fitly qualified through education and experience. Exact knowledge of the nature of material, its behavior under loads, the competent design of main members and details, and a multitude of other similar matters, must be intelligently considered." Obviously, engineers were the only people that the editors of the *Record* considered qualified to judge building-related disasters. Further, the efforts of government officials to create a safe building environment were indirectly criticized: "The recommendations of one such competent jury [composed of engineers] would furnish a ready means of clearing the Building Ordinance of numerous absurdities and of introducing efficient provision for the attainment of creditable building construction, so that buildings, in the process of erection at least, could not fall down like a child's block-house." ("Ireland Building Inquest," 1895) This criticism was not entirely fair, as the jury heard testimony by expert builders and engineers and arguably was therefore no less well-informed than any jury examining a technical issue with the help of expert witnesses, but it did highlight a gap between engineers and the public. Engineers saw vindication of their work in the fact of a disaster that occurred from un-engineered design, in the method of determining blame through professional investigation, and in their preference for steel over cast iron; while the public saw a building of "modern" construction collapsing catastrophically and without warning.

Investigation and Forensic Analysis

Like the Ireland Building, the Darlington had no engineer formally involved with structural design; like the Ireland, there were contributing causes that suggested negligence on the part of the general contractor; but most importantly, like the Ireland, the root cause of collapse was eventually identified as the use of a cage frame with cast-iron columns. By this date, that combination of material and system was used in New York only for medium-rise hotels and apartment houses, having been otherwise superseded by steel-frame technology.

Increased sophistication of both the public and city officials can be seen in the first article in the *New York Times* describing the collapse. ("Death in Collapse...," 1904) In addition to a narrative description of the collapse and rescue efforts on site, the article contained notice of arrest of the site foreman for negligence in overloading the ninth floor with materials to be used in construction of the tenth floor; mention of the beginnings of investigation by the coroner's office, the district attorney's office, and the Building Department; and three distinct theories of the cause of collapse: overloading, faulty construction, and the explosion of a hoist steam engine. The article compares the overload theory to the collapse of the Ireland Building and quotes the Fire Department's Chief Croker as stating that "complaint after complaint" had been filed with the Building Department during the preceding months for faulty construction. There was a coordinated effort by various city agencies to assign blame by assessing the collapse for specific technical causes, and the *Times* assumed that the reading public had enough understanding of tall buildings to know the context of the Ireland collapse and the dangers of specific poor practices.

One of the disturbing details found was the extreme flexibility of the frame. Since the exterior walls were meant to provide lateral bracing to the building, it is no surprise to hear that the frame wobbled with iron erection having so far outpaced masonry, but reports that the frame was 18 inches out of plumb when erected to the 9th floor and had to be pulled back erect with a block and tackle, or that it swayed noticeably in the breeze despite the lack of significant sail area, raise the question of damage from secondary stresses. ("Concerning the Fall...," 1904; "Further Notes...," 1904)

Over the next few days, the public was given detailed explanations of the pre-collapse Building Department violations and the methods used to cure them as well as descriptions of the difference between inadequate construction (where the fault would lie primarily with the contractor) and inadequate design (where the fault would lie primarily with the architect and engineer). In both cases, there would be secondary blame on those responsible for review: the building department inspectors and, for inadequate construction, the architects and engineers. Interestingly, the first technical opinion stated by an official – that the building code had been violated by the engineers through the use of too-small floor beams and too few bolts between beams and columns – came from an Assistant District Attorney, not one of the experts on construction. His logic was simple: "But it is all foolishness for any one to think that there was nothing wrong when a building like that falls in on itself." ("Six More Bodies...", 1904; "Darlington Owner Held...," 1904; "Darlington's Builder Surrenders...," 1904) With its emphasis on the failure of a common type of building, that is a statement of betrayal of the lay-person by technology.

The formal results of the coroner's inquest were mixed. Unlike the Ireland jury nine years earlier, experts were consulted by the city agencies to ensure that technical information was not reviewed unaided by jurors and lawyers. In addition to technical discussion, the coroner's jury heard testimony concerning confusion over the meaning of architectural and engineering consultation. Guy Waite, a consulting engineer formerly employed by the Building Department, described his refusal to reduce the size of floor beams smaller at the owner's request. Julius Tomiek, an employee of the general contracting firm Pole & Schwandtler and an immigrant who testified in German, stated that he had no contact with the city officials or direct knowledge of the work performed on site, and that he had been told by his employers that the plans (showing the lighter floor beams Waite thought incorrect) had been approved by the building department. Similar testimony followed, focusing to a large extent on the lack of continuity among the designers. One juror was quoted saying that "If there is a system by which one architect comes into a job and starts and stops and then another and another starts and stops so that no responsibility can be fixed, we want to know it, so that the laws may be amended so as to put an end to it." This statement is technically an attack on architects, but the legal blame was found to be the owner who hired and fired the architects. The jury found the owner, Eugene Allison, and Pole & Schwandtler guilty of criminal negligence and recommended two new laws: that only the architects responsible for a design or experienced contractors be allowed to supervise construction, and that the building department hire engineers as inspectors of "all buildings requiring engineering skill." Four jurors added as unofficial recommendations a requirement for licensing all architects and engineers working in the city. ("Darlington's Builder Surrenders...," 1904; "Warned Allison...," 1904; "Darlington Verdict...," 1904; "Architects Condemn Law," 1904) The District Attorney indicted Allison but apparently never tried him, saying that he had disappeared. In 1911, Allison (who had been in new York the entire time) had the charges dropped on account of his poor health. (*Engineering News* editorial, 1911)

The separate investigation conducted by the district attorney's office was quite simple in concept: Harry de Berkeley Parsons, a leading consulting engineer of the time, was hired to investigate the wreckage and drawings to see if the design and construction violated any laws. Parsons's conclusion was that in a narrow sense the failure was the result of poor beam-to-column connections that allowed the columns to buckle sideways without restraint, but that in a broader sense the cage frame type was inherently flawed. This was a significant conclusion for engineers, but his statement that the collapse was the result of "faulty design and carelessness and neglect in the erection of the members" was vague in a legal sense, allowing the designers, the contractors, and the building department inspectors to share the blame. ("Darlington Disaster Report," 1904; Parsons, 1904) The existence of his report was itself a statement of the professional method of conducting forensic engineering investigations, regardless of whether that was perceived; the report was quoted in the popular press and an abbreviated version published in both the *Proceedings of the American Society of Civil Engineers* and the *Engineering News*.

Parsons was quite specific in his report, identifying a column near the center of the building as the first to fail. He explained how this argued against various other causes that had been suggested, including the explosion of the boiler for the on-site steam engine used for hoisting, column overload caused by storing plasterer's gypsum and unerected iron and steel on the topmost floor, and foundation movement. ("Further Notes...," 1904; "Concerning the Fall...," 1904) Interestingly, at least one more recent engineer, Jacob Feld, is of the opinion that the trigger for the failure was that some of the column footings rested directly on rock while others were on clayey soil. (Feld, 1968, p. 42) However, this analysis focuses on the trigger for the specific failure rather than the general cause; had the same foundation conditions existed below a structural steel frame, any resulting failure would have been limited to settlement and minor member deformation. Similarly, pre-collapse sway may have cracked the column-splice flanges, but this does not constitute a separate overall cause of failure as much as reinforce the case against the use of cast iron and cage frames.

Conclusion

Various problems that may have led to the collapse – overloading of the top floor, cracks in the columns caused sway and truing of the frame, the possibility of an isolated footing failure – are unrelated in their origins, but share a common theme: none of them could cause catastrophic failure of a steel skeleton building. The complete loss of the Darlington, regardless of the importance attributed to the different causes in forensic investigation, was the result of an inadequate structural system – one that had no resilience to prevent progressive failure when subjected to temporary overloading, isolated connection failure, or differential settlement.

The shortcomings of the cage building system were known and had been described in detail in the engineering press before the collapse. Prominent engineers in the 1890s and 1900s criticized the cage frame, the use of cast-iron columns, and the use of integrally cast connections – for example, William Sooy Smith describing in 1902 the use of cast iron "flimsy lug and bracket connections" and eccentric loads on cast-iron columns as "radical defects" – but had only managed to eliminate the use of cast iron and cage frames from the commercial buildings where they were regularly employed. In other buildings, particularly hotels and apartment houses, where consulting engineers were rare, the older and less safe construction types persisted. (Sooy Smith, 1902; "Current Practice...,"1904)

In that sense, this disaster was foreseeable and preventable and hardly seems worthy of forensic analysis. However, the shock of the utter failure of a formerly popular building type, the shock of the great loss of life, and the sense that structural engineers, as a group, had allowed an unsafe type to continue in use combined to exaggerate the importance of this one incident. It is difficult overstate the importance of the Darlington collapse when sources as knowledgeable as the builder William Starrett and the steel expert Robins Fleming later cited it as the cause for the final abandonment of cast-iron structure. (Starrett, 1928, p. 41; Fleming, 1934)

One of the articles in the engineering press following the collapse expressed the engineers' view clearly: "Steel skeleton...construction...gives us a rigid steel framework connected and braced like a bridge in all directions, carrying walls, floors, and finish merely as so much clothing, and able to safely bear all loads which come on either the completed building or the framework alone. Structures of the 'office-building' type are practically all of this kind. But very many other buildings, from five or six-story warehouses and stores to twelve (and in at least one case seventeen-) story apartment buildings, are radically different in structural respects, though masquerading under the general name of 'skeleton' or 'steel-frame' construction. They have floor beams and girders of steel, columns of either steel or cast-iron designed for vertical, central loading only; they are provided with no bracing, and the connections between the beams and columns are unfitted to resist any calculable bending or twisting moments. In such buildings the walls and floors give lateral strength and stiffness to the structure, while vertical loads are carried by the frame. Evidently this is quite different from the condition in an office-building structure, both after and during erection. Of course, it is known that when the walls are in place and firmly set, they supply sufficient lateral resistance, though the amount of this resistance cannot be calculated, especially when the wall is pierced by windows in every panel." ("Concerning the Fall...," 1904) In other words, even though cage buildings might be generally safe when the walls were completed, they were inherently inferior because they contained undesigned structure. Engineers were looking for more responsibility, not less: the owner has the responsibility "to employ competent architects and engineers to design and supervise his work, and honest contractors to execute it; and it is with those parties that the real responsibility for safe work must lie because *they do the work.*" [Emphasis in original.] (*Engineering News* editorial, 1904)

An informal result of the Darlington collapse seems to have been the end of cast-iron-column cage construction. By 1904 this type had been technically obsolete for at least a decade and it was absent from commercial buildings because the professional engineers involved with them no longer used cast iron. As William Starrett wrote in his memoirs, "the suspicion [of cast iron] was ruinous," leaving steel frames as the only acceptable new technology for most buildings. (Starrett, 1928, p. 41) In other words, the public perception of new building technology had caught up with engineers in distinguishing between cast iron and steel. Public concerns over safety and liability put an end to cast-iron use in a way that engineers' concerns over the technical issues of safety had not. The cage frame, which was associated with but not dependent on iron columns, disappeared from use at roughly the same time; steel-frame buildings were generally free from a reputation for catastrophic collapse. This collapse marked the end of acceptance of unanalyzed structure.

The lessons from the Darlington were available after the Ireland collapse, but the context was different. In 1895 there were only a handful of steel-frame buildings in New York and Chicago, often visually indistinguishable from their cast-iron cage-frame neighbors. By 1904, steel-frame construction, identified as such and visible during construction, was an accepted part of the cityscape and had spread across the country. The difference between this technology and its predecessors was dramatized by such buildings as the Fuller (Flatiron) Building at the triangular intersection of

Fifth Avenue, Broadway, and 23rd Street; the combination of a 285-foot height and a triangular plan that narrows to a 6-foot apex was not achievable using cast-iron technology and created a dramatic appearance that heightened public awareness. Compared to the increasingly-tall steel-framed buildings, the design of the Darlington was seen as backward, even though that technology was itself only thirty years old. Because the transition to modern technology was largely complete by the time of the Darlington collapse, the press coverage focused on process – who was responsible? – and regulation of how to encourage the use of the best technology available.

References

"Architects Condemn Law," *The New York Times*, March 4, 1904, p. 3

Bijker, W., Hughes, T., and Pinch, T., editors (1987) *The Social Construction of Technological Systems*, The MIT Press, Cambridge.

"Bodies All Taken Out," *New York Times*, August 15, 1895, p. 9

Bogardus, J. (1856) *Cast Iron Buildings; Their Construction and Advantages*, J. W. Harrison, New York

"The Collapse of the Darlington Apartment House in New York City," *Engineering News*, March 10, 1904

Condit, C. (1982) *American Building*, 2nd edition, University of Chicago Press, Chicago.

"Concerning the Fall of the Darlington Building," *Engineering News*, March 24, 1904

"Current Practice in Apartment House Construction in New York City," *Engineering News*, April 14, 1904

"Darlington's Builder Surrenders to Coroner," *The New York Times*, March 17, 1904, p. 7

"Darlington Disaster Report," *New York Times*, March 20, 1904, p. 20

"Darlington Owner Held by Coroner," *The New York Times*, March 5, 1904, p. 3

"Darlington Verdict Holds Three Guilty," *The New York Times*, March 23, 1904, p. 1

"Death in Collapse of Hotel Skeleton," *New York Times*, March 3, 1904, p. 1

Engineering News editorial, untitled, March 17, 1904

Engineering News editorial, untitled, June 15, 1911

Feld, Jacob, *Construction Failure*, New York: John Wiley and Sons, 1968

Fleming, Robins, "A Half-Century of the Skyscraper," *Civil Engineering*, December 1934

Freyer, W. (1967 [1898]) *History of Real Estate, Building, and Architecture in New York City*, Arno Press, New York

"Further Notes on the Collapse of the Darlington Building," *Engineering News*, March 17, 1904

"Imprisoned in a Wreck," *New York Times*, August 9, 1895, p. 1

"The Investigation of Building Failures," *Engineering Record*, 9/14/1895

"Ireland Building Inquest," *Brooklyn Eagle*, August 21, 1895, p. 1

"Mr. Constable Inspects," *New York Times*, August 13, 1895, p. 6

Parsons, H. De B., "Collapse of a Building During Construction," *Engineering News*, May 12, 1904

"Six More Bodies From Hotel Ruins," *The New York Times*, March 4, 1904, p. 3

Sooy Smith, W., "The Modern Tall Building – Corrosion and Fire Dangers," *Cassier's Magazine*, May 1902

Starrett, W. (1928) *Skyscrapers and the Men Who Build Them*, Charles Scribner's Sons, New York

"Steel Frames," *The New York Times*, March 4, 1904, p. 4

"To Go Before the Grand Jury," *Brooklyn Eagle,* August 30, 1895, p. 12

"Visited the Ruins," *Brooklyn Eagle*, August 23, 1895, p. 10

"Warned Allison of Darlington's Peril," *The New York Times*, March 18, 1904, p. 14

Learning From Failure:
Teaching a Course on Building Performance and Forensic Techniques

M. Kevin Parfitt, P.E.[1]

Abstract

University courses in the areas of building performance and building failures can take a variety of successful formats. This paper describes a course in building failures and forensic techniques offered primarily to architectural and civil engineers at Penn State University. The course incorporates a broad architectural / engineering definition of failures concentrating on topics ranging from water penetration resulting from roofing defects and poorly installed building envelope flashing, to a review of both major and lesser known historic structural collapses. A discussion of the use of industry partnerships and various forms of student centered learning techniques used to enhance the educational experience is included. The paper also describes the incorporation of case history class examples as they relate to teaching failures topics in an educational setting.

Background on Failures Curriculums in the US

The value and need for failures courses and integrated failures information in civil engineering or related curriculums such as architecture and architectural engineering is well documented in the literature. A number of educators and practitioners have discussed the need and value of integrating failures information in both undergraduate and graduate programs, particularly in Civil Engineering (Bosela 1993; Rendon-Herrero 1993a, b; Baer 1996; Delatte 1997; Rens and Knott 1997; Pietroforte 1998; Delatte 2000; Jennings and Mackinnon 2000; Rens, Rendon-Herrero, and Clark 2000; Rens, Clark and Knott 2000; Delatte and Rens 2002). These same educational needs and advantages apply to any students, including those in architectural engineering, who concentrate their education in the area of building design.

Specific topics incorporated in a course on failures vary depending on the needs of the individual schools and their curriculums. In general, topics cover a broad range of items including, but not limited to, the study of historic collapses, failures of buildings and civil engineering structures, ethical considerations, the construction and performance of various architecture and engineering systems, durability and lessons on how to become a better designer.

[1] Associate Professor, Department of Architectural Engineering, Penn State University, 104 Engineering Unit A, University Park, PA 16802, mkp@psu.edu

Challenges and difficulties that exist in developing failures related courses or implementing significant failures modules in existing courses have been discussed in the literature by a number of researchers and educators (Delatte 2000; Rens, Clark and Knott 2000; Reynolds 2003; Sutterer 2003). In general terms, a representative list of the roadblocks to successful implementation of failures information in a curriculum are summarized in the list below:

- Failures information and case history studies or modules are not readily available for academic use due to factors such as difficulties in obtaining permission for the dissemination of confidential client material or the need to condense large volumes of raw documentation for class room use. In addition, older case history information can sometimes be difficult to obtain.

- There is often a lack of course instructors with significant practical engineering or failures investigation experience available to develop and teach failures and performance related courses.

- Guest lecturers and practitioners with extensive forensic experience are often not available for frequent or extended periods of time to assist in course development or delivery.

- Finding room in the curriculum is difficult given ABET requirements and all the other demands on an engineering program to adequately cover existing and developing topics as well as to provide a breadth of experiences for engineers.

The development and availability of quality instructional materials and case history examples is one of most important items needed for offering a failures related course. (Rens, Rendon-Herro, and Clark 2000; Parfitt 2001)

Through a number of committees, including the Committee on Education and The Committee on Dissemination of Failure Information, the ASCE Technical Council on Forensic Engineering (TCFE) has as one of its goals to improve the practice of Civil Engineering by promoting the study of failure case histories in educational activities. TCFE believes failure case histories can be used as a tool that has the potential to reduce the frequency and severity of failures (Carper 2003)

A number of educators and practitioners have taken the approach that the teaching of failures topics in university programs should not focus exclusively on classic structural related failures and collapses. Failure topics, case history studies, and general failure course studies focusing on performance related issues such as general deterioration of materials and water infiltration of building envelopes (roofs, doors and windows, facades, flashing and sealants, etc.) represent an important class of frequently encountered performance failures that should be included in many failures curriculums (Reynolds 2003).

Introduction and History of the Course

Many of the problems and roadblocks to implementing and maintaining a failures based course discussed in the previous section were also experienced by Penn State AE. Fortunately, the initiation of an integrated bachelor's and master's degree (BAE/MAE) program in AE generated an opportunity for a few additional graduate level practice based courses. In conjunction with industry support in the form of visiting practitioner lectures, the course was first offered in fall semester of 1999. Up until that time and continuing today, some failures topics, particularly in the areas of structural performance, construction issues, fire protection, and materials, are imbedded or incorporated into regular courses in the five year Bachelor of Architectural Engineering (BAE) undergraduate program curriculum.

Although structural failure and performance issues were a regular part of a number of undergraduate courses, a more comprehensive course in building failures had long been contemplated by a number of the faculty in AE. A heavy demand for AE structural option students, who like all Penn State AE students also had a breadth background in materials, mechanical systems, lighting/ electrical, and construction management, was developing due to expanding interest from a number of forensic and building technology firms in the industry.

The Penn State AE course described in this paper, AE 537 - Building Failures and Forensic Techniques (Building Failures), was initiated in the fall of 1999. It is offered once each year to graduate students and upper level (4th and 5th year) undergraduate students who are in the integrated BAE/MAE program in architectural engineering and with permission for qualified upper level undergraduate BAE students. While most of the students taking the course are in AE, students from Civil Engineering, and to a lesser degree, Architecture (a separate program in the College of Arts and Architecture at Penn State) often enroll in the course as a technical elective.

As a course, Building Failures focuses on structural and architectural building systems and incorporates a simple yet broad definition of failures related to the design, construction, and operation of buildings that states: "a failure is any system that does not perform as intended." As such, the course covers topics that incorporate a variety of structural and architectural components as noted below:

- Building envelope failures (facades and roofs)
- Flashing, waterproofing, sealants, and related issues
- Structural failures – special loadings, damage assessment, full and partial collapses, and structural performance
- Historic preservation issues for buildings
- Durability, deterioration, maintenance, and repair of building materials
- Failures due to communication and procedural issues or errors
- Legal issues and the role / responsibilities of an expert witness
- Ethics in engineering practice

Course Offered in Partnership with Practitioners

In order to keep the topics of the course in line with current industry issues and practices, and to provide students access and interaction with practicing professionals, the building failures course takes advantage of an ongoing educational / industry partnership within the AE Department. This partnership results in a yearly visiting lecture series related to technical topics incorporated into the course, an overview of industry practices, and interaction between students and potential employers in the building technology and forensic fields. This lecture series was initiated the first year the course was offered as a way of jump starting the number of case history studies available to the students and as an example of how the profession can become involved in the education of students related to building failures (Parfitt 1999).

Over the years, a number of firms and individuals have participated in this educational partnership. The industry participants represent a cross section of forensic and building technology firms, structural engineers, repair and restoration contractors, and attorneys selected to expose the students to a balance and cross section of activities within the industry. The following list is representative of those who have participated on a frequent basis:

- Facility Engineering Associates (FEA)
- Gerard J. Pisarcik, Esq.
- Ryan-Biggs Associates, P.C.
- Simpson Gumpertz & Heger, Inc. (SGH)
- Structural Preservation Systems, Inc. (SPS)
- The LZA Technology Division of Thornton-Tomasetti Group, Inc
- Wiss Janney Elstner Associates, Inc. (WJE)

Techniques for Teaching Failures Topics

A number of different techniques are available to teach failures. Many educators have had considerable success with the use of the case study method. Curriculum improvement surveys completed by students in AE 537 have shown that no one teaching technique is considered best for all aspects of the course. Students in the course have indicated that they prefer a variety of techniques as appropriate to the topic or particular teaching goal with student centered and active learning techniques as the most popular.

It was also quickly discovered that one of the best tools for active learning in AE 537 was to use the extensive building infrastructure resources of the Penn State campus and the local community, State College, Pennsylvania, as a learning laboratory. While structurally damaged and collapsed buildings seldom occur in this laboratory, not to mention timed to coincide with the class schedule, Penn State like most major institutions, has a considerable amount of deferred maintenance and ongoing capital improvement projects that result in many opportunities for building envelope performance observation and investigation. Available examples include historic (and sometimes more recent) masonry maintenance and repair issues, deterioration and repair of concrete, roofing and façade waterproofing maintenance and repairs, and occasional storm damage. The Penn State Office of Physical Plant (OPP) has been very cooperative relative to providing file information and site visits for students in the class. In some cases, they have even created invasive inspection openings in building assemblies of interest to the students and OPP alike.

Historic buildings on campus are often reviewed to show the differences between mass wall masonry construction, and the resulting lack of masonry movement joints, as compared to buildings with more recent masonry cavity wall facades. Students are asked to observe the number and spacing of movement joints in different types of masonry façade buildings, as well as to note how well those joints are performing (sealant condition and quality, sealant pushed out of joint, etc.) as an active learning exercise.

Exploring campus by the students has led to the discovery of several cavity wall facades that are lacking the proper number and/or size of movement joints resulting in cracking and damage to the masonry. Figure 1 is a photograph of a student identified example of this type from a building that contains few movement joints overall, and none near the building corners.

Figure 1
This vertical crack near a building corner was identified by student campus surveys.

Older buildings on campus are also frequently used to point out conditions related to historic preservation issues such as the proper repair and maintenance of masonry, slate roofs and other period materials. Students are shown that observations of historic structures can also provide clues as to how even newer materials will weather and perform over time. One excellent example on campus used for the class is the Penn State Obelisk (Figure 2). Constructed in 1896, this mass masonry monument-like structure contains 281 stones arranged in natural geologic order. The Obelisk was originally erected to demonstrate the weathering properties and subsequent value of Pennsylvania stones. While some of the mortar has been repaired, the stones have continued to weather and the obelisk is still being used in Building Failures and several other classes for demonstrating the long term impacts of weathering on building stones, including variations due to orientation and bedding (Figure 3).

Severe structural collapses are rare, especially those that occur within the class time frame. Occasionally students have the opportunity to view the aftermath of a large scale collapse, however, typical up-close failure evaluations and reviews by students are limited to minor structural damage or building performance conditions. Figure 4 is a photograph showing the aftermath, including search and rescue markings, from a local parking deck collapse. Experience has shown that students often identify more closely with local failures and collapses since they know the buildings or their subsequent reconstructed replacements. As such, they are often surprised to find out that even a small town like State College has experienced a number of structural or performance failures. Local failures and performance problems have been documented and are included in the course as examples and mini case histories to reinforce the point that failures can happen anywhere. Figure 5 shows an example of a performance issue related to roof leaks and minor concrete deterioration at an auditorium on campus that has been studied as a class example.

Figure 2
Penn State Obelisk: Used to study weathering of building stones.

Figure 3
The Photograph above is a close up of one segment of the Obelisk. Students are alerted to different weathering patterns of the stones relative to type of stone, exposure and bedding placement.

Figure 4
Parking Deck Collapse

Figure 5
Penn State Forum Re-roofing

Studying failures in a classroom can not always be accomplished first hand. As a result, in order to target specific learning goals and to provide the students with a variety of learning experiences, a mixture of teaching methods have been implemented in AE 537. A representative list is provided in Table 1 below:

Instructional Activity - Example	Goals of Activity	Comments
Traditional Classroom Lectures	A,B,G,H	By course instructor and visiting professionals
Case History Presentation by with document package and figures and/or video	B,D,G,H	By course instructor and visiting professionals
Condition Surveys and Reports – campus and local buildings	B,C,D,E,F,H,I,K	Professional level writing stressed
Case History Research Project and Report – building is individually selected by student	B,D,G,I	Professional level writing stressed
Independent Research and Report on a system, type, or category of failure	B,D,G,H,I	Professional level writing stressed
Written Proposal for Engineering Services	I,J,K	Professional level writing stressed
Classroom Practicum – includes discussion and collaboration among students	J,K	Differing viewpoints are encouraged and discussed
Campus survey "blitzes" with follow up poster presentation and discussion	B,C,E,I,K	Initial observation time is intentionally limited in some of these exercises.

Key to Goals of Activity:

A. Quick dissemination of knowledge, provide background material necessary for projects, case history study, follow up to guest lectures.
B. Develop and Improve Investigative Techniques
C. Improve powers of observation
D. Develop ability to analyze information and draw conclusions
E. Ability to document conditions with figures, graphs and/or sketches
F. Supplement an instructor or guest practitioner lecture by reinforcing key points with direct field observation.
G. Learn about failure and performance from a historical perspective
H. Identify characteristics of specific building types or systems
I. Improve written, graphic and oral communication skills
J. Develop ability to identify and convey all aspects of investigation work
K. Investigative team work, coordination, and ability to summarize information in a concise manner.

Table 1 – Instructional Activities and Goals

The specific class exercises and assignments vary slightly from year to year based on the availability of practitioner guest lecturers, investigation and repair projects in progress or available on campus, and natural hazard events that may take place during a particular semester such as severe wind or snow storms.

Case histories of significant and well known failures remain an important aspect of the course. Each student is required to investigate through study and research, a major historical failure that can be examined using conventional literature of various types, as well as online archives and failure sites. A variety of resources are required to be used as documentation including obtaining original reports and investigation records and contacting the original investigators as appropriate.

Challenges in Offering the Course

Once the decision has been made to offer a failures course, a number of challenges remain relative to preparation and execution. Challenges such as finding adequate materials and case history information for AE 537 were resolved by implementing a partnership with the industry. A number of firms provided guest lectures the first several years while at the same time providing the instructor permission to use much of the material in subsequent years as appropriate, especially if the guest professional was not available due to scheduling conflicts. Industry representatives provided a number of files, photographs, drawings, material samples, and related resources to supplement those from the course instructor's research and practice experience. For example, the American Institute of Steel Construction (AISC) donated their entire original historic file from the Kansas City Hyatt walkway collapse while a consulting firm provided access to a photographic library of roofing projects. Despite the extensive outside help, the time investment in preparing and updating the course remains extensive.

Difficulties related to the timing of recent collapses that might be desirable for class study has been resolved by preparing extensive representative collections of raw data on collapses (photographs, depositions, reports, etc.) that permit the instructor to offer a case study of the final outcome, or take a step back and provide raw information so the students can simulate parts of a real investigation.

Access to campus buildings is provided by Penn State, but scheduling can be difficult, especially when you are trying to accommodate large numbers of students in what sometimes is a small space. Some field study situations can place students on roof tops and ladders making safety the highest priority. At the same time in keeping with the active learning role, these types of field trips and construction tours provide an opportunity for an introductory safety lecture or a discussion on OSHA requirements for certain types of construction.

The use of service teaching, where a building owner gets something in return, (results of the survey, photographic documentation, recommendations etc.) has been found to be a potential source of liability as documented by others teaching courses of this nature (Sutterer 2003). In AE 537, service assignments are limited to Penn State buildings that are, or will be, followed up by investigations by OPP or outside consultants hired by the university. The studies are therefore handled as initial investigations to identify or classify problems and to suggest possible solutions. They are not considered or represented as final reports, even though the investigations are monitored by licensed professional engineers.

Summary

Active or student centered learning and the use of case history studies is not without challenges and drawbacks. Experience by AE in this course, and other educators throughout the country, reveals that although these types of learning experiences are well received by students, facilitating these methods and related course projects can be time consuming, a potential source of liability, and difficult to find in the right stage at the right time to satisfy a particular course objective. The rewards are excellent, however, as students enter the work force with direct exposure to the building failures and performance issues that will confront them in the years to come.

References

Baer, Ruben J. (1996). Guest editorial, "Are civil engineering graduates adequately informed on failure? A practitioner's view." ASCE *Journal of Constructed Facilities*, 10(2), 46.

Bosela, Paul A. (1993). "Failure of Engineered Facilities: Academia Responds to the Challenge." ASCE *Journal of Constructed Facilities*, 7(2), 140-144.

Carper, Kenneth L. (2003). "Technical Council on Forensic Engineering: Twenty-year Retrospective Review." *Proceedings of the Third Forensic Engineering Congress,* P. A. Bosela, N. J. Delatte, and K. L. Rens, eds., ASCE, Reston, VA, 280-296.

Delatte, Jr., Norbert J. (1997). "Failure Case Studies and Ethics in Engineering Mechanics Courses." ASCE *Journal of Professional Issues in Engineering Education Practice*, 123(3), 111-116.

Delatte, Norbert. J. (2000). "Using failure case studies in civil engineering education." *Proceedings of the Second Forensic Engineering Congress*, K. L. Rens, O. Rendon-Herrero, and P. A. Bosela, eds., ASCE, Reston , VA., 430-440.

Delatte, Norbert J., Rens, Kevin, L. (2002). "Forensics and Case Studies in Civil Engineering Education: State of the Art." ASCE, *Journal of Constructed Facilities*, 16(3), 98-109.

Jennings, Alan and Mackinnon, Pauline. (2000). "Case for Undergraduate Study of Disasters." ASCE, *Journal of Constructed Facilities*, 14(1), 38-41.

Parfitt, M. Kevin (1999). "New Course in Building Failures Offered with Industry Support." *Architectural Engineering Newsletter,* Fall / Winter, 9-10.

Parfitt, M. Kevin (2001). Discussion of "Failure of Constructed Facilities in Civil Engineering Curricula" by Kevin L. Rens, Oswald Rendon-Herrero, and Melissa J. Clark, ASCE, *Journal of Performance of Constructed Facilities*, 15(1), 38-39.

Pietroforte, Roberto. (1998). "Civil Engineering Education through Case Studies of Failures." ASCE *Journal of Constructed Facilities,* 12(2), 51-55.

Rendon-Herrero, Oswald. (1993a). "Too Many Failures: What can Education Do?" ASCE *Journal of Constructed Facilities,* 7(2), 133-139

Rendon-Herrero, Oswald. (1993b). "Including Failure Case Studies in Civil Engineering Courses." ASCE *Journal of Constructed Facilities,* 7(3), 181-185.

Rens, K. L. and Knott, A. W. (1997). "Teaching Experiences: A Graduate Course in Condition Assessment and Forensic Engineering." *Proceedings of the First Forensic Engineering Congress*, K. L. Rens, ed., ASCE, New York, 178-185.

Rens, Kevin. L., Clark, Melissa, and Knott, Albert W. (2000). "Development of an Internet Failure Information Disseminator for Professors." *Proceedings of the Second Forensic Engineering Congress*, K. L. Rens, O. Rendon-Herrero, and P. A. Bosela, eds., ASCE, Reston , VA., 441-453.

Rens, Kevin. L., Rendon-Herrero, Oswald., and Clark, Melissa J. (2000). "Failure of Constructed Facilities in Civil Engineering Curricula." ASCE *Journal of Constructed Facilities,* 4(1), 27-37.

Reynolds, Christine (2003). "Rewriting the Curriculum: a Review and Proposal of Forensic Engineering Coursework in U.S. Universities", *Proceedings of the Third Forensic Engineering Congress,* P. A. Bosela, N. J. Delatte, and K. L. Rens, eds., ASCE, Reston, VA, 307-315.

Sutterer, Kevin G. (2003). "Service Learning and Forensic Engineering in Soil Mechanics." *Proceedings of the Third Forensic Engineering Congress,* P. A. Bosela, N. J. Delatte, and K. L. Rens, eds., ASCE, Reston, VA, 297-306.

Vibrations of Cable-Stayed Bridges

S. M. Palmquist[1]

[1]Department of Engineering, Western Kentucky University, 1906 College Heights Blvd. 21082, Bowling Green, KY 42101-1082; PH (270) 745-7190, FAX (270) 745-5658; email: shane.palmquist@wku.edu

Abstract

Recently, cable-stayed bridges have become very common in the United States. However, the cables of many of these structures have experienced unforeseen vibrations causing significant damage to what were relatively new structures. The damage in some cases has cost state departments of transportation millions of dollars to inspect, maintain, and repair. Perhaps even more perplexing than the cost of the repairs is the question: why were the vibrations unforeseen? For many years, vortex shedding, occurring around the cables such as what occurred with the Tacoma Narrows Bridge (except around the girder deck system), was thought to have been causing the problem. Recently, another phenomenon termed rain-wind induced vibration was found to be causing the problem. But was rain-wind induced vibration really causing the problem or were there other factors involved that could have eliminated the problem from the beginning?

This paper examines two such bridges that have experienced vibrational problems requiring major repair. These bridges are the Cochrane Bridge in Mobile, Alabama and the Talmadge Memorial Bridge in Savannah, Georgia. An understanding of structural vibrations and the importance of the need to properly design critical connections is vital, when designing and constructing cost effectively flexible structures, such as cable-stayed bridges.

Introduction

Excessive cable vibrations due to wind have been observed on many cable-stayed bridges around the world (Gu and Du 2005; Jones and Main 1998). Occurrences of such events have happened throughout the United States, including: Burlington Bridge in Iowa; Clarke Bridge in Alton, Illinois; Cochrane Bridge in Mobile, Alabama; Dame Point Bridge in Jacksonville, Florida; East Huntington Bridge in West Virginia; Fred Hartman Bridge in Baytown, Texas; Houston Ship Channel Bridge in La Porte, Texas; James River Bridge in Richmond, Virginia;

Veterans Memorial Bridge in Port Author, Texas; and Weirton-Steubenville Bridge in West Virginia.

Wind related excitation, that can potentially cause cable vibrations, include the following phenomena: vortex excitation, wake galloping, and rain-wind excitation. On many bridges, the combination of rain and wind has caused excessive cable excitation. This phenomenon was first observed and described by Hikami and Shiraishi (1988) on the Meikonishi cable-stayed bridge in Nagoya Harbor, Japan. Cables with a 14 cm diameter and lengths varying between 65 to 200 meters in length began oscillating, when weather conditions were light rain and wind.

Rain-wind excitation of the cables can begin at low wind speeds, and only involve the lowest modes shapes. Modes up to the sixth mode have been report (Irwin 1997). Large amplitude vibrations on cable-stayed bridges can result in unintended stresses and fatigue in the cables as well as in the connections at the bridge deck and tower. Large amplitude vibrations have been observed on cables with a yawning angle of 20° to 60° over a range of wind speeds from 6 to 40 m/s causing amplitudes of up to 2 meters (Gimsing 2000).

The weather conditions that cause rain-wind cable oscillation have very specific characteristics. The rain and wind have to be light to moderate. When this occurs, water collects along the length of the stay cable forming rivulets that are nearly diametrically opposite to each other, as shown in Figure 1. The rivulets that form along the length of the cable can be seen with the naked eye even though the size is relatively small as compared to the cable diameter, as observed on the Cochrane Bridge (Telang and Norton 1999). The two rivulets, running along the length of the cable, cause a change in the cross-sectional profile of the cable resulting in aerodynamic instability. The aerodynamic instability can develop into excitation of the cables in the vertical plane of the cable stays. Another factor that contributes to the development of the water rivulets is the polyvinyl tape that wraps around the polyethylene pipe containing the high strength steel strands. The polyvinyl tape provides ultraviolet protection; however, it is extremely smooth, promoting movement of water around the stays to the location of the rivulets. Cable-stayed bridges with steel pipes have rough surfaces that are not conducive to the movement of water in forming the rivulets.

Cables with potential for rain-wind induced vibrations can be experimentally tested to determine the value of the Scruton number, S_C (Irwin 1997). The nondimenional number has been found to be an important parameter in aerodynamic oscillation phenomenon. The Scruton number, S_C, is defined by,

$$S_c = \frac{m\zeta}{\rho D^2}$$

where, m is the mass per unit length of the cable (kg/m); ζ is the damping ratio of the cable; ρ is the air density (kg/m^3), which has a standard value of approximately 1.225 kg/m^3; and D is the outside diameter of the cable (m). The damping ratio of some cables has been reported to be as low as 0.0005 with a corresponding Scruton number, S_c, of 1 (Irwin 1997). According to the Post-Tensioning Institute (PTI) Guide Specification, 4th edition, "Recommendations for Stay Cable Design, Testing and Installation," the Scruton number for cables with smooth surfaces should be

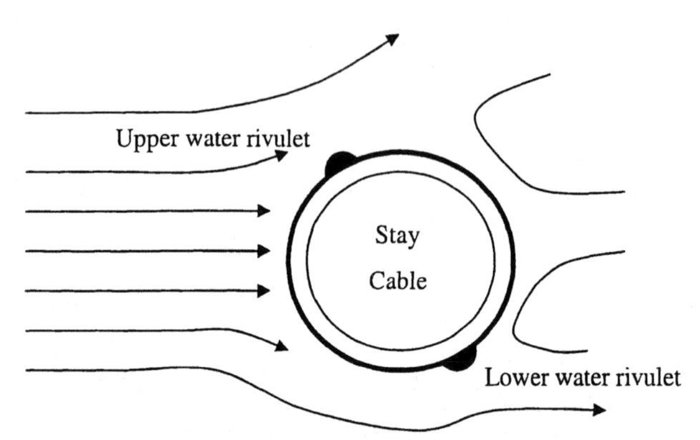

Figure 1. Aerodynamics of rain-wind induced vibration

greater than 10 to suppress the potential of aerodynamic instability caused by rain-wind induced vibration.

Based on the equation, reducing the diameter of the cable will have a significant effect on raising the Scruton number since the parameter is squared in the denominator. Use of high density grout in the cable stays would also increase the Scruton number but to a much lesser degree than changing the cable diameter. For existing cable stayed bridges, reducing the cable size or replacing the cable grout are not feasible, and other methods must be considered such as: shape modification of the cables, raising the natural frequencies of the cables, or raising the damping ratio of the cable stays. Cables with Scruton numbers less than 10 have been found to be prone to vibrations caused by wind-rain events. For such cables, the connections near and at the anchorage are critical and proper analysis and design is required.

Cable guide pipe connections

Near the cable end connections, the stay cable passes through a neoprene or rubber damper into the anchorage at the deck or tower. The dampers are small and relatively short in length in comparison with the cable length. Many cables are 60 to 200 meters in length, while the dampers are only 25 cm in length. For a cable oscillating, this can be thought of as three children playing jump rope, one at each end and one jumping. Should the two children, at the ends, hold the rope with their fingertips or with their hands? If the children hold the rope with their finger tips, their wrists, hands and fingers will quickly become fatigued and sore!

At the location of the damper, which is usually a meter or so above the deck or away from the tower, the potential exists for significant stresses to be developed if the cable begins to oscillate. The stresses will need to be mitigated and transferred from the cable, into the damper, into the steel guide pipe, through the guide pipe and into the superstructure. Proper design of the guide pipes and washers for transfer of these stresses is critical especially since large moments where the guide pipes connect with the deck or tower can develop. Unfortunately, the design of the guide pipe and washer connection was developed from "rules of thumb" based on anticipated lateral loads (Poston 1998). The purpose of the damper was to mitigate stay vibrations arising from low amplitude, high frequency vibrations associated with buffeting and vortex shedding associated with wind. The dampers were not designed for low frequency, high amplitude vibrations associated with rain-wind induced vibrations.

While the low amplitude, high frequency argument appears to validate the material selection and structural design of the guide pipe and washer connections based on the information available at the time, an entirely opposite argument could have been proposed based on the fundamental principles of structural dynamics, as follows. Flexible structures such as cable-stayed bridges tend to oscillate at the lowest natural frequencies, corresponding to the lowest modes. Cables of bridges of this type are very flexible by design and behave accordingly, vibrating at the lowest modes when circumstances create the potential for such a condition. Whether the structure is the girder system of the Tacoma Narrow Bridge, which experienced a violent resonance condition in 1940, resulting in collapse of the structure, or a cable of a cable-stayed bridge, the lowest modes will be targeted and excited, unless conditions favor a specific higher mode or modes. Given this, more attention to the design of the guide pipe and washer connections was warranted. The design should have adequately accounted for the potential of low frequency, high amplitude vibrations.

Methods for vibration mitigation of existing cable-stay bridges

A number of methods are available to mitigate cable vibrations on existing cable-stay bridges. For rain-wind vibrations to occur, the cable must have a low Scruton number, less than 10, and water rivulets must develop along the length of the cable. Several methods used to mitigate cable vibrations meet these criteria by either eliminating the formation of the water rivulets, increasing the Scruton number to greater than 10, or a combination of the two. Some of the common methods are: shape modification of the cable stays, raising the natural frequencies of the cable stays, and raising the damping of the cable stays. With each method, there are advantages and disadvantages when compared to each other.

Altering the shape or profile along the cable length of the stay can be achieve by either attaching helical fillets to the length of the cable or increasing the roughness of the surface of the cable. Helical fillets have been used in France to stabilize cables against rain-wind induced vibrations. The helical fillets stabilize the cable during rain-wind events by disrupting the flow of water along the length of the cable. In Japan, protrusions along the cable surface have been used to deter the formation of the water rivulets. However, careful attention is required when considering shape modification of an existing cable-stayed bridge since the effects on aerodynamic stability based on the new shape of the cables will have to be evaluated.

Increasing the natural frequencies of a cable directly raises the wind velocity needed to excite the cable. In terms of a taught flexible cable, the natural frequencies for any flexural mode, n, is,

$$\omega_n = \frac{n\pi}{L}\sqrt{\frac{Tg}{m}}$$

where m is the weight per cable unit length of cable (N/m); g is the acceleration due to gravity (m/s^2), which is 9.81 m/s^2; T is the tension applied to the cable by the anchorages (N); and L is the effective cable length (m).

Tension and mass are not quantities that can be readily changed on an existing cable-stayed bridge. However, the effective length, L, can be changed by using cross ties. The effective length, L, is the most significant parameter in this equation. Reducing this parameter will have a significant effect on the corresponding natural frequencies of the cables. The lowest natural frequencies susceptible to aerodynamic excitation can be greatly increased by the addition of cable ties. While this method will help dampen out vibrations, the amount of increased damping by the cross ties is unknown and is difficult to quantify.

Attaching dampers on cable stays can also be used as an effective method for minimizing the potential for cable vibration. This method hinders the development of cable oscillation by substantially increasing the duration required for an event to occur. In practice, this has become the preferred method despite the significant cost associated with their use. Existing cable-stayed bridges, which have been retrofitted with external dampers, include: Fred Hartman Bridge in Baytown, Texas; Veterans Memorial Bridge in Put Author, Texas; Cochrane Bridge in Mobile, Alabama; Talmadge Memorial Bridge in Savannah, Georgia; and many others.

Cochrane Bridge

The Cochrane Bridge majestically crosses the Mobile River in Mobile, Alabama, as shown in Figure 2. The two semi-harped arrangements of cables on H-shaped pylons complement the vast flat landscape of the area. The structure carries US Routes 90 and 31, and spans a total of 460 meters over the Mobile River. The main center span is 240 meters with two flanking spans of 110 meters. The total length of the bridge including all approach spans is approximately 2200 meters in length. The bridge was completed in 1991 at a cost of $70,000,000.

Figure 2. South elevation of the Cochrane Bridge

The cable stayed spans carry two 12 meter wide roadways on a 28 meter wide by 2.3 meter deep prestressed concrete deck. The deck consists of longitudinally and traversely post-tensioned and segmentally constructed twin cast-in-place single cell trapezoidal concrete box girders supported along each side by 48 cable stays in a semi-harped arrangement. The cable stays consist of a varying number of 1.5 centimeter diameter parallel seven wire high strength strands encased in polyethylene pipes. The strands, which vary from 26 on the short cables to 72 on the long cables, are embedded in a cement grout within the polyethylene pipe. At the anchorage, the stay cable passes through a steel guide pipe, and a neoprene washer is provided in the annular space between the polyethylene pipe and the steel guide pipe, as shown in Figure 3.

Since the bridge opened in 1991, several instances of excessive cable oscillations have been reported. In one instance, an event occur when the weather was a combination of light rain and wind with a sustained wind speed of approximately 47 km/hr and gusts of up to 65 km/hr (Telang and Norton 2000). Vibration in many of the 48 cables in the vertical plane of the stays was observed. The maximum reported amplitude near the midspan of the cables was approximately 1.5 meters. This event in conjunction with previous events, prompted the Alabama Department of Transportation (ALDOT) to perform an emergency response inspection of the deck level cable anchorages. The inspection revealed locations of

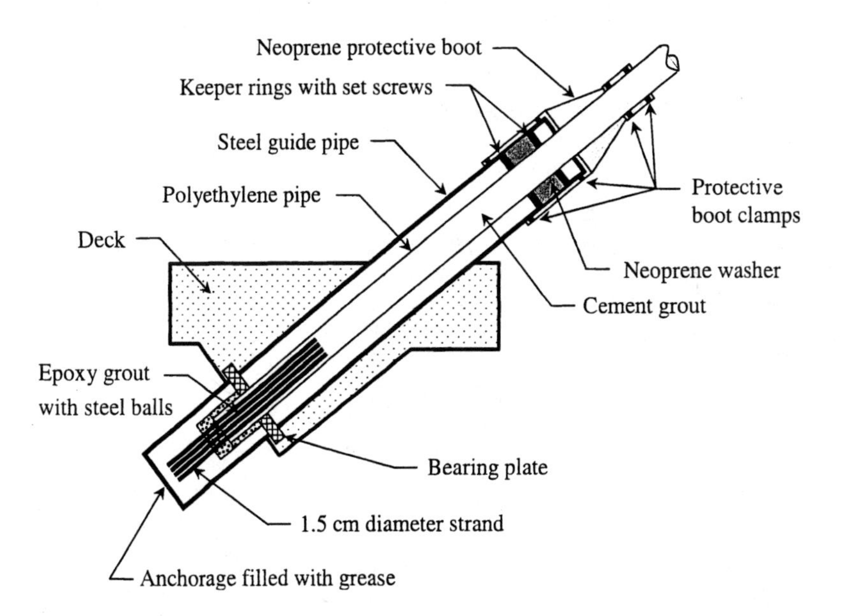

Figure 3. Guide pipe and washer connection at the anchorage

damage, including: dislodged protective boots around the guide pipes, broken or stretched clamps around the protective boots, dislodged and misaligned keeper rings and washers, and sheared off bolts that attached the keeper rings to the steel guide pipes.

Later, during an in-depth inspection of the cables in March of 1998, engineers witnessed an event where large amplitudes in many of the cables began to occur. The event occurred during a period of light rain and wind, where the wind velocity was between 32 to 48 km/hr. Large amplitude cable vibrations in the vertical plane of the stays were observed in seven of the longest cables on west side of the southern pylon towers. The maximum amplitude was estimated to be 1.2 to 1.5 meters, and the frequency of vibration was approximately 0.9 to 2.0 hertz (Telang 1999). Water rivulets located approximately diametrically opposite to each other and running down the cable stays were observed. As the wind speed increased and the rainfall became heavier, the water rivulets on the cable surface were "disrupted or blow away," and the vibration of the cable mitigated quickly (Telang 1999). A similar event with nearly the same weather conditions occurred again in April of 1998.

Based on the significant rain-wind event March of 1998, a plan of inspection, instrumentation, and testing was implemented to determine the nature of the problem and to determine what options were available to mitigate the problem. Inspections revealed localized damage to the cable stay connection where the polyethylene pipe connects to the guide pipe. After extensive field inspection, two conditions were routinely found throughout the bridge. The first condition was minimal to no bearing contact between the cables and washers at numerous locations throughout. Many cables did not pass concentrically through the guide pipe at the location of the washers, so the thickness of the washer materials around the cable was not uniform. In some cases, the thickness of the washer varied from 6 millimeters on one side to more than 150 millimeters on the other side. In addition, there was a gap from 6 millimeters to as much as 25 millimeters between the guide pipe and the washer. Minimal damage to the superstructure box girders and to the pylon towers at the guide pipe connections was observed. The second condition was dislodging of the washers from the initial location, which was caused by failure of the steel keeper rings due to excessive movement of the cables. The displaced washers provided limited to negligible damping to the cable stays. In addition, at some locations, the proximity of the cable to the anchorage components allowed the cables to violently impact with the guide pipe during a rain-wind event. The impact caused damage in the form of gouges on the surface of the polyethylene pipe, and cracks were observed on the guide pipe.

Field measurement were performed to quantify the damping in each cable and to calculate the corresponding Scruton number, S_C. Damping ratios were in the range of 0.1 to 0.9 percent (Telang and Norton 1999). For 86 percent of the cables, the Scruton number was less than 10. As a result, nearly all of the cables were susceptible to rain-wind induced vibration. To mitigate potential cable vibrations, external viscous dampers were installed at the deck level to all of the cables. In addition, new washers and keeper rings were also installed in the guide pipes at the deck level and at the tower locations. The cost for inspection, repair/rehabilitation of all of the guide pipe washer connections, and installation of the external dampers was in the many millions of dollars.

Talmadge Memorial Bridge

The Talmadge Memorial Bridge is gracefully poised over the Savannah River in Savannah, Georgia, as shown in Figure 4. The structure is a twin-pylon cable-stayed bridge with two semi-harped arrangements of cables on H-shaped pylons that compliment the local landscape. The bridge carries US Route 17, and spans a total distance of 620 meters over the Savannah River. The main center span is 335 meters with two side spans of 140 meters. The total length of the bridge including all approach spans is 2300 meters. The bridge was completed in 1991 at a cost of $26,000,000 for the cable-stayed spans. The total cost of the bridge including all approach spans was $71,000,000.

The cable stayed spans carry two 105 meter wide roadways on a 24 meter wide post-tensioned segmental deck consisting of 28 centimeter thick concrete slab with 1.4 meter deep concrete edge girders. The deck is supported transversely by concrete floorbeams, which are up to 2.1 meters deep and located at each deck level cable anchorage. The edge girders are continuous with no expansion joints for the full length of the main span. There are 72 cables per pylon (36 stays on each side), which support the deck.

During construction, movement of the cables occurred (Liles 1999). In 1998, excessive cable vibrations were reported by maintenance workers, and one (1) of the guide pipes had fractured (due to excessive movement of the cable stay). Soon thereafter, an inspection by the Georgia Department of Transportation (GADOT) and Federal Highway Administration (FHWA) was performed, and a number of problems associated with cable vibrations were reported, including: (1) partial fracture of three additional guide pipes (see Figure 5), (2) missing dampers or misaligned cables at the guide pipe entrance, and (3) further spalling and cracking at the guide pipe/bridge

Figure 4. West elevation of the Talmadge Memorial Bridge

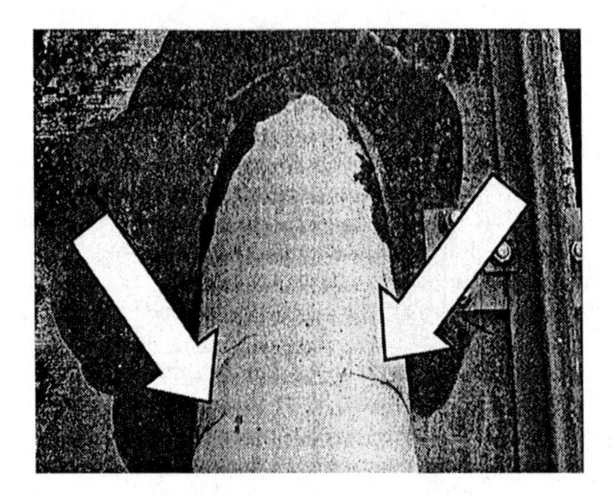

Figure 5. Fractured steel guide pipe

deck surface (Liles 1999). As a result, an extensive plan of inspection, testing, and major rehabilitation of the guide pipe washer connections was required, similar to the Cochrane Bridge. The repairs included installation of external dampers at the deck level, and the total cost of the work was substantial, well into the millions of dollars.

Cost Effective Vibration Mitigation

Excessive vibrations in cable-stayed bridges have significantly increased the cost of inspections, maintenance, and repairs. The cost of this work resulting from a single cable-stayed bridge is in the millions of dollars (Poston 1998). For the Cochrane Bridge and the Talmadge Memorial Bridge, the costs were well into the millions of dollars (and those costs only included evaluation and installation of the cable damper systems). On the Fred Hartman Bridge, cross ties were added as a temporary solution for vibration control. The cost for installing the cable ties was $573,000 (Poston 1998). Unfortunately, cross ties are not considered as a permanent solution since the cross ties alter the aesthetic appearance of the structure (Poston 1998). In actuality, cross ties do not noticeably influence the appearance of the structure (Gimsing 2000).

Cross ties offer a more cost effective approach to mitigation of cable vibrations than installing expensive external dampers. Failures of cross ties have resulted due to improper details and materials selection. Aesthetics should not govern over practicality. The aesthetic impact of cross ties can be minimized by proper selection such as thin flexible wire rope or a similar system with adequate fatigue resistance and wear characteristics.

Conclusions

The design and details of the steel guide pipe and washer connections were overlooked and considered less important in the overall complexity of the structure on the Cochrane Bridge and Talmadge Memorial Bridge. The potential flexibility of the cables based on the fundamental principles of structural dynamics should have been considered more closely in the design of these details. The deficiencies of these details have greatly increased the maintenance cost of these bridges and many like it with similar details. On both of these bridges, expensive external dampers were installed on all of the cables to mitigate the potential of future vibrations due to rain-wind events. Cross ties, despite being cost effective as compared to the external dampers, were not considered as a viable option, since they were thought to detract from the aesthetic appearance of the bridges.

References

Bosch, H. R., (1999). "Wind-Induced Vibration of Stay Cables," *Proceeding of Workshop on Rain-Wind Vibrations in Cable Stayed Bridges*, FHWA Southern Resource Center Area, August 30 – Septermber1, Atlanta, Georgia.

Cioldo, A. T., and W. P. Yen, (1999). "An Immediate Payoff," Public Roads, May/June, pp. 10-17.

Gimsing, N. J., (2000). "Chapter 3: Cable System," Cable Supported Bridges, Wiley, New York, pp. 270-281.

Gu, M., and X. Du, (2005). "Experimental Instigation of Rain-Wind-Induced Vibration of Cable in Cable-Stayed Bridge and Its Mitigation," *Journal of Wind Engineering and Industrial Aerodynamics*, 93, pp. 79-95.

Hikami, Y., and N. Shiraishi, (1988). "Rain-Wind Induced Vibrations of Cable in Cable Stayed Bridges," *Journal of Wind Engineering and Industrial Aerodynamics*, 29 (1988), pp. 409-418.

Irwin, P. A., (1997). "Wind Vibration of Cables on Cable-Stayed Bridges," *Building to Last*, pp. 383-387.

Jones, N. P., and J. Main, (1998). "Stay Cable Vibration Measurements, Mechanisms and Mitigation," *Proceedings of 12th ASCE Engineering Mechanics Conference*, La Jolla, CA, May.

Norton, P. G., Darling, S. M., and K. J. Bean, (2003), "Design Considerations for Cable-Stayed Bridges," *Proceedings of the 20th International Bridge Conference*, Pittsburgh, June.

Poston, R. W., (1998). "Cable-Stay Conundrum," *Civil Engineering*, August., pp. 58-61.

Post-Tension Institute, (2001), "Recommendations for Stay Cable Design, Testing and Installation," *PTI Guide Specification*, 4th edition.

Telang, N. M., and P. G. Norton, (1999). "Evaluation of Rain-Wind Vibration of the Cochrane Bridge," *Proceedings of the 16th International Bridge Conference*, Pittsburgh, June.

Telang, N. M., (2000). "Rehabilitation for Mitigation of Rain-Wind Vibrations on the Cochrane Bridge," *Proceedings of the 17th International Bridge Conference*, Pittsburgh, June.

Human Induced Vibration Monitoring of a College Football Stadium

Dilip Choudhuri, P.E., M.ASCE[1] and Prasad Samarajiva, Ph.D., P.E., M.ASCE[2]

[1]Principal, Walter P. Moore and Associates, Inc., Structural Diagnostic Services Group, 3131 Eastside St, 2nd Floor, Houston, TX 77098, Phone: (713) 630-7300, email: DChoudhuri@walterpmoore.com.

[2]Associate, Walter P. Moore and Associates, Inc., Structural Diagnostic Services Group, 3131 Eastside St, 2nd Floor, Houston, TX 77098, Phone: (713) 630-7300, email: PSamarajiva@walterpmoore.com.

Abstract

Individuals sitting in the press box and suite levels of a college football stadium had noted significant structural vibrations during games, specifically when the home crowd sings their sports hymn. Synchronized side sway of tens of thousands of spectators during the chanting of the sports hymn appeared to be the prime reason of the vibrations. This reinforced concrete stadium has gone through several stages of expansions and renovations over the years. Walter P. Moore and Associates, Inc. (WPM) instrumented the various section of the stadium with accelerometers to evaluate the levels of vibrations during three games of a recent season. Our testing revealed that the lateral vibration levels in the press box and suite levels to be strongly perceptible, mostly occurring around 1.4 Hz.

Background Information

Walter P. Moore and Associates was requested by the institution to conduct an evaluation of vibrations induced by human activities at their football stadium structure during use for three home game of a recent football season. Anecdotal evidence suggested that in the past there were several occupants that had complained about the perceived lateral movement at the press box and suite levels, specifically during the chanting of their sports hymn.

The stadium stand of interest consists of three concrete deck levels with a pre-engineered steel box on top of the third deck housing the press box and suite levels. The stadium stand structure consisted of typical concrete column elements, concrete raker beams, and precast seating risers. Figure 1 show a cross section of the stadium stand of interest. A schematic layout of the stadium is shown in Figure 2.

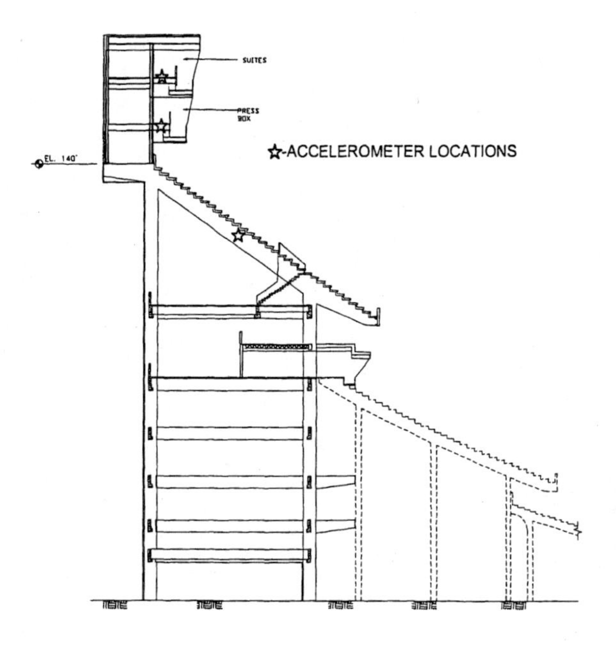

Figure 1. Cross Section of the Stadium Stand

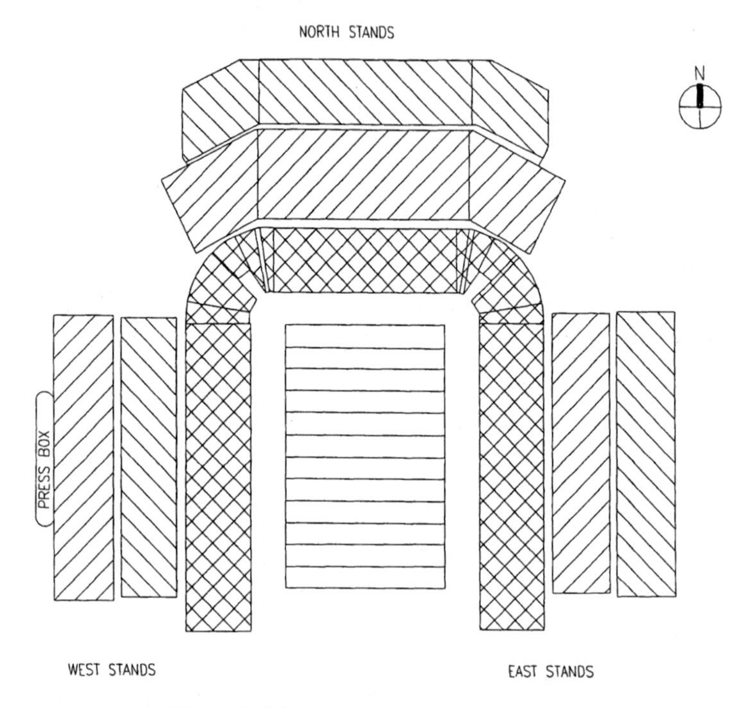

Figure 2. Schematic Layout of the Stadium

Vibration Monitoring Techniques

The monitoring was conducted using capacitive accelerometers (Fraden) at preselected locations in the stadium (Figure 1). The accelerometers were either the uniaxial type or the triaxial type. In addition, Instantel BlastMate III blast monitors were used as back-up instrumentation for the uniaxial and triaxial sensors. The uniaxial accelerometers measures acceleration in one direction – longitudinal. The triaxial accelerometers measures acceleration in all three directions – longitudinal, transverse and vertical. The blast monitors measure peak particle velocity (PPV) for a defined duration (15 seconds) after a preset threshold is exceeded. Data acquisition systems for the uniaxial and triaxial accelerometers were automated to acquire accelerations continuously; while the blast monitors were preset to trigger at a PPV of 0.05 inch/sec threshold.

Human Response To Floor Motion

The response of humans to floor motion is a complex phenomenon, involving the magnitude of motion, the environment around the sensor, and the human sensor. A continuous motion can be more annoying than a transient motion caused by impact loads. The threshold of perception of floor motions in a busy workplace can be significantly higher than in a quiet apartment. Human reaction to the same induced motion on a structure can vary depending on the age and the location of the individual in the structure. A standard that can be used to compare the acquired vibration data qualitatively is provided by Siskind et al. (Figure 3). In this standard particle velocity is plotted against exposure time and shows that shorter transient pulses are less perceptible and thus less potentially annoying than are long transient and continuous motions.

The reaction of people who feel the vibrations depend strongly on what they are doing. The threshold for human comfort is lower in the 4 -8 Hz vibration frequency (Figure 4); outside this frequency range people accept higher vibration accelerations. This shows the acceleration limits as recommended by the International Standards Organization (ISO) adjusted for intended occupancy. People in offices or residences do not like "distinctly perceptible" vibrations. In the 4-8 Hz frequency range "distinctly perceptible" vibrations would be an acceleration of 0.5 percent of gravity for offices and residences and approximately 5.0 percent of gravity for people participating in rhythmic activities (Murray et al.).

The ISO standard suggests limits in terms of root mean square (rms) acceleration as a multiple of the baseline shown in Figure 4. The multipliers for the proposed criterion which is expressed in terms of peak acceleration is 10 for offices, 30 for shopping malls and 100 for rhythmic activities. For structural design purposes the limits can be assumed to range between 0.8 times and 1.5 times the recommended values depending on duration and frequency of vibration events (Murray et al.).

The vibration acceptance criteria would be different for the Press box structure and the rest of stadium stands vis-à-vis spectator areas. The Press box Level of the football stadium is an area occupied by personnel from the working press from both home and opposing teams, as well as the network television crew reporters that typically file reports on the game. It is common to see people work on laptops, write data on sheets, and review documents while they observe the proceedings of the game. This level would hence be regarded as an 'Office Space' for the criteria of recommended accelerations for human comfort. The spectator areas, where people are participating in the rhythmic activities, have significantly higher threshold as described previously.

The current vibration monitoring study of the football stadium uses two criteria to evaluate the vibration data. It compares the recorded accelerations due to people induced loads to the limits recommended by the ISO 2631-2, 1989 adjusted for the intended occupancy (Murray et al.). It qualitatively compares the vibration data for selected areas to the standard provided by Siskind et al. on the human response to continuous and transient motions.

Field Observations

This study included monitoring the last three homes games for the recent football season. The game details are presented in Table 1 below:

Table 1. Key Information of Monitored Games

Game-Date	Winner	Attendance
Game 1	Visitor	80,000
Game 2	Home	70,000
Game 3	Visitor	90,000

The spectator induced motions during the monitored games occurred during the rendition of the sports hymn at specific times of the game. The duration of motion was approximately 30 seconds during the chanting of a verse typically performed at the end of the sports hymn.

This current study focused on vibrations induced due to the spectators during the chanting of the sports hymn; effects of ambient wind conditions and gusts, and/or codified lateral loads are not included in this study.

Game 1

Four triaxial accelerometers for this game were located at the West Stand – press-box beam at level 10, third deck raker; North Zone – third deck raker; East Zone – third deck raker. In addition, two blast monitor were located at the west stands on level 9 and level 8.

The data from the various accelerometers were analyzed in the frequency domain using Fourier Transfer Functions. The Fourier response spectrum (Bendat and Piersol) revealed that the random accelerations during the sports hymn are banded in the 1- 2 Hertz frequency range. The maximum acceleration recorded during the sports hymn during the game was at the Press box. The accelerations were approximately: 1% Gravity (Figure 5) in the longitudinal direction and 0.8% Gravity in the transverse direction. The maximum resultant accelerations (during the sports hymn) at the Press box Level in the west stands were recorded to be 1.3% Gravity at the dominant frequency of 1.43 Hertz (Figure 6).

The Press box Level of the football stadium houses personnel from the working press as well as the network television crew reporters. People working on laptops, filling data sheets, and reviewing documents while they observe the proceedings of the game is a common occurrence at this location. This level, as discussed in the previous sections, would hence be regarded as an Office Space. Figure 4 plots the peak accelerations recorded during the chanting of the sports hymn on the ISO standard curve adjusted for the intended occupancy. This comparison would suggest that the peak accelerations for the Press box exceeded the threshold specified by ISO. The induced vibrations at the press-box would be considered to be "strongly perceptible" as defined by Siskind et al. (Figure 3).

Vertical accelerations recorded during this game were below the sensitivity of the measurement systems (< 0.001g). The results show that there were no appreciable accelerations in the east stand. The data from blast monitors are not presented since the frequency of induced motions on the structure were at the lower limit of the equipment's sensitivity vis-à-vis 2 Hz.

Game 2

The attendance for this game was significantly lower than the previous home game.

Four triaxial accelerometers, three uniaxial accelerometers and two blast monitors were used to collect vibration data.

The data from the various accelerometers were analyzed in the frequency domain using Fourier Transfer Functions. The Fourier response spectrum revealed that the random accelerations during the sports hymn are banded in the 1- 2 hertz frequency range. The maximum acceleration recorded during the sports hymn during the game was at the Press box. The accelerations were approximately - 1% Gravity in the longitudinal direction and 0.7% Gravity in the transverse direction. The maximum resultant accelerations (during the sports hymn) at the Press box Level in the west stands was recorded to be 1.2% Gravity at the dominant frequency of 1.4 Hertz. The peak acceleration recorded for the North Zone was 0.6% Gravity at a dominant frequency of 1.45 Hertz. There were no significant accelerations recorded on the East Stands during the chanting of the sports hymn. The induced vibrations at the press-box would be considered to be "strongly perceptible" as defined by Siskind et al. (Figure 3).

Vertical accelerations recorded during this game were below the sensitivity of the measurement systems (< 0.001g). The results show that there were no appreciable accelerations in the east stand. The data from blast monitors are not presented since the frequency of induced motions on the structure were at the lower limit of the equipment's sensitivity vis-à-vis 2 Hz.

Game 3

The attendance for this final game was the highest of the three home games monitored.

Seven triaxial accelerometers, one biaxial (combination of two uniaxial) accelerometer and two blast monitors were used to collect vibration data.

Data analysis in the frequency domain using Fourier Transfer Functions revealed that the random accelerations during the sports hymn are banded in the 1- 2 hertz frequency range. The maximum acceleration recorded during the sports hymn during the game was at the Press box. The maximum acceleration recorded were approximately - 1% Gravity in the longitudinal direction and 0.8% Gravity in the transverse direction. The maximum resultant accelerations (during the sports hymn) at the Press box Level in the west stands was recorded to be 1.2% Gravity at the dominant frequency of 1.52 Hertz. The peak acceleration recorded for the North Zone was 0.8% Gravity at a dominant frequency of 1.41 Hertz. The peak acceleration recorded for the East stand was 0.3% Gravity at a dominant frequency of 1.52 Hertz. The induced vibrations at the press-box would be considered to be "strongly perceptible" as defined by Siskind et al. (Figure 3).

Vertical accelerations recorded during this game were below the sensitivity of the measurement systems (< 0.001g). The data from blast monitors are not presented since the frequency of induced motions on the structure were at the lower limit of the equipment's sensitivity vis-à-vis 2 Hz.

Conclusions

This study primarily focused on the effects of spectator induced vibrations specifically during the chanting of the sports hymn during the last three home games of a recent football season. Our conclusions are as follows:

1. The intensity of the vibrations increases with increased participation of a greater number of spectators. The peak acceleration was recorded in the Press box structure during the chanting of the sports hymn. These accelerations exceed the standards prescribed for human comfort by the ISO.

2. The comparison of velocities in the Press box, during the sports hymn, with Siskend's standards suggests that the induced vibrations can be characterized as distinctly to strongly perceptible.

References

Bendat, J. S. and Piersol, A. G. (1993), "Engineering Applications of Correlation and Spectral Analysis," 2nd Edition, John Wiley and Sons, Inc. ISBN 0-47157-055-9

Dowding, C. H. (2000), "Construction Vibrations (2nd Ed.)," Prentice-Hall, ISBN: 0-96443-131-9

Fraden, J. (1996), "Handbook of Modern Sensors: Physics, Design, and Applications," Springer-Verlag, New York Inc., ISBN 1-56396-538-0

Murray, T.M., Allen, D. E., and Ungar, E. E. (2003), "Floor Vibrations Due to Human Activity," Steel Design Guide Series 11, American Institute of Steel Construction.

Siskind, D. E., Staghura, V. J., Stagg, M. S., and Kopp, J. W. (1980), "Structure Response and Damage Produced by Airblast from Mining," Report of Investigations 8485, U.S. Bureau of Mines.

Wiss, J. and Parmelee, R. (1974), "Human Perception of Transient Vibrations," Journal of the Structural Division, ASCE, Vol. 100, No. S74, pp.773-787.

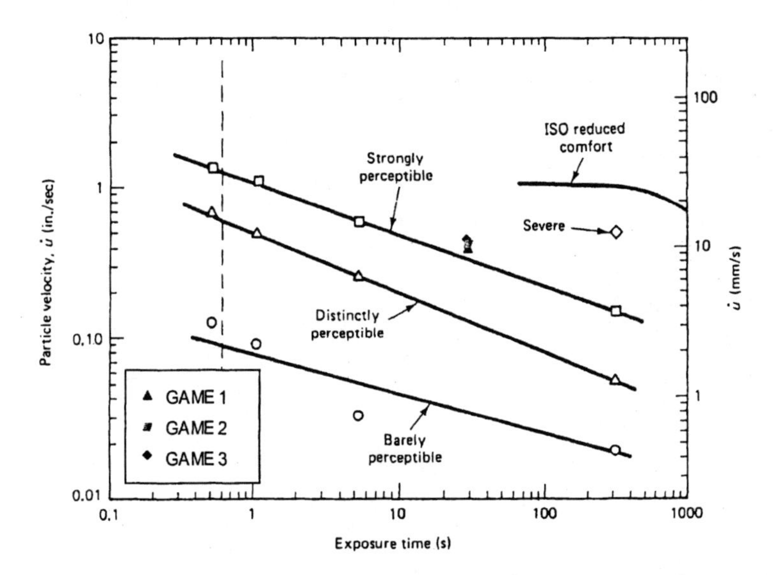

Figure 3. Human Response to Transient Pulses of Varying Durations After Wiss and Parmelee (1974) As Replotted by Siskind et al. (Extracted From Construction Vibrations by C.H. Dowding)

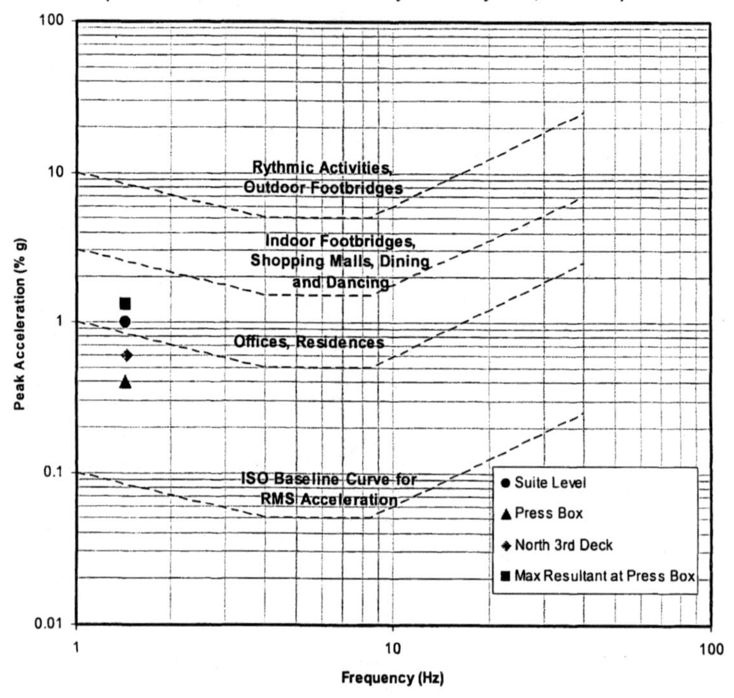

Figure 4. Vibration Response of the Stadium On Game 1

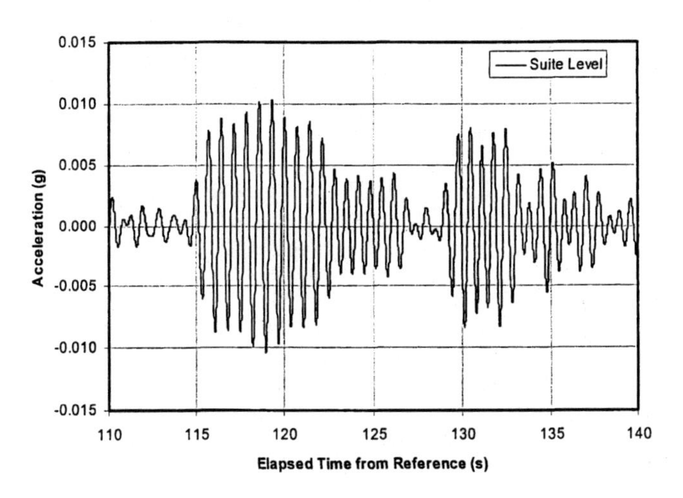

Figure 5. Lateral Acceleration at Suite Level on Game 1

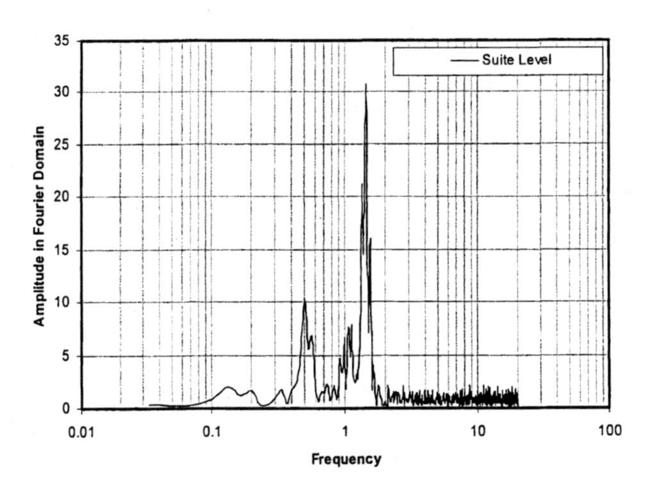

Figure 6. Fourier Spectrum of Lateral Acceleration at Suite Level on Game 1

Forensic Engineering of Intolerable Structural Vibrations and Damage from Construction and Industrial Dynamic Sources

Mark R. Svinkin, Member, ASCE[1]

[1]VIBRACONSULT, 13821 Cedar Road, #205, Cleveland, OH 44118-2376; PH (216) 397-9625; FAX (216) 397-1175; msvinkin@stratos.net

Abstract

Intolerable vibrations and damage to structures from construction and industrial dynamic sources are subjects of the investigations in forensic engineering. There are some typical vibration problems and different ways can be used to diagnose building conditions and understand the causes of high vibrations or failure. Analysis of the structural response and a choice of vibration limits are very important. Practical solutions are demonstrated in four case histories.

Introduction

Construction and industrial sources with impact and vibratory forces generate diverse vibrations of structures which levels depend on values and frequency components of dynamic forces, soil conditions at a site, soil-structure interaction, and susceptibility of structures. Not all vibrations are detrimental. Some of them are benign. Intolerable structural vibrations are the reasons for concern because such vibrations can adversely affect structures. Their effects range from annoyance of people, disturbance of sensitive devices, and disruption of some businesses to visible structural damage.

There are various combinations of dynamic loads, wave propagation paths, soil and structure responses. When sources such as machines with dynamic loads do not properly operate and/or function, the goals are to diagnose building conditions, judge severity, find the cause of vibration problems, and perform remedial actions. For blasting, pile driving, and other construction dynamic forces, the major priority is determination whether dynamic loads or environmental forces were the cause of diverse structural damage.

This paper evaluates some queries in analysis of intolerable vibrations and finding the causes of structural failures, elaborates problems of soil and structure responses to dynamic loading and allowable vibration limits, and describes four case histories.

Forensic Engineering Approach to Structural Vibration Problems

In the field of structural vibrations, forensic engineering is the application of engineering principles to investigate intolerable vibrations and failures of structures.

There are different objectives in the forensic research in Soil and Structural Dynamics: interpretation of measured data, searching the causes which can trigger extreme structural vibrations and performance of remedial actions; resolving inverse problems to find the actual cause of structural damage without measured vibrations; correct measurement and assessment of ground and structural vibrations; and others.

Several queries are considered below and two topics are separately presented in detail.

Unknown Causes of Intolerable Vibrations from Known Dynamic Sources

Machines with dynamic loads installed on structures and foundations are often the sources of perceptible structural vibrations which can exceed the tolerable levels. Such vibrations can provoke excessive wear and cracking of some machine parts, equipment malfunctions, damage to structures, and dangerous soil deformations. The beginning of perceptible vibrations of machine supportive structures can occur in one or a few places of the machine-structure-soil system. It can occur because of suddenly increased dynamic loads transmitted from the machine to the supportive structure, structural damage of the supportive structure, or changes of soil properties under influence of vibrations and/or production process.

Effects of Dynamic and Environmental Forces on Structures

Damage to structures from construction and industrial dynamic sources has been described in a number of technical publications. Woods (1997) summarized dynamic effects of pile installation on adjacent structures. Siskind et al. (1980) suggested damage classification with description of damage. Siskind (2000) presented accumulated results of ground vibrations measured near 1-2 story houses and actual observations of structural damage conducted by the U.S. Bureau of Mines (USBM).

Nevertheless, buildings, houses, and other structures usually have cracks and other damage from non-vibration environmental forces such as atmospheric humidity changes, weather cycles, human activities and others.

Dynamic or environmental forces can be the causes of similar structural damage. It is important to find a solution of an inverse problem: there is structural damage and the cause of damage has to be found. This problem becomes more complicated without the proper survey of structures.

To discern effects of dynamic and environmental forces on structures, it is necessary to provide a condition survey of structures surrounding a construction site prior to the beginning of construction operations with the use of dynamic forces. The condition survey report has to present descriptions and photographs of areas with existing damage in houses and buildings. It is a significant step for proper assessment of vibration effects on structures and minimizing claims.

Non-Vibration Damage to Structures from Dynamic Sources

Dynamic loads from construction sources can damage structures without vibration effects. During pile driving in clay, dynamic loads force piles to penetrate into the ground that result in displacements of soil surrounding a pile. The soil movements may produce heave, settlement and lateral displacement toward the existing nearby foundations of buildings.

According to D'Appolonia (1971), pile penetration into clay produces an increase in lateral stress, in pore pressure and heave of the ground surface. During pile driving, the excess pore pressure increases with each driven pile and may reach big values at large distances beyond the pile group area. This excess pore pressure can be much larger than the initial effective overburden stress. After the completion of pile driving and the dissipation of excess pore pressure, the soil reconsolidates and ground surface settles. The soil settlement is usually greater than the heave during pile driving because soil compressibility is significantly increased by soil remodeling after pile installation.

Svinkin (2005a) has analyzed the effects of several factors on soil heave and settlement, and presented mitigation measures of soil movements from pile driving in soft to medium clay on adjacent buildings.

Damage to Structures from Non-Dynamic Construction Activities

In some projects, pile driving or blasting for foundation construction are usually conducted after excavation and dewatering at a construction site. These construction activities may trigger significant ground movements and produce damage to nearby buildings. Therefore, it is necessary to separate damage to structures from construction activities and from dynamic forces. Because of that the pre-driving or pre-blasting condition survey has to be provided when excavating and dewatering have been completed at the site.

Vibration Measurement Equipment

Seismographs are applied for measurement of a vibration velocity which is considered as the criterion of allowable vibrations. Records of soil and structural vibrations are evidences of vibration activities at a site and they should be used with confidence. Therefore, quality and accuracy of vibration records are a primary concern.

Obviously, the vibration measurement system has to be calibrated. Manufacturers usually supply calibration curves along with seismographs. Calibration curves should have a flat part at the low frequency between 4-12 Hz for recording oscillations in the range of natural frequencies of horizontal vibrations for most houses and buildings. Moreover the codes in several states and counties such as Florida, Ohio, etc. demand a flat part of a calibration curve starting from 2 Hz. It is particularly significant in Florida where, according to Siskind and Stagg (2000), some areas have wave paths with low attenuation and ground vibrations manifest the dominant frequency of 2-4 Hz.

In measurement of ground vibrations, it is necessary to keep in mind that the condition of resonance can exist for the seismograph – soil system because of elastic properties of soils. This system can have different natural frequencies in vertical, longitudinal, and transverse directions. If these frequencies are within the frequency range of the measurement system, development of system resonant vibrations can be reflected on the records. Such distortions of vibration records have been described in the publications (Pasechnik 1952, Bycroft 1957, Fogelson and Johnson 1962, Bollinger 1971, Levin and Svinkin 1973). Natural frequencies of the seismograph-soil systems should be verified in the field.

If more than one seismograph is employed at the site, coupling in time is essential for analysis of all records of seismographs used for measurements. In practice of vibration measurement, it seems that a restricted number of people obtain records with coupling in time which is very helpful for understanding measured vibrations.

People with various experiences make measurements of soil and structural vibrations. In order to meet requirements of quality records, it would be beneficial to use the sample rate of at least 1000 samples per second.

Expert Witness Presentation

For successful expert witness work, it make sense to say using mathematical terminology that well prepared arguments are the necessary condition and psychological preparation to answer any questions of an opposing counsel is the sufficient condition. Litigation support is stressful but stimulating and challenging job.

Soil and Structure Responses to Ground Vibrations

Structural damage can be classified depending on the structural response to ground vibrations. Siskind et al. (1980) suggested the following damage classification with a description of damage. Threshold: loosening of paint, small plaster cracks at joints between construction elements, lengthening of old cracks. Minor: loosening and falling of plaster, cracks in masonry around openings near partitions, hairline to 3 mm cracks (0-1/8 in), fall of loose mortar. Major: cracks of several mm in walls, rupture of opening vaults, structural weakening, fall of masonry, load support ability affected.

The USBM accumulated results of structural damage from ground vibrations generated by surface mine and quarry blasting are shown in Figure 1 which was modified from Siskind (2000). The data were obtained from 718 blasts and 233 documented observations of threshold, minor and major damage. Non-damaging blasts are not shown, although some of them produced relatively high levels of ground vibrations even exceeding 51 mm/s (2 in./s), the maximum allowable limit of ground vibrations. Analysis of these data indicates different vibration effects on structures depending on the dominant frequency and the peak particle velocity of ground vibrations, Svinkin (2004a, 2005a).

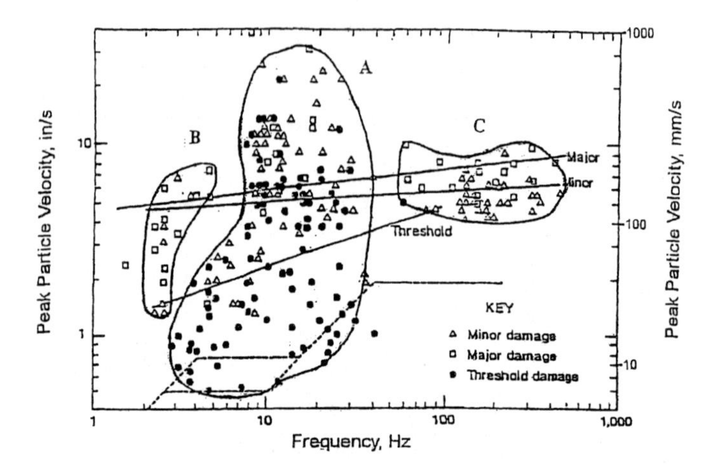

Figure 1. Three zones with closely grouped structure responses and damage summary from ground vibrations generated by blasting, and USBM recommended safe limits-dashed lines. Data were modified from Siskind (2000)

Ground vibrations from construction dynamic sources may produce direct damage to houses and buildings, trigger structural and soil layer resonant vibrations, and be the cause of non-uniform dynamic settlements.

Direct Structural Damage

Soil-structure interaction determines structural response to the ground vibrations that may produce direct damage to structures when frequencies of ground vibrations do not match natural frequencies of structures. This kind of damage, mostly minor and major, is predominant in zones B and C and partially reflected in zone A (Figure 1). Velocity values of ground vibrations should be substantial for direct structural damage. Thus, the velocity range is 33-191 mm/s (1.3-7.5 in/s) for frequencies of 2 to 5 Hz in zone B and 102-254 mm/s (4-10 in/s) for frequencies of 60 to 450 Hz in zone C. These velocity levels are substantially higher than the USBM vibration limits.

It is necessary to point out that there are a number of case histories that demonstrate no structural damage at relatively small distances from blasts though velocities of ground vibrations were higher than the threshold levels for the wide frequency range.

Because of conservatism of vibration limits which can make obstacles for implementation of the blasting projects, Oriard (2002) has monitored vibrations of critical structures or other items of interest and demonstrated high vibration levels acceptable for structures.

It is important to underline that high vibration limits can be used only for direct vibration effects on structures when frequencies of ground vibrations do not match natural frequencies of structures. At relatively small distances from blasts, transient ground vibrations with short durations cannot trigger resonant structural vibrations. However, these criteria cannot be applied for the condition of resonance and dynamic settlements.

Resonant Structure Vibrations

The proximity of the frequencies of ground vibrations to building's natural frequencies may generate the condition of resonance in the building. Resonant structural vibrations are predominant in zone A and partially presented in zone B. Amplitudes of ground vibrations are substantially less important then a coincidence of frequencies. Increasing structure vibrations starts after the first cycle of ground vibrations with the dominant frequency near the natural structure frequency. According to Siskind (1980) and Quesne (2002), a dynamic magnifying factor at resonance was measured in limits of 2-9 for one-two stories residential houses. This factor can be substantially higher for steel and concrete structures. Resonant structural vibrations can be triggered at large distances from a construction site and even more than one mile from quarry blasting. Structure resonance of horizontal building and wall vibrations and also vertical floor vibrations can result in damage to structures. Resonant horizontal building and house vibrations occur most frequently. Resonant vertical floor vibrations are important when precise and sensitive devices and also fragile objects are installed on the floors, Svinkin (2004a).

The following are two examples of the condition of resonance in the buildings generated by dynamic sources with impact loads. Rausch (1950) described a case history where intolerable vibrations occurred in an administrative building located 200 m from the foundation of a forge hammer with a falling weight of 14.7 kN. Svinkin (1993) reported a similar situation with resonant building vibrations of a five story apartment building located at approximately 500 m from the foundation under a vibroisolated block for a sizable forge hammer with a falling weight of 157.0 kN.

Resonance of Soil Layers

Waves from dynamic sources travel through the soil medium, and they can be amplified as a result of resonance of soil layers. In most cases, analysis of site responses is focused on the motion at the free ground surface. However, resonant effects may occur at any point within a layered soil profile. It is possible to consider two locations with the same soil within the same site excited by the same dynamic source, and these locations could respond quite differently because of the nature and dimensions of surrounding soil layers, Davis and Berrill (1998).

Bodare and Erlingsson (1993) described an example of strong structure vibrations due to resonance of a soil deposit. At the time of a rock concert held in the stadium, more than twenty-five thousand people were standing on the field close to the stage. During the concert, the audience jumped in time to the music. In this way, the audience excited from the surface a clay layer (a thickness of 25 m) that had the same frequency (about 2.4 Hz) as the beat of the rock music. Resonance of the clay deposit excited violent vibrations that damaged stadium structures. Also, residential buildings 400 m away experienced vibrations. After that, rock concerts were not permitted at the stadium.

Dynamic Settlements

Ground and foundation settlements as a result of relatively small ground vibrations in loose soils may happen at various distances from the source. Densification and liquefaction of soils can occur under the vibration effects of blasting and pile driving operations. At short distances from the dynamic sources, densification is expected, but ground and foundation settlements extend beyond the zone of densification for account of joint influence of static and dynamic loads. Dynamic settlements could be partially presented in zones A, B, and C.

Practical Application of Existing Vibration Limits

There are no general regulations of vibrations from various construction and industrial sources. Some branches of industry, state and county authorities have developed their own safe criteria of ground vibrations to prevent structural damage and annoyance of people.

For 1-2 story houses surrounding surface mining facilities of coal industry, the USBM has developed the frequency-based safe limits of cosmetic cracking threshold shown in Figure 2, Siskind et al. (1980). It was a great achievement that provided safety of low-rise residential structures from vibrations generated by mining blasting. The Office of Surface Mining (OSM) included the derivative version of the USBM safe limits as the Chart Option into the OSM surface coal mine federal regulations, OSM (1983). In several years later, the USBM safe limits also received a status of federal regulations, NFPA 495 (2001).

The USBM safe limits were built up on the basis of a correlation between ground vibrations and observations of cracking damage in low-rise houses, and these limits are applied for ground vibrations as the criteria of possible structural damage. As a matter of fact these vibration limits can be successfully used for structures and ground conditions similar to those for which they were developed. However, the safe vibration criteria are used for indirect assessment of vibration effects on various super and underground structures even though this approach is not founded for them. AASHTO (1996) stated the application of the USBM limits to markedly different types of structures is common and inaccurate.

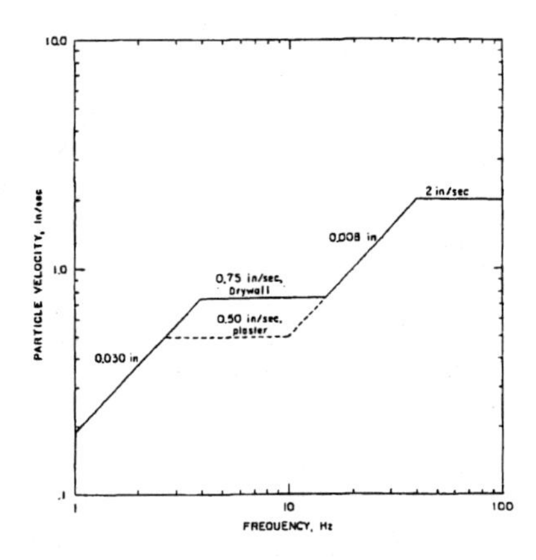

Figure 2. USBM safe levels of blasting vibrations for houses using a combination of velocity and displacement, after Siskind et al. (1980)

Oriard (1999) has made the summary comment on vibrations: "Routine blasting operations which conform to standard regulations and standard criteria would not be expected to have any adverse effect on soils on a properly prepared building site". Such an approach is broadly used in practical assessment of vibration effects on structures though it could not prevent damage to structures in a number of cases. On the one hand this approach is correct for the direct vibration effect on structures when ground vibrations do not trigger plastic soil deformations. At the same time, the results from Figure 1 and Oriard (2002) demonstrated that structures can withstand substantial higher vibration levels from direct vibration effects than the USBM and OSM vibration limits. It means these limits used for assessment of direct vibration effects on structures can restrict construction operations with the low allowable vibrations. On the other hand such an approach is wrong and misleading for structural and soil layer resonant vibrations and non-uniform soil deformations caused by soil densification and liquefaction. Obviously, Siskind (2000) made the point that most regulatory bodies, including OSM, recognize "that a more rigorous treatment may be needed in special cases, such as that outlined in RI 8507".

Departments of Transportation, state and local authorities use their own vibration limits of the peak particle velocity. These limits are applied independently of soil conditions and soil-structure interaction. Also, they do not take into account type, age and stress history of structures. If structures receive even cosmetic cracks from blasting, the authorities try to decrease the existing vibration limits. It is a wrong policy because such a step cannot prevent new damage without analysis of the causes of cosmetic cracking. Also, this action can negatively affect production blasting, pile driving, and other construction operations. What vibration limits can be used for 3-10 and higher story buildings? Such criteria are not available. This is the reason why some researchers and practitioners measure vibrations of structures.

The structural response is an indication of vibration effects on structures, and there is a growing need for measurement of the structural response as the criterion of vibration monitoring and control. OSM has funded research projects for evaluation of structural responses to blasting, Crum (1997). Measurements of structural vibrations were made in different projects for determining vibration effects on historic buildings, curing concrete, sensitive devices and equipment, e.g. Elliott and Goumans (2004), Lucca (2004), and Svinkin (2004b).

Furthermore, direct measurement of structural responses to ground vibrations from construction sources is more accurate and has the substantial advantages in comparison with ground vibration measurement because structure responses ensure the flexibility of implicitly considering the variety of soil deposits and the condition of structures. Such an approach provides measurement of structural response after soil-structure interaction, takes into account distinction of type, configuration, age and stress history of structures in the wide frequency range. As a result of these advantages, tolerable limits of structural vibrations can be substantially increased.

Each site is unique because of soil conditions and various structures at sites. Therefore, specifications for construction operations with dynamic sources have to consider specifics of each site, and sometimes preliminary tests should be done to determine tolerable levels of structural vibrations. It is necessary to make direct measurement of structural vibrations accompanied by observation of the results of dynamic effects. For multi-story residential, commercial and industrial buildings, the frequency-independent safe limit of 51 mm/s (2 in/s) can be chosen for the PPV of structural vibrations, Svinkin (2003, 2004a). This criterion automatically takes into account soil-structure interaction for the whole building frequency range. The proposed criterion does not exclude higher allowable vibration levels. There are two reasons which confirm truthfulness and expediency of this criterion. First, in the middle of 1940s, the safe vibration limits of 30-50 mm/s (1.18-1.97 in/s) for sound structures were found by the Moscow Institute of Physics of the Earth. These limits were successfully used for years in former USSR. Second, according to Siskind (2000), the PPV of 51 mm/s (2 in./s) is the highest vibration level generated inside homes by walking, jumping, slamming doors, etc.

There are some general rules for measurement of structural vibrations. Horizontal building vibrations are most important for measurements. Therefore, transducers should be installed on a windowsill at the upper floor and a foundation for vibration measurements in the perpendicular direction to the building wall. Such measurement is necessary to detect resonant structural vibrations. Vertical floor vibrations should be measured when precise and sensitive devices and also fragile objects are installed on the floor. The preconstruction condition survey of structures surrounding a construction site may determine certain locations for vibration monitoring and control.

There are no regulations of ground and structure settlements triggered by soil densification and liquefaction, but information about these phenomena is available in some publications (Barkan 1962, Clough and Chameau 1980, Sanders 1981, Lacy and Gould 1985).

Case Histories

Case 1. Extreme Vibrations of Cone Crasher Foundation

Two cone fine crushers with the unbalanced horizontal forces of 25 kN were used for coal fragmentation. Both crusher were mounted on two separate identical concrete foundations consisting of walls, top and bottom slabs. The distance between crushers' axes was about 8 m. During a long time, vibrations of crasher foundations were in allowable limits, i.e. less than 0.3 mm, Svinkin (1993).

All of a sudden, vibrations of one foundation increased and the displacement amplitude reached 1 mm. Such vibration level is dangerous for the machine and its foundation. The problem was to find the source of these vibrations. Changes in any part of the whole machine-foundation-soil system can be the cause of extreme vibrations. Consequently, a few logical steps were done in a search of the vibration source.

The crusher. The unbalanced horizontal centrifugal force, horizontal and vertical impulse loads are transmitted from the operative crusher to its foundation. Impulse loads could not suddenly rise because of permanent technological conditions. The unbalanced force can only increase gradually in time. Because of these reasons, the crusher was excluded as the source of extreme vibrations.

The crasher foundation. A condition survey of vibrating foundation structures revealed no visible cracks and other damage in spite of the extreme foundation vibrations. It meant the foundation itself could not be the cause of arisen vibrations.

The soil. There were good soil strata under the crusher foundation: limestone with sand and clay interlayers. Such soil conditions usually create no troubles for machine foundations. However, crushing is a wet technological process. Therefore, it was assumed that water could penetrate under the crasher foundation and reversed soil properties. Boring made near the foundation proved that water turned a weak soil interlayer into slash. After the water was pumped out, foundation vibrations returned back to the allowable limit.

Case 2. Perceptible Vibrations of Foundation under Sizeable Exhaust Fan

A few sizeable exhaust fans were placed at the plant for preparation of metallurgical raw materials. The fans were mounted on the identical elevated pedestal foundations which columns had a cross-section of 1 m^2. Perceptible foundation oscillations appeared during launching of one fan. In order to reduce these oscillations, the design company worked out a project to reinforce the foundation structures. Unfortunately, in a number of cases design engineers consider foundation reinforcement as the best remedy against perceptible vibrations, Svinkin (1993).

In diagnostic work it is necessary to analyze possible causes of intolerable vibrations. Therefore, a condition survey of foundation structures and vibration measurements were made for two foundations with allowable and perceptible vibrations. Both foundation structures had no cracks or other damage, but identical machines exerted different vibration effects on those foundations. A comparison of acquired data provided information to draw the conclusion that an unbalanced exhaust fan was the cause of perceptible vibrations. After the fan was rebalanced, foundation vibrations came back to the allowable limit.

Case 3. Damage to Houses from Quarry Blasting

Siskind and Stagg (2000) studied the same 10 typical residential houses selected at five areas in Miami-Dade County for investigation of cracks in various structures of the houses. Distances between the test sites and active quarries were in the 3.2-4.8 km range. Vibration measurements of ground and house vibrations were made about a 2-month period from late February to late April 2000. Pre- and post-blasting condition surveys were made before and after vibration monitoring.

The report findings are briefly presented in Svinkin (2005b): (a) structure oscillations are superposition of vibrations with various frequencies including low frequencies of about 8 Hz, which are close to the natural frequencies of house horizontal vibrations, and very low frequencies of 2 to 4 Hz; (b) structural responses in South Florida are sufficiently different from responses of frame structures studied elsewhere; (c) vibrations are of long duration at the houses with maximum duration about 17 second; (d) the highest dynamic structure amplification exceeded 6x; (e) ground vibrations with the PPV of 19 mm/s (0.75 in./s) with structure response of 6.1x would produce a global wall strain that is 1.5 times higher of the strain limit for crack formation.

It was found that 5 of 10 inspected houses had some wall cracks, mostly exterior, which could be from dynamic sources (blasting and wind). For house damage other than wall cracks, blasting was excluded. However, the authors have stated they would need to have additional tests done to determine the exact causes for every crack and fault.

For three homes built in 1995, 1997, and 1998, wall cracks could be from quarry blasting and wind, but in the writer opinion, the wind from hurricane Andrew (1992) could not be the cause of wall damage in these homes. It makes sense to separate structural damage made by blasting from similar damage obtained at the time of hurricanes using a comparison of the condition survey presented in the Report with a survey of houses in the Miami-Dade County where there are no complaints on blasting.

Case 4. Decreasing Vibrations of Apartment Building

A five story apartment building was located at a distance about 500 m from the foundation under a vibroisolated concrete block and a sizeable forge hammer with a falling weight of 157 kN. The building had two perpendicular parts. A major part of the building was oriented in a radial direction from the hammer foundation and had large stiffness in this direction. There were no vibration problems with this building part. A minor part of the building had low stiffness and harmful vibrations in the radial direction, Svinkin (1993).

Vibrations of the hammer foundation induced transversal horizontal structural vibrations with the frequency of 3.1 Hz. Obviously, this frequency was closed to the natural frequency of horizontal building vibrations. A change of the frequency of vertical vibrations of the vibroisolated block with the hammer is the simplest and economical way for diminishing building vibrations. To rich this goal, it is necessary to decrease stiffness of vibroisolators by eliminating part of them. Steel springs are chosen in accordance with the condition of strength to support the concrete block. Therefore, it is possible to diminish the quantity only of dashpots.

Buildings usually have a narrow resonant zone. In the described case history, the natural frequency of vibroisolated concrete block with hammer was decreased from 3.1 to 2.9 Hz due to elimination of a few dashpots. It was sufficient to diminish structural vibrations to the acceptable limits.

Conclusions

Construction and industrial dynamic sources may generate vibrations that harmfully affect structures. Intolerable vibrations and damage to structures are subjects of the investigations in forensic engineering.

There are different problems in forensic research in Soil and Structural Dynamics: unknown causes of intolerable vibrations from known dynamic sources, effects of dynamic and environmental forces on structures, non-vibration damage to structures from dynamic sources, damage to structures from non-dynamic construction activities, and others.

Sensible choice should be provided between measurement of ground and structural vibrations. The most important question in assessment of vibration effects on structures is determination of the structural responses that are a final goal of vibration monitoring and control. There is a growing need for measurement of the structural vibrations. The existing vibration criteria should be used as guidelines, and structure conditions must be taken into account in a choice of the safe vibration limits. For multi-story residential, commercial and industrial buildings, the frequency-independent safe limit of 51 mm/s (2 in/s) can be chosen for the PPV of structural vibrations.

It is beneficial to make direct measurement of structural vibrations accompanied by observation of the results of dynamic effects for determining vibration limits to combine construction activities with application of dynamic forces and for prevention of cosmetic cracking and other damage to structures.

References

AASHTO (1996). AASHTO DESIGNATION: R 8-96. SECIFICATIONS FOR MATERIALS, Standard Recommended Practice for Evaluation of Transportation-Related Earthborn Vibrations, 909-916.

Barkan, D.D. (1962). Dynamics of Bases and Foundations. McGraw Hill Co., New York

Bodare, A. and Erlingsson, S. (1993). "Rock music induced damage at Nya Ullevi Stadium." Proceedings of the Third International Conference on Case Histories in Geotechnical Engineering, UMR, V. I, 671-675.

Bollinger, G.A. (1971). Blast Vibration Analysis. Southern Illinois University Press.

Bycroft, G.N. (1957). "The magnification caused by partial resonance of the foundation of a ground vibration detector." Trans. Amer. Geophysics. V. 38, No. 6, 928-931.

Clough, G.W. and Chameau, J-L. (1980). Measured effects of vibratory sheetpile driving. American Society of Civil Engineers, ASCE Journal of the Geotechnical Engineering Division, Vol. 106, GT10, 1081-1099.

Crum, S.V. (1997). House Responses from Blast-Induced Low Frequency Ground Vibrations and Inspections for Related Interior Cracking. S-Wave GeoTech, Minneapolis, MN. Final Report to Office of Surface Mining, Contract ID: 143868-PO96-12616.

D'Appolonia, D.J. (1971). Effects of foundation construction on nearby structures. Proceedings of Fourth Panamerican Conference on Soil Mechanics and Foundation Engineering, ASCE, Vol. 2, Puerto Rico, 189-235.

Davis, R.O. and Berrill, J.B. (1998). "Site-specific prediction of liquefaction." Ceotechnique, 48(2), 289-293.

Elliott, R.J. and Goumans, C. (2004). "Drilling and blasting of a small diameter shaft next to historic structures", Proceedings of the 30th Annual Conference on Explosives and Blasting Technique, ISEE, New Orleans, Louisiana, USA, V. I, 15-23.

Fogelson, D.E. and Johnson, C.F. (1962). Calibration Studies of Three Portable Seismographs. USBM, RI 6009.

Lacy, H.S. and Gould, J.P. (1985). Settlement from pile driving in sand. American Society of Civil Engineers, Proceedings of a Symposium on Vibration Problems in Geotechnical Engineering, ASCE, G. Gazetas and E.T. Selig, Editors, New York, 152-173.

Levin, G.E. and Svinkin, M.R. (1973). "About spectrum analysis of ground vibrations excited by impacts and blasts." Dynamic Properties of Soils and Seismic Resistance of Hydro-Technical Structures, Energiya, 108-111 (in Russian).

Lucca, F. (2004). "Municipal blasting: blast design, vibration monitoring and control", Proceedings of the 30th Annual Conference on Explosives and Blasting Technique, ISEE, Nee Orleans, Louisiana, USA, V. II, 479-488.

NFPA 495 (2001). Explosive Materials Code. National Fire Protection Association, An International Codes and Standards Organization, Quincy, Massachusetts.

Office of Surface Mining Reclamation and Enforcement (1983). Surface Coal Mining and Reclamation Operations; Initial and Permanent Regulatory Programs; Use of Explosives. Federal Register, V. 48, No. 46, March 8, 9788-9811.

Oriard, L.L. (1999). The Effects of Vibrations and Environmental Forces. International Society of Explosives Engineers, Cleveland, Ohio.

Oriard, L.L. (2002). Explosive Engineering, Construction Vibrations and Geotechnology, International Society of Explosives Engineers, Cleveland, Ohio.

Pasechnik, I.P. (1952). "Results of experimental study of resonance phenomena in oscillation seismograph-soil system." Proceedings the USSR Academy of Science, Geophysical Series, No. 3, 34-57.

Quesne, J.D. (2001). "Blasting vibration from limestone quarries and their effect on concrete block and stucco homes", Vibration problems, Geo Discussion Forum, www.geoforum.com (May 6, 2002).

Rausch, E. (1950). Maschinen Fundamente (in German), VDI-Verlag, Dusseldorf, Germany.

Sanders, S.G. (1982). Assessment of the Liquefaction Hazards Resulting from Explosive Removal of the Bird's Point New-Madrid Fuze Plug Levee, paper No. GL-

82-5, U.S. Army Corps of Engineering Waterways Experiment Station, Vicksburg, Mississippi.

Siskind, D.E. (2000). Vibration from Blasting. International Society of Explosives Engineers, Cleveland, Ohio.

Siskind, D.E., Stagg, M.S., Kopp, J.W., and Dowding, C.H. (1980). Structure response and damage produced by ground vibrations from surface blasting. RI 8507, U.S. Bureau of Mines, Washington, D.C.

Siskind, D.E. and Stagg, M.S. (2000). "The fourth report". Blast Vibration Damage Assessment Study and Report. Miami-Dade County, C3TS Project No.: 1322-01.

Svinkin, M.R. (1993). "Analyzing man-made vibrations, diagnostics and monitoring". Proceedings of the Third International Conference on Case Histories in Geotechnical Engineering, St. Louis, Missouri, June 1-4, 663-670.

Svinkin, M.R. (2003). "Drawbacks of blast vibration regulations." International Society of Explosives Engineers, Proceedings of the 29th Annual Conference on Explosives and blasting Technique, Nashville, Vol. II,. 157-168.

Svinkin, M. R. (2004a). "Minimizing construction vibration effects." American Society of Civil Engineers, Practice Periodical on Structural Design and Construction, ASCE, Vol. 9, No. 2, 108-115.

Svinkin, M.R. (2004b). "Vibration effects of coal pit blasts on tower pithead concrete structures", Proceedings of the 30th Annual Conf. on Explosives and Blasting Technique, ISEE, Cleveland, V. II, 295-304.

Svinkin, M.R. (2005a). "Mitigation of soil movements from pile driving." Underground Construction in Urban Environments, A Specialty Seminar Presented by ASCE Metropolitan Section Geotechnical Group and the GEO-Institute of ASCE, May 11-12, New York.

Svinkin, M.R. (2005b). Closure to "Minimizing construction vibration effects," by Mark R. Svinkin. American Society of Civil Engineers, Practice Periodical on Structural Design and Construction, ASCE, Vol. 10, No.3, 202-204.

Woods, R.D. (1997). Dynamic Effects of Pile Installations on Adjacent Structures, NCHRP Synthesis 253, Transportation Research Board, National Research Council, Washington, D.C.

90 Feet - The Difference an Avenue Makes: Protection of Existing Historic Structures during Adjacent Construction

by

Keith Kesner[1]
Eric Hammarberg[2]
Derek Trelstad[3]

Abstract

The protection of existing structures during adjacent construction is a common problem confronted by engineers and architects in urban environments. The problem is considerably more complicated when the existing structures are older, historic structures or structures in historic districts. This paper reviews the potential impact of adjacent construction on existing structures as well as published guidelines for the protection of existing structures during adjacent construction. Case studies are presented that show how these guidelines can be incorporated into construction protection plans tailored to specific projects.

[1] Senior Engineer, WDP & Associates, 8832 Rixlew Lane, Manassas, VA 20101; PH 703-257-9280; FAX 703-257-9281; email: kkesner@wdpa.com
[2] Vice President, Thornton Tomasetti, 51 Madison Ave., New York, NY 10010; PH 917-661-7800; FAX 917-661-7801; email: ehammarberg@thorntontomasetti.com
[3] Associate, Thornton Tomasetti, 51 Madison Ave., New York, NY 10010; PH 917-661-7800; FAX 917-661-7801; email: dtrelstad@thorntontomasetti.com

Introduction

The construction of new structures adjacent to existing structures is common, particularly in urban environments. Construction of this type is also potentially problematic because of the need to minimize the impact of the new construction on the existing construction. Historic structures and structures within historic districts are potentially more susceptible to construction damage. This is due to the brittle materials (brick masonry, terra cotta, and plaster) traditionally used in their construction, non-redundant types on construction, the age of the structures, and uncertainties in the current condition of the structures. All of these factors, combined with the potential for undocumented modifications to older structures, make these structures potentially more susceptible to damage from adjacent construction than new structures.

Construction can result in damage to the structure, envelope and/or finishes in adjacent buildings. "Structural" damage affects the capacity of the primary or secondary support for the building. Damage to the "envelope" is typically non-structural in nature and includes cracking of exterior finishes, mortar joint cracking / spalling / debonding, extension of exising cracks and others forms of damage. Because the envelope is typically the primary waterproofing system for the building, this type of damage may develop into structural damage if left untreated. Corrosion of embedded iron or steel structural components and deterioration of masonry through freeze-thaw cycling are typical where envelope damage is not addressed. Significantly, on many historic structures, the building envelope also is the building structure; structural damage may include loss of bearing, failure of members, and building collapse. Cosmetic damage adversely affects the appearance of a structure without a decrease in the structural performance of the building or its weather resistance. Older, heavily ornamented structures are particularly susceptible to cosmetic damage. Most importantly, both envelope and cosmetic damage negatively affect the historic character of these structures.

Adjacent Construction Effects on Existing Buildings

The most significant impact of adjacent construction will typically occur during demolition, excavation and foundation construction operations. Based on our experience the following foundation construction activities will likely produce the most significant effects on the existing buildings:

- Installation of piles using vibration or impact methods around the site
- Excavation of the site using large equipment; rock removal; demolition of existing building foundations; etc.
- Underpinning of existing foundations
- Installation of new deep foundations
- Dewatering of the site (if required)

The following sections describe the effects of demolition, construction vibrations and underpinning on existing structures.

Construction Vibrations

Construction vibrations are created when construction materials or equipment impact soil or rock creating localized displacements in the material. These displacements propagate through the ground, and result in the vibration (excitation) of buildings. The propagation and attenuation (decrease in intensity over time) of the vibrations through a material (soil or bedrock) is a function of the material type, material density, degree of compaction and distance from the source of the vibrations. Generally, greater attenuation of vibrations occurs in soil when compared to bedrock.

In terms of vibrations, both the magnitude of ground vibration and frequency content (dominant frequencies) are of interest. The magnitude of the ground motion is typically expressed in terms of ground velocity (distance / time). Both the frequency of the ground motion and the ground motion velocity affect the displacement of the ground at the point of measurement.

The vibration of a structure is primarily determined by the ground motion frequency. Both the ground vibrations transmitted through the soil and underlying rock will contribute to the vibrations of the structure. When the ground motion (input frequency) matches a natural frequency of a structure, a condition known as resonance occurs. Resonant or near resonant conditions will result in large amplitude displacements of a structure. In a complex structure the individual components of the structure (such as the towers, machine rooms, or other components) all have natural frequencies which can be excited by ground motion. Additions or other alterations (common in historic structures) also may complicate the response of a structure.

Setting Limits

To prevent damage from construction-induced ground vibration, limits are typically placed on the maximum ground velocity as a function of the frequency content of the ground motion. These values are developed from experience with protection of existing structures, and are therefore empirical in nature. The values also depend upon the vibration source, soil type(s) and structure condition (Bachman et al, 1995).

Acceptable vibration limits vary significantly depending upon the type and character of the structure under consideration. Nearly all of the commonly used vibration standards recognize the need for lower vibration limits when dealing with older and historic structures. There is no single standard criterion for allowable vibration levels. While this makes the process of identifying a standard for a project more difficult, it allows rational standards to be adopted for the particular project and probable extent of damage to adjacent structures.

Table 1 provides a summary of allowable construction vibration levels from a variety of references. As seen in the table, significant variations exist in the allowable vibration levels. Some of the standards include different allowable vibration limits depending upon the expected duration of the construction operation. This accounts for the greater damage produced by long-term (steady-state) vibrations when compared to short-term vibrations of similar magnitude. Examples of construction operations producing long-term / steady-state vibrations include vibratory installation of sheet piling, pile driving, and vibratory compactors. Short–term vibrations can be produced during blasting operations, dynamic soil compaction or by other means.

The most appropriate standards for the protection of historic structures are the standards which have been specifically developed for the protection of older, damaged or vulnerable structures. These include the recommendations contained in the CALTRANS technical advisory, the standards developed for the Central Artery / Tunnel project and the German and Swiss Standards. The USBM recommendations were developed from extensive data obtained on residential buildings subject to vibrations from blasting (Siskind, 2000). Significant differences may exist between the residential structures examined by USBM (disbanded in 1995) and fragile historic structures. There are also differences in the frequency and duration of blast versus construction induced vibrations. Accordingly, these are the least appropriate standards for protection of historic structures.

The lower vibration limits contained in the Swiss, German and Central Artery / Tunnel standards have been shown to be effective in preventing damage in existing structures from construction vibrations in urban environments (Kelley, et al, 1998 and Glatt, et al, 2004). In each of these research reports, no damage was observed in structures when the construction-induced vibration levels were less than the values contained in the German and Swiss Standards.

The New York City guideline (TPPN #10/88) contains specific recommendations for monitoring of historic structures located within 90 feet of new construction—the typical width of the major north-south roadways in Manhattan. While the recommendations contained in the guideline are generally appropriate for protection of existing structures, the vibration limit of 0.5 inches per second may not be sufficient to preclude damage in relatively fragile historic structures. Further, while the research supporting the codification of the vibration limit (and the limits on settlement, see below) appears to be sound, the very codification of these limits without specific follow-up study to confirm their appropriateness for a broad range of historic structures makes adopting project-specific limits that differ from the TPPN difficult.

Underpinning

The primary concern associated with underpinning (and dewatering) of existing buildings is the settlement of the structures. Some amount of settlement occurs in nearly all underpinning projects. Any settlement, particularly differential or nonuniform settlement, can result in damage both to structural and non-structural elements of a building. The amount of settlement can vary extensively depending upon the building conditions, soil types, construction practices and workmanship. The rate of settlement can also vary significantly between structures, with the settlement potentially continuing after the underpinning work has been completed. The variation in rate of settlement occurs as a consequence of the shifting of foundation loads to different soil strata, changes in soil stresses on the newly loaded strata, and variations in time-settlement properties of soils.

To monitor the settlement of buildings, surveying methods are typically used. Prior to the start of underpinning, a baseline survey of the building is performed. The baseline survey will include the installation of monitoring targets on the buildings. During the subsequent construction operations, the elevation surveys can be repeated to determine the extent of building settlement. More frequent surveys may be required at the onset of the underpinning. The surveys should also be repeated whenever construction operations are halted because of excessive vibrations.

A maximum horizontal and vertical movement of 0.25" is allowed in NYC TPPN #10/88. However, absolute limits on allowable displacements are difficult to establish. Settlement limits need to be based upon the in-situ condition of the structure, and should reflect the type of construction present. Large settlements, significant changes in crack sizes or other types of building displacement will require investigations to determine the causes.

Protection of Adjacent Structures during Construction

The protection of existing construction structures requires a balance between providing safe limitations on vibration levels without precluding cost-effective construction. Construction vibration limits must be established for specific structures based upon the fragility of the existing structure and expected construction vibrations. Effective protection plans include limits on vibrations as well as recommendations for preconstruction surveys, monitoring frequency during construction, reporting and notification procedures, response for a sliding scale of damage and other considerations.

Preconstruction Condition Survey

Prior to starting vibration producing construction operations, preconstruction surveys are required. These surveys are used to establish a baseline condition of the structures prior to the start of construction. Typical information recorded during preconstruction surveys includes the following:

- Description of the building, building layout and site plan
- Type of construction
- Approximate age of construction
- Type of foundations
- General condition of the building
- Location, width and orientation of visible defects/cracks
- Location of loose materials
- Location of previous repairs
- Distance to construction operations

The preconstruction surveys focus on documentation of existing defects and cosmetic damage in both the common areas and individual building units. Accurate preconstruction surveys of existing damage are essential to prevent spurious claims of new damage occurring as a result of the new construction. The results obtained during preconstruction surveys can be used for comparison purposes in the event of reported damage during construction.

During the preconstruction surveys, any potentially loose or damaged materials should be identified. If possible, these materials should be made safe prior to starting construction. Additionally, consideration should be given to retrofitting hanging mechanical equipment to prevent damage from sway.

Visual Surveys During Construction

During the construction period, routine visual surveys are recommended to allow for any changes in condition to be documented. These surveys should focus on existing conditions identified during the preconstruction surveys to see if any changes have occurred, and evaluating if new damage is present. Depending upon the ongoing construction activities, the visual surveys should be completed on at least a weekly basis. Additional surveys will be required if critical vibration thresholds are exceeded. These "priority" surveys are recommended to be completed within 12 hours of the recorded vibration events. A survey report should be prepared after completion of each survey with copies distributed to the contractor, building owners and developers.

Construction Vibration Monitoring System

To allow for a rapid reaction to excessive vibrations, an integrated monitoring system is needed to allow for notification of the monitoring engineer, contractor and owner when critical vibration thresholds are exceeded. The integration will require the connection of all seismographs to a central data acquisition system which has the capability to immediately notify the monitoring engineers and contractor via telephone/email when critical vibration levels are exceeded. The vibration monitoring system should be installed and operational at least thirty days prior to the start of construction. This will allow for baseline monitoring results to be established.

Crack Monitoring

To accurately determine if changes in existing cracks occur, the installation of crack monitors is recommended. The exact locations of the crack monitors should be determined during the preconstruction surveys of the buildings. These monitors can be either visual (tell-tale type) monitors or electronic monitors that are linked to the vibration monitoring system. Additional crack monitors may be required in the event of new crack formation during construction.

The crack monitors should be examined during all routine surveys of the building condition. Any changes in crack size should be included in the survey report. Significant changes in crack size will warrant further investigation to determine the cause of the changes.

Human Vibration Perception

Effective construction protection plans require the education of owners/tenants to discern the difference between perceptible vibrations and potentially damaging vibrations. Human perception to vibrations begins at much lower levels than those shown in Table 1. Vibrations with a peak velocity as low as 0.02 inches per second are perceptible to a standing person, while a vibration velocity of 0.26 inches per second may be considered disturbing or unpleasant (Bachman, 1995). The human perception of vibration, at 0.02 inches per second, is one-sixth the 0.12 inches per second value shown in Table 1 as a recommended vibration limit.

The low threshold for human perception of vibration highlights the need for automated monitoring of these vibrations, and the need to establish baseline vibration levels in the buildings. To establish baseline vibration levels in the buildings, the monitoring should be started a minimum of thirty days prior to construction. During the baseline period, the monitoring system should be used continuously to allow for all sources of "normal occupancy" vibrations to be documented. These sources may include subways, traffic, elevators and other sources. Results from these baseline studies should be presented to the building owners to allow for a feel of "typical" vibration levels to be developed.

Case Studies

Construction protection plans have recently been developed for two projects in the New York City area. One of the projects involved the protection of two historic structures adjacent to a proposed high rise tower; the other involved the protection of existing buildings in a historic district.

Historic Buildings

The existing buildings represent well preserved examples of early skyscraper construction in New York City. Building 1 (Figure 1), built from 1878 to 1880, was originally ten stories tall, two additions increased the height to 15 stories. Building 2 (Figure 2) is 21 stories tall, and was originally constructed in the early 1890s. Both buildings have been converted from their original commercial office use to residential.

To assist in the development of the protection plans, limited surveys were completed in both buildings to assess the construction types, current condition and other factors. The surveys indicated that the buildings were generally in good condition and were well maintained. However, potentially susceptible details were observed in both buildings. These details included the prior underpinning of portions of Building 2 to accommodate subway construction, cast iron structures, and remnants of previous common wall construction.

Building 1 was largely constructed using brick masonry construction for the exterior walls. The floors appear to be constructed using closely spaced wrought iron or steel beams with barrel vault construction between the beams. Decorative features on the roof include ante-fixes on the parapets. These features were recently replaced using replicas of the original construction, substituting synthetic materials for the original terra cotta.

Building 2 was constructed using a combination of steel and masonry construction. The building is clad with a mix of gray granite, brick and terra cotta. The three level penthouse is wrapped with an ornate copper fascia with terra cotta caryatids, in the shape of winged angels, at the corners of the penthouse (Figure 4). The caryatids are topped with gold spheres. The caryatids are currently strapped to the building (Figure 5) and show signs of cracking and spalling.

The historic nature of the buildings and the highly ornate exterior facades lead us to recommend conservative vibration limits. Table 2 shows the recommended vibration limits for the buildings. These limits were largely based upon the German, and Swiss Standards shown in Table 1, and are also consistent with NYC TPPN # 10/88. Table 2 also contains recommendations for visual surveys of the buildings based upon the recorded level of vibrations, and an upper vibration limit at which all construction operations should cease until an inspection of the buildings can be performed. If new damage is observed during inspections of the buildings, more stringent vibration limits will need to be imposed (consistent with the recommendations contained in NYC TPPN # 10/88). In addition to the monitoring program, the following additional recommendations were made:

- Detailed preconstruction surveys should be completed at both buildings
- Preconstruction "baseline" vibrations should be recorded for at least 30 days prior to the start of construction
- Installation of crack monitors at select locations
- Installation of survey targets to allow for evaluation of building settlement
- Installation of a gas leakage monitoring system
- Installation of braces to prevent sway of hung mechanical systems
- Vibration and noise monitoring during the construction period

Historic District

The construction of a new mezzanine and stairway has been proposed to add to an existing mass transit system. The proposed new construction will be located within a historical district recognized by both New York City and the National Register of Historic Places. To insure the protection of the buildings within the project area, and to assess the need for possible limitations on construction, a construction protection plan was commissioned by residents of the district.

The buildings in the area were generally federal style buildings (Figure 6), and were constructed starting in the 1820s. Signs of modifications and alterations from the original construction were visible in the majority of the surveyed buildings. Typical modifications included changes to the front facades to allow for retail space on the ground level, and conversion of the buildings into multi-unit apartments.

During the site evaluation, the foundations of the federal style buildings were observed to be highly susceptible to damage. These foundations are constructed with loose stone, rubble type materials and can be easily damaged by construction induced vibrations, or settlement of the surrounding soil (Figure 7). The potential for damage is exacerbated by the extensive erosion of mortar observed during the survey.

To prevent construction damage to the existing buildings in the district, limitations are required on both the vibrations and tolerable settlement produced by the construction. These limitations should apply to all buildings within 90 feet of the construction area. Table 3 shows specific vibration limits from the construction protection plan developed for the historic district. These limits require damage surveys at lower thresholds than those shown in Table 2 due to the more fragile nature of the buildings. The remaining facets of the program were similar to those discussed above.

Summary

Protection of historic structures and structures within historic districts during adjacent construction is frequently required in urban areas. The protection of these structures is more demanding than protection of newer structure. This is because of both the fragile nature of these structures, and the need to prevent both structural and cosmetic damage from occurring. Monitoring and protection plans for these structures require an accurate assessment of the structures current condition and the development of appropriate construction vibration limits. The implementation of monitoring and protection plans will not preclude damage from occurring during construction. The monitoring and protection plan allow for a rational quantification of the impact of the construction on the buildings. The efficacy of the protection measures will be discussed in subsequent papers.

References

Bachman, H., et al, *Vibrations Problems in Structures: Practical Guidelines*, Birkhauser, Berlin, 1995, 234 pp.

CALTRANS, "Transportation Related Earthborne Vibrations," TAV-023-01-R9601, Prepared by Rudy Hendriks, 31 pp.

Deutsches Institut for Normung, *Structural Vibrations in Buildings, Effects on Structures*, DIN 4150

Glatt, J., et al, "Sheetpile-Induced Vibrations at the Lurie Excavation Project," Geotechnical Engineering for Transportation Projects, Yegian, M.K., and Kavazanjian, E., Eds., Special Geotechnical Publication #126, ASCE, July 2004.

Kelley, P.L., et al, "Building Response to Adjacent Excavation and Construction," Effects of Construction on Structures, Geotechnical Special Publication No. 84, ASCE 1998, pp. 80-97.

Konon, W. and Schuring, J., "Vibration Criteria for Historic Buildings," Journal of Construction Engineering and Management, Vol. 111, No. 3, September 1985.

Massachusetts Highway Department – Central Artery / Tunnel, "Design Policy Memorandum No. 1 (Revision 6) Construction Impact Mitigation."

Nichols, H.R., et al, *Blasting Vibrations and Their Effects on Buildings*, U.S. Bureau of Mines Bulletin 656, 105 pp.

Siskind, D.E., *Vibrations from Blasting*, International Society of Explosive Engineers, Cleveland, 2000, 120 pp.

Swiss Standards Association, SN 604312, "Criteria for Construction Vibrations."

Technical Policy and Procedure Notice # 10/88, "Procedure for the Avoidance of Damage to Historic Structures Resulting from Adjacent Construction when Subject to Controlled Inspection by Section 27-724 and for any Existing Structure Designated by the Commissioner," New York City Department of Buildings, June 6, 1988.

Figure 1. Front elevation of Building 1.

Figure 2. Front elevation of Building 2.

Figure 3. Masonry arch windows on south elevation of Building 1.

Figure 4. Terra cotta caryatids in the form of winged angels at penthouse corners

Figure 5. Straps used to support caryatids

Figure 6. Front elevation of building in Historic District

Figure 7. Rubble foundation in basement of historic district building.

Table 1 – Summary of Recommended Construction Vibration Limits

Standard	PPV[1] Limits in. / sec.	Frequency (Hz)	Vibration Duration / Building or Soil Type	Regulatory Agency	Comments
U.S. Bureau of Mines - 8507	0.18 to 0.5	1 to 2.5	Blast	US Dept. of Interior	Linear slope
	0.5	2.5 to 10			
	0.5 to 2.0	10 to 40			Linear slope
	2.0	> 40			
NYC TPPN # 10/88	0.5	none	Historic Buildings	New York City Dept. of Buildings	¼" max. movement
Caltrans TAV- 02-02-R9601	2	blast	Residential	California Dept. of Trans.	Technical advisory
	0.2	continuous	Ruins and Ancient monuments		
	0.08	continuous			
	0.2 – 2.0	pile driving	Well engineered structures		
	0.3	pile driving	Normal dwellings		
Central Artery / Tunnel	0.2	1 – 30 Source M	Type III Building – Non-engineered with plaster finishes	Mass. Highway Dept.	Source M includes vibratory pile installation
	0.12	1 – 30 Source M	Type IV Building – Extremely susceptible to damage		
Swiss Standard SN 604312[2]	0.12	10 - 30	Vulnerable to vibration / Steady state	Swiss Standards Association	Buildings which are especially sensitive or worthy of protection
	0.12 – 0.5	30 - 60			
	0.3	10 – 30	Vulnerable to vibration / transient		
	0.3 – 0.5	30 - 60			
DIN 4150	0.12	< 10	Short term	German Institute of Standards	Building particularly sensitive to vibrations
	0.12 – 0.3	10 – 50			
	0.3 – 0.4	50 - 100			

1. PPV – peak particle velocity (1 inch / second = 25.4 mm /second)
2. Referenced in Kelley, P.L., et al, "Building Response to Adjacent Excavation and Construction," Effects of Construction on Structures, Geotechnical Special Publication No. 84, ASCE 1998, pp. 80-97.

Table 2 – Recommended Vibration Limits for Historic Buildings

Peak Vibration Level (inches/second)[1]	Reporting	Engineer Action	Contractor Action
≥0.05	Engineer	Routine inspection	None
≥0.12	Engineer/Contractor	Daily inspection	None
≥0.50	Engineer/Contractor/Owner	Priority[2] survey	Cease vibration producing work[3]
≥2.0	Engineer/Contractor/Owner	Priority survey	Cease all work[4]

1. Vibration level recorded at building site (1 inch / second = 25.4 mm /second).
2. Survey within 12 hours of vibration limit being exceeded.
3. Vibration producing work should be stopped and alternate construction methods used.
4. All vibration producing work should be stopped until after completion of engineers survey and alternate construction methods used.

Table 3 – Recommended Vibration Limits for Historic District Buildings

Peak Vibration Level (inches/second)[1]	Reporting	Engineer Action	Contractor Action
≥0.05	Engineer	Daily inspection	None
≥0.12	Engineer/Contractor	Priority[2] survey	None
≥0.50	Engineer/Contractor/Owner	Priority[2] survey	Cease vibration producing work[3]
≥2.0	Engineer/Contractor/Owner	Priority survey	Cease all work[4]

1. Vibration level recorded at individual building (1 inch / second = 25.4 mm /second).
2. Survey within 12 hours of vibration limit being exceeded.
3. Vibration producing work should be stopped and alternate construction methods used.
4. All vibration producing work should be stopped until after completion of engineers survey and alternate construction methods used.

Roof Collapse – A Forensic Analysis Years After

W.W. Small, P.E.[1] and P.G. Swanson, P.E.[2]

[1]Facility Engineering Associates, P.C., 11001 Lee Highway Ste. D, Fairfax, VA 22030; PH (703) 591-4855; FAX (703) 591-4857; email: small@feapc.com
[2]Facility Engineering Associates, P.C., 11001 Lee Highway Ste. D, Fairfax, VA 22030; PH (703) 591-4855; FAX (703) 591-4857; email: swanson@feapc.com

ABSTRACT

In March of 2002, a construction company specializing in light steel construction was awarded a contract for the construction of a fire and rescue facility located in rural Virginia. The project was completed and put into service in October of 2002. Between February 15, 2003 and February 17, 2003, the site area experienced a snow storm resulting in measured snow accumulations of approximately 20 inches. On February 17, 2003, the roof structure of the building failed, collapsing the roof and portions of the second floor walls.

The paper describes the process of analyzing the collapse from the perspective of three years later. A review of construction documents, site photographs, and communication documents were used to establish the mechanism and cause of the failure. The lack of available direct site evidence required that the analysis primarily utilize pre-failure information as a means of determining cause of failure.

INTRODUCTION

The building consists of a two-story combination garage and office/warehouse structure approximately 60 feet by 100 feet in plan. The first story was approximately 16 feet high with a system of cold-rolled light-gage steel exterior framing and hot-rolled interior columns and beams. The second floor was 9 feet high with predominantly light-gage steel framing throughout. The roof system consisted of continuous rafters from eave to ridge with a 4:12 pitch and gable ends (refer to Figure 2). The roof rafters were supported by light-gage steel at the exterior walls and along a light-gage beam line and light-gage steel built-up columns to either side of the centerline of the roof (at approximate third points from the front to the rear of the structure). The rafters spanned approximately 20 feet from the exterior bearing wall to the beam line and cantilevered 10 feet to the ridgeline. The rafters were spaced at 24 inches on centers.

The building was constructed in the summer and fall of 2002 and was completed in September of that year. According to the structural drawings, the building was designed in accordance with the Virginia Uniform Statewide Building Code with reference to the 1996 BOCA National Building Code. The 1996 BOCA code references 1986 AISI code with 1989 Addenda for issues related to cold-formed steel.

Based on the Codes in force at the time of design and construction, the designer used a flat roof snow load of 25 psf (pounds per square foot) and a minimum roof design live load of 30 psf. Available National Weather Service data and site observations taken the day of the failure confirmed a snow depth of 20 inches for the site area. Measurements (by weighing 1 foot x 1 foot x accumulated depth of actual snow) taken on site the day following the collapse revealed an actual snow load of approximately 20 to 21 psf could be attributed to the storm.

SCOPE OF INVESTIGATION

Since the failure occurred over three years prior to our investigation, there was no opportunity to visit the site or observe failed members. The collapse occurred during the early morning hours of February 17, 2003. Consequently, there were no witnesses to the collapse. Information used in the investigation consisted primarily of the historical records of the construction. Several forensic investigations were conducted immediately and shortly after the time of the collapse representing the various parties involved. Although these were reviewed and considered during the course of our investigation, our analysis focused also on the documents created during construction as a means of establishing an objective viewpoint.

Over 3,000 pages of written documentation were available consisting of:

- Forensic reports prepared by engineers representing the parties involved.
- Communications between the general contractor and the steel erection contractor.
- Construction drawings and design computations documenting the original design intent.
- Shop drawings prepared by the steel erector.
- Meeting minutes between the owner, contractor, and design team.

ANALYSIS STRATEGY

Objective

The objective of the analysis was to formulate an objective opinion regarding the probable cause of collapse.

Approach

In approaching the objective, knowing that the analysis would only rely on documentation without the aid of direct observation of the failed structure, it was necessary to establish goals that would need to be met on a step-by-step basis. In this fashion, the analysis would proceed until such time as it became clear that the objective could not be met based on objective evidence available from the documentation provided. If, at any point, it was determined that the overall objective could not be met based on the documentation available the analysis would have been discontinued; or, at least, would have been postponed until such time as additional documentation was produced that allowed for the analysis to continue toward meeting the objective.

Goals

The step-by-step goals of the analysis were as follows:

1) Conduct preliminary document review to establish if significant documentation elements deemed critical were provided.
    a. If no to step one, request the desired outstanding documents
    b. If documents are received, continue to Goal 2, otherwise cease analysis.
2) Review documentation to establish codes governing design and construction and overall design intent.
3) Review documentation to establish contract duties for fabrication and erection of the light-gauge steel framing.
4) Review documentation to establish if what was intended by design was fabricated and installed and if not, to establish structural deviations.
5) Review documentation to establish responsibility for the structural deviations and to determine if the Engineer of Record was party to the changes.
6) Review documentation and perform calculations to establish if structural elements that deviated from the design intent were weaker than those called for in the design.
7) Review documentation to establish if weaknesses in the as-built deviations appeared critical to the collapse of the structure.
8) Formulate opinion of the probable cause of collapse based on this objective review of the documentation provided.

Execution

The first goal of the analysis was to establish if significant documentation elements deemed critical were present and if not, to secure them. This was accomplished by conducting a preliminary review of the documentation followed by a request for additional critical documentation (such as design calculations and construction photographs), followed by a period of waiting until such time as the additional documentation was provided. When it was deemed that enough documentation was available for analysis, the second goal was pursued. Likewise, each successive goal was pursued to success based on the objective review of the documentation provided and ultimately, an opinion of the probable cause of collapse was formulated.

ANALYSIS DISCUSSION

Our analysis consisted of review of documents provided relative to the design, construction and subsequent failure of the subject fire and rescue facility. Based on the codes that governed the design of the building, as stated in the documents provided, various sections of codes that applied to the structural design, fabrication and erection of the light-gauge steel framing system utilized for the building were also reviewed.

Over the course of the analysis, approximately 3000 pages of documents and drawings were reviewed. Although numerous copies of the drawings were provided, it appeared that the structural design of the building, including the light-gauge steel system, was not materially changed on the drawings from any one set to another. Therefore, what appeared to be the most current full set of drawings that were stamped and signed by the designers of record were relied on in the analysis. The drawings conveyed the design intent and enumerated the codes that governed the design and construction of the building.

In addition to the written documentation provided, various photographs were provided that captured the building and many of the structural elements during construction and of failed structural components and the overall building following the collapse. These included both digital photographs (from the fabricator) taken during construction and copies of printed photographs taken by various parties following the collapse of the structure. The photographs confirmed many, but not all, of the structural members used in the as-built light-gauge steel system. Although the actual gauge of the members was not discernible from the photographs, this did provide confirmation of the component numbers and geometry of some of the critical framing members and made possible the comparison of as-built conditions to design intent.

Applicable Code and Structural Design Review

The structural drawings identified the applicable building code as the Virginia Uniform Statewide Building Code (BOCA 1996 as amended). The BOCA National Building Code/1996 (BOCA) adopts, as a referenced standard, the American Iron and Steel Institute 1986 Specification for Design of Cold-Formed Steel Structural Members with 1989 Addendum (CFSD-ASD-86).

The structural drawings also documented the design live loads for the building. The ground snow load, as required by the local authority having jurisdiction (AHJ), was indicated to be 35 psf. The drawings indicated use of a snow exposure factor of 0.7 and a snow load importance factor of 1.0, resulting in a flat roof snow load of 24.5 psf. Although indicated as 1.0 on the structural drawings, the snow load importance factor required by BOCA/1996 for "fire, rescue, and police stations" is 1.2. In actuality, the snow load importance factor documented in the design calculations provided was 1.2, not 1.0, which resulted in a calculated flat roof snow load of 29.4 psf. Although the structural drawings documented a flat roof snow load, they also documented that the "minimum" roof live load was 30 psf.

Pertinent to our analysis was that the live load utilized for the roof design was indicated on the structural drawings to be 30 psf, minimum, and that the structural calculations documented the use of the "minimum" roof live load of 30 psf. The design calculations also indicated the design roof dead load to be 15 psf. The design calculations also enumerated the design wind loads, which appeared consistent with the BOCA code requirements.

As-built Construction Review

The documentation provided revealed that the as-built construction deviated from the Construction Drawings and the original design intent. The most significant of these departures were the following:

1) The 2nd floor to roof columns were originally designed as built-up members consisting of (4) 600S162-54 light gauge steel structural studs configured as two pairs of back-to-back structural studs wrapped (continuous from top of floor to underside of roof beams) on two sides (across the stud flanges) with (2) 800-T-100-54 light gauge steel track sections; (1) on each side. The original design recognized the columns to be unbraced from the top of the second floor to the elevation of the roof beams in the columns' weak direction (oriented left-to-right or side-to-side in the building) and unbraced from the top of the second floor to the elevation of the roof rafters in the columns' strong direction (oriented front-to-back in the building). It appears to be for this reason that the steel track sections were required. From the photographs (taken both during construction and after failure), it was evident that the as-built construction omitted the track sections and substituted (3) 2" steel straps screwed across the structural stud flanges (see figures 1 and 2 taken during construction). Three straps were located at roughly 6 feet, or greater, on centers, with one located near the top of the second floor, one near mid-height, and one near the underside of the beam framing into the column.

Figure 1. Overall configuration of the 2nd floor-to-roof columns.

Figure 2. Close-up photograph documenting the omission of the continuous track sections intended to be fastened across the outside flanges of the built-up columns. This was important objective evidence of a critical departure from the intended structural design. This was also confirmed in the photos taken after the failure.

2) From the photographs, it was evident that some of the connections utilized fewer fasteners than were required by design. This was specifically documented for the rafter-to-rafter connections at the ridge. It was also apparent, from the photographs, that the fastening pattern at the beam-to-column connections was inconsistent from column to column.

3) The design calculations appeared to assume that the second floor would be constructed with a finished 5/8-inch thick gypsum board ceiling applied to the underside of the roof rafters. This was specifically indicated in the list of assumed roof dead loads and was also apparent in the design of the roof rafters. The calculations for the roof rafters indicated that the rafters were designed with continuous top and bottom chord lateral bracing. The top chord (or flange) lateral bracing was provided by the roof sheathing; however, without installation of a ceiling at the underside of the rafters, the bottom chords were not braced at all. This was significant due to the fact that the rafters were constructed as continuous members, which exposed them to compression in both the top and bottom chords.

It was not clear from the documentation provided that the structural engineer of record was involved in the decisions for the above items to deviate from the original design. However, these deviations appeared to contribute to the collapse of the structure.

CONCLUSIONS

From the analysis of the materials provided, the collapse of the building involved the light-gage steel that was designed and furnished by the fabricator. We also understand that the original structural design for the light gauge steel system was provided to the fabricator, stamped and signed, by the structural Engineer of Record. However, the furnished light-gauge steel was erected by the General Contractor.

From the documentation provided, the collapse appeared generally to be the result of the failure of the light-gauge steel framing system (columns particularly) supporting the roof structure. The second floor framing and the exterior walls from the ground level to the second floor generally remained intact throughout and following the collapse. Therefore, the analysis could be focused on the structural design and as-built construction of the roof framing. The documentation provided an understanding of the nature of the design intent and allowed objective comparison of the design intent to what was actually constructed.

The structural design notes/calculations provided demonstrated that the design recognized the minimum design loads required by code and that the design ground snow load satisfied the requirements of the local AHJ. The design notes/calculations provided also documented the assumptions for the weight of materials anticipated to be permanently applied to the roof structure, which totaled 13 psf. However, the design notes then documented that a 15 psf minimum was utilized for the design dead load for the roof. Finally, the design notes/calculations documented that the appropriate wind loads were recognized in the structural design of the building. Therefore, the design loads for the building were consistent with code requirements.

From the analysis, it was determined that:

a) **What was designed was not constructed** – the deviations were enumerated in the previous section;

b) **What was constructed was weaker than what was designed** – each of the deviations indicated introduced a weakness in the constructed product as compared to design. In the case of the built-up columns, the design stiffness of these members was critical in resisting the loads imparted on them. Similarly, the structural connections that deviated from design were also weaker, specifically where fewer fasteners than specified were used. The omission of lateral bracing in the plane of the bottom flange of the roof rafters both weakened the rafters themselves and the overall roof diaphragm.

c) **The weakest of the deviations failed** – The collapse of the structure precipitated from the failure of the columns due to their reduced stiffness.

Although it was difficult to ascertain from the documentation provided if there was a single, specific point of failure initiation, it was clear that the only members that completely failed were the interior columns from the second floor to the roof framing (refer to Figure 2). These as-built columns were weaker in both directions (their strong and weak directions) than the designed columns (refer to Table 1). Based on photographs of the failed columns, this appeared critical in the weak direction. The straps utilized in lieu of the continuous tracks were negligible with regard to increased stiffness of the built-up columns and were incapable of reducing the unbraced length of the columns.

d) **What was not constructed per design could have been repaired at any time** – although it is unclear why the deviations identified were implemented, it is clear that decisions were made to deviate from the design intent. However, these conditions were such that the correction of the deviant conditions could have been made at any time during or after substantial completion of the building.

e) **The structural engineer of record did not acknowledge deviations from design** – It is customary to consult the engineer of record when either the design cannot be implemented or if deviations from design are made for any reason. It is unclear in this case if the deviations were joint decisions involving the structural engineer of record or if they were done as a matter of convenience based on frame geometries, material handling and erection, cost, etc. The "Suggested Cold-formed Steel Structural Framing Engineering, Fabrication, and Erection Procedures for Quality Construction" outlined in Section 6 of the CFSD-ASD-86 indicate that "Periodic inspections during construction should be made by the professional structural engineer who either is or represents the design engineer of record" and "At the conclusion of the cold-formed steel construction process a final inspection should be made by the engineer of record who should certify that the cold-formed steel framing has been constructed in accordance with the plans and specifications and in accordance with all applicable building codes and regulatory requirements." Although the General Contractor requested the "sign-off" several times from the fabricator, it did not appear (from the documentation provided) that this sign-off was issued. Also it was not stipulated in the contracts reviewed that the engineer of record was retained to provide the services suggested by the CFSD-ASD-86.

OPINION ON CAUSE OF ROOF FAILURE

The stiffness of several components of the building was reduced due to as-built conditions. Each deviation from design introduced a weakness to the system. The opinion formulated regarding the probable cause of collapse was the weakness introduced by the omission of the track sections from the built-up light-gauge steel columns supporting the roof structure. The as-built geometry of the built-up columns was well-documented, their failure was predictable and probable by calculation (refer to Table 1), and their failure was substantiated by photographs included with the documentation provided.

Table 1. Section Properties and Load Comparisons

Column Section	Area (in.^{2})	Moment of Inertia (weak axis, in.^{4})	Allowable Load (kips)	Est. Actual Load (kips)	Percent Overstress
As-Built	4.4	2 @ 2.0 ea.	11.6	15.0	30
Design	>4.4	>13.7 (as unit)	>30	15.0	n/a

Table Notes: The above is approximated to simplify the comparison and illustrate the relative difference in capacities between the as-built member and design member configurations. The above is based on gross section properties and concentric loading (ignoring track sections of design configuration, which allow built-up sections to behave as a unit). Impact of effective section properties and moments induced by actual member loading and framing eccentricities are not considered.

The stiffness of the columns was critically reduced by the omission of the track sections. With the weaker (less stiff) columns, the unbraced length would have to have been reduced in some fashion in order to support the combined axial and lateral loads. The 2" intermittent steel straps used in lieu of the omitted continuous track sections were incapable of providing lateral bracing in any direction and specifically not in the weak direction. Based on the photographs taken after failure, the columns appeared to fail due to buckling in the weak direction.

SUMMARY

By application of the approach outlined at the start of the project, the objective was met. The documentation was sufficiently complete and was able to be reviewed in an objective manner. The analysis process remained objective throughout, which allowed for the formulation of an objective opinion regarding the probable cause of collapse.

REFERENCES

American Iron and Steel Institute. (August 19, 1986 Edition with December 11, 1989 Addendum). *Cold-formed Steel Design Manual,* Computerized Structural Design, S.C., Milwaukee, Wisconsin.

Building Officials & Code Administrators International, Inc. (BOCA). (1996). *The BOCA National Building Code/1996,* Country Club Hills, Illinois.

The Kinzua Decoded, the Forensic Investigation of the July 21, 2003 Collapse

Thomas G. Leech, P.E., S.E., M.ASCE[1], Robert J. Connor, Ph.D.[2], Eric J. Kaufmann, Ph.D.[3], and Jonathan McHugh, P.E., M.ASCE[4]

[1] Vice President, Gannett Fleming, Inc., 601 Holiday Drive, Pittsburgh, PA 15220. Email: tleech@gfnet.com
[2] Assistant Professor of Civil Engineering, Purdue University, 550 Stadium Mall Drive, West Lafayette, IN, 47907. Email: rconnor@purdue.edu
[3] Senior Research Scientist, ATLSS Engineering Research Center, Lehigh University, 117 ATLSS Drive, Bethlehem, PA 18015
Email: ek02@lehigh.edu
[4] Project Engineer, Gannett Fleming, Inc., 601 Holiday Drive, Pittsburgh, PA 15220. Email: jmchugh@gfnet.com

Abstract

On July 21, 2003, The Kinzua Viaduct, a 91m (300 ft) tall, 626 m (2,053 ft) long railroad bridge in northwestern Pennsylvania, located approximately 26 km (16 mi) south of the border of New York and Pennsylvania, collapsed during a severe wind storm. The fingerprints of the collapse were evident within the debris field which extended over a distance of 320 m (1050 ft). A *Board of Inquiry* investigation of this historic Civil Engineering Landmark was conducted on August 12, 2003 by a team of forensic engineers, metallurgical specialists, meteorological scientists and government engineers, given the assignment to decode the secrets of the collapse. The findings of this forensic investigation are presented in the final *Board of Inquiry* Report dated December 31, 2003. This paper will highlight the systematic processes employed in the investigation and present the specific findings attributable to the tragic collapse.

Overview

A series of unfavorable weather conditions produced a severe weather event on July 21, 2003. This event was the largest severe weather outbreak in the Commonwealth of Pennsylvania for the 2003 season. An intense weather system, called a mesoscale convective system (or MCS), formed in the afternoon over a wide area including eastern Ohio, western Pennsylvania, western New York, and southern Ontario. At the leading edge of the front, the contribution of wind shear and moisture with afternoon instability initiated intense thunderstorms. Tornado activity appeared due to intense smaller vortices within the larger MCS system. At Kinzua State Park, an F-1 tornado, with associated wind speeds varying between 33 and 50 m/s (73 and 112 mph), touched down at approximately 3:20 pm local time. (Grumm, 2003) Wind impinged the structure initially from the east due to the counter clockwise rotation of winds within the tornado vortex. As the vortex trailed northward moving past the structure, the structure was attacked a second time from the south by a strong inflow of air spiraling into the tornado vortex. (Markowski 2003)

The spectacular collapse of twenty-three of the structure's forty-one spans resulted. (Photo 1) The *Board of Inquiry* investigation ultimately concluded that due to a specific hidden deterioration within the anchorage system, the structure was especially vulnerable from high winds attacking from an easterly direction. (Leech et al. 2003)

Photo 1 - Kinzua Viaduct, Mt. Jewett, PA, July 2003 – looking southeast

Historical Setting

At the time of its first construction (1882), the Kinzua Viaduct was the tallest bridge in the world. The slender wrought iron viaduct enabled steam locomotives to carry coal and other natural resources high above the deep stream valley, eliminating 13 km (8 mi) of steep grades and treacherous terrain. However, excessive vibrations and high crosswinds prevented trains from traveling faster than 8 km/h (5 mph) while crossing the bridge. With the advent of heavier rail loads in a time of rapidly changing technology, the original bridge was dismantled in 1901. (Shank 1980) The Kinzua Viaduct was rebuilt as a steel structure with a span arrangement, tower geometry, and grade line identical to those of the original bridge. The new steel superstructure was secured to the existing pedestals using the original wrought iron anchor bolts from the first viaduct. The western tower legs were secured directly to the existing anchor bolts, while the eastern tower legs for towers 4 thru 14, were fitted with roller expansion bearings so that the structure could expand laterally in response to temperature. Because the existing wrought iron anchor bolts were too short to accommodate the fitted expansion bearings used in the new design, the wrought iron anchor bolts were spliced to steel anchor bolt extensions by means of threaded, steel collar coupling assemblies. (Grimm 1901) The collar coupling assemblies were encapsulated by large diameter circular washers, and were consequently obstructed from view prior to the collapse. (Figure 1)

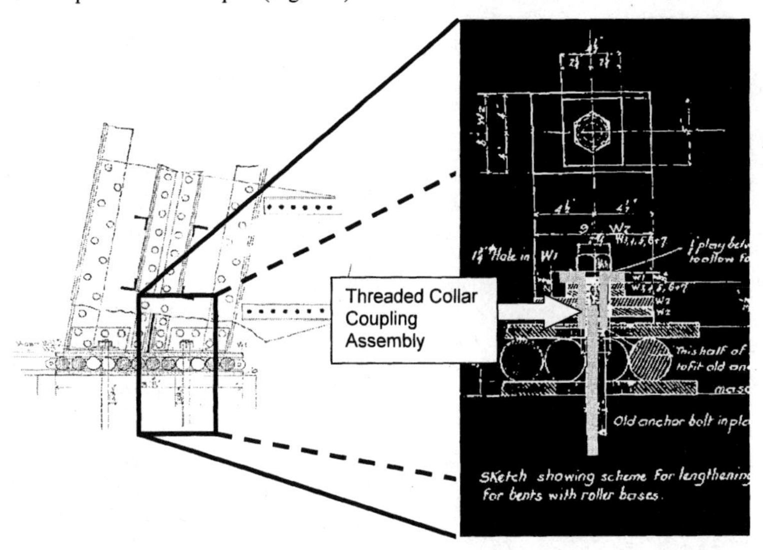

Figure 1 – Wrought Iron Anchor Bolt Extension via Threaded Collar Coupling Assembly

As the 20th century progressed, nearby coal resources were depleted, and rail traffic diminished. The viaduct remained in service but activity was light. Finally, in 1959 the railroad company sold the structure for scrap to the owner of a private salvage company. Realizing its value as a historic resource, the owner chose not to dismantle the structure, but instead, sold it in 1963 to the Commonwealth of Pennsylvania, who then established at the site a state park that featured the bridge as its centerpiece. Listed in the National Park Services National Register of Historic Places in 1977 and designated a civil engineering landmark by ASCE in 1982, the viaduct was used by a private railroad concessionaire from 1980 through early 2002. That year, Pennsylvania Department of Conservation and Natural Resources closed the structure after a routine inspection uncovered severe deterioration in observable structural elements of the towers. In 2003, repair work that focused solely on the restoration of these visibly deteriorated tower elements began and was progressing at the time of the collapse.

Board of Inquiry of Investigation

The Commonwealth of Pennsylvania has established guideline procedures, in Pennsylvania Department of Transportation Publication 220 dated July 2001, which govern the investigation of a catastrophic collapse. The intention of these procedures is to provide a thorough forensic and analytical investigation of a catastrophic collapse by a team of specialists, who are entirely independent from any ongoing design, maintenance, construction, or rehabilitation activities associated with the structure. Within the framework of these guideline procedures, the following methodology was established for conducting the forensic investigation.

Methodology of Investigation

Strict protocols were established to support the field and analytical phases of the investigation. These protocols were established within 24 hours of the incident and set in motion a rapid chain of events. Within one week of the collapse, [1] a team of specialists, skilled in forensics, meteorology, and fracture interpretation were assembled and placed under contract to the Pennsylvania Department of Conservation and Environmental Resources, [2] all arrangements were made to conduct a one day field investigation including all logistical support necessary to freely transport all investigative personnel over the rugged terrain, and [3] high resolution aerial photography of the 1200 m (4000 ft) wind impacted area was obtained by the Commonwealth of Pennsylvania, Department of Transportation. (Photograph 2)

Photograph 2 – High Resolution Aerial Photograph – tower numbers indicated

The field investigation phase relied upon the assembly of a team comprising a wide range of engineering and scientific disciplines, who conducted a single day examination of the site. The team was delegated with specific investigative inspection responsibilities to complete the investigation in a single day as required by the Commonwealth's guidelines. The site was cordoned and the debris field was treated as evidence which was permitted to be observed but otherwise was to remain untouched, with the exception of select material samples taken off site for laboratory examination. The team was subdivided into three squads which were individually tasked with (a) the examination of the 1200 m (4000 ft) roughly circular wind impacted area of the park surrounding the structure, to assess the gross scale meteorological implications, (b) the examination of the 320 m (1250 ft) debris field on a span by span basis, to systematically record all damaged elements of the structure, observe and record obvious forensic markers and develop initial hypotheses regarding the sequence of collapse, and (c) the examination and selection of limited fractured samples for further laboratory investigation, to support the follow-on analytical phase of the investigation. Additionally, eyewitness testimony, was recorded from all park maintenance personnel and all construction laborers who were present on the site during the storm event and subsequent collapse. The eyewitness testimony was taken with the understanding that accounts likely were emotionally charged and were taken from individuals without scientific backgrounds.

Physical and mechanical properties as well as chemical composition and metallurgical microstructure were determined through laboratory investigation for the fractured wrought iron anchor bolts, the fractured steel collar coupling assemblies and the steel anchor bolt extensions.

Subsequent analytical investigation included back calculation of wind velocity necessary to induce a rotational failure of a tower element, based on visual observations evidenced at the site during the investigative phase. The back calculation relied on the strength of connection between substructure and superstructure determined from laboratory testing and utilized published wind coefficients (ASCE-7). The back calculation additionally considered inertial effects. In addition, the analytical investigation relied on careful examination of the high resolution aerial photography as well as eye-witness testimony to synthesize a complete and consistent explanation of the initiation and sequence of the collapse. The analytical investigation concluded with the preparation of a comprehensive report accompanied by animated computer renderings demonstrating the sequence of collapse. Upon presentation to responsible government agencies, the complete *Board of Inquiry* report and animations were made available to the general public on the Pennsylvania Department of Conservation and Natural Resources' web site. (http://www.dcnr.state.pa.us/info/kinzuabridgereport/kinzua.html)

Findings

In the course of a forensic investigation, there are many indicators, that when taken collectively, decode the event. (Leech et al. 2004) At the Kinzua site, the following four distinct forensic markers were apparent:

Order markers – including the ordering of materials clustered within a debris field. The inversion of clustered materials within a debris field allows the reconstruction of the direct order of collapse. Order markers were apparent during the field inspection enabling the determination of the precise order of tower collapse.

Directional markers – including both the direction of fallen trees and collapsed towers. (Figure 2) Directional markers were evident through high resolution photography and revealed wind directed in two orthogonal directions (initially from the east, then from the south). The observation of these markers confirmed the sequence of collapse as well as the contribution of both "vortex leading edge" (easterly) winds accompanied by "inflow" (southerly) winds occurring in rapid sequence in this extreme weather event.

Figure 2 – Depiction of Wind Streamlines (Directional Markers) with "leading edge" wind streamlines – directed westward; and "inflow" wind streamlines – directed northward

Separation markers – including all evidences of "clean" breaks. (Figure 3) During the field investigative phase, it was apparent to the investigative team that the initiation of failure occurred at boundary between superstructure (of trestle bent configuration) and substructure, most likely at the specific boundary of the 1882 (original construction) and the 1901 (reconstruction).

At this boundary, the original wrought iron anchor bolts and masonry were preserved in the reconstruction. Fractures were observed within certain wrought iron anchor bolts (1882 construction) and a majority of expansion bearing, steel collar coupling assemblies (1901 construction), that provided connectivity between the original masonry and reconstructed superstructure. Consistently "clean breaks" were observed at the interface of the 1882 and 1901 construction. The separation markers were observed during the field inspection. The separation markers evidenced distinctive separation of the superstructure from its roller bearing assembly at the base of the superstructure for all eleven collapsed towers.

Figure 3 – Illustration of separation failure at substructure/superstructure interface

Fracture markers – including steel evidence of consistent patterns of small sub-critical fractures within members. (Photograph 3) Consistently, small sub-critical fractures were observed in the steel collar couplings connecting the 1882 construction and the 1901 construction on the windward side of the towers. Subsequent fractographic examination using a light microscope and microscopic examination using a scanning electron microscope, revealed evidence of long term fatigue crack propagation. The visual appearance of the fractures was characteristic of fatigue fracture with a flat, smooth surface without any evidence of plastic deformation in the fracture region. Microscopically, the fracture morphology also appeared fatigue-like with smooth featureless regions, typically found in high sulfur steels. (Photograph 4) No evidence of fatigue beach marks or striations were observed in these regions, however, the absence of evidence of brittle fracture by a cleavage mechanism or ductile fracture mechanism by dimple fracture suggested that the flat regions were most likely the result of stable crack propagation by a fatigue mechanism.

Based on the four forensic markers and subsequent analysis, the *Board of Inquiry* investigation concluded that failure initiated at the "weak-link" of the system – the anchor bolt system on the eastern faces, which was initially installed in the 1882 construction and subsequently modified in the 1901 construction. The 1901 construction provided a collar coupling assembly which included a series of washers that surrounded the anchor bolts and couplings. Consequently, the through wall-cracking of the collar couplings was hidden from view during routine condition inspections.

Photograph 3 – Delineating Through-Wall Cracking

The *Board of Inquiry* investigation concluded that the circumstance of a nearly north-south structure alignment and fractures within the collar couplings at the eastern tower legs resulted in a structure which was specifically vulnerable to winds from the east but not otherwise vulnerable to prevailing westerly winds. (Leech et al. 2003)

All eleven collapsed towers were fitted in the 1901 reconstruction with expansion bearings secured to the existing masonry via (1882) wrought iron bolts and (1902) collar coupling assemblies. Based on site observation that for the majority of these locations, three out of four of the anchor bolt assemblies at each tower leg displayed complete separation of the superstructure from the substructure at the collar coupling connection, the immediate failure at the expansion bearings may be characterized as a separation failure. As this separation failure resulted in a rotational failure of each tower about the fixed (or opposite) tower bearings, a back calculation of the force effects necessary to overturn the structure became the basis for prediction of the limiting wind velocity. (Figure 4) This back calculation derived uplift capacity based on the fracture of one anchor bolt out of four (with no strength attributed to the remaining three anchor bolts within a typical four bolt tower location) and utilized limit strengths derived from specimen samples. This back calculation subsequently derived an applied wind velocity of 42 m/s (94 mph) accompanied by a wind pressure of 1.9 kPa (39 psf) and recognized two distinct modes of failure of the anchor bolt system.

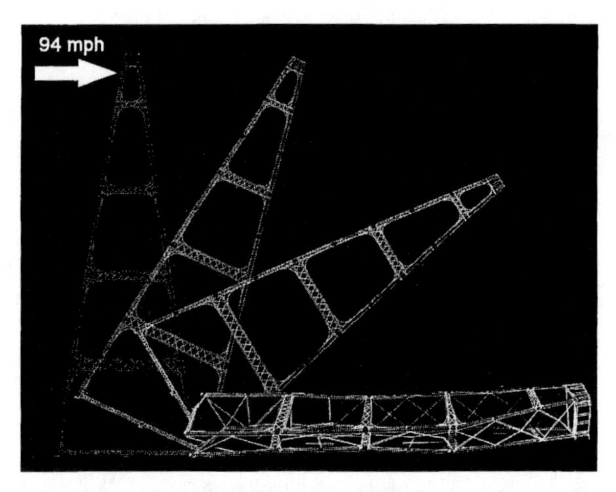

Figure 4 – Illustration of Rotational Failure Mechanism

Failure Mode 1 – Coupling Failure at the Boundary of 1882 and 1901 Construction – Expansion Bearing Anchor Bolt Collar Coupling. This mode accounts for approximately 3/4 of observed separation failures. All collar couplings observed at the site exhibited a radial cracking pattern, along with multiple longitudinal "splits" within the anchorages. The equiangular "splits" completely penetrated the collar couplings. The collar couplings found throughout the debris field exhibited similar fracture indications. Coupling failures showed evidence of fatigue fracture with secondary fractures occurring by overload presumably during the collapse. (Photograph 4) The *Board of Inquiry* investigation concluded that the couplings, which separated from the bearing assemblies and which were strewn within the debris field, experienced long term fatigue crack propagation prior to the time of the collapse incident. Because the cracks propagated through the entire thickness of the coupling, these collar couplings were judged to be ineffective for the transmission of uplift forces to the substructure. (Kaufmann and Connor 2003)

Mode 2 – Ductile Failure within the existing 1882 anchor bolts – Expansion Anchor Bolts. (Photograph 5) This mode accounts for approximately 1/4 of observed failures. Fractographic examination of fractured original 1882 anchor bolts showed that the fracture resulted from tensile overload and was a fully ductile fracture. (Kauffmann and Connor 2003) The estimated tensile capacity of a single (1882), 31.75 mm (1-1/4 in) anchor bolt at failure, considering a 20% corrosion loss, was determined to be 13 kN (30 tons) based on Brinnel Hardness evaluation. Based on the observed, 3:1 ratio of collar coupling failure to ductile anchor bolt failures, an uplift capacity of 13 kN (30 tons) was attributed to each tower. This capacity established a lower bound, critical wind speed of 42 m/s (94 mph), which was sufficient to initiate failure. (Leech et al. 2003) The failure was sudden and catastrophic.

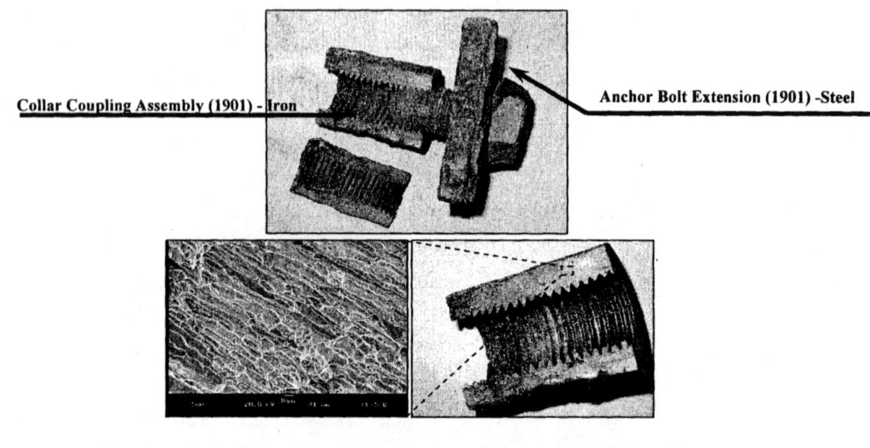

Photograph 4 – Laboratory Investigation of Through-Wall Cracking
(demonstrating long term fatigue crack propagation)

Photograph 5 – Wrought Iron Anchor Bolt – Ductile Fracture

The collapse of eleven supporting towers and twenty-three of the forty-one structure spans was rapid and proceeded in three distinct and separate episodes as illustrated on the accompanying figures. All girders and towers between towers 3 and 15 collapsed. Separation of the structure into three distinct collapsing segments is attributable to the arrangement of the wind locks within the girder system and to the nature of the wind event. (Figure 5) The 1901 design introduced expansion joints (and accompanying wind locks) at irregular locations within the structure. The collapse of the structure in three distinct episodes was controlled by the location of these wind locks.

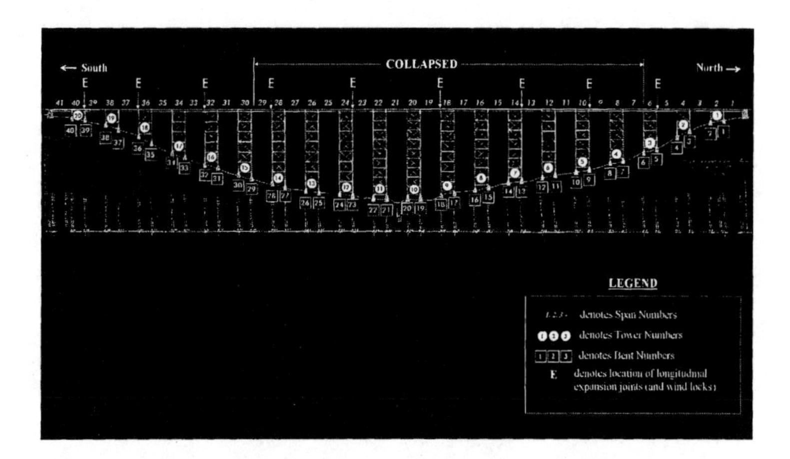

Figure 5 – Structure Elevation View Demonstrating Wind Lock Location

The following occurred in Episode 1 in the sequence indicated. (Photograph 6)

1. Tornado touched down – easterly (or "vortex leading edge") winds grew rapidly – local wind speeds (from the east) exceeded 40 m/s (90 mph) – as wind speeds grew, the towers oscillated laterally in response to their natural frequency.
2. "Separation" failures occurred within the "expansion" anchor bolt system of Towers 10, 11, 12, 13 and 14.
3. Rotational failure accompanied by collapse of Towers 10 and 11 and adjoining spans occurred.
4. Towers 12, 13 and 14 initially become airborne and "jumped" a small distance north and westward. Towers 12, 13 & 14 momentarily came to rest in the upright position on the ground, not initially collapsing. The rails and wooden decking for a brief period of time remained affixed to several spans and held the three towers in a vertical position, initially preventing immediate catastrophic collapse.

Photograph 6 - Debris Field – Episode 1 – tower numbers indicated

The following occurred in Episode 2 in the sequence indicated. (Photograph 7)

1. Tornado moved northward – easterly ("vortex leading edge") winds grew rapidly – local wind speeds (from the east) exceeded 40 m/s (90 mph). As wind speeds grew, all towers oscillated laterally in response to their respective natural lateral frequencies.
2. Wooden decking and rails (spans 1 – 18) separated from the structure.
3. "Separation" failure occurred in sequence within the "expansion" bearings of Towers 9, 8, 7, 6, 5 and 4.
4. In sequence, rotational failure of Tower 9 occurred, shortly, followed by rotational failure of Towers 8, 7, 6, and 5. Collapse was progressive from South to North.
5. Tower 4 was initially restrained by Tower 3 and the connecting girder span. However, after elongation of the girder span's connection to Tower 3, rapid collapse and clockwise twist of the tower occurred. Tower 3, although standing, was visibly distorted.

Photograph 7 - Debris Field – Episode 2 – tower numbers indicated

The following occurred in Episode 3 in the sequence indicated. (Photograph 8)

1. Tornado moved northward – rapid and confined southerly ("inflow") winds attack from the south.
2. The wooden decking and rails, momentarily connected to Towers 12, 13 and 14 during Episode 1, separated from the structure.
3. Towers 12, 13 and 14 and adjoining girder spans, having separated from the bearings during Episode 1, twisted and subsequently collapsed in a southerly direction.
4. The final remaining girder span, momentarily affixed to Tower 15, oscillated laterally several times at the Tower 15 connection, eventually separating and rotating upside down before impact. The rails remained attached and "hung" from Tower 15.

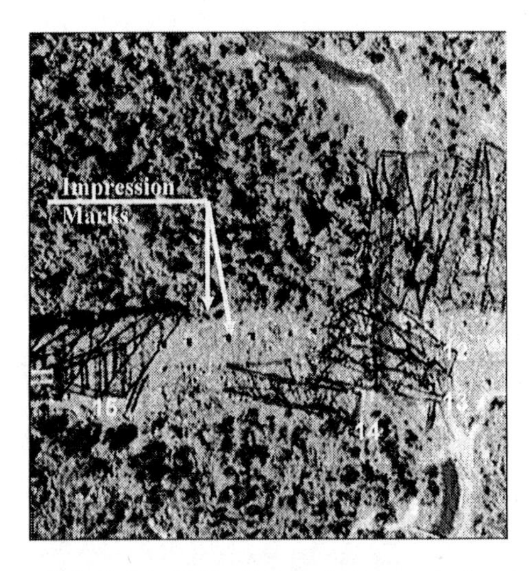

Photograph 8 - Debris Field – Episode 3 – tower numbers indicated

From a forensic perspective, the most puzzling aspect of the debris field at the time of field investigation was the location and orientation of Towers 12, 13 & 14 after collapse. The towers clearly were affected by the winds during Episode 1, but were positioned on top of all Episode 1 debris and were oriented in a direction opposing all other structure debris. Interpretation of the debris clearly indicates the effects of "inflow" wind throughout the collapse cycle. Examination of impression marks on the ground surface displayed on the high resolution photography depicted evidence of initial motion westward during Episode 1, attributable to easterly ("inflow") winds. Evidence of the effects of southerly ("inflow") winds included the overall southerly vector orientation of a significant region of the debris field.

Conclusions/Recommendations

The *Board of Inquiry* investigation and subsequent report serve as a model for the development of a comprehensive forensic investigation within a compressed time frame, typically associated with a catastrophic collapse. Logistical support, high resolution aerial photography and fractographic examination are necessary components of any catastrophic collapse investigation. The *Board of Inquiry* investigation demonstrates the need to identify potential and applicable forensic markers in the earliest stages of an investigation. Eyewitness testimony, often an overlooked component of forensic investigation, was invaluable for this situation. The most revealing eyewitness account was that of the field superintendent who recalled the number of independent concussion sounds during the collapse. This testimony confirmed the otherwise inferred episodes and sequence of collapse ascertained by the investigative team.

The *Board of Inquiry* investigation disclosed unusual wind phenomena peculiar to extreme wind events associated with tornados. The moving tornado vortex with its strongest "leading edge winds" may be accompanied by "inflow" winds. In certain geographical settings such as the mountainous region of the Kinzua site, the "inflow winds" can produce concentrated, straight line winds in distinctly differing vector direction from the "leading edge winds". Thus, a structure can be attacked with design strength wind pressures from two differing directions within a small increment of time.

The *Board of Inquiry* investigation concluded that the ongoing repair of the visibly deteriorated tower elements was not influential in any respect and did not otherwise affect the order or sequence of the collapse. The debris field contained many structural members which were strengthened during the ongoing repair program.

The *Board of Inquiry* investigation underscores the need for owners of tall viaduct, trestle structures to critically examine all anchor bolt tie-down components of the superstructure system. The *Board of Inquiry* comprehensive report identifies these structures as "wind susceptible" and offers specific recommendations to owners of these "wind susceptible" structures to [1] recognize that wind induced tensile forces within the tie-down system may be critical to the survival of the structure in an extreme wind event, [2] inspect all elements of anchor bolts systems, [3] determine if "weak link" or otherwise hidden components are present, [4] re-evaluate capacity of the tie-down systems with respect anticipated wind speed under extreme events and [5] schedule tie-down retrofits in advance of all other rehabilitation activities. (Leech et al. 2003, Leech et al. 2004, Leech et al. 2005)

Epilogue

The Commonwealth of Pennsylvania has no immediate plans to restore the twenty-three collapsed spans, but intends to strengthen the towers supporting the remaining eighteen spans and offer the site as an interpretive center to the public.

Acknowledgements

The authors wish to acknowledge the following Board of Inquiry Members whose assistance was invaluable in this forensic investigation including:

Mr. James Epply, P.E., Mr. George DiCarlantonio, Ms. Denise Kelly, P.G., Ms. Ginny Davison, Mr. Greg Sassaman, Mr. Barrett Clark – Pennsylvania Department of Conservation and Natural Resources; Mr. Craig Beissel, P.E., Mr. Jim Surkovich, P.E., Mr. Jerry Bruck, P.E. - Pennsylvania Department of Transportation; Mr. Joseph Portelli, Paul Markowski, Ph.D., Pennsylvania State University; Mr. Russell Ricker, III, P.E., Mr. Merl Steimer, P.E.; Mr. Vincent LaCross, P.E. - Gannett Fleming, Inc.

References:

ANSI/ASCE 7-98, American Society of Civil Engineers, *Minimum Design Loads for*

Building and Other Structures, 1998 (American Society of Civil Engineers: Reston, VA).

AREMA *Manual for Railway Engineering*, 2000 (American Railway Engineering and Maintenance of Right-of-Way Association: Landover, MD).

Erie Railroad Company, Bradford Division, Design Drawings, Kinzua Viaduct, (Elmira Bridge Co., Ltd: Elmira, NY), March, 1900.

Grimm, C. R., The Kinzua Viaduct of the Erie Railroad Company, Paper No. 899, *Transactions of the American Society of Civil Engineers*, January, 1901.

Grumm, R. H. The Mesoscale Corrective Vortex Severe Weather Event of 21 July 2003, *NOAA Local Research Paper*, August 2003(NOAA: Washington, DC).

Kaufmann, E., Connor, R., Evaluation of the Anchor Bolt Components, Kinzua Bridge Collapse, ATLSS Engineering Research Center Report (published as an appendix to Board of Inquiry Investigative Report), Bethlehem, PA, December 2003.

Leech, T. G., Eppley, J.A., et al., Board of Inquiry Investigation, Report on the July 21stCollapse of the Kinzua Viaduct, McKean County, PA, Pennsylvania Department of Conservation and Natural Resources, Harrisburg, PA, December 2003 <http://www.dcnr.state.pa.us/info/kinzuabridgereport/main.html>.

Leech, T.G., Ricker, III, Russel L., Comoss, E.J., Eppley, J.A., Anatomy of a Collapse – The July 21st Collapse of the Kinzua Viaduct, IBC 04-61, *Proceedings of the Twenty First International Bridge Conference*, Engineers Society of Western Pennsylvania, Pittsburgh, PA, June 2004.

Leech, T. G., McHugh, J. G., DiCarlantonio, Lessons from the Kinzua, *Proceedings of the 3rd New York City Bridge Conference,* Bridge Engineering Association, September 2005.

Markowski, P., *Meteorological Aspects of the 21 July 2003 Kinzua Viaduct Storm*, The Pennsylvania State University Report (published as an appendix to Board of Inquiry Investigative Report), University Park, PA, December 2003.

Shank, W.H., Historic Bridges of Pennsylvania, 1980 (American Canal and Transportation Center: York, PA).

Pennsylvania Department of Transportation, *Publication 220*, Harrisburg, PA, July 2001.

US Department of Congress, National Oceanic and Atmospheric Administration, National Weather Service, A Guide to F-Scale Damage Assessment, Silver Spring, MD, April 2003.

Wrigley Field: Forensic Investigation and Structural Analysis of the Friendly Confines

Jonathan E. Lewis[1], Associate Member, ASCE
Arne P. Johnson[2]
Gary J. Klein[3], Member, ASCE

Abstract

During the 2004 baseball season, instances of concrete falling from the upper stands of Wrigley Field were reported, generating local and national media coverage. Several engineering firms were retained by the Chicago Cubs to determine the cause of the falling concrete, develop appropriate remedial measures, and review the general condition of the stadium structure. This paper describes the investigation and structural analysis carried out by the authors. Structural analysis of the grandstands, on-site strain and vibration measurements during a crowded baseball game, and laboratory testing did not reveal any systemic structural deficiencies. The cause of the falling concrete was attributed primarily to corrosion of metals embedded in chloride-contaminated concrete near joints, thermal movements restrained by welded connections, and unintended concrete-to-concrete contact between precast sections. Recommendations for remedial action were provided. The case study underscores the value of forensic engineering techniques to isolate structural performance problems and to develop appropriate and practical remedial solutions.

Introduction and Background

Wrigley Field in Chicago, Illinois is a 41,000 seat outdoor stadium that is the current home of the Chicago Cubs baseball team. Originally constructed in 1914, Wrigley Field is the second oldest ballpark in the major leagues. The ballpark's signature features, including its ivy-covered outfield walls, hand-operated center field

[1] Engineer, Wiss, Janney, Elstner Associates, Inc., 330 Pfingsten Road, Northbrook, IL 60062, Phone: 847.272.7400, Fax: 847.291.9599, jlewis@wje.com

[2] Consultant, Wiss, Janney, Elstner Associates, Inc., 330 Pfingsten Road, Northbrook, IL 60062, Phone: 847.272.7400, Fax: 847.291.9599, ajohnson@wje.com

[3] Principal and Executive Vice President, Wiss, Janney, Elstner Associates, Inc., 330 Pfingsten Road, Northbrook, IL 60062, Phone: 847.272.7400, Fax: 847.291.9599, gklein@wje.com

scoreboard, and neighboring residential rooftop bleachers, make it both a local and national landmark, synonymous with the national pastime.

During the 2004 baseball season, three separate incidences when a small piece of concrete fell from the underside of the upper stands (locations indicated with white markers in Figure 1) generated a flurry of local and national media coverage. Each incident was investigated by structural engineers retained by the Cubs, and protective netting was hung above public areas to catch any new spalls that might occur during the remainder of the season.

Following these incidents, local building officials questioned the cause of the concrete spalling as well as the general structural integrity of the stadium's structural frames. Also questioned was the structural significance of crowd-induced vibrations at the cantilevered portion of the upper deck framing. The Cubs commissioned a structural investigation to respond to these concerns.

Figure 1. Main grandstands of historic Wrigley Field, Chicago, Illinois, showing approximate locations of spalled concrete during 2004 season

To determine the cause of the spalling concrete and to respond to the concerns expressed about the performance of the stadium structure, the authors conducted investigative studies, including the following:

- General inspections of the overall stadium structure
- Close-up examination and material analysis at the three locations from which material fell
- Structural analysis of typical grandstand frames to verify structural integrity and review the effects of structural behavior and crowd-induced vibrations on the upper deck concrete
- Field instrumentation of the grandstand structure and collection of static and dynamic strains and accelerations during a crowded baseball game

By synthesis of the results from these various studies, appropriate and practical remedial measures were recommended, as will be described in this paper.

Description of Structure

Wrigley Field occupies a full city block (approximately 600 feet by 600 feet in plan) on the near north side of Chicago. The main roof atop the upper deck stands is about 100 feet above the playing surface. Original structural drawings for the stadium are no longer available, but based on various historical reports, the stadium's facade, lower deck, and an original lower roof were constructed circa 1914. In the late 1920s, the original roof was removed and the upper deck, ramps, and higher roof were constructed. The bleacher stands and bleacher ramps were added in the 1930s. At some point, reportedly in about the 1960s, the original cast-in-place tread-and-riser system for the upper deck was replaced with a precast concrete tread-and-riser system. The mezzanine suites, press box areas, and lights atop the roof were added in 1988-89 (apparently to the chagrin of the baseball gods as the first night game was rained out).

Structural framing for the lower stands consists of cast-in-place concrete tread-and-riser sections that span between radially-oriented lines of raker beams and columns. Near the playing field, the raker beams and columns are cast-in-place reinforced concrete. Elsewhere, the raker beams and columns are built-up structural steel sections. The stadium columns are reportedly founded on spread footings. The concourse floors at grade level are typically concrete slabs or brick pavers supported on-grade. Figure 2 shows a portion of the main grandstand framing along the third base line.

Figure 2. View of main stands and mezzanine suites along third base line

The upper stands consist of lightweight precast concrete tread-and-riser sections that span between radially-oriented steel frames and trusses. There are 59 trusses in total, each about 96 feet long and typically spaced approximately 19 feet apart. Each truss has a 33 foot cantilever at the end closest to the playing field. In order to improve spectator sightlines, not every truss is supported by a column at the end closest to the playing field. Rather, a second truss running parallel to the playing field transfers the reaction from every other truss to the columns supporting the adjacent trusses. Such a configuration was typical for steel-framed stadiums of this vintage. A similar truss configuration supports the timber roof structure and rooftop lighting. The individual members of the main and transfer trusses are typically double angles with riveted connections. Figure 3 is an overall cross section through the main grandstand framing.

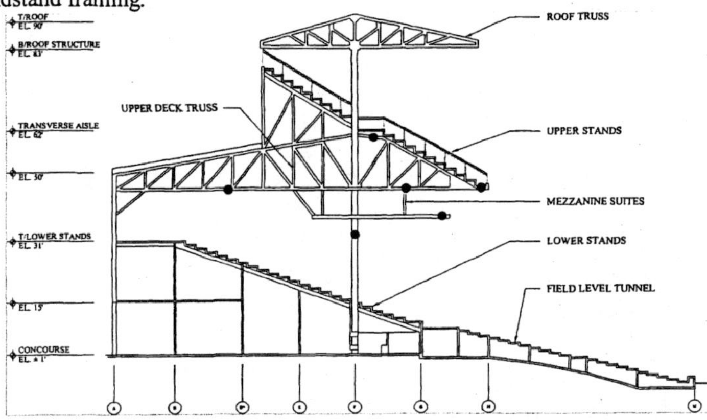

Figure 3. Typical section through grandstands showing locations of field instrumentation (described below)

The concourse level, lower stands and upper stands are interconnected by eight ramps located at the perimeter of the main stands. The ramps are constructed of cast-in-place concrete slabs that span between reinforced concrete or concrete-encased steel beams. The ramps are supported by the building columns or suspended from the upper deck trusses, and provide an added means of lateral support for the superstructure.

Also visible in Figure 2, the roof over the upper stands is constructed of structural steel trusses that support heavy timber wood purlins and wood decking. The roof at the back of the upper deck trusses (between ramps) is also constructed of wood purlins and decking.

Perhaps the best-known portion of the structure is the outfield bleachers. The bleacher stands are composed of a cast-in-place reinforced concrete tread-and-riser system that spans between concrete or structural steel raker beams and columns. Two cast-in-place concrete ramps framed with steel beams service the bleacher stands. A multi-story steel-clad manual scoreboard above the centerfield bleachers is supported by steel columns extending down through the stands below.

Many improvements and modifications have been made to the structure over its lifetime. In 1988-89, the original (circa 1926) steel frames were strengthened and extended with additional structural steel elements to support the new mezzanine suites and press box areas, an additional row of seating at the bottom of the upper stands, and rooftop lighting. Visible in Figure 2, the mezzanine suites are suspended via hangers from the underside of the main trusses. Many built-up steel columns were also strengthened in the late-1980s.

Investigation of Falling Concrete

Close-up examinations. The authors carried out close-up examinations of the precast concrete at each of the three reported spalls, the general locations of each noted on Figure 1. At Location No. 1, a piece of concrete had spalled from the lower corner at the end of a precast tread-and-riser section, as shown in Figure 4. The spall was shallow (maximum depth of about 1/2 inch) and had a total length of about 12 inches. The piece that fell was likely only a small portion of this spall. A heavily corroded reinforcing bar was exposed along the length of the spall. Concrete clear cover over this bar was 1/4 to 1/2 inch. On the top side of the precast tread at this location, there was a crack in the tread parallel to the joint at the end of the precast section. A few inches away from the end of the precast section, the concrete was sound. The spall location is largely obscured from view from below, because it is directly above the top flange of the upper deck truss top chord.

Figure 4. Precast concrete distress at Location No. 1

At Location No. 2, a thin piece of concrete, approximately 1/8-inch thick by 4 inches square, had spalled from the underside of a precast tread-and-riser alignment bracket. The spalled area is shown in Figure 5. The remaining portion of the alignment bracket was sound, though heavy water staining was present on and immediately surrounding the alignment bracket. Also, the face of the alignment bracket was in contact with the back side of the next precast riser. Typically, there is a gap between the alignment bracket and adjacent riser. No reinforcing steel or corrosion staining was exposed at the surface of the spall.

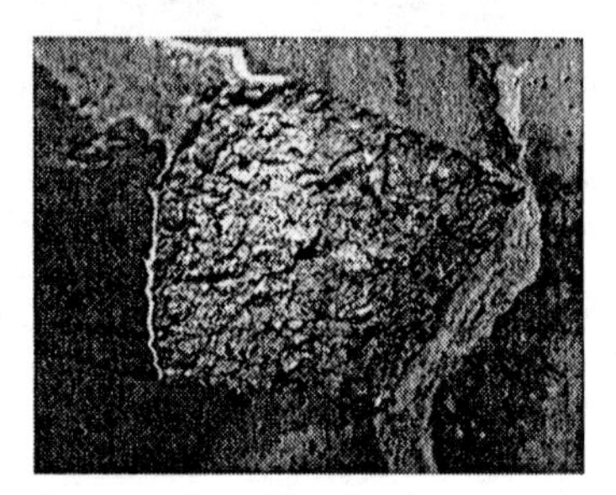

Figure 5. Precast concrete distress at Location No. 2

At Location No. 3, a small piece of concrete had spalled around a threaded rod that was anchored into the underside of a precast concrete tread. The threaded rod suspended a plumbing pipe below. At the time of the authors' inspection, the concrete at this location had been excavated much deeper than the spall in order to expose the embedded reinforcing steel in preparation for a concrete patch. The condition of this location is shown in Figure 6.

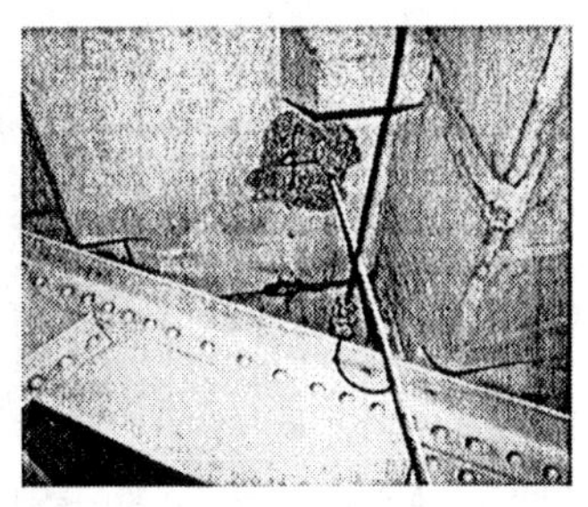
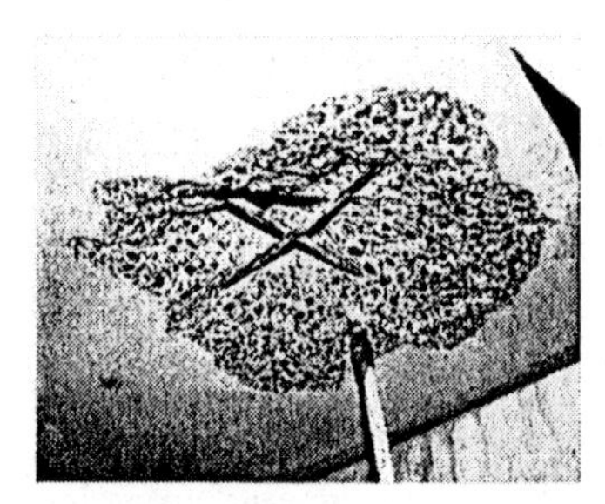

Figure 6. Precast concrete distress at Location No. 3

Laboratory Analysis. The authors performed laboratory analyses on cores and chunk samples removed from the three spall locations and from other general areas of the upper deck concrete. Samples were analyzed petrographically for depth of carbonation, and for chloride content, with the following results:

- The upper deck concrete consists of expanded shale lightweight aggregate with a maximum nominal size of 3/8 inch. The fine aggregate is a mixture of expanded shale and natural sand. The water to cement ratio and air content were estimated to 0.48 to 0.54 and 5 to 6 percent, respectively. No inherently unstable or incompatible materials were detected in the aggregate or paste systems.
- The depths of carbonation typically extended near or past the depth of the embedded reinforcing steel, which creates an environment favorable for corrosion.
- Chloride content of the samples removed from the distressed regions near joints was very high, from 0.26 to 0.53 percent by weight of concrete. Values in excess of approximately 0.03 percent indicate a probable corrosive environment for embedded steel. The base level of acid-soluble chloride in the concrete was very low (0.007 percent by weight of concrete, just at the lower detection limit), indicating that chloride was not purposely added to the concrete during original construction.
- The distress at Location No. 1 where concrete fell was attributed to chloride-induced corrosion of embedded reinforcement with shallow cover.
- The cracking in the sample removed from Location No. 2 was judged to be the result of horizontal bearing stress on the edge of the spalled surface (from unintended concrete-to-concrete contact with the next riser).
- Analysis of the samples removed from Location No. 3 indicated that the concrete had been damaged to some degree when a power-driven anchor for suspending a pipe was originally installed at this location. Also, the anchor had been driven a

slight angle to the suspended load, such that vertical loading on the anchor induced lateral forces on the concrete at the anchorage. The damage went undetected, and eventually a small piece of the concrete around the anchor spalled.

General Structural Inspections

Upper Deck Precast Concrete. In addition to the detailed examinations at the three spall locations, representative areas of the upper deck concrete were accessed for close-up examination and hammer sounding, and the remaining areas were visually inspected from the closest vantage point. Concurrently, other engineers retained by the Cubs were conducting a comprehensive "hands-on" inspection of all overhead areas to identify and remove all loose material.

In general, the "field" of the precast sections (that is, away from the end bearing regions and away from the alignment brackets) were in good condition and exhibited no significant distress. Water staining was present on many of the members, but corrosion staining was present only at isolated locations. Most of the deterioration observed was at the ends of the sections or at the alignment brackets. Some unsound previous patches were also present. Various stages of all of these categories of distress were present, indicating that the distress had been developing over a long period of time. The distress observed can be categorized as follows:

Category 1 - Cracks above embedded steel plates at bearing connections. The distress is caused by volume change movement of the concrete (shrinkage and thermal expansion/contraction), which is retrained by the embedded plates and welded connections at both ends of the members.

Category 2 - Distress at alignment brackets. A gap is typically present between the bracket and the face of the adjacent riser. However, where the gap was tight, the unintended concrete-to-concrete contact had often caused bearing failure of the unreinforced bracket.

Category 3 - Distress at end joints. Leakage through the joints over time has resulted in corrosion of embedded steel elements at the panel ends, including reinforcing steel with shallow cover and steel inserts presumably used for lifting during erection. Corrosion

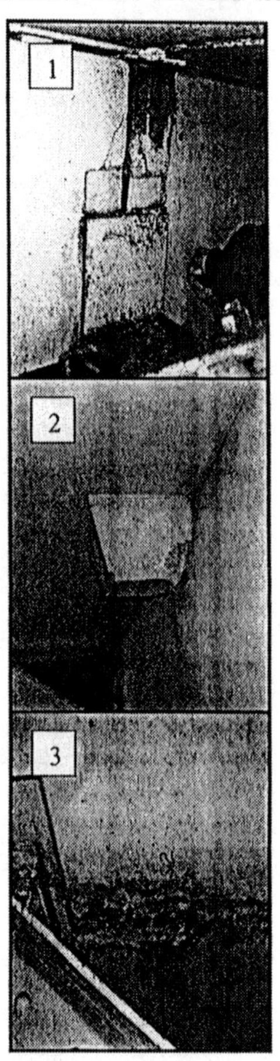

Figure 7. Examples of typical precast distress

of the embedded steel has led to cracking and spalling of the bottom corners of some of the precast members.

Category 4 - Unsound previous patches. Some previous patches were found to be unsound due to being installed in a manner that did not include chipping behind reinforcing bars or provision of mechanical anchorage. These patches were removed during the hands-on inspection.

Two of the three pieces that fell were of categories 2 and 3 above; the third came from an unusual condition (at an anchor for a pipe support).

Steel Framing at Upper Deck and Roof. The riveted steel trusses and framing members appeared to be in reasonably good condition, except for localized areas of peeling paint and light to moderate surface corrosion. Steel repairs addressing the localized corrosion were being installed at the time of this investigation. Non-destructive testing on previous welded strengthening repairs revealed no indications of fatigue-related cracking.

Ramps. At the time of this investigation, repairs to the ramps throughout the stadium were underway. The present investigation included visual examination of the ramps, review of the test results and repair design performed by others, and structural analysis of a typical ramp. Spalling of the cast-in-place concrete of the ramps was attributed primarily to corrosion of embedded reinforcing steel due to high chloride content in the concrete, most likely from application of deicing salts over the years. Full-depth concrete patches were being installed to address the spalling. Numerous inspection openings revealed that appreciable section loss in the embedded steel framing was limited to the areas where the members exited the concrete encasements. Welded repairs were being installed to address this localized corrosion.

Structural Analysis

In order to address expressed concerns about the overall structural integrity of the ballpark, the authors performed a limited structural analysis of typical steel framing supporting the seating and roof levels, as well as a typical pedestrian ramp. A combination of computer modeling (finite element analysis) and hand calculations was used to evaluate the adequacy of typical structural components with respect to the Chicago Building Code. Information on existing structural members was obtained from several sources, including miscellaneous drawings for previous structural improvements and additional field measurements made by the authors. Where no existing information was available for material strengths, reasonable values were assumed based on historical data.

Methodology. A three-dimensional finite element model of a portion of the steel superstructure along the third base line was created to evaluate the adequacy of typical steel framing. The analysis included the effects of various strengthening repairs employed over the years and also reflected the added loads imposed on the main trusses by the mezzanine suites and associated pedestrian ramps. Loads applied in the model included an estimated weight of the structural and architectural components, code-specified live loading for seating areas and ramps, and snow and wind loads per the Chicago Building Code. The effects of unbalanced snow loading on the upper roof and unbalanced live loading on the upper deck stands were also considered.

For purposes of the analysis, the yield strength of the steel was originally assumed to be 30,000 psi, typical for 1915-1930 vintage carbon steel of Grades A7 or A9. When preliminary results of the structural analysis showed that some members were overstressed for the assumed minimum yield strength, steel samples were removed from the stadium structure in order to determine actual steel properties. Material testing of twelve steel coupons showed a mean yield stress of 39,000 psi. A reliable yield stress of 35,000 psi was calculated from the sample data following customary steel design procedures. All analyses were modified to incorporate this higher yield stress of 35,000 psi. The ultimate strength of the rivet material was assumed to be 46,000 psi, the typical minimum strength for rivet steel of this vintage. Relevant load combinations were obtained from AISC-ASD 1989 and ASCE 7. Despite exposure to an aggressive climate for at least 75 years, very few structural steel members exhibited any significant deterioration. As such, full section properties were used in the analysis.

Typical framing components of a representative pedestrian ramp (concrete slab, steel stringers and hangers) were also analyzed to verify that they could sustain the 100 psf live load required by the code. The Cubs also requested that the analysis consider the effects of forklifts used to service various upper deck vendor stands.

Strength Evaluation of Grandstand Framing. Imposed demands in several of the truss members were found to slightly exceed the code-specified limits when a full 100 psf live load was applied to the stands, mezzanine suites, and ramps simultaneously, which is the governing case per the Chicago Building Code. However, since the main stands consist of fixed seating, for which ASCE-7 prescribes a 60 psf live load, and the spatial layout of the mezzanine suites makes a 100 psf live load exceedingly unlikely, additional analyses with a characteristic live load were performed. When a 60 psf live load was applied to the grandstand framing, the stress levels in the most heavily loaded members were within or very near the code-specified limits. Using forces from the finite element analysis, several typical connections between the various truss members were evaluated and found to be adequate for the imposed demands. Live load deflections of about 1-1/2 inches at the tip of the cantilevered upper deck stands were computed (using 100 psf live load in stands).

Cases of unbalanced live loading on the upper stands were included in the analysis. While these unbalanced cases were found to govern for some members of the main trusses, no overstress conditions were identified for the unbalanced load cases. Deflections at the field edge of the upper stands and mezzanine suites increased only slightly upon consideration of unbalanced live loading on the main truss cantilever.

Several of the welded strengthening repairs of the main truss members contained fatigue-sensitive details. The calculated forces from the finite element model were used to determine the live load stress ranges at several critical locations.

Uniform live loading of 100 psf in the stands and ramps creates maximum tension and compression stress values of approximately +1.0 ksi and -4.3 ksi, respectively. Under unbalanced live loading conditions, the maximum live load tensile and compressive stresses increased to approximately +3.2 ksi (ramps loaded) and -6.4 ksi (cantilever loaded), respectively.

Framing components of the ramps were found to be adequate for the required 100 psf pedestrian loading. Steel framing components were also found to be adequate for the suite of forklifts employed by the Cubs, but the concrete slab lacked sufficient strength to sustain the forklift loadings—most likely explaining the areas of significant cracking on various ramps throughout the park. The Cubs were subsequently informed that forklift traffic should be prohibited on the ramps, and repairs to deteriorated portions of the ramps were made.

Vibration Analysis. Mode shapes and periods of vibration for the first several modes of the structure were obtained from the finite element model. The primary mode shape involving a vertical vibration of the stands, similar to what would be expected from crowd-induced motion, was found to have a natural frequency of approximately 3.6 Hz. The measured natural frequency of this mode (as determined via accelerometers during an actual game), was somewhat higher (approximately 8 Hz), most likely due to the stiffening effects of the precast stands and non-structural elements.

Comparison of Computed and Measured Stresses in Instrumented Members. In order to determine an estimate of the actual live load in the stands during a nearly-sold out game, the computed stresses from the finite element modeling (considering 100 psf live loading) were compared to the stresses obtained from the instrumented structural members. Based on the measured information, an actual live load in the upper deck stands of approximately 25 psf was obtained. Similarly, the actual live load in the mezzanine suites was determined to be approximately 15 psf. Both these loads are significantly less than the live loads used in the structural models discussed above, but appear reasonable based upon the geometry of the fixed seating in the upper deck stands. Assuming the average fan weighs 175 pounds and occupies about 5 square feet, a live load of around 35 psf in the fixed seating areas is obtained. Accounting for vacancies in the aisles (which would not be fully loaded concurrently with the seating areas) yields a live load of around 25 psf, similar to that determined from the instrumentation.

Field Instrumentation

Vibration and Dynamic Strain Measurements. In order to better understand the structural behavior of the main stands, a typical grandstand frame was instrumented and continuously monitored during a nearly sold-out home game during the 2004 pennant race. Seven strain gages and three accelerometers were installed on the upper deck steel trusses and a supporting steel column along the third base line. This portion of the structure was also subsequently analyzed with a finite element computer program. Strain gages were 1/4 inch long, bondable foil gages with high strain sensitivity. The accelerometers measured the crowd-induced dynamic response of the structure in terms of acceleration (g's) and frequency (Hertz). Accelerometers had a 100 mV per g sensitivity, ±50 g amplitude range and a 0.7 Hz to 6.5 Hz frequency range. The vibrations measured were compared to human perceptibility thresholds. The locations of the accelerometers and strain gages are shown on Figure 3.

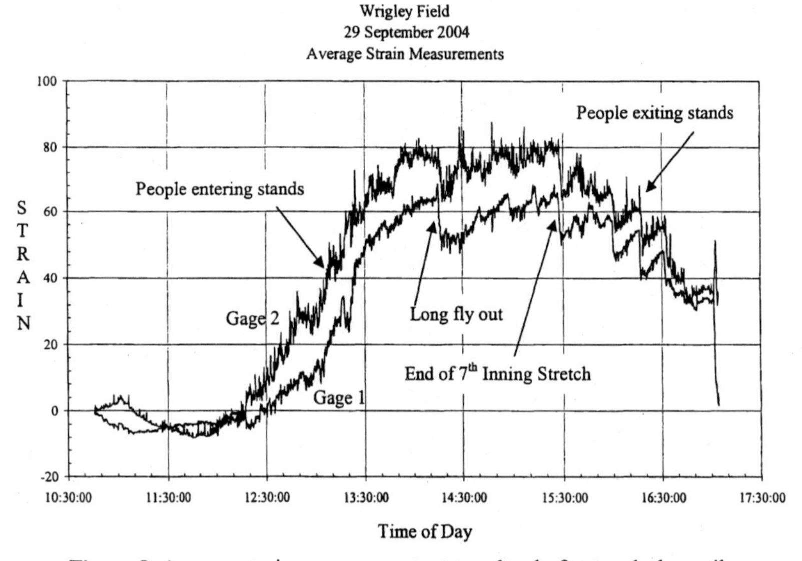

Figure 8. Average strain measurements at top chord of upper deck cantilever

A centrally located data acquisition system analyzed and displayed the data in real-time and stored detailed data for further analysis and comparison to results predicted by the finite element modeling. Data collection began well before the game and continued until the crowd had vacated the stadium.

The instrumented game was a very close contest that went to extra innings and provided several exciting moments which were reflected in the acquired data. With only a few seats vacant in the instrumented area, the maximum static live load strain measured in the upper deck truss members was 2.6 ksi. Dynamic strains, that is, short duration peaks in strain due to sudden or rhythmic crowd movements, were low, corresponding to an additional truss member stress of up to 0.4 ksi. The maximum dynamic stress occurred when the Cubs pitcher—of all people—hit a home run in the bottom of the third inning. Figure 8 shows the average strains recorded at two locations on the upper deck cantilever throughout the afternoon with significant game-related events denoted. It should be noted that all of the strain gages did not return to their starting position when people vacated the stands. This effect was attributed to thermal changes over the course of the day.

Peak accelerations measured due to vibration of the stands occurred during the third inning home run, a long fly-out in the fourth inning with the bases loaded, and during the seventh inning stretch where the crowd engaged in singing the traditional "Take Me Out to the Ballgame." A slightly perceptible side-to-side swaying of the stands was detected by the authors during the seventh inning stretch. Overall, peak accelerations of up to about 0.008g, with a corresponding frequency of about 3 Hz, were recorded at the tip of the upper deck cantilever. Displacement of the structure due to dynamic excitation (that is, in addition to the static deflection) was up to 1/32 inch, again very low. Based on comparison with published standards, the measured vibrations should not be perceptible to people involved in the crowd activities. Measured stresses in the upper deck structure due to the near-capacity crowd corresponded to an actual live loading of about 25 psf on all public areas (stands, aisles and ramps).

Findings and Recommendations

Static and Dynamic Response of Structure. In general, the structural integrity of the typical upper deck trusses can be described as adequate, but not robust. When code-level design loads are considered (100 psf in the stands; 100 psf or 60 psf in the mezzanine suites; and 25 psf snow load on the roof), the calculated stresses in the truss members are generally within or very near the limits allowed by the Chicago Building Code.

The actual live load during the September 29 game was calculated by comparing the measured stresses to those predicted by the analysis. When almost all the seats were occupied, the effective crowd load was about 25 psf in the main stands and even less in the mezzanine suites.

In summary, no abnormalities in the static or dynamic behavior of the trusses were detected that would initiate distress in the upper deck precast stands.

Cause and Significance of Spalling Concrete. Several details of the precast tread-riser section were found to be vulnerable to damage and spalling, although none of the observed damage impairs the load-carrying performance of the stands. The general categories of the distress and the underlying causes for the damage and spalling are as follows:

- Cracks and spalls along joints at ends of precast sections, caused by water leakage at the joints, chloride contamination of the concrete at the joint and corrosion of embedded steel elements with shallow cover. Chloride content at the joints, apparently due to past use of deicing salts, was ten times that needed to initiate corrosion.
- Cracks above embedded steel plates at bearing connections, caused by volume change movement of the concrete, which is restrained by the steel plates welded to the bearing connections at both ends of the members. Strength loss at these locations is not an issue, but further damage will develop due to restrained thermal movements.
- Cracks and spalls at bottom surfaces of alignment brackets, caused by bearing stresses induced by unintentional concrete-to-concrete contact. Additional damage may develop at these locations due to normal structural deflections and volume change movement.
- Previous patches that have become unsound.

In addition to the above, there are a variety of other locations where damage has or may occur. For example, one of the three incidents of concrete spalling during the 2004 season occurred where a threaded rod was anchored (power-driven) into the underside of a precast concrete tread. Concrete in the vicinity of the anchor was damaged when the anchor was originally installed, and the anchor had been installed at a slight angle.

Recommended Measures for Structural Safety. The distress conditions observed in the upper stands did not impair the load-carrying ability of the precast sections. As such, the only safety hazard posed by the distress conditions was the potential for non-structural pieces of concrete to fall.

Concurrent with this investigation, the hands-on inspection by others identified and removed all potentially loose concrete over public areas of the stadium. As such, the inspection and removal program mitigated the spalling hazard to a point where the safety of the stadium with respect to the potential for falling concrete was comparable to that of other similar, well-maintained, exposed concrete structures. Mitigation of spalling hazards by regular inspection and removal of unsound materials is an acceptable, standard practice for both buildings and bridges. To maintain the stadium in a safe condition with respect to the potential for falling concrete, the hands-on inspection must be repeated regularly.

Optional Measures for Increased Safety. Though the regular inspection and removal program provide a level of safety with respect to falling concrete judged acceptable and comparable to that of other well-maintained, similar structures, this level of safety does not mean that the risk of concrete cracking and spalling is absolutely zero. There are thousands of potential sites from which spalling can occur and the deterioration mechanisms are ongoing. It is therefore possible that small pieces of concrete could become dislodged fairly soon after a thorough hands-on inspection. An optional measure to provide an even greater degree of safety with respect to falling material would be to install an engineered protective barrier system (such as custom netting, solid or perforated screening, mounted panels, or the like) below the upper stands. Such a protective barrier system must not be viewed as a substitute for the regular hands-on inspections, and the barrier system must be removable to allow for the close-up inspections in all areas.

Conclusions

1. Overall, the investigation revealed that the structural system at Wrigley Field is in reasonably good condition considering its age, construction type, and exposure. As with any aging structure of this type, maintenance repairs need to be identified and implemented on an ongoing basis.
2. The typical grandstand frames were found to have adequate strength to safely support the 100 psf code-level design loading.
3. The level of crowd-induced vibrations measured at Wrigley Field was very low, well within the acceptable range for occupants, and well below levels that could initiate structural damage.

4. Investigative studies concluded that the cause of spalling concrete is mainly corrosion of steel embedded in chloride-contaminated concrete, thermal movements restrained by welded connections, and unintended concrete-to-concrete contact between precast sections. None of these conditions affected overall structural integrity. Inspection and removal of loose material were judged to have mitigated the concrete spalling hazard to a level comparable to other similar, well-maintained, exposed concrete structures. Nonetheless, minor damage of this type will continue to develop and normal deflection of the structure makes it more likely that pieces of unsound concrete would dislodge during a ballgame than at other times. At the discretion of the Cubs, additional measures (protective barrier systems) were implemented to further mitigate the risk of spalling concrete, thereby providing an even higher degree of protection to the public.
5. The investigation underscores the need for regular inspection of all public structures with overhead concrete. However, it also underscores the need for the public, engineers, and building officials to recognize that, even for buildings designed and maintained according to codified practices, there is some risk that concrete will dislodge from exposed structures. This concept is similar to the approach taken in standard structural engineering design, in which there is a finite, although very small, probability that a properly designed structure will fail during its service life.
6. Testing performed as part of this investigation reveals the conservatism in code-specified live loads for grandstands. Measurements indicated an at-capacity crowd weight of about 35 to 40 psf within the fixed seating areas, which was only 25 psf when averaged over the grandstand area including aisles. These values are much less than the 100 psf required by the Chicago Building Code for all grandstands and the 60 psf required by ASCE-7 for grandstands with fixed seating. In the opinion of the authors, the Chicago Building Code is unduly conservative for fixed seating areas and the ASCE-7 provisions should be viewed as an upper bound.

To preserve this historic stadium for decades to come, a long-range plan coordinated with ongoing annual inspections was recommended, which is appropriate for all exposed concrete structures of this type.

A Case Study of Granite Cladding Distress

Deepak Ahuja, P.E., M.ASCE[1] and Matthew D. Oestrike, P.E., M.ASCE[2]

Abstract

Distress of building cladding is either a symptom of an underlying problem within the support structure or is a sign that the cladding is deficient in some way. Cladding distress typically affects the appearance of the building, may allow unwanted intrusion from the elements to affect and degrade interior materials, and/or may allow the potential for unsafe conditions to develop if conditions are left uncorrected. Like other building components that fail, cladding distress develops from a number of factors that often involves more than just a single cause. The reason why cladding fails and the extent of this failure is a continued topic of debate among engineers, architects, contractors, owners of buildings, and our courts.

Presented herein is a case of exterior cladding distress at a building that includes a search for reasons why the distress occurred based on forensic engineering methods and evaluation. This search concluded that the likely causes of distress for the subject structure were related to corrosion of embedded steel bars, differential movement due to dissimilar materials, excess water infiltration in conjunction with inadequate drainage at the exterior walls, changes in the as-built construction not represented on the plans, and inadequate design coordination/supervision prior to and during construction.

Introduction

The case involves a temple structure located in Texas that was built in 1994-1995. The majority of the temple consists of one and two-story cast-in-place columns and floors (or similar) with granite cladding covering its exterior wall, fascia, and soffit areas. At the back of and connected to the main portion of the temple structure is a shrine that is similarly constructed with cast-in-place concrete and granite cladding with decorative accents. Built above the roof of the shrine and at the back portion of

[1] Vice President, Nelson Architectural Engineers, Inc., 2740 Dallas Parkway, Suite 220, Plano, Texas 75093, 469-429-9000.

[2] Branch Manager, Nelson Architectural Engineers, Inc., 13231 Champion Forest Drive, Suite 112, Houston, Texas 77069, 281-453-8765.

the temple rests a dome-like structure. This dome structure consists of a reinforced gunite shell with exterior granite cladding. This dome structure and its exterior cladding distress is the focus of this paper.

The paper will discuss background information related to the structure's construction, details of the structure, and the steps that were undertaken to evaluate the structure and arrive at the most probable causes of cladding distress. These steps included considering and/or ruling out typical causes, making site/structure observations, comparing as-built construction to the plans, evaluating the effects of water within the cavity, evaluating the effects of corrosion, evaluating differential movement of dissimilar materials, field and material testing of the as-built construction, and arriving at conclusions based on the information available.

Background

For discussion purposes, this dome structure is considered a part of the temple, but will be indicated as "dome" within the context of this paper. The dome portion of the temple was designed by an architect based in India and a structural engineer based in Texas. The contractors were based in Texas, and the fabricator of the granite was based in Italy. The dome was built in stages with separate contractors for the shell of the dome structure, the granite cladding, and for decorative accents at localized areas of the dome. No architect or other design professional was involved during the construction of the separate stages of the dome construction. Figure 1 shows an elevation view of the dome, shrine, and remaining temple structure.

Figure 1. General elevation view.

The owners began to see separations at granite cladding joints and cracks at the granite pieces several years after the dome structure was completed. Most of the damage was located at the dome whereas the remaining areas of the temple sustained only minor damage.

Details of the Structure

The dome was constructed as a gunite reinforced concrete shell structure that extended approximately 43 feet above the top of the shrine and roof of the temple. Thus, the dome is more than 70 feet above the finished floor of the temple. The dome is a curved structure having two intermediate ring beams between the base and the top. The top of the dome is flat with an opening in the middle. The drawings indicate the dome to have a 1'–5" wall thickness at the base of the dome (above the shrine roof); it tapers to a wider thickness between the base and the first ring beam, and then tapers down in thickness at the first ring beam elevation. The wall thickness between the first ring beam and the top is indicated as 8". The dome walls are supported on a thickened structural two-way slab that makes up its floor. Four main cast-in-place columns that bear on straight-shafted piers support the floor slab. Three step-like projections, or "leaves," that follow the curved shape of the dome are located above the temple roof level on each of the four (4) faces of the dome and were constructed monolithically with the dome walls. An elastomeric coating was specified to cover the entire exterior of the gunite dome. After the curved gunite dome was completed, projections at the base of the dome were constructed out of brick and CMU block (neither of which were indicated on the drawings). Figures 2 and 3 show the exterior and the interior of the dome, respectively.

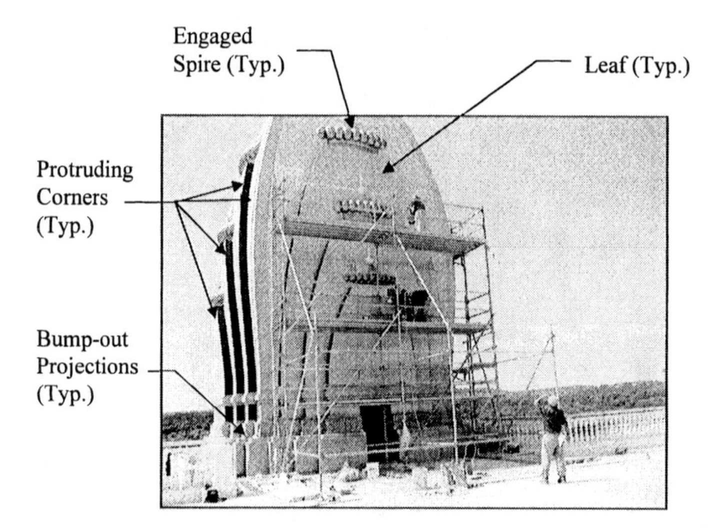

Figure 2. Dome exterior.

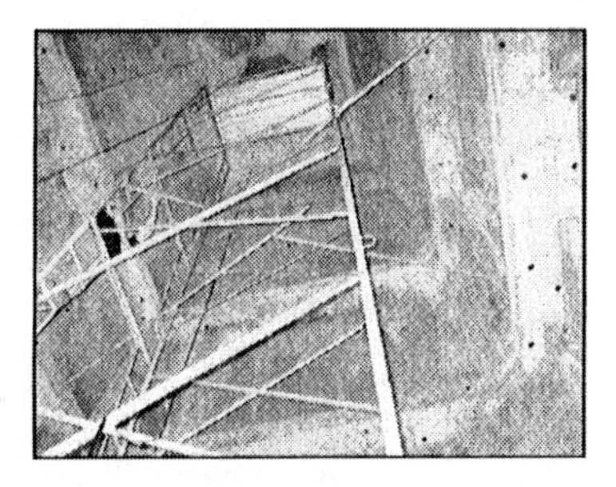

Figure 3. Dome interior (looking up).

The cladding on the dome as well as the temple soffits and fascia consists of 1 1/4" thick granite of varying widths and lengths with 1/4" wide by 1" deep continuous kerfs (slots at edges of granite pieces). Stainless steel anchors set (anchored) into the exterior side of the concrete shell of the dome are utilized to independently support the individual granite pieces. Mortar daubs at each anchor location were used during granite installation to facilitate construction of the veneer. The granite joints are comprised of sealant with backer rods used as the joint backing material. Figures 4 and 5 show typical plan and section details of the granite support at the protruding corners of the dome.

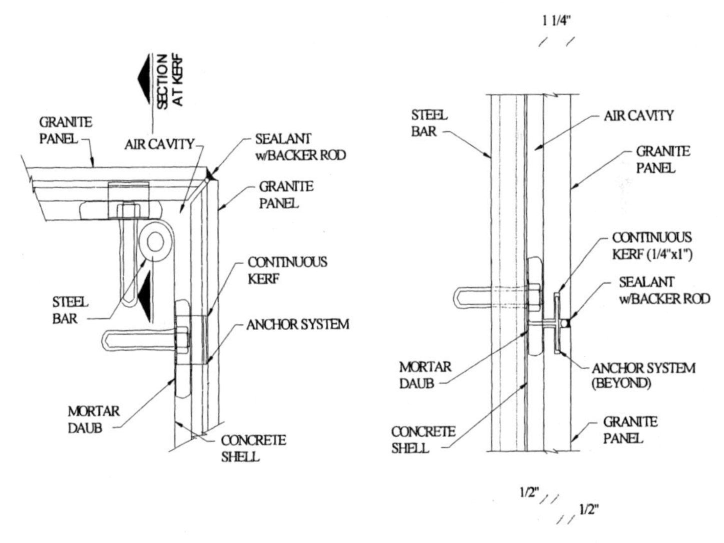

Figure 4. As-built plan at protruding corner. **Figure 5. Section at kerf.**

The sealant joints act as both a moisture barrier for the structure and as control joints for the granite cladding. An air space, indicated on the drawings to be 3/4" to 1 3/4" for the dome and 3/4" for the shrine, separates the exterior face of the concrete shell and the granite veneer. This air space is continuous around the perimeter of the dome except at the top of each of the "leaves," where the air space is interrupted by mortar being utilized to help support the flat granite pieces at these locations. Thus, the exterior wall is intended to behave as a cavity wall system allowing water that penetrates through the exterior veneer or water that condenses within the cavity to drain downward within the cavity space of the wall system. Drainage from the top of each leaf is allowed to occur within the cavity at either of the sides and/or at the front of each leaf. The drainage is intended to continue to the base of the wall where a continuous weep slot (horizontal gap at the base of the cavity running the length of the wall) was specified, so that excess water in the cavity would be allowed to drain. However, this weep slot is covered with a decorative horizontal banded trim that consists of beaded glass stones set in grout around the perimeter of the base (the horizontal banded trim was not indicated on the drawings, Figures 7-9 on next page).

After the granite cladding was installed, a decorative fiberglass "finial" (a crowning ornament or detail, Figure 1) was installed to cap the opening at the top of the dome. Additionally, decorative fiberglass engaged-spires were attached to the granite veneer exterior at the top of the "leaves," located on each face of the dome (Figure 2). A "catwalk" surrounds the base of the dome on three sides and overhangs the shrine that lies directly below the dome. Granite cladding installed at the fascia and soffit bands runs along the face of and below the catwalk.

Typical Causes of Cladding Distress

Cladding problems develop in a number of ways and can be caused by a combination of factors. Some of these include, but are not limited to, the following:

- Differential foundation movement
- Superstructure framing movement (unrelated to effects from the foundation)
- Storm effects from severe wind including windborne debris impact, rain, flood, lightning, hail, ice, etc.
- Movement of the various materials due to temperature, moisture, freeze-thaw, seismic, etc.
- Remedial repairs during construction or after construction completion
- Impact from vehicles, humans, equipment, etc.
- Construction defects of the cladding or components of the supporting structure including inferior or improper materials
- Age and deterioration of the materials
- Volumetric changes in the backup support material due to internal chemical reaction

All of these factors were initially considered prior to visiting the site. Some of these factors were ruled out after initial site observations were conducted as discussed below.

Initial Observations

Dome: An initial visual walk-through of the site confirmed that most of the damage was indeed located at the dome with little distress observed at the remaining portions of the temple. Most of the distress was noted at the protruding corners of the dome. The granite pieces at these corner joints were displaced outward on each side of the joint. Figures 6 and 7 show separations at the protruding corners.

Figure 6. Separation at protruding corner.

Figure 7. Separation at protruding corner near projection.

Additionally, localized areas of granite developed fractures in the pieces themselves. Evidence of accumulated and excess moisture including staining, efflorescence, and spalled glass beads at the horizontal banded trim at the base of the dome were noted. Figures 8 and 9 show typical areas exhibiting distress.

Figure 8. Moisture staining at base projection.

Figure 9. Granite fracture.

Additionally, moisture staining and efflorescence was noted throughout the interior walls (exposed gunite) of the dome. Hairline cracks were observed primarily at the upper half at the interior of the dome shell. The sealant was weathered with signs of alligatoring, cracking, and holes noted. Furthermore, corrosion was observed at the protruding corners of the dome shell at areas of the granite pieces that were removed prior to our site visit.

Based on the reported information and initial walk-through of the structure, there were no significant signs-of-distress that could be related to movement of the foundation or to differential movement between the dome and foundation. Hairline cracking at the top of the dome appeared to be related to normal shrinkage or not significant enough to cause the noted cladding distress. Additionally, the cladding distress was not noted after any significant storm event that one might correlate to severe wind, hail, lightning, etc. The site is located in a temperate climate zone, where freeze-thaw problems of masonry are not an issue. Additionally, there were no significant remedial repairs, seismic events, impact events that could have caused the noted distress. Furthermore, the cladding distress was on all sides of the dome; however, appeared to be localized to certain areas of the dome shell.

Shrine (located below Dome): The shrine cladding is protected from water intrusion by the presence of the catwalk at the roof elevation. This catwalk protects the top portion of the shrine cladding from any direct rain and any potential water infiltration that may enter the top of the cavity. Additionally, a continuous weep slot at the base of the shrine was observed allowing any water penetration through the cladding to drain out of the cavity. No fractured or displaced granite was noted. Some of the sealant appeared to have weathered/degraded.

Remaining Temple (excluding Dome and Shrine): The remainder of the Temple beyond the Dome and Shrine structure was evaluated for signs of distress. The majority of the granite cladding installed at the soffits and fascia on all terrace/roof levels, Porte-cochere, porch and at perimeter of the Temple building, revealed little to no distress. The granite in almost all of the locations was in good condition. The primary area of distress was at a few of the sealant joints at the protruding corners of the terrace fascia. These separations ranged from hairline to approximately 1/4" in width. Additionally, some weathering of the sealant was observed without any adhesion failure at the interface between the sealant and the granite substrate. Some moisture stains were noted at the base of the granite cladding walls.

As-Built Construction

The as-built construction of the dome revealed differences and/or modifications from the design drawings. The concrete shell appeared to have been constructed in accordance to the structural drawings. However, the structural drawings did not show the as-built bump-out projections that were observed at the base of the dome (Figure 8). Samples taken at these projections, considered along with the noted observations, indicated that the projections were constructed of various materials including CMU block and at least two different types of brick masonry. Additionally, steel hollow sections (bar/tube) were noted at the edges of the protruding corners of the dome shell (Figures 4-5). There was no indication on the structural drawings that steel hollow sections (bar/tube) were to be installed. The horizontal blue and pink beaded stone bands were built after the granite was installed (Figures 7-9). No indication of this type of installation and construction was noted in the shop drawings or in the architectural schematic drawings. Details and/or specifications for the horizontal bands were not prepared. Therefore, the contractor installed the horizontal beaded bands under no design supervision, did not conduct the work based on any specific details provided by a design professional, and was not aware of the necessity of allowing the cavity wall to weep excess water. Furthermore, the structural drawings indicated that the dome shell was squared off and open at the top. Observations indicated that the top of the dome consisted of an additional concrete shell, above the specified opening, that was not indicated on the structural drawings.

Water in Cavity

From observations and testing, it was clear that water was entering into the exterior wall cavity between the granite cladding and the dome shell. Entry points for water to infiltrate into the cavity existed at many of the sealant joints primarily at the protruding vertical corner joints where separations were observed. Additionally, there are openings at the "curb" located at the top of the dome where the top finial was attached. These openings are in the form of gaps, holes, and hairline separations. At the interior of the dome, dripping water was seen coming from the top of the shell. Efflorescence, mineral deposits, and moisture staining were prevalent throughout the exposed concrete dome interior (Figure 3). Additionally, the moisture readings taken at the surface of the shell interior indicated that the shell was wet. All of these factors provide evidence that water is penetrating the dome envelope by either entering through the cavity and/or from around the top finial and then leaching through the shell to the interior of the dome. Furthermore, separations at the horizontal butt joints at the outer edge of the flat surfaces of the dome (top of dome and top of "leaves") and unsealed screw holes that were drilled through the face of the granite to aid in attaching the engaged-spires to the faces of the dome, are other points of water infiltration.

Evidence of the effects of this water intrusion is noticed at the dome exterior especially at the blue beaded stone bands that surround the base of the dome (Figure 8). Additionally, upon removing the granite panels and at locations where the granite had already been removed, bubbling and blistering of the paint/coating that was applied over the CMU block was observed. Additionally, mineral deposits were observed to be leaching through the coating from the bump-out projections/shell locations. The mineral deposits and bubbling/blistering is a typical affect of water transpiring from the shell and/or CMU block through the applied paint/coating. Rust streaks and corrosion of the embedded steel hollow sections were observed at the protruding corners of the shell. Corrosion and the effects thereof are discussed in detail in the following section.

Corrosion

A hollow section steel bar was embedded (cast-in-place with the concrete) at the protruded corners of the shell without an effective cover or clearance. It was noted that the cover over the hollow steel bar was either non-existent at corroded portions along the bar or very thin at other locations. Due to the lack of an effective cover, the bar was exposed to the elements in the cladding cavity. This exposure resulted in significant corrosion at the hollow steel sections. No signs of protective coating in the form of galvanization or paint on the hollow steel section was noted on the steel bar. The embedded bar does not perform a load bearing or structural purpose and was not shown on any of the drawings. It is not clear what function the exposed steel bar served other than to aid in forming the shell. If a bar was required for reasons, which were not structural, then it is not clear why a steel hollow section without any protective coating was used.

Water and oxygen are essential for corrosion to occur. The lack of any cover at these protruding corners in conjunction with the abundance of water affecting the exterior of the shell, allowed corrosion to develop and affect these metal bars/tubes. Unhydrated ferric oxide (Fe_2O_3), or rust, when fully dense has a volume of about two times that of the steel it replaces. Broomfield (Broomfield, 2003) reports that when ferric oxide gets hydrated it swells even more and becomes porous. Hydrated ferric oxide ($Fe_2O_3.H_2O + 2H_2O$) is approximately two to ten times the volume of original steel. Kaminetzky (Kaminetzky, 1991), reports that the increase in volume of rusted embedded steel (typically reinforcing bars) produces internal pressures in the concrete that split the concrete in the path of least resistance. Figures 10 and 11 are typical views of the effects of the corrosion at the protruding corners of the shell.

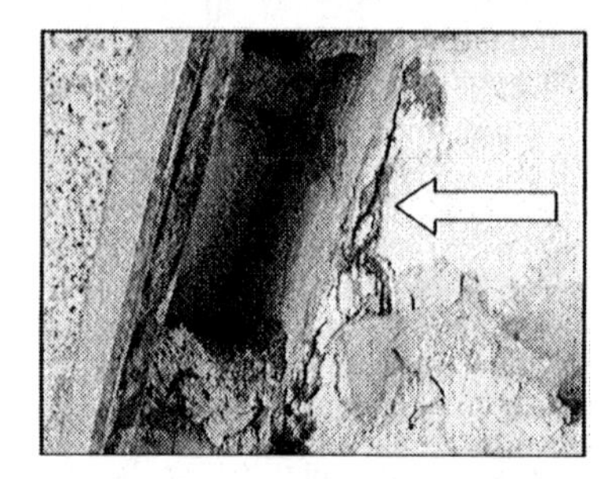

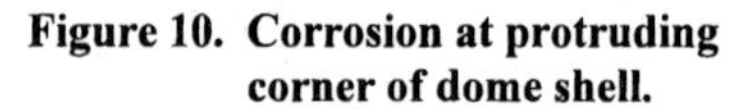

Figure 10. Corrosion at protruding corner of dome shell.

Figure 11. Similar as Figure 10 at another location.

The stainless steel anchors supporting the granite cladding at the sides of the leaves are centered on the individual granite panels. The center of the side panel and attachment of the stainless steel anchor along with the surrounding mortar daub is proximate to the embedded hollow section steel bar in the concrete shell and is not centered on the side face of the shell. The corrosion of the embedded steel bar has created rust products which have increased the volume of the original steel. This volume change due to the formation of rust has created expansive forces onto and through the mortar daubs and the anchors, which in effect have pushed the granite panel at the side of the leaves outward (away from the shell). This results in opening of the sealant joints at the protruding corners of the cladding. On the other hand, the flat faces of the leaves (front granite panels) have mortar daubs/anchors located away from the protruding corners of the shell and, thus, have not been subjected to any expansive forces from the corroding hollow steel sections. This creates a condition where one side of a cladding corner is moving outward causing a separation, while the other side of the corner joint is relatively stable. This also explains why there is no separation at any of the re-entrant corners. It was observed that the corroded hollow section steel bar has moved in the order of ¼" from its original position.

Furthermore, fractures at some of the granite panels (black in color) located at the sides of the leaves, correlates with the area of the stainless steel anchor attachment. As discussed above, the stainless steel anchors were attached to the shell, proximate to the corroding steel bar. The expansion creating by the corroded bar led to outward pressure at the anchor. This resulted in the split-tail portion of the anchor, which was attached to the granite panel, to induce an outward force on the sides of the kerf located at the center of the granite panel. Figures 12 and 13 show how the corroded bar affected the granite.

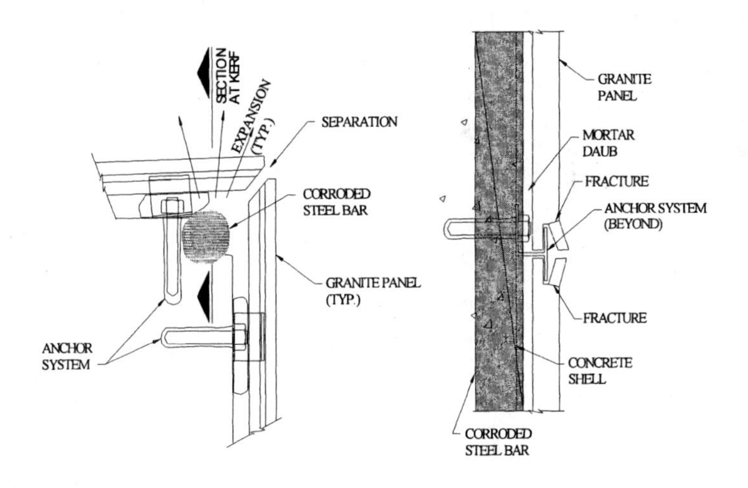

Figure 12. Plan at protruding corner after corroded bar expansion.

Figure 13. Section at kerf after corroded bar expansion.

The panel is 1 1/4" thick with a 1/4" wide kerf. The granite thickness on each side of the kerf is approximately 1/2" thick. The section at the exterior face of the kerf is less than half the thickness of the panel (approximately 1/2") and is weakest at this location. The outward movement of the anchor creates a localized bending and shear at the exterior face of the granite panel. Where the capacity of the granite panel has been exceeded, it is exhibiting fractures.

The corrosion of the embedded steel hollow section is proximate to the observed distress. Most of the distress at the dome is in the form of separations at the protruding corners and at the granite pieces at the sides of the leaves where the stainless steel anchors are attached to the kerfs. The expansive effects created by the corrosion of the embedded steel at the protruding corners, explain the type of cladding distress observed.

Differential Movement due to Dissimilar Materials

As reported by Beall (Beall, 1993), bricks (clay) expand from exposure to moisture and to variations in temperature. Conversely, concrete shrinks as it cures. At the base of the dome projections, which exist above the roof of the temple, brick and CMU block were used. The projection at the base of the dome consists of CMU block forming the outer most projection, with brick masonry above the CMU to form the next step of the dome projection. Brick masonry is adhered to CMU without a bond breaker. The expansion of the masonry and the opposing CMU shrinkage has caused separations in the joints at the protruding corners of the dome projections (Figure 7). The reason the separations happen at the protruding corners and not the re-entrant corners is that the protruding corners have the least restraint to movement. This explains the separations at the southwest corner of the dome where the dome projects outward. At this location, the embedded steel bar/tube is not present and the brick is supported on CMU.

Field Testing

Field-testing of the sealant at the granite joints was conducted to see if there were any deficiencies in the adhesion of the sealant that may have led to the noted separations at the granite joints. The in-situ field adhesion tests were conducted in substantial accordance with "Method A of Field-Applied Sealant Joint Hand Pull Tab" (ASTM C 1193-00, 2000). The tests on the dome sealant concluded that both types of sealant utilized were not defective and that, in general, the adhesion was acceptable. However, the alligatoring and cracking of the sealant indicated that there were signs of weathering from normal age and exposure.

Elevated surface moisture readings and humidity readings within the dome verified the moisture staining, efflorescence, and mineral deposits observed at the surface of the gunite/shotcrete dome interior. Moisture readings were as high as 100% (relative) at many of the locations.

Material Testing

Material testing was conducted to further evaluate the conditions at the structure. Additionally, materials testing was used to confirm whether or not specifications were met and/or if deficiencies in the materials exist that may have led to the observed cladding distress. Since the majority of the granite appeared to be in relatively good condition (except at localized areas), it was important to verify the characteristics of the material that was either attached to or in back of the granite. Additionally, material testing was used to verify conclusions reached by others from prior material testing results. Prior chemical analysis and petrographic examination testing of the mortar daubs by others concluded that the primary cause of the cladding distress was related to the expansion of the mortar due to sulfate attack (expansive reaction that can cause cracking in the mortar).

Additional field sampling and testing of the brick, mortar, and gunite (concrete) shell materials at the dome was conducted in order to perform chemical analysis and petrographic examination on the material samples. Additionally, sampling was also conducted at the terraces of the Temple building. Sampling locations were chosen in order to verify laboratory results of samples that were previously taken by others as well as to evaluate laboratory data of the material at other representative locations.

Twenty samples taken from the site were sent to an independent qualified laboratory (CTL[3]) so that chemical and petrographic testing could be performed. The purpose of the laboratory testing was to verify the amount of sulfur/sulfates and ettringite that were present within the material through chemical analysis and to provide other information related to the material (as required) through petrographic examination.

Laboratory Testing: Chemical analysis included X-Ray Diffraction (XRD), X-Ray Fluorescence Spectrometry (XRF), and LECO Total Sulfur testing on the samples. The hardened mortar and concrete samples were analyzed by XRF for 14 chemical elements and multi-step loss on ignition. Additionally, 3 samples were analyzed for total sulfur and expressed as SO_3 by LECO Total Sulfur testing. Chemical analysis results of the samples are summarized below.

<u>Chemical Analysis Results from XRD Analysis:</u> The XRD analysis of Mortar – Samples #2, #3, #4 and #8 indicated the presence of calcite, aragonite, quartz, gibbsite, calcium aluminate, and brownmillerite. Gypsum was also found in Mortar – Sample #5, #6, and #7. The coating on the glass beads shown on Core Sample #9 was predominantly calcite. Secondary ettringite was not detected.

<u>Chemical Analysis Results from XRF Analysis and LECO Total Sulfur Testing:</u> The sulfur content was determined by XRF and by LECO Total Sulfur (as noted) for each

---

[3] Construction Technology Laboratories, Inc. (CTL) performed chemical analysis and petrographic examination of the samples and prepared an unpublished report (dated November 2, 2004) of their laboratory results.

of the samples and in the unused mortar. The chemical analysis by XRF resulted in SO_3 contents between 0.45 – 1.67 (weight %). The LECO total sulfur analysis of mortar samples yielded results between 0.82 – 2.0 (weight %). Previous testing by others indicated an average of 2.26% SO_3 (by mass of sample) by LECO total sulfur analysis. This is compared to CTL's results that determined that the mortar at the dome had a total sulfur content of 0.82% and 2.0% SO_3 (by weight %) respectively, by LECO total sulfur analysis. The differing sulfate amounts are explained by CTL as being due to the ettringite (normal under hydration process) converting into gypsum, gibbsite and calcite/aragonite in the presence of carbon dioxide, moisture and facilitated by high temperature. The cracks in the mortar were due to shrinkage, which occurs with such a chemical reaction. This finding was contrary to assertions made by the previous laboratory.

Petrographic analysis on the mortar, grout, and gunite (concrete) was conducted in accordance with "Standard Practice for Petrographic Examination of Hardened Concrete" (ASTM C 856-04, 2004). ASTM C 856 is often used for petrographic examination of mortar and/or cementitious grout as well as concrete. The purpose of the petrographic examination was to gain insight into the composition, curing history, condition, and usage of the mortar/grout/concrete by viewing the samples under a microscope. Petrographic examinations of hardened concrete/mortar can be used to estimate cement content/type, water-cementitious material ratio, the aggregate content and grading, color, microcracking, depth of carbonation, nature of the air void system, secondary deposits, etc. These methods can assist in determination of the quality of the mortar/grout/concrete when originally cast, causes of distress, and the degree to which damage has occurred within the material.

Petrographic Results of Mortar: In general, CTL found that the cementitious matrix of the mortar primarily consisted of calcium-aluminate cement and lesser amounts of Portland cement and calcium sulfate compounds (originally either gypsum and/or plaster). The mortar samples have undergone one or more alteration stages that appear to have contributed to non-uniform degradation of paste properties and localized cracking/microcracking. Carbonation of the cementitious matrix was found to be extensive in most of the examined sample fragments and appears to have contributed to changes in the mineral composition of the mortar including the formation of secondary compounds and deposits. Carbonation is a "normal reaction in most Portland cement-based construction materials, with the degree and rate of carbonation mainly dependent on the relative permeability of the cementitious matrix." The secondary compounds and deposits consisted of calcium carbonate, gypsum, ettringite, and other crystalline compounds.

Petrographic Results of Gunite (Concrete): The gunite (concrete) samples were in generally good condition. Traces of alkali-silica reaction gel were found in at least one core; however, no significant cracking was found around the sparsely affected aggregates and no evidence of significant distress was found in the concrete. Outer surfaces of the samples revealed remnants of finish mortar and textured, elastomeric paint. Many of the cores exhibited secondary deposits of primarily calcium carbonate at the outer edge of the elastomeric paint and/or beneath the paint. Some secondary deposits of ettringite were observed in voids throughout most of the examined core samples. The cementitious binder of the samples is extensively hydrated, and considered along with the presence of secondary deposits (as described above), suggests that the wall was at least periodically kept in a moist to saturated condition while in service and that some of this moisture has migrated through the gunite (concrete).

CTL concluded that the subject mortar was not expansive and might have undergone a reversal of the hydration reaction explaining the difference in the sulfate amounts. CTL concluded that cracks of the sampled mortar were due to leaching, heating and cooling, and wetting and drying of the mortar. Additionally, CTL concluded that the subject mortar was not expansive by mixing the subject mortar with water and then setting it in a glass jar while it hardened. They also did the same for a known expansive mortar. The results revealed that the expansive mortar fractures the glass jar as it expands whereas the subject mortar does not.

Conclusions

Based on the information available, documents reviewed, analysis, site observations, and testing conducted; we found that the **primary causes** of distress are related to the following:

- Volume change due to corrosion of the embedded steel bar/tube at the protruding corners of the dome shell.
- Differential movement due to dissimilar materials at the base of the dome.
- Excessive amount of water infiltration into the wall cavity of the dome without a proper means to drain the excess water. Entry points for water were noted at the sealant joints between granite panels primarily at the protruding corners, at the curb where the top finial was attached, improper horizontal (butt) joint at top of dome, and unsealed screw holes for the engaged fiberglass spire attachments.
- As-built construction was not represented on the plans/drawings.
- No design coordination between the different design disciplines or trades prior to the construction.
- No design professional involved onsite to conduct any construction administration and/or supervision during the building of the dome.

The majority of the cladding distress is proximate to the corrosion observed at the protruding corners of the dome. Other contributing factors that led to isolated granite fractures are related to the presence of mortar in the kerfs in conjunction with chronic water penetration, likely fissures in some of the granite pieces, and isolated areas of improper construction at the face of the leaves. Furthermore, differential foundation movement, wind, wind-borne debris or impact, lightning, hail, or seismic activities were identified as having no contribution to the cladding distress.

References

American Concrete Institute (ACI). (1995). *Building Code Requirements for Structural Concrete (ACI 318-95)*, Detroit, MI, 1995.

American Society for Testing and Materials (ASTM). (2000). ASTM C 1193-00 *Standard Guide for Use of Joint Sealants,* West Conshohocken, PA, 2000.

American Society for Testing and Materials (ASTM). (2004). ASTM C 856-04 *Standard Practice for Petrographic Examination of Hardened Concrete,* West Conshohocken, PA, 2004.

Beall, Christine. (1993). *Masonry Design and Detailing For Architects, Engineers, and Contractors,* Third Edition, McGraw-Hill, Inc.

Broomfield, John P. (2003). *Corrosion of Steel in Concrete*, Spon Press.

Kaminetzky, Dov. (1991). *Design and Construction Failures: Lessons from Forensic Investigations*, McGraw-Hill, Inc.

LATERAL STRENGTH EVALUATION OF EXISTING ORIENTED STRANDBOARD WALL SHEATHING

Kenneth B. Simons[1], F.ASCE

Abstract

The lateral strength of thirty-six, 7/16" thick oriented strandboard samples taken from two existing multi-story apartment buildings deemed to be in a state of substantial impairment in Western Washington state, USA were evaluated. These samples, although stained and partially deteriorated, met the minimum requirements for structural wood panels set forth by governmental and industry standards. A discussion of imminent collapse and substantial structural impairment and their relationships to the 1997 Uniform Code for Abatement of Dangerous buildings and the 2003 International Existing Building Code is presented as well as discussion on the economic burden created by the complete removal (stripping) of exterior wall claddings to replace oriented strandboard sheathing that still meets the minimum requirements for a structural wood panel.

Introduction

Deterioration of wood framed buildings has become problematic in North America. Many of these deterioration problems are caused by water intrusion through the building envelope. According to the Barrett Report (Barrett, 2000), approximately 50,0000 condominium units in British Columbia, Canada were estimated to be affected by premature building envelope failure with an estimated total cost of repair of $1,000,000,000. During the last 20 years, the combination of modern architectural design, energy conservation methods, new building materials, and new wall cladding systems have reduced redundant water resistant features previously present that provided protection for the structure (Simons, 2001). Wood decay (rotting) of a wood structure, which previously may have taken many years or even centuries in some cases (Sack, 1989) to cause a building to collapse, currently initiates shortly after construction and manifests itself to the extent that portions of a building may be significantly reduced in strength within a few years. Included in the new building materials referred to above is the continued popularity of oriented strandboard sheathing (OSB). OSB is a panel product that is made of aspen, poplar or yellow pine in bands or wafers, bonded together under heat and pressure using a waterproof

[1] Principal Engineer, Damage Consultants, Inc., 9725 SE 36th Street Suite 102, PO Box 1336, Mercer Island, Washington 98040-1336

phenolic resin adhesive or equivalent waterproof binder. Like most manufactured wood products, its primary challenge is dimensional stability; that is, the OSB swells irreversibly, especially near the edges, when it gets wet. The swelling of the panels can cause uneven and partially laminated conditions. In addition, the fasteners (nails) can appear to be overdriven due to the increased thickness of the panel from swelling (APA, 2002).

Figure 1. Swollen and deteriorated OSB.

Figure 2. Location of OSB sample #6.

General Background Information

The oriented strandboard evaluated was located on exterior walls of two large, multi-story wood framed apartment buildings located in western Washington State, USA. The buildings were constructed of 2 x 6 wood framed walls covered with 7/16" thick oriented strandboard sheathing, which in turn was covered with 7/8" thick Portland cement stucco cladding over an asphalt-impregnated felt weather resistive barrier. Windows and sliding doors of the building consisted of insulated glass set within extruded vinyl frames. The penetrations for the windows at this type of building are likely sources for water infiltration (Nicastro, 1997). Repairs had been provided to address water infiltration at various locations on these buildings and subsequently a collapse claim had been made to the first party insurance carrier. Prior to this claim, controversy had arisen between insureds and insurers of buildings in Washington State; specifically multi-family, multi-story, wood framed buildings. The insurers typically exclude coverage on losses caused by fungus, mold, decay, and deterioration, continuous water seepage that occurs over a period of time, faulty, inadequate or defective construction, design, and collapse, unless it is caused by hidden decay.

Local jurisdictions in Washington State have assigned a more relaxed standard of "substantial impairment of structural integrity for the term "collapse" in insurance policies. During 2001, the Washington State Supreme Court decided that "hidden" decay causing collapse, means decay obscured from view, and the policy limitations begin to run when the decay is no longer hidden (Panorama, 2001). Subsequently, at these two apartment buildings, the cement stucco wall cladding was completely removed at the recommendation of the building owner's consultant to reveal all of the exterior wall sheathing. Localized portions of the framing at the three elevations of the buildings investigated, revealed some minimal decay to the framing that varied up to 1" in depth. In addition, it was discovered that the majority of the 7/16" thick oriented strandboard sheathing was stained to some degree. Much of the OSB sheathing varied in thickness between 1/2" and 5/8" in thickness where it had appeared to partially delaminate. Although dark discoloration and some decay were noted at the OSB sheathing at several areas, the OSB appeared to retain strength. However, these conditions caused the building owner's consultant to deem these exterior walls to be substantially structurally impaired.

At the subject buildings, the exterior wood framed walls that are sheathed with OSB are indicated to be shear walls. These exterior walls are also sheathed at the interior face with gypsum sheathing boards. There are also numerous interior walls that are sheathed with combinations of plywood and gypsum sheathing board at each floor level; thus, even in a worst case scenario where one might theorize that all of the OSB on the exterior elevations of the buildings has deteriorated completely (Kent, et.al., 2004), the total reduction in lateral strength in each direction for both buildings would be less than 30%.

Discussion of Applicable Building Codes and Standards

The 1997 Uniform Code for the Abatement of Dangerous Buildings as presented by The International Conference of Building Officials (International, 1997) defines dangerous buildings. In section 302 there are several categories; however, a dangerous building is defined in section 302-8 as "Whenever the building or structure or any portion thereof, because of (i) dilapidation, deterioration or decay; (ii) faulty construction; (iii) the removal, movement or instability of any portion of the ground necessary for the purpose of supporting such buildings; (iv) the deterioration, decay or inadequacy of its foundation; or (v) an other cause is likely to partially or completely collapse."

Section 302-11 states: "Whenever the building or structure, exclusive of the foundations shows 33% or more damage or deterioration of its supporting member or members, or 50% damage or deterioration of it non-supporting members, enclosing or outside walls or coverings."

Section 302-14 states: "Whenever any building or structure which whether or not is erected in accordance with all applicable laws, ordinances, has any non-supporting part, member, or portion less that 50% or in any supporting part, member, or portion less that 66% of the (i) strength; (ii) fire resisting qualities or characteristics; (iii) weather resisting qualities or characteristics required by law in the case of a newly constructed building of like area, height and occupancy in the same location." In summary, these three code requirements define a dangerous building as one that may collapse from 33% or more damage or deterioration of its supporting members.

The 2003 International Existing Building Code as presented by the International Code Council (International, 2003), in Section 202 on page 11 defines **Dangerous,** as

"Any building or structure or any individual member with any of the structural conditions or defects described below shall be deemed dangerous:

1. The stress in a member or portion thereof due to all factored dead and live loads is more than one and one third the nominal strength allowed in the *International Building Code* for new buildings of similar structure, purpose, or location.

2. Any portion, member, or appurtenance thereof likely to fail, or to become detached or dislodged, or to collapse and thereby injure persons.

3. Any portion of a building, or any member, appurtenance, or ornamentation on the exterior thereof is not of sufficient strength or stability, or is not anchored, attached, or fastened in place so as to be capable of resisting wind pressure of two thirds of that specified in the *International Building Code* for new buildings of similar structure, purpose, or location without exceeding the nominal strength permitted in the *International Building Code* for such buildings.

4. The building, or any portion thereof, is likely to collapse partially or completely because of dilapidation, deterioration or decay; construction in violation of the International Building Code; the removal, movement or instability of any portion of the ground necessary for the purpose of supporting such building; the deterioration, decay or inadequacy of it foundation; damage due to fire, earthquake, wind or flood; or any other similar cause.

5. The exterior walls or other vertical structural members list, lean, or buckle to such an extent that a plumb line passing through the center of gravity does not fall inside the middle one third of the base."

The 2003 International Existing Building Code as presented by the International Code Council (International, 2003) in Section 202 on page 12 defines **Substantial Structural Damage** as "[a] condition where:

1. in any story, the vertical elements of the lateral-force-resisting system, in any direction and taken as a whole, have suffered damage such that the lateral load-carrying capacity has been reduced by more that 20 percent from its pre-damaged condition, or

2. the vertical load-carrying components supporting more than 30 percent of the structure's floor or roof area have suffered a reduction in vertical load-carrying capacity to below 75 percent of the International Building Code required strength levels calculated by either the strength or allowable stress method."

It should be noted that the term substantial structural damage as presented by the International Code Council in the International Existing Building Code is interchangeable with "substantial structural impairment" after researching the definition of impairment in Webster's New World Encyclopedic Dictionary, (Webster, 2002) and Blacks Law Dictionary (Blacks 2001) and substituting it with it's synonym "damage".

Removal and Testing of OSB Samples

In order to determine if the stained and deteriorated portions of OSB panels met the criteria set forth by the applicable codes, 36 samples approximately 4" square each were removed from the building at areas that were stained and deteriorated. The OSB varied in thickness between approximately 1/2" and 5/8" thick (refer to Figure 1). Four of the 36 OSB sample locations and their subsequent testing are shown in Figures 2 through 14.

Figure 3. Sample #6 in testing apparatus.

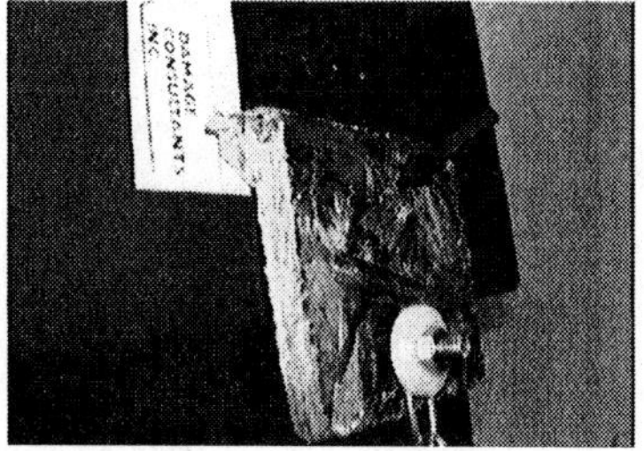

Figure 4. Close-up view of OSB sample #6 fastened with 8d nail.

The testing apparatus consisted of a wood frame with a metal nailing plate at the top where one of the 8-penny sheathing nails retrieved from the site was inserted into the OSB sample, approximately 3/4" from the edge. The bottom portion of the OSB was fastened with 3/8" diameter bolts and a washer that was connected to a chain to a spring loaded mechanical scale that was connected to an eyebolt at the base of the frame. As the nut on the eyebolt was tightened, tension would be applied to the sample and subsequently be recorded on the scale. Testing was ceased at approximately 90 pounds which was the minimum lateral strength requirement set forth by the Department of Commerce Voluntary Performance Standard PS2-04 authored by the APA-engineered Wood Association (Performance, 2004). The samples were not tested to failure in order to preserve and present the samples at settlement conferences and to use the samples for further independent testing should the results be disputed. Only one sample (#29) did not meet the requirement. This sample would no longer accept additional load at 75 pounds where the 8-penny nail failed at an existing nail hole. It should be noted that this 75-pound load exceeded the 1991 UBC 63 pound lateral load capacity for an 8-penny nail (International, 1991). The results of the lateral load tests are shown in Table 1.

Figure 5. 90 pound tension force applied to OSB sample #6.

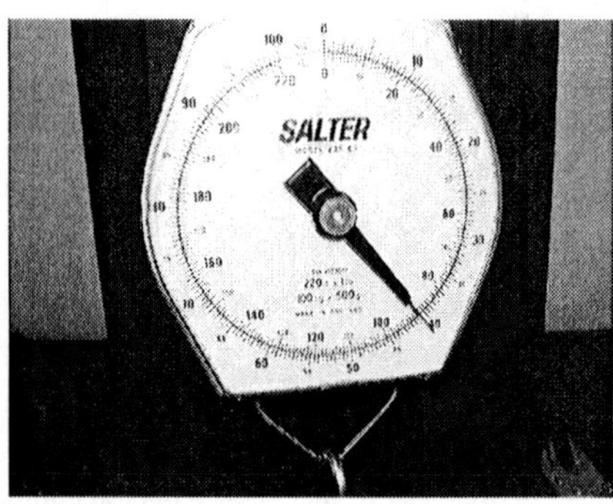

Figure 6. Close-up view of scale face.

Figure 7. Stained, partially deteriorated OSB sample #3 beneath environmental exhaust vent.

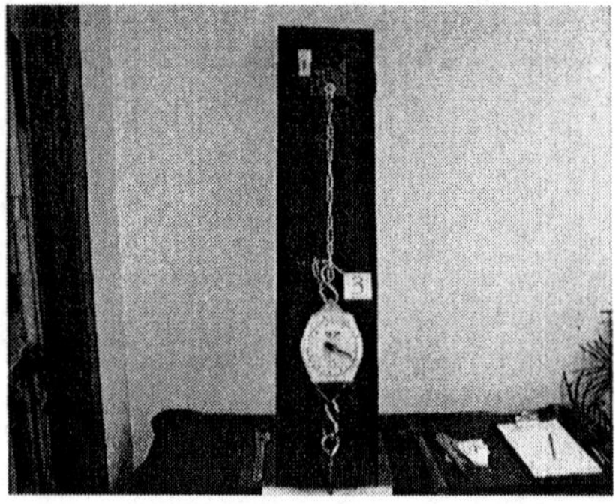

Figure 8. Sample #3 in test apparatus with partial loading.

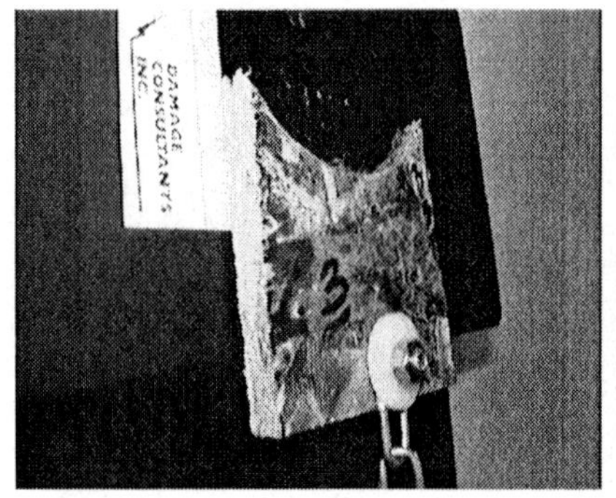

Figure 9. Close-up view of OSB sample #3 fastened to testing apparatus.

Figure 10. Close-up view of scale at conclusion of testing OSB sample #3.

TABLE 1

LATERAL LOAD TEST OF OSB SAMPLES

Sample No.	Load (lbs)	Notes
1	92	1 No Failure
2	92	1 No Failure
3	90	1 No Failure
4	91	1 No Failure
5	92	1 No Failure
6	90	1 No Failure
7	93	1 No Failure
8	90	1 No Failure
9	90	1 No Failure
10	94	1 No Failure
11	92	1 No Failure
12	92	1 No Failure
13	91	1 No Failure
14	92	1 No Failure
15	94	1 No Failure
16	90	1 No Failure
17	91	1 No Failure
18	93	1 No Failure
19	90	1 No Failure
20	90	1 No Failure
21	92	1 No Failure
22	90	1 No Failure
23	94	1 No Failure
24	92	1 No Failure
25	92	1 No Failure
26	92	1 No Failure
27	94	1 No Failure
28	90	1 No Failure
29	75	2 Failure at existing nail hole with 8d nail
30	94	1 No Failure
31	93	1 No Failure
32	88	1 No Failure
33	90	1 No Failure
34	91	1 No Failure
35	90	1 No Failure
36	90	1 No Failure

1 Meets APA/The Engineered Wood Association requirements for a structural wood panel
2 Exceeds UBC/IBC Lateral capacity for 8d nail

Figure 11. Location of stained and deteriorated OSB sample #12.

Figure 12. OSB sample removed revealing 1" of decay at 2 x 6 corner stud.

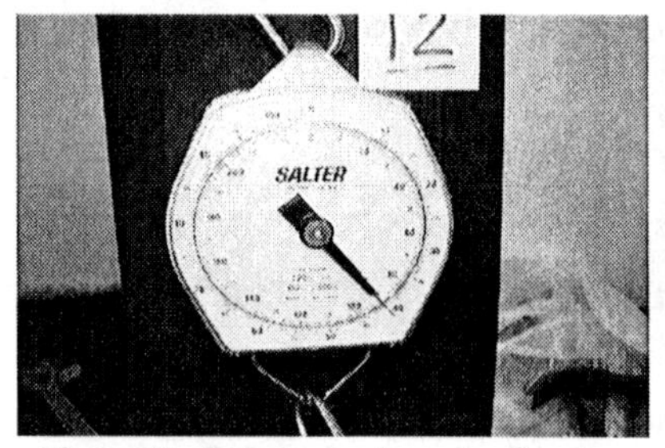

Figure 13. A close-up view of scale when testing of OSB sample #12 was completed.

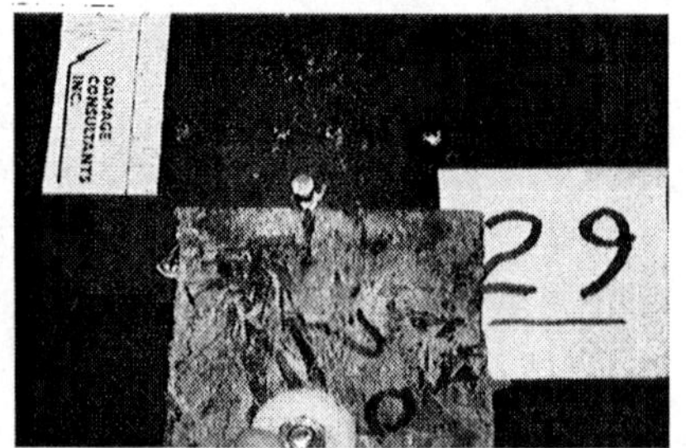

Figure 14. Failure at 75 pounds of an existing nail hole in OSB sample #29.

Concluding Remarks

The determination whether "imminent collapse" and/or "substantial structural impairment" due to hidden decay is present at a wood frame building is a challenge in Western Washington where conditions are conducive to decay formation. The extent of decay can vary from building to building or from different parts of the same building and from the discussion above, the stained, swollen and partially deteriorated oriented strand- board sheathing that was discovered still met the requirements for a structural wood panel set forth by governmental and industry standards. Although these conditions are not desirable, the OSB still meets the requirements for structural wood panels without any reductions based upon the 1997 UBC Abatement of Dangerous Building Code and the 2003 International Existing Building Code. Both of these code requirements do not come into effect until an approximate 20% to 33% reduction in strength of these connections. The code reductions would equate to approximately 50 to 72 pounds for the nail to OSB connection based on the 90 pound requirement from the US Department of Commerce or 42 to 50 pounds for the connection based upon the UBC's 63-pound maximum lateral capacity for an 8-penny nail. Removal and replacement of large exterior wall claddings in order to remove and replace oriented strandboard sheathing that still meets the requirements of structural wood panels appears to be economically wasteful. If this type of practice is continued, it will create an increasing demand on the economy.

References

APA – The Engineered Wood Association, *Effect of Overdriven Fasteners on Shear Capacity*, Technical Topics Form No. TT-012,Tacoma, Washington, 2002

APA – The Engineered Wood Association, *Performance Standard for Wood Based Structured Use Panels*, Tacoma, Washington, December 2004.

Barrett Report, *The Renewal of Trust in Residential Construction, Part II, Volume One* Commission of Inquiry into the quality of Condominium Construction in British Columbia, Government of British Columbia, February 2000.

Black's Law Dictionary, 8th Edition, St. Paul, 2001, page 332

International Conference of Building Officials, *Uniform Building Code*, ICBO, Whittier, California, 1991.

International Code Council 2003 *International Existing Building Code*, ICC, Country Club Hills, Illinois, 2003.

International Council of Building Officials, *Uniform Code for the Abatement of Dangerous Buildings*, ICBO, Whittier, California, 1997.

Kent, Scott M., Leichti, Robert J., Rosowsky, David V., Morrell, Jeffrey J., *Effects of Wood Decay by Postia Placenta on the Lateral Capacity of Nailed on Strandboard Sheathing and Douglas-fir Framing Members, Wood and Fiber Science*, 2004, Vol. 36, Number 4, pages 560-572.

Nicastro, David H., Simons, Kenneth B., *Fenestration Frustration: Failure Mechanisms in Building Construction*, ASCE, New York, 1997.

Panorama Village Condominium Owners v. Allstate Insurance Company, 2001 WL776175 (Washington Supreme Court July 12, 2001) Bullivant (Bulletin July 2001.

Sack, Ronald L. and Aune Petter. *The Norwegian Stave Church A Legacy in Wood, Classic Wood Structures*, ASCE, New York, 1989.

Simons, Kenneth B., *Imminent Collapse of Wood Structures Affected by Decay, Forensic Engineering*, Thames Telford Publishing, London U.K., November 2001, pages 149-158.

Webster's New World Encyclopedic Dictionary 2002 Edition.

Performance of a Design-Build Project

Ross J. Smith, P.E.,[1] and Rochelle C. Jaffe, S.E., Ar., C.C.S., S.M.I.[2]

[1]NTH Consultants, Ltd., 4635 44th Street SE, Suite C-180, Grand Rapids, MI 49512; PH (616)-957-3690; FAX (616)-575-1000; email: rsmith@nthconsultants.com

[2]ASCE Professional Member, NTH Consultants, Ltd., 38955 Hills Tech Drive, Farmington Hills, MI, 48331-3432; PH (248)-553-6300; FAX (248)-324-5187; email: rjaffe@nthconsultants.com

Abstract

A small district library retained the authors' consulting firm to perform a condition assessment of a single story facility that had been renovated from a grocery store some eight years prior to the assessment. To perform the renovation, the library staff entered into a design-build arrangement with a local contractor and their affiliated architect. The document review uncovered evidence of design errors, code violations and a general lack of appropriate detailing. The condition assessment revealed improper material selections, incorrect system installations and overall low quality workmanship. The inexperience of the library board, coupled with a contractor-architect partnership that did not provide for checks and balances, fostered a situation where the owner's interests weren't represented and industry standards of care and quality were not met. Retaining an independent professional consultant to review project documents, assist with contractor procurement, and provide quality assurance during construction would have better served the owner's interests and likely saved them money in the long run.

Introduction

As the construction industry continues to evolve, the number of companies trying their hand in the design-build arena grows. Many Architect/Engineer (A/E) design firms, construction companies and other entities have implemented the design-build model and successfully delivered on their promises of lower costs, single-point-of-contact service, and fast-track construction timelines. Recently, however, the authors had the opportunity to review a building that had been renovated under a design-build type contract and found the results disheartening. As participants in the construction industry, we were concerned to discover a situation in which the owners, *our* clients, did not have their interests represented and ultimately, were taken advantage of. Unfortunately, this case is not isolated and is part of a disturbing trend.

In this specific situation, the staff of a small district library in southern Michigan retained the authors' consulting company to provide a condition assessment of a facility they had occupied for less than ten years. The assessment was requested in order to establish a baseline condition report for the facility and to provide repair recommendations and budgeting strategy for future maintenance.

Building History

The district library relocated to its new facility in 1998 following an extensive renovation of a building that formerly served as a grocery store. The new home of the library is a single story building featuring a structural steel frame that was in-filled with light-gauge metal stud framing. Newly added cladding features included horizontal bands of colored, split-face concrete masonry units (CMU), clay face brick, and exterior insulation and finish system (EIFS) near the roofline. All façades contain sections of metal-framed curtainwall, which includes operable panels. (Refer to Figure 1)

The structure of the library building is comprised of two nearly independent systems. An original structural steel frame with traditional steel joists and steel deck serves as the primary system and supports a low-slope, ballasted built-up roof. As part of the renovation, a second system was added to support a new sloped-roof architectural feature. Plate-connected wood trusses frame the new roof and bear on a steel frame, which is supported by six retrofitted steel columns that penetrate the original roof surface. To interface between the two roof systems, a metal stud knee-wall was constructed around the perimeter of the newer sloped roof.

Since the renovation, a new built-up roof was installed on the remaining exposed low-slope areas. No other significant exterior maintenance work has been performed since the renovation. In the past few years, and as recently as December of 2005, the building experienced leakage in several different areas. As a result of these incidents and a growing concern for the sustainability of the facility, the library staff chose to retain a consultant to conduct a condition assessment and develop a maintenance budget.

Project Problem 1: Non-compliant drawings, Incomplete design

As part of the assessment, the authors reviewed of a set of "as-built" renovation drawings that was provided by the library. A small architectural firm, that is reportedly no longer in business, prepared these documents. In general, the drawings were incomplete and in some cases, in violation of applicable building codes.

Code Violations - At the masonry veneer system, the drawings call for a 3/8-inch air space between the masonry and the sheathing. This is a direct violation of the requirements of the current Michigan Building Code (MBC) (a local variation of the International Building Code), and the 1996 BOCA (Building Officials & Code Administrators) Code, the governing code at the time of the renovation. Both the 2003 MBC and the 1996 BOCA codes reference the ACI 530/ASCE 5/TMS 402 *Building Code Requirements for Masonry Structures*, a document produced by the Masonry Standards Joint Committee (MSJC), for masonry veneer requirements. The MSJC standard requires a minimum of 1-inch air space in a masonry veneer system backed by metal stud framing. The purpose of the air space is to permit water that penetrates the masonry veneer to run down the back face of the masonry to the bottom, where it can be collected and directed to the building exterior. A 3/8-inch air space cannot be kept clear of mortar protrusions that bridge between the masonry and the sheathing. Water that penetrates the masonry veneer can travel across the mortar protrusions to the sheathing, where it can enter the building interior.

Another code violation in the drawings is the omission of through-wall flashing and weepholes to collect and drain water that penetrates the veneer system. The MSJC standard requires flashing and weepholes be designed and detailed "to resist water penetration into the building interior". Further, the standard requires a minimum of 3/16-inch diameter weepholes spaced at 33 inches or less. These materials are not shown or called out in the drawings or details.

Incomplete project documents –The as-built drawings were also incomplete in many areas. Several specific detail sheets are referenced but not included in the drawing package. The numbering system for the referenced detail sheets differs from that used for the elevation, plan, and wall section sheets, leading one to suspect that the numbers related to a different project. In some cases, detail bubbles are not filled out at all. (Figure 2) The drawing set contains nothing more detailed than overall wall sections.

Details vital to proper completion of the building envelope were, apparently, never provided. Specifically, no detailing of EIFS base and edge conditions, or the related isolation joints between EIFS and other façade materials, were provided. No details were provided for the new sloped roof system, its integration with the existing built up roof, or the eave and gable terminations. The masonry veneer system was similarly undefined with no details describing the base condition, a flashing or weep system, wall ties, or expansion joints. In fact, the wall sections don't even show ties to connect the masonry veneer to the metal stud framing. These omissions, as well as many others, left the nearly schematic, sparsely labeled wall sections and plans as the only documents from which to construct the building. During field observations, many of these drawing omissions resulted in as-built construction problems. With no clear direction from the construction documents, the contractor's field personnel were left to make important architectural design decisions, despite being untrained and inexperienced in such design.

Industry Standards Ignored – In many aspects, the architect's design ignored industry standards. Besides providing no requirements for a water collection system behind the masonry veneer and no expansion joints within the masonry veneer, the drawings did not address the method of accommodating the differential movement between the clay brick (irreversible moisture expansion) and the concrete block (irreversible shrinkage) that are shown to be bonded together within the veneer system. The National Concrete Masonry Association (NCMA) TEK Note #5-2A describes two methods of addressing this situation: either providing additional joint reinforcement within the concrete masonry courses to resist tensile stresses generated by the clay brick expansion, or providing a slip plane between the two materials to prevent transfer of those stresses. Failure to address the predictable material volume changes led to cracking in the CMU head joints and may lead to infiltration of water into the veneer system.

The sheet metal gutter system at the eaves of the newer sloped roof appears to serve no functional purpose. The presence of the eave gutters suggests that runoff from the sloped roof may be greater than the capacity of the original low-slope roof drains. However, water that is collected in the gutter system is channeled to downspouts that deposit directly onto the built-up roof below, rather than into an independent drainage system. (Figure 3) This accomplishes the same result as allowing the water to run off the eave on to the built-up roof. Furthermore, the presence of the gutters contributes to an ongoing ice-damming problem occurring at the eaves. The gutters are substantially undersized, according to the Architectural Sheet Metal Manual by Sheet Metal and Air Conditioning Contractors' National Association, Inc. (SMACNA) and are easily filled with ice buildup. Ice that backs up under the shingles enters the building interior due to faulty ice and water shield installation at the eaves.

The failure to detail the knee wall interface between the two roof systems indicates that the designer did not understand the behavior of attic spaces under sloped roofs relative to heat and vapor movement. Because the drawings did not indicate the need for continuity of the vapor barrier nor for continuity of the thermal insulation, gaps in these systems resulted from the construction. Inadequate ventilation of the attic space resulted in interior condensation and exterior ice damming. Recommendations for venting heated and unheated attics is commonly available, and can be found in publications such as Architectural Graphic Standards.

Problem 2: Improper applications, Poor workmanship.
Following the document review, the authors performed an on-site visual assessment of the building. During the survey, the lack of directions and details in the drawings became readily apparent in the observed (poor) quality of the construction work. Most of the material systems employed in the building envelope have fundamental problems resulting from improper detailing, inappropriate application, or a combination thereof.

EIFS– In general, EIFS is a specialized proprietary system that requires trained installers who pay close attention to detail. Terminations of the EIFS are of particular concern. The EIFS system installed around the library facility has multiple installation deficiencies. The as-installed EIFS terminations were without proper back-wrapping (Figure 4) and no isolation joints were provided between the EIFS and dissimilar façade materials. Furthermore, no movement joints were provided within the field of the EIFS where the geometry of the design would normally dictate such a joint. In some locations, the EIFS is terminated below grade and is in contact with landscaping mulch and soil. Penetrations through the EIFS, such as for hose bibs, were left unsealed. As a result of these construction deficiencies, cracked coating, exposed mesh, and system deterioration were observed. Each of these conditions individually and collectively contributes to the premature deterioration and failure of the EIFS system.

In addition to the system-wide deficiencies, the EIFS was not detailed to properly integrate with the other building envelope systems. For example, on one elevation where EIFS covers the full height of the exterior wall, several window pan flashings were installed first. When the EIFS application followed, it was installed over the leading edge of the flashing (rather than under it), causing the window sill pans to drain directly into the EIFS (behind the lamina). The gable ends of the sloped roof system provide another example. At that location, the raked edge metal fascia was covered by the EIFS. A poorly executed "fix" included a fascia extension piece that was not tightly fit to the original metal fascia. Where the EIFS interfaced with the masonry veneer, the tight joint was covered with a smeared coating of sealant (which has already torn and failed) rather than placing sealant and backer rod into a recessed movement joint. In general, the quality of the EIFS installation, detailing and interfacing does not meet EIFS industry standards or general levels of acceptable workmanship.

Masonry veneer - Despite the lack of direction in the contract drawings, the masonry veneer system was apparently built with a through-wall flashing at the bottom of the walls. However, no weepholes were constructed at the flashing line. In many locations, the veneer extends below grade, further hindering proper drainage. Common masonry construction includes a means of draining through-wall flashing and maintaining the flashing level above grade. Neither properly trained masons nor an experienced general contractor should have allowed a masonry veneer system to be installed with no means of draining the space behind the veneer. One also wonders whether ties were installed to connect the veneer to the metal stud framing, despite none being shown in the drawings. If ties were provided, they should have been adjustable type, as required by the building code.

Sloped roof – As noted in a previous discussion, the contract drawings provided no direction on how to construct and enclose the newer sloped roof system or how to detail the interface between roof systems. These omissions left crucial detailing decisions to the discretion of the contractor's field employees.

The soffit on the underside of the eaves was enclosed with gypsum sheathing board and covered with a layer with plywood. These materials were fastened in an overhead application with common nails. Nail fasteners provide little pullout resistance and subsequent gravity-induced pullout failures allowed the plywood to fall and expose the gypsum sheathing board. (Figure 5) The gypsum sheathing is delaminating and falling in some locations, causing the soffit area to be completely exposed to the elements. Two of the eave ends were not enclosed at all during original construction, leaving the soffit area (and access into the attic space) exposed to wind, rain, snow, flying debris and birds. (Figure 6) The lack of details notwithstanding, eaves and soffits are common even in residential construction, and are not unusually difficult to properly construct. The as-built condition suggests that the installer used scrap materials and performed a half-hearted installation with no attention to common-sense details. Leaving the ends of the eaves open, thereby exposing untreated wood truss elements and the interior face of gypsum sheathing, is purely inexcusable.

To permit installation of the knee wall on the existing low-slope roof, existing ballast was relocated, a strip of the roofing membrane was cut and pulled back, rigid insulation was removed, and the top surface of the corrugated steel deck was exposed. The base track of the stud wall was fastened directly to the steel deck. On the exterior face of the wall, the existing roof membrane was turned up wall, but was not adequately anchored. Inside the attic space and adjacent to the knee wall, no vapor barrier was installed and batt insulation was haphazardly placed over the exposed portions of the deck. Without a vapor barrier, the ridge vent does not provide enough upper area ventilation to meet the requirements of the MBC.

In addition to the discontinuous vapor barrier, inadequate insulation around the perimeter of the new attic compromises the intended cold air space between the original flat roof and new sloped roof systems. The random insulation placement also allows a substantial amount of heat to radiate from the warm interior space through the roof deck. Escaped heat warms the attic air and facilitates ice damming at the eaves. (Figure 7) The ice-damming becomes more problematic due to the presence of gutters (as discussed) and the inadequate extension of ice and water shield on the sloped surface. As the ice accumulates and advances up-slope of the roof, thawing allows water to infiltrate the shingles and pass through the seams in the sheathing. In common roofing practice, the "ice and water shield" is extended up the surface of the roof for a minimum of 24 inches past the knee wall to reduce the amount of water infiltration resulting from the ice-damming (instead of 24 inches overall length). Though the drawings provide no direction on appropriate components or proper assembly of this area, general framing guidelines and appropriate selection of materials, which are not beyond the expertise of an experienced roofing contractor, could have dramatically improved the quality and durability of the installation.

Structural deficiencies –A cursory review of the structural systems revealed that one truss near a gable end had broken splice plates at both the top and bottom chords as well as a completely disengaged diagonal web member in the next panel of the truss. (Figure 8) The nature and grouping of the broken elements suggest that the truss was damaged during shipment or placement. Sandwich repairs with as few as three nails were installed at the chord splice plates and the web diagonal was attached to the nearby connection plate with a single bent nail. Based on inspection alone, it is clear that none of the repairs provide adequate structural capacity or meet the requirements of NDS (National Design Specification for Wood). The very presence of the poorly attempted repairs on the truss signals there was knowledge of the damage. It is equally clear that the truss manufacturer and/or design professional of record were not notified so that an appropriate repair could be designed. Such disregard for safety is indefensible.

What Went Wrong?

From a global perspective, the entire building presents evidence of an incomplete effort both in design and execution. To the common passerby at grade, the building may appear sound and weather-tight with no distinctly obvious shortcuts taken. But, to anyone on the roof level and especially to those with the technically trained eye, an endless list of improper details, inappropriate applications and noncompliant conditions highlight a project that was severely lacking in quality and workmanship. Based on the authors' observations and understanding of the situation, a few simple key elements led to this travesty.

Inexperienced Owner – The small town library board likely had no construction experience and didn't really know what they were getting into. As such, they blindly trusted the architect and/or contractor with whom they became involved. Knowing that municipal budgets are typically limited, the contractor/architect likely convinced the decision makers that a design/build type approach would be more cost effective. They may have stressed that bringing an "inspector" into the project would burden the project with unnecessary costs. (As an aside, the population of the board has turned over since the renovation project, so the atmosphere surrounding the original construction decisions is not completely clear.) It is highly probable that no member of the library board knew enough about construction to be able to critically inspect the work as it was occurring, or even after substantial completion. It is also likely that because the board was responsible for the project, no one individual had authority over the contractor/architect and, therefore, no one monitored the construction progress.

A/E partnered with Contractor – Whether or not it was officially defined as "design-build", it is clear the contractor and architect began working together at some point early in the project, based on the evidence provided. This arrangement, coupled with the naiveté of the district board, quickly shifted the focus of the document development process from the goal of producing a detailed set of construction drawings to the goal of simply attaining a building permit. The eventual result was threefold: First, an incomplete set of drawings, with inconsistent and inadequate details was developed and used as a basis for construction. Second, since the drawings lacked the appropriate level of detail, several design decisions and material selections were left to the discretion of the contractor. It is likely that the architect, whose name is on the drawings, had no involvement during construction. Therefore, there was no entity to provide checks and balances on the contractor's work. Most importantly, the interests of the owner went unrepresented. These circumstances sent the project down the road to premature failure and will end up costing the owner more money over the long term.

Unjustified reliance on building official review – The library board may have been under the impression that the city building officials were responsible for inspecting the construction. After all, the city issues the permit, inspects some aspects of construction, and issues the certificate of occupancy. However, the reality is that the city building inspectors probably visited the site only to look at plumbing and electrical components, and maybe to look at the foundations for the retrofitted columns to support the sloped roof. City inspectors generally do not review installation of façade materials, because they are not required to do so by code and because they do not have the expertise to do so.

What Should Have Occurred?

Without question, this situation could have been avoided. The contractor and architect could have operated under the same partnered arrangement, but proceeded responsibly, keeping the interests of their mutual client, the owner, in mind. Likely, a more realistic circumstance would have seen the owner hire a professional consultant, who was independent of the design/build team, to provide technical expertise and quality assurance throughout all phases of the project. Given the competitive, fiscally restricted environment in the current construction industry, the last scenario would have been the most appropriate in order to protect the interest of the owner. This concept holds true for several reasons.

Retaining a qualified consultant in the initial phases of the design-build project to perform a peer review of the construction documents helps assure the design architect produces quality drawings, compliant with applicable codes and standards, and suitable for use in construction. A consultant can also provide input to the owner on material selections and perform a subsequent review of the contractor's product submittals to verify that the owner's desires will be met.

Perhaps most importantly, an independent consultant can provide field observations. An educated, on-site presence who represents the owner is the most direct and effective way to hold a contractor and subcontractors accountable for the quality of their workmanship. By observing the work, verifying appropriate installations, promoting correct detailing and requiring adherence to governing codes and standards, the consultant serves as a constant reminder of an engaged owner, who is intimately connected to the project and who expects the quality for which he/she is paying. In comparison to other alternatives, the quality attained by using a professional consultant is unmatched.

Budget Implications

However strong the argument for doing the "right" thing by retaining a consultant, the reality remains: the bottom line still drives the final decision. But, where additional costs are cited as a barrier for many owners, the financial repercussions of ***not*** using a consultant far outweigh the initial outlay. Forgoing a third-party representative may allow the owner to realize the savings of typical consultant fees - ranging from five to fifteen percent of the total project cost. On the other hand, unsupervised construction may lead to a low quality product which will require substantially higher annual maintenance and large-scale repairs/replacements much sooner than would normally be anticipated. Furthermore, damage to interior finishes and furnishings could be avoided. These unexpected increases could collectively escalate to 50 percent or more of the original construction cost.

For the library, the need to completely reclad the building envelope is a potential reality, only eight years after initial construction. Furthermore, they have sustained damage to books and ceiling tiles that require additional expenditure to replace. If long term financing is considered for the construction project, engaging a consultant may very well save the owner money over time. Simply put, doing it right the first time is cheaper than fixing mistakes later.

Conclusion

As in any industry, the rules in construction are not steadfast nor are the patterns without their exceptions. Sometimes the conditions and the involved parties in a design-build contract can provide a winning deal for the owner as well for the contractor. However, as the design-build giants continue to grow and absorb all architectural, engineering and construction management functions, the traditional owner-architect-contractor project relationships have begun to change. With this change should come a renewed sense of caution by owners. Before entering into any project, retaining a qualified consultant to represent owner interests should be considered. More often than not, when the new package deal seems just too good to be true, it probably is.

Figure 1. Overall view of library building and newer sloped roof feature.

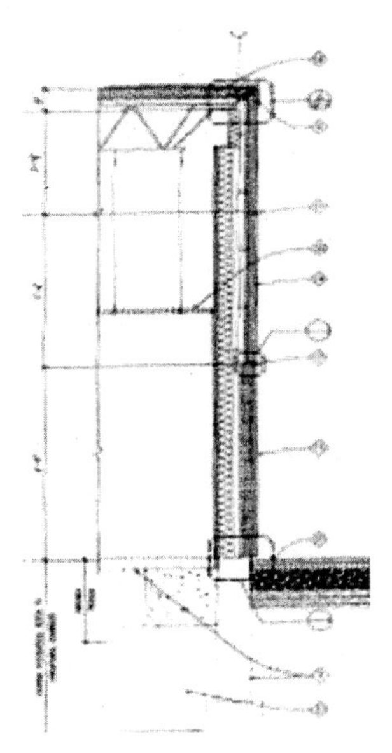

Figure 2 – Excerpt wall section from construction drawings with unfilled reference bubbles.

Figure 3. Overall view of sloped roof feature on library building.

Figure 4. EIFS system is improperly terminated without backwrapping.

Figure 5. Pullout failure and fallen soffit materials at sloped roof eave.

Figure 6 – Open end of soffit at gable eave for sloped roof. The poorly executed fascia extension is also seen at the raked edge of the slope roof.

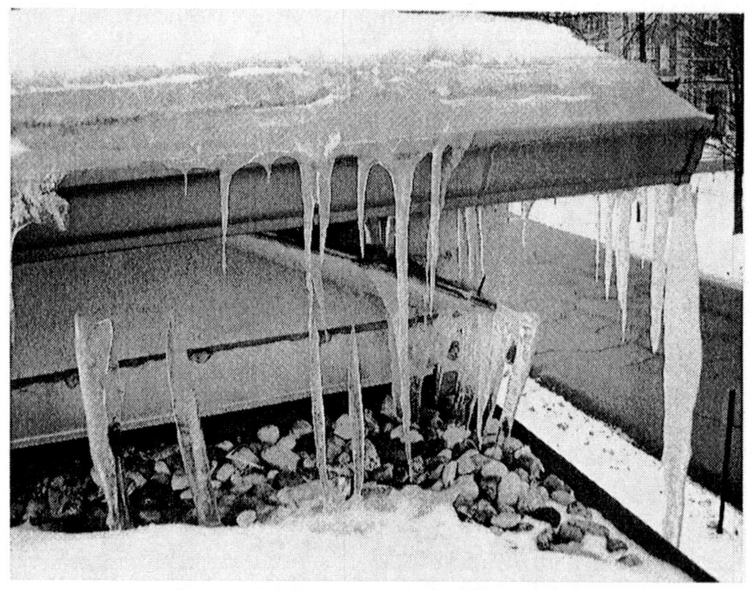

Figure 7. Ice damming at the sloped roof eave.

Figure 8 – Disengaged web diagonal at damaged roof truss.

Confidence in Expert Opinion: A Structural Engineering View

Gary C. Hart[1], Ayse Hortacsu[2], Stephanie A. King[3]

[1]Emeritus Professor, Department of Civil Engineering, University of California, Los Angeles, and Principal, Weidlinger Associates, Inc., Santa Monica, CA, hart@wai.com; [2]Senior Engineer, Weidlinger Associates, Inc., Mountain View, CA, hortacsu@wai.com; [3]Associate Principal, Weidlinger Associates, Inc., Mountain View, CA, sking@wai.com.

Keywords: expert opinion, confidence, damage estimation, structural engineering

ABSTRACT

This paper first briefly describes a method for using consensus-based publicly-available models and computer software to estimate earthquake-induced structural damage. Next, and the primary focus of the paper, is the presentation of a technique for quantifying and increasing the confidence in the damage estimate with the use of the professional structural engineering culture and the principles of probability theory.

INTRODUCTION

Over the past several decades, the field of Bayesian Decision Making has become a well-developed application within probability and statistics, spanning many different disciplines. For example, the Wall Street Journal recently described the retired chairman of the Federal Reserve, Alan Greenspan, as a Bayesian Decision Maker. In simple terms, Bayesian Decision Making entails the updating of a decision by the incorporation of new data or information. When we are making decisions, or more specifically, offering an expert opinion, it is a fact of life that we have uncertainty in our opinion and that we can use more information to reduce our uncertainty and increase our confidence.

In the structural engineering field, we are often asked to offer opinions with very limited information and typically in a very short time frame. For example, consider the green, yellow or red tagging of buildings immediately after an earthquake. It is in the culture of structural engineering to make decisions with limited information and with little time or budget for information gathering, and we do it often in the area of post-earthquake and post-hurricane damage estimation.

One historical example of a decision made in structural engineering tradition and culture is the definition of the design basis earthquake, first published in the ATC 3-06 [1] document in 1978. The definition focused on answering the question: What is the maximum earthquake ground motion that can occur at a specific building site, assuming that the building being designed will be exposed to future earthquakes for 50 years? No single number can be used to answer this question because the answer is a random variable. However, using the science of probability theory and Bayesian Decision Making the design basis earthquake was defined to be an earthquake ground motion that has a 90% probability of not being exceeded in the 50-year design life of the building. (The 50-year design life was used based on the expectation that the building would likely be replaced in approximately 50 years.) Therefore, using Table 1 shown below, this means that the ATC3-06 authors were *Almost Certain* that the maximum earthquake ground motion in the 50-year design life would not exceed the design basis earthquake defined in the document.

It is possible that a structural engineering expert is asked to state his or her opinion under not the best of conditions and thus will not state an opinion unless they have a minimum level of confidence in that opinion. Table 1 shows different levels of confidence in both qualitative and numerical terms that are used to describe the confidence in an opinion [2].

Table 1 Qualitative to Numerical Transformation of Confidence, adapted from [2] and [3]

Expression	Single-numbered (median) Probability Equivalent (%)	Range (%)
Almost Certain	90	90-99.5
Very Likely	85	75-90
Likely	70	65-85
Even Chance	50	45-55
Possible	40	40-70
Unlikely	15	10-25
Very Unlikely	10	2-15
Almost Impossible	2	0-5

It is often the case that after a hurricane or earthquake a speaker will give a slide presentation based only on field observations he or she has made without consulting with others. The other extreme often occurs as well: the results from the FEMA/HAZUS [4] computer program or its derivatives will be presented and financial or emergency decisions are made without ever seeing a building. This paper presents an approach to quantify the confidence in an expert's opinion in, for example, calculating a dollar estimate of natural hazard damage for a given building.

ILLUSTRATIVE EXAMPLE

Consider the case where the opinion that is sought is the expected dollar loss for a building experiencing a hurricane or an earthquake.

Using Only Consensus-Based Publicly-Available Models

In the mid-1990's FEMA funded the National Institute of Building Sciences (NIBS) to develop a methodology and computer software for estimating the damage to a building due to the occurrence of an earthquake (HAZUS). The methodology was later extended to address multiple hazards, such as hurricane and flood (HAZUS-MH). The methodology enables us to obtain a best estimate of the damage to a building using default values for all of the variables in the models. These default values were obtained for different standard building types using the experience and results of analytical studies performed by experts in the appropriate areas of the methodology.

As a professional engineer one can use the HAZUS method with its default variable values to estimate a value of damage to a given building. Prior to providing this estimate to the client, the expert must be able to say, for example, that this estimate is *Likely* to be correct. In describing this confidence, the term *Likely* is from Table 1 and says in effect that he or she is 70% confident that the opinion, based on the HAZUS default results, is correct.

Improving Confidence with Additional Effort

But what if the engineer wants a better estimate of the damage that he or she can state with more confidence? The client can pay for a more detailed engineering assessment and the expert can use the same steps defined in the HAZUS methodology, but now he or she will not use the default values provided in the computer software. Instead, the expert will improve upon the accuracy of the default values. For example, the engineer might be paid to do the following: (1) a site-specific estimation of the earthquake ground motion at the building site using the measured ground motions in the vicinity of the building site obtained from the USGS, CSMIP, and other public sources; (2) a site visit to observe the damage to the building; (3) a computer model of the structure to estimate the building's roof displacement and wall deflections during the earthquake; and (4) a head-to-head discussion with a knowledgeable construction/cost estimation expert on what they both observed at the building site. It is reasonable to expect that the expert's new estimate of damage will be different from the HAZUS damage estimate using the default values. In addition, this revised estimate will be presented to the client with more confidence that it is correct. But the question is: How much more confidence did the client get for the expenditure of the money for the expert to do this additional work?

METHOD FOR QUANTIFYING CONFIDENCE

The proposed method for quantifying confidence in an expert opinion first considers each component that goes into preparing a damage evaluation and related cost calculations. The confidence associated with each component is then propagated through the decision-making process to determine the confidence in the final opinion, as described below.

In general, there are two components a structural engineer uses to form an opinion: Observation and Simulation, as illustrated in Figure 1. In the case of an earthquake damage evaluation, the Observation component refers to the field investigation performed in order to document and measure damage to the building in question. The Simulation component refers to the structural analysis that is performed in order to estimate the effects of ground shaking on a building model, created according to details obtained about the building. The confidence in the opinion of the damage is a function of the completeness and competence of the work in each of these two primary sources of information, which can be viewed as branches on an information tree. The variation of scope within each branch is important because it offers a spectrum of tasks that can performed to meet the required service for the project under consideration. It is possible to obtain a professional estimate of damage using information from any one of these information sources alone but the confidence the structural engineer has in an opinion rises with an increase in the scope of services that are performed.

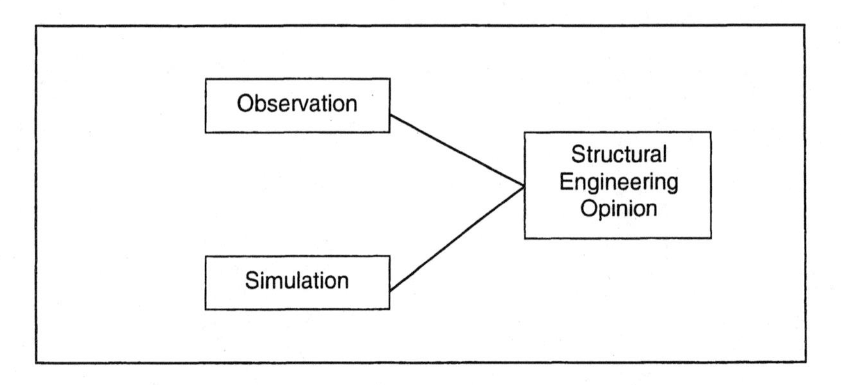

Figure 1. Information tree of forming a Structural Engineering Opinion.

Consider the Observation branch, which may be comprised of a field investigation of the exterior, interior, and/or common areas of the building. In addition, for each of these areas, the scope may range from none to destructive testing, as shown in Figure 2. For each path on the information tree for the Observation branch, a confidence is assigned, based on the scope of the field investigation and the information in Table 1. For example, we may state, "If the field investigation consisted of destructive testing an all three areas of the building, I am Almost Certain [90%] that the statements made in the damage evaluation report are correct." This initial confidence, based on the scope alone, is designated as F_{obs}.

Continuing with the Observation branch, in the next step, the field investigation of the three areas (exterior, interior, and common areas) is evaluated to assign a detail/completeness (d) and quality/competence (q) score to each area. These scores are used to quantify, in terms of a confidence level, the thoroughness of the investigation (d) and how well it was carried out (q). The d and q scores are combined and used to update, in a Bayesian Decision Making-type process, the initial confidence value (F_{obs}) to produce a new confidence for the Observation branch (F_{obs}').

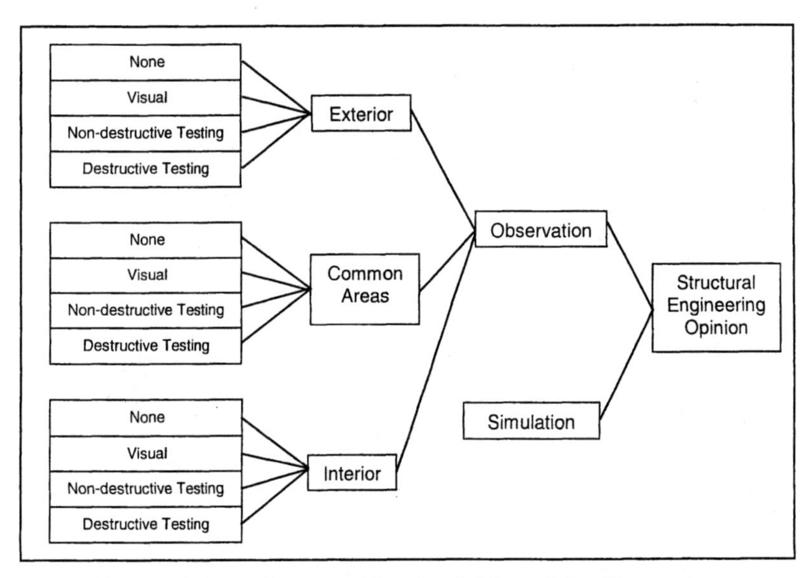

Figure 2. Information tree with expanded branch for Observation.

For the Simulation branch of the information tree, the procedure is essentially the same as for the Observation branch. The initial or prior confidence value in this main component of a structural engineering opinion is termed F_{sim} and is based only on the scope of a Simulation or analysis. Figure 3 shows how the Simulation branch of the information tree can be expanded to comprise linear and non-linear analysis as well as engineering judgment.

Similar to the Observation branch, each analysis type is assigned a detail/completeness (d) and quality/competence (q) score. These scores are used to update the initial Simulation branch confidence value (F_{sim}) to produce a new confidence value (F_{sim}') that incorporates the new information on how complete the simulation component is and how well it was carried out.

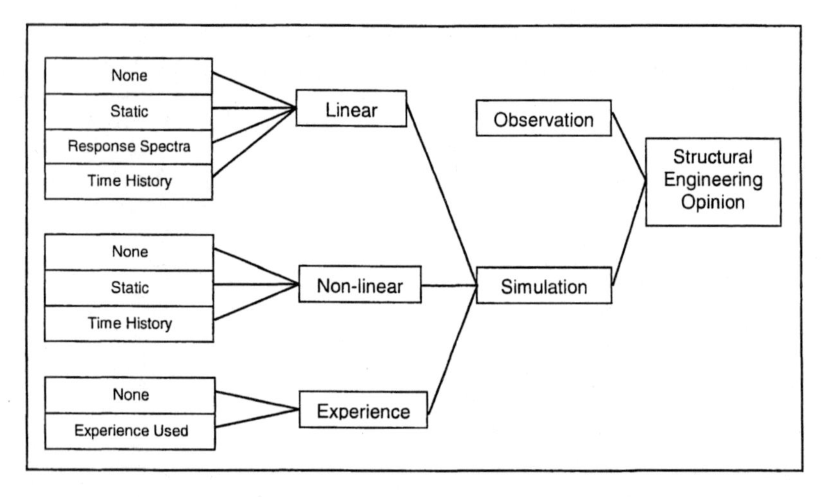

Figure 3. Information tree with expanded branch for Simulation.

In the final step, the confidence values associated with each of the two main components, F_{obs}' and F_{sim}', are combined to yield the overall confidence in the structural engineering opinion, F_{opi}. The information in Table 1 can be used to relate F_{opi}, as well as the other numerical confidence values in the method (F_{obs}, F_{obs}', F_{sim} and F_{sim}') to the qualitative descriptions of confidence that are more easily understood by the client.

CONCLUSION

The method for quantifying confidence in a structural engineering opinion described in this paper is straightforward and relies on the professional culture of structural engineering. The method is based on an information tree approach, whereby the initial confidence in the opinion is quantified only as a function of the scope of the investigation. This initial confidence estimate is updated, through a Bayesian Decision Making-type approach, by incorporating the additional information on the completeness and quality of the given investigation.

This method provides a comprehensive and objective technique for a third-party reviewer to evaluate two or more reports about the same building and safely state his or her degree of confidence in each report. In addition, the method provides a means for determining the increase in confidence in an opinion that would be expected to occur for an increased investigation scope – essentially a cost (fees for additional tasks) versus benefit (reduced uncertainty and increased confidence) analysis.

REFERENCES

1. Applied Technology Council (ATC). "Tentative Provisions for the Development of Seismic Regulations for Buildings." Redwood City, California: ATC, 1978.
2. Vick SG. "Degrees of Belief: Subjective Probability and Engineering Judgment." Reston, Virginia: American Society of Civil Engineers Press, 2002.
3. Reagan, R., Mosteller, F., and Youtz, C. "Quantitative Meanings of Verbal Probability Expressions." Journal of Applied Psychology, 1989.
4. Federal Emergency Management Agency (FEMA). "HAZUS99 Earthquake Loss Estimation Methodology, Technical Manual." Prepared by the National Institute of Building Sciences for the Federal Emergency Management Agency, 1999.

Bolt Failure at Gable Roof of 42-Story Building

William D. Bast, M.ASCE[1] and Ken R. Maschke, M.ASCE[2]

[1]Thornton Tomasetti, 14 E. Jackson Boulevard, Suite 1100, Chicago, IL 60604; PH (312) 596-2000; FAX (312) 596-2001; email: wbast@thorntontomasetti.com
[2]Thornton Tomasetti, 14 E. Jackson Boulevard, Suite 1100, Chicago, IL 60604; PH (312) 596-2000; FAX (312) 596-2001; email: kmaschke@thorntontomasetti.com

Abstract

Following reports of a "loud noise," the building engineer for a 20-year-old, 42-story high-rise in Chicago, Illinois, found the head of a sheared bolt in the mechanical room on the 42nd floor. Structural engineers subsequently discovered a connection missing five of its six 7/8-inch (22mm) diameter ASTM A325 bolts.

This paper provides an account of the events at the building and describes the analysis through which engineers concluded that differential elastic rebound between the concrete shear walls and the exterior steel framing was likely responsible for the connection failure.

Introduction

Unusual things often seem to happen on Friday afternoons! The afternoon of Friday, June 24, 2005, confirmed this statement when at 1:10 p.m. an entire building shook and several tenants called the Fire Department and the Emergency 911 line to report the incident. Tenants located between the 7th and 42nd floors reported a loud noise and shaking, describing it as similar to "a heavy piece of equipment dropped on the floor directly above us." In fear, many tenants immediately evacuated the building. Building management performed a quick check of the vital building systems, in particular the rooftop mechanical equipment, and found nothing unusual. No earthquake activity was determined to have occurred in the Midwest that afternoon.

The 42-story building combines a composite steel frame with an interior concrete core. Construction of the building was completed around 1985. Steel members frame the outside wall and floors of the building, and a central concrete core provides both gravity load support and lateral stability. The concrete core continues to the level of the 42nd floor, but only on the north part of the building. The top of the high rise features a unique gabled roof.

Building management reported that over the course of three weekends in June 2005, a major tenant of the building, occupying 20 floors, moved out. On the afternoon of June 24, approximately one week after the "move out" concluded, tenants between the 7th and 42nd floors reported the shaking and loud noise as described above.

Later that weekend, a building engineer discovered large amounts of flaked-off fireproofing and one-half of a bolt on the floor of the mechanical room on the 42nd floor. The equipment in the room was found to be in good order. Upon further inspection, building engineers noticed that three of the doors leading to the north stairwell had been racked and no longer closed properly.

Observations

On July 1, 2005, structural engineers visited the site to investigate the disturbance and the broken bolt (Bast 2005). Closer inspection of the mechanical space on the 42nd floor revealed four additional bolt halves lying on top of a mechanical duct. These bolts were found beneath the connection between column lines 9-10 and A-D, as located in Figure 1 and shown with greater detail in Figure 2. A small "pin-hole" of light was observed at the gabled peak near this connection.

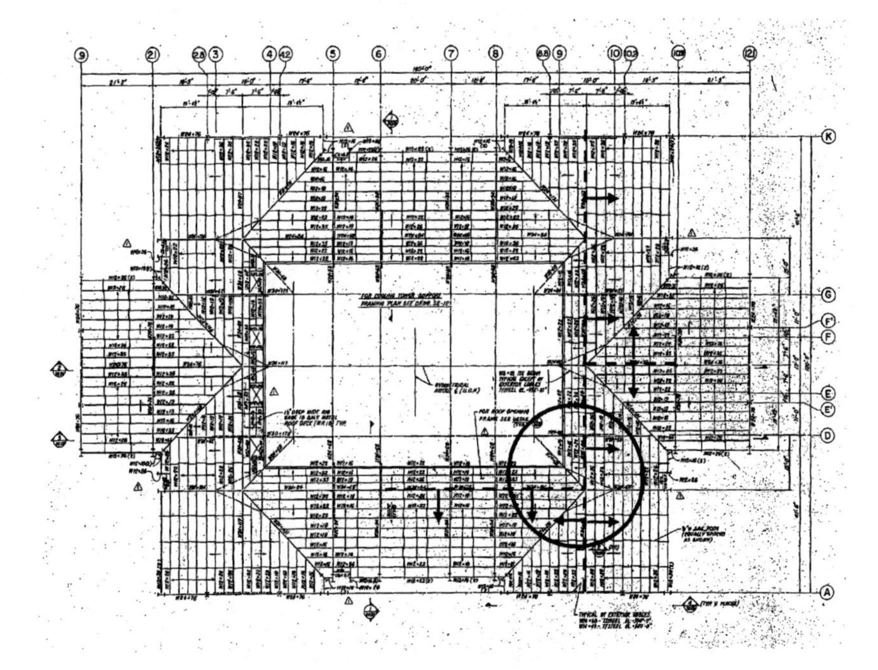

Figure 1 - Roof Plan (CBM 1985) (North to Right)
Arrows Indicate Roof Slope, Area of Distress Circled

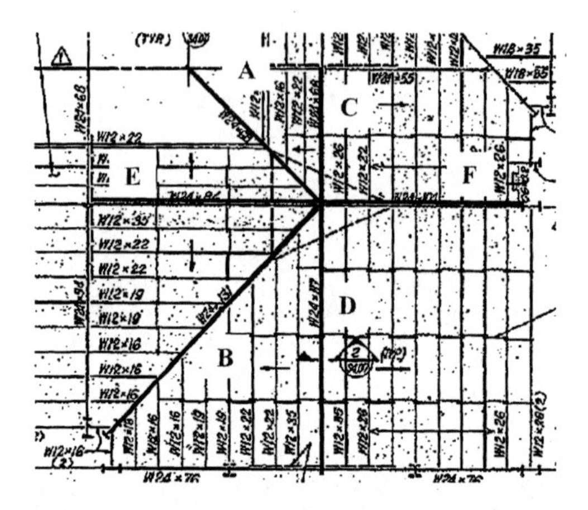

Figure 2 - Roof Partial Plan (CBM 1985) (North to Right)

Following removal of additional fire proofing from the steel framing, the structural engineers located a beam connection missing five bolts. The connecting W24x68 beam (noted "A") spanned diagonally upward from the core wall of the 42nd floor to the peak of the northeast corner of the room. The upper end of the beam framed into a connection with five other steel beams ("B" to "F"). The lower end rested on top of a corbel attached to the concrete core wall, as illustrated below (Figures 3 and 4).

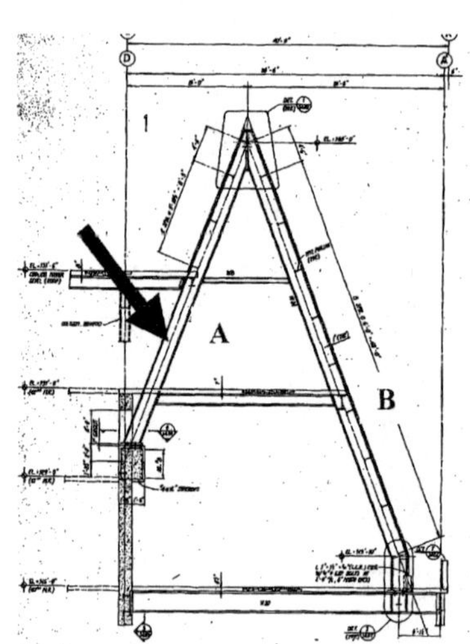

Figure 3 - Gable Detail (CBM 1985) Arrow Indicates Beam Framing into Shear Wall

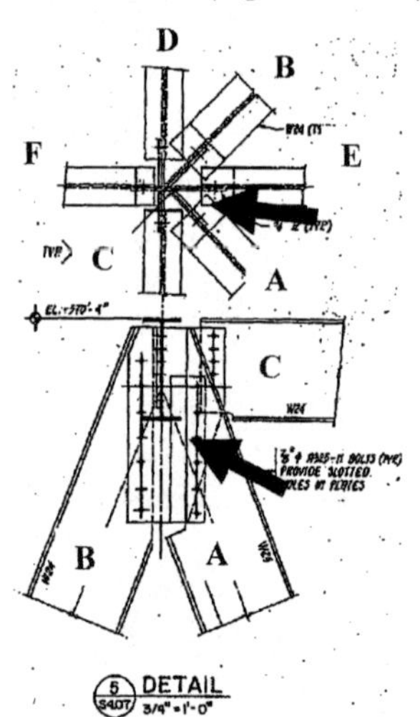

Figure 4 - Detail of Connection (CBM 1985) Arrow Indicates Distressed Connection

Figure 5 is a photograph of the connection missing five of the six bolts. In this photo, the holes of the backing plate are not visible. Observed scrape marks on the backing plate, visible through the bolt holes and the interface between the fireproofing and connection plate (Figure 6), indicate some longitudinal movement at the top of the connection.

Figure 5 - Failed Connection
Arrow Indicates Unsheared Bolt

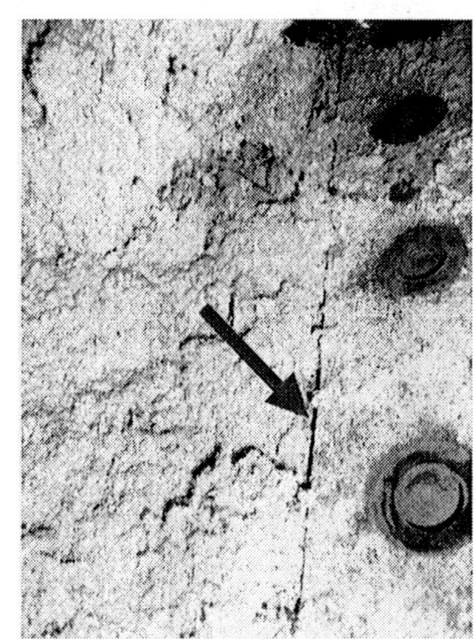

Figure 6 - View of Fireproofing at Connection
Arrow Indicates Edge of Connection Plate

The bolts were marked as type A325 and measured 7/8 in. (22mm) in diameter. All discovered bolts were sheared across the shank, not the threads. No sheared bolts were discovered at similar locations in the other quadrants of the symmetric roof system.

Analysis

The analysis of the cause of the bolt failure focused on the possibility that significant differential movement occurred between the core and the exterior as a result of the large tenant "move out". The structural engineers noted that the load imposed on the building by the aforementioned tenant was distributed between the concrete core wall and exterior steel columns, though not evenly on both sides of the generally symmetric building. This heavy loading elastically compressed both the concrete core and the steel columns. However, based upon stiffness properties and load distribution, it is believed that more total elastic shortening occurred in the exterior columns than the interior concrete core.

The "move out" of the tenant removed much of the load that was distributed to the gravity system. To compute the differential shortening of the structural elements the full design live load (50 lb/ft^2 (2.4 kN/m^2) plus an additional 20 lb/ft^2 (1.0 kN/m^2) for partitions) was compared to the fully unloaded live load condition (0 lb/ft^2) on floors 19 through 39. Reductions in live load were applied only as permitted by the Chicago Building Code for the design of bearing walls and columns. No further reduction was taken for unused space because the tenant, a law firm, reportedly stored large volumes of paper in high density filing units. The Chicago Building Code requires that a density of 65 lb/ft^3 (1040 kg/m^3) be assumed for the weight of books and shelving.

Based upon the structural engineers' calculations of building shortening under gravity loading conditions, the exterior steel columns may have elastically rebounded as much as 0.51 in. (13mm) at the level of the 42nd floor (Table 1). Conversely, the calculations indicate that the concrete core may have elastically rebounded as little as 0.05 in. (1mm). The differential elastic rebound of 0.46 in. (12mm) likely induced significant stresses in the roof framing system, enough to cause the failure.

Table 1 - Structural Frame Elastic Shortening (inches (mm))

	Total Elastic Shortening	**Elastic Shortening After Move**	**Elastic Rebound**
Core Wall	0.71 (18)	0.66 (17)	0.05 (1)
Steel Column A-9	2.28 (58)	1.77 (45)	0.51 (13)
Difference			0.46 (12)

The steel framing at this area of the roof, as shown in Figures 2 and 3, is quite complex. Since the connecting W24x68 steel member ("A") framed directly into a corbel attached to the concrete core wall and was oriented at a steep upward angle, it was essentially constrained by the core wall. Beams "B" and "F" framed into the exterior steel columns. After the unloading of the structure, the rebound pushed on these members, inducing a force at the distressed connection. The orientation of the six framing members prohibited translation of the connection. However, the presence of one unsheared bolt suggests a rotational component to the forces acted on the connection.

The structural engineers noted that the connection bolt failure occurred with little observed deformation in the bolt holes and connection plate. Calculations indicated that each 7/8-inch (22mm) diameter A325 bolt had an ultimate shear capacity of approximately 36 kips (160 kN). Based on the observed bolt spacing, the connection material could resist a maximum load of 42 kips (187 kN) per bolt before significant deformation would occur. The total capacity of the sheared connection could resist a maximum of approximately 216 kips (961 kN) (6 bolts at 36 kips (160 kN) per bolt) before failure. The unrestrained axial capacity of the W24x68 beam was 730 kips (3,249 kN). Therefore, the axial compression of the beam would cause failure of all six bolts prior to significant deformation of the connection material or buckling of the beam.

Thermal effects were not believed to be the primary cause of the failure. Although the roof was reportedly "very hot," the interior members were insulated from outside weather conditions. Additionally, in the building's 20-year history no loud unidentifiable noises or damaged connections had previously been observed, regardless of weather cycles. The engineers, therefore, concluded that shearing of the bolts occurred all at once and subsequent to the described tenant "move-out."

Conclusions

It was concluded that uneven loading and unloading of the concrete core caused differential rebound of the building frame and subsequent connection failure. This behavior was initiated by the removal of large live loads imposed by the tenant previously occupying the top 20 floors of the building. Thermal expansion of members may have played a small role, but was not likely the primary cause of the bolt failure.

This investigation revealed the importance of considering the influence of non-local loading as it specifically relates to the potential for non-uniform elastic shortening in structural elements.

References

Bast, Cho, Kurth (Bast 2005). "Bolt Failure Investigation," Client Report (July 20, 2005).

Chicago Building Code. 2004. *Minimum Design Loads*. Group 16, Chapter 13.

Cohen, Barreto, Marchertas Inc. (CBM 1985). Original Structural Drawings, for Building Permit. S2.14, S4.06, S4.01. (April 1, 1985).

Strengthening of Fire Damaged Concrete Joists at a High School

Jerome F. Prugar, P.E. [1]

Scott M. Osowski, P.E. [2]

Joshua Brighton, P.E. [3]

[1] M. ASCE, BSCE, Case Western Reserve University, MSCE Case Western Reserve University, Principal Engineer and President, Prugar Consulting, Inc. 7550 Lucerne Drive, Suite 409, Middleburg Heights, Ohio 44130, PH (440) 891-1414; FAX (440) 891-1454; email: jf.prugar@prugarinc.com.

[2] BSCE Cleveland State University, Project Engineer, Prugar Consulting, Inc. 7550 Lucerne Drive, Suite 409, Middleburg Heights, Ohio 44130, PH (440) 891-1414; FAX (440) 891-1454; email: scott.osowski@prugarinc.com.

[3] BSCE The Pennsylvania State University, MSCE The Pennsylvania State University, Project Engineer, Prugar Consulting, Inc. 7550 Lucerne Drive, Suite 409, Middleburg Heights, Ohio 44130, PH (440) 891-1414; FAX (440) 891-1454; email: josh.brighton@prugarinc.com.

Abstract

Cast-in-place concrete joists at a high school building were severely damaged by an intense fire. Three of the joists exhibited spalled concrete, cracking indicative of shear and bending failure, buckled reinforcing, and a color change in the concrete. The damaged concrete was removed. The bending and shear capacities were reestablished by strengthening the joists with reinforcing steel encased in shotcrete and by clamping the cracked joists with threaded steel rods. The repairs were performed in a relatively short period of time with only minor disruption to the operations at the school due to the planning and teamwork of the school district, building department, contractor, and engineer.

Introduction

Concrete structures are known to be fairly resistant to damage from the excessive heat of a fire. However, experience has shown that when plastic, commonly used for children's toys and furniture, provides the fuel for the fire, the heat from the fire can become intense and the concrete members can be damaged in a relatively short period of time. Such was the case at a high school building.

Prugar Consulting, Inc. (Prugar) was contacted in February 1994 by an insurance company to evaluate the extent of fire damage to the concrete structure of a local high school. The high school had been constructed in the late 1960's with a cast-in-place concrete frame comprised of joists, beams, columns, and a floor slab (Figure 1). The concrete slab was 4 inches thick. The joists, which were spaced at approximately 5 feet, measured 30 inches deep by 15 inches wide. The beams measured 30 inches deep and approximately 48 inches wide. The joists, beams, columns, and the underside of the floor slab were exposed and painted in many areas of the school, including the area that was damaged. The joists were designed and constructed with a two span condition. The spans were not equal with one being 56 feet and the other 42 feet. The joists in the 42 foot span were observed to be sagged on both sides of an intermediate concrete block wall. Thus, the joists in the 42 foot span were inadvertently supported by the intermediate partition wall.

Figure 1 Overall view of fire damaged area.

It was reported that a fire occurred in a storage area on the lower level. The storage area was a fenced area located near a column, directly under four concrete joists spanning 56 feet. It was reported that the fire burned for only one half hour and the storage area was predominantly used to contain children's plastic furniture and plastic toys.

Site Observations

Examination disclosed smoke damage that extended throughout most of the lower level of the school. A portion of the concrete in the area of the reported fire exhibited a pink hue. A large area of the concrete material on three of the joists exhibited a dull thud when sounded with a carpenter's hammer. Some of the concrete was easily dislodged from the joists and a corner of a column when struck by a moderate blow with a hammer. The longitudinal bottom reinforcing was exposed where some of the concrete had been dislodged from the joists (Figure 2). Cracking was observed along the length of the longitudinal reinforcing bars at the bottom of the joists: the cracking extended from one end of the joists to near the midspan. One of the bottom bars was elongated and buckled out near the support of one joist. Three of the joists exhibited diagonal cracking within 4 feet of their support. The diagonal cracking extended from the bottom of the joist to the bottom of the slab and along the intersection between the joist web and the slab. All of the cracking on these three

Figure 2 Fire damage at bottom of joist.

joists exhibited sharp edges and relatively clean fracture surfaces indicating that they occurred recently. The corner of a nearby column exhibited spalled concrete and a slightly buckled steel reinforcing bar.

The three damaged joists were shored immediately so that the classrooms could be used on the upper level directly above the damaged areas. The shoring consisted of a series of heavy adjustable steel posts which were placed under the fire damaged concrete joists.

The school district provided a copy of the original design drawings. No project manual could be located, but the drawings contained sufficient design and material information for an analysis of the concrete floor framing.

Analysis of the Damaged Structure

The joists were analyzed in the two span condition with the material properties specified on the drawings and the loads required in the 1992 Ohio Basic Building Code (Board of Building Standards, 1992). The analysis of the bending and the shear capacities showed that an undamaged joist was capable of safely supporting the dead load and the expected live loads. However, a closer review of the joist reinforcement specified on the drawings revealed that the joists were not designed or constructed with the minimum shear reinforcement required by the 1992 concrete code (ACI, 1992). The minimum shear reinforcement was required by the code because the wider pans resulted in a joist spacing that was greater than the spacing allowed for a bending member to be considered a joist.

Review of the joists with the 1963 concrete code (ACI, 1963) requirements disclosed that the design and construction were in compliance with the code in effect at the time of the original design and construction. The minimum shear reinforcement requirement had been added in the 1971 concrete code (ACI, 1971) to reduce the risk of a catastrophic shear failure. This later requirement would need to be considered in the design of the repair.

A second analysis was performed to consider the effect of the inadvertent support provided by the intermediate block wall. The second analysis considered the full live load and a portion of the dead load to be supported by the intermediate wall. The intermediate support reduced the shear at the interior support of the two span condition, but increased the bending moment in the damaged 56 foot span.

The bending capacity was computed for the joists in their damaged state. The development length of the buckled longitudinal bar was considered to be reduced by the length of the visible buckle plus twelve inches on each end. The development length of the exposed bars was reduced by the exposed surface of the bar plus an additional surface area of 1/2 inch wide along the length of the exposed surface. With these reductions, the bending capacity of the joists was significantly reduced.

The lack of shear reinforcement and the sharp-edged diagonal cracking in the joists indicated that the shear capacity of the joists had been compromised. Furthermore, the loss of development length for the tensile reinforcement at the bottom of the joists indicated that the bending capacity had been compromised. Therefore, the joists required repair in compliance with 1992 standards.

Repair Development Phase

The school district requested that the local building department closely oversee the repair work, especially since the fire damage was widely reported by the local news media. The building department required they be kept informed of design development and design changes during the repair process, be provided engineering drawings for their review prior to starting the repairs, and be updated regularly about the progress of the construction.

A local concrete repair contractor, Exterior Services of Ohio, Inc. (ESO), who had experience in the repair of fire damaged concrete, was contacted to assist with development of the repair method. ESO and Prugar discussed the possible methods of strengthening the joists. These included the installation of a post-tensioning cable, clamping the joist together with steel rods, the addition of steel plates along the bottom of the joists for bending strength and the addition of steel members along the sides for shear strength, and adding reinforcing steel and concrete around the joist. After some preliminary analysis it was determined that the addition of reinforcing steel and concrete around the joists would allow the joists to be exposed with little change in appearance. This method was considered because it had been described in an article written by John F. Vincent, P.E. (Vincent, 1993) of Construction Technologies Laboratories. It was also decided that the new concrete for this method would be shotcrete. The use of shotcrete omitted the need for forms, time to construct forms, and the need to pump and consolidate the concrete in a relatively blind form.

A meeting was held between the school district, building department, engineer, and contractor to discuss expectations and limitations. The possible repair methods were reviewed; the deadlines, the schedules, and the logistics of the repair were discussed; a preliminary plan of action was developed; and the duties of each party were agreed upon for the next step of action.

Two methods of repair were predominantly discussed at the meeting: replacement of the entire floor in the area of the damaged joists or the strengthening of the damaged joists with a reinforced shotcrete. Prugar described the scope for each of these methods. ESO outlined a preliminary schedule for each of these methods. It was decided that strengthening was a better option to pursue, as it allowed the school to continue to operate with the least disruption. Within a day ESO provided preliminary estimates for each method and a preliminary schedule. The information was provided to the school district, building department, and insurance company. The method of repair and reinforcement was agreed upon by all.

The repair of the joists required access to the floor above the fire damaged area on occasion. Since this area continued to be used by the school, the school district required that the work be performed between 6 PM and 4 AM for the convenience of school operations. A removable platform was constructed in the hallway to cover the open holes and slots required for the repair work. The platform had ramps in compliance with the Americans with Disabilities Act. The two classrooms affected by the repair work were to be closed for only two weeks. The school district provided overhead signage warning of the ramp.

Design of the Repair and the Document Preparation

Analysis of the new joists considered the additional weight of the concrete, affect of the support provided by the intermediate concrete block wall on the distribution of the bending moment and shear, severe failure of the existing joists in shear, and transfer of the existing loads to the new portion of the joists. ESO began to submit material information during the preparation of the drawings to reduce the time for approval. The materials were evaluated and included in the specifications. The drawings showing the demolition and the repair details were prepared by Prugar within one week. While the drawings were being prepared, ESO also began to prepare the construction site for selective demolition.

The analysis considered not only the bending moment and the shear for the clear span condition, but also the re-distribution of the bending moment and shear due to the inadvertent support provided by the intermediate walls. The greatest bending moments and the greatest shears were used in the design of the new reinforced portion of the joists even though they could not have occurred concurrently.

The diagonal shear cracking indicated that the concrete joists had failed in shear and were resisting shear by shear friction and the dowel action of the longitudinal bottom reinforcing steel. The shear strength resulting from shear friction was calculated and considered. It was decided to include the shear friction in the shear capacity of the joists. To include the shear friction, the two separated sections of the joist needed to be held together. It was decided that pretensioned threaded rods would be installed to hold the cracked sections of the joists together. See Figures 3 and 4.

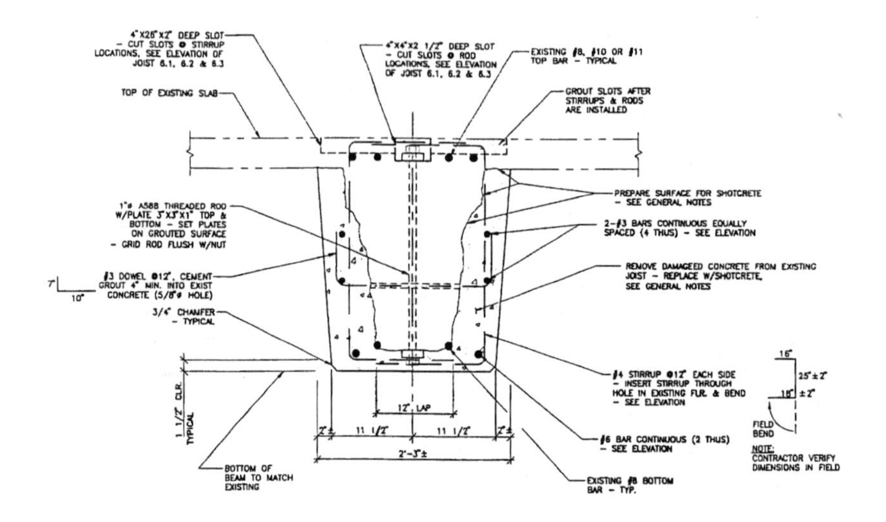

Figure 3 Cross section through joist showing repairs.

Figure 4 Bearing plates for clamping rods set in a grout bed.

Due to installation limitations, it was thought that the new stirrups would be lapped on the sides of the joists. However, ESO suggested that they could field bend the stirrups across the top the joist. Prugar observed their bending method and the condition of the bar at the bend. The diameter of the bend was such that the steel was not kinked. Thus, new L-shaped stirrups were installed from below and the vertical legs bent over the top of the joists and the bottom legs of the L-shaped stirrups were lapped at the bottom. The top legs of the stirrups were concealed in a slot cut in the top of the slab. See Figures 5 through 8.

L-shaped dowels were also grouted into holes drilled into the sides of the joists to provide a mechanical means of holding the shotcrete to the joists. Longitudinal steel was installed in the new portion of the joists for additional bending strength and to reduce shrinkage cracking on the sides. See Figures 3 and 7.

Figure 5 Floor above with slots for joist ties.

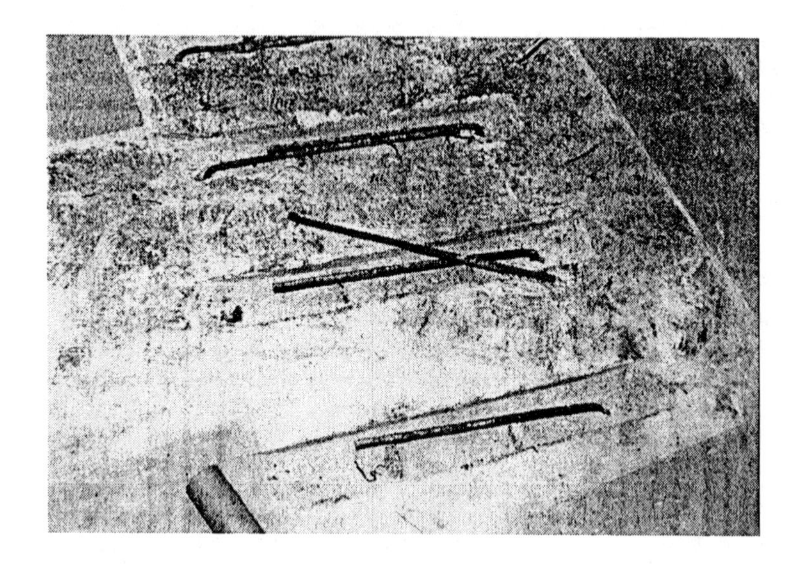

Figure 6 Joist ties bent over into slot.

Figure 7 Joist ties viewed from below.

Figure 8 Side of joist exhibits ties. Bottom exhibits welding plates for longitudinal bars.

The buckling of the longitudinal steel reinforcing in one of the joists caused it to lose a significant portion of its development length. The longitudinal bar was re-secured to the joist by welding plates to the side of the bar: the plate had studs which were embedded and grouted into the side of the joist. See Figures 8 and 9.

Since the joists were supporting the additional weight of the repair, additional bending strength was required. Additional bending strength was obtained by adding longitudinal steel reinforcing bars at the bottom of the new shotcrete. See Figure 3.

The analysis also considered how to transfer the loads from the existing joists into the new portion of the joist. It was decided that the existing joists would be shored and lifted along their length with the purpose of removing as much of the existing live and dead loads as possible. The elastic deflection of the joists was calculated for the actual dead and live loads. It was determined that a portion of this deflection, and thus, a portion of the actual load, would be removed by lifting the joists. A shoring scheme was provided on the drawings showing the amount of lifting required for the joists. When the repair was completed the joists were to be lowered slowly.

The joists in the adjacent 42 foot span exhibited diagonal cracking with paint coverage and paint in the cracking. The paint coverage and paint in the cracking indicate that it was present prior to the fire. The diagonal cracking in the shorter undamaged spans and the analysis of the two span joists disclosed that the lifting of the shorter span might need to be limited to prevent extension of the existing cracking. Thus, the jacking was to be reviewed and adjusted as needed at the site by Prugar and ESO based on observations of the existing cracking.

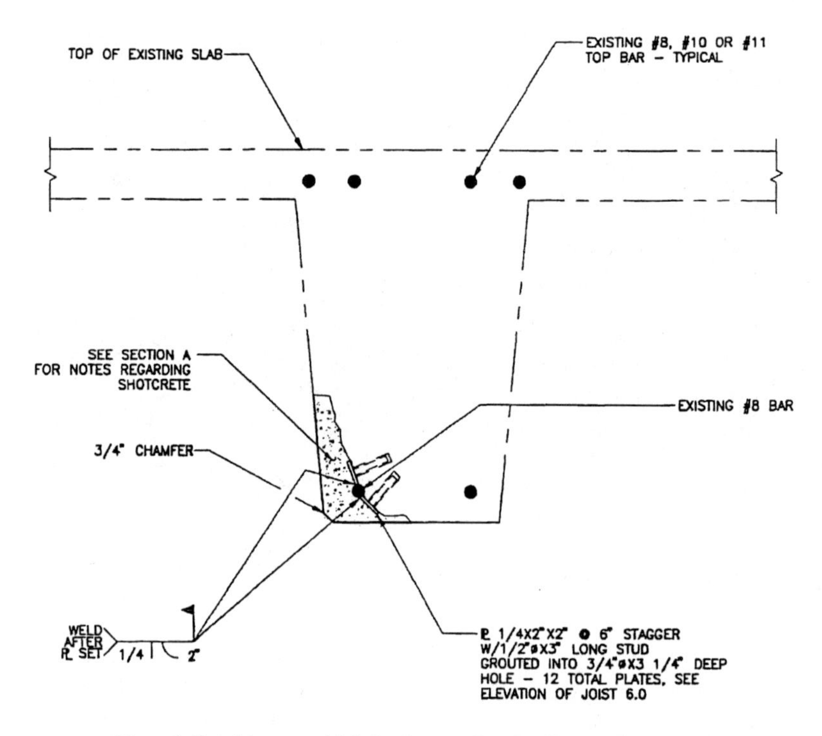

Figure 9 Detail to re-establish development length of longitudinal bar.

For simplicity, all three joists were to receive the same scope of repair required for the most severely damaged joist. The scope would then be reduced or modified in the field at the direction of Prugar.

A fourth joist that exhibited minor damage was also analyzed and determined to have enough extra capacity so that it did not warrant repairs. Due to the great number of columns, and their relatively large size, it was also determined that the fire damaged concrete column only required removal of the loose and damaged concrete and patching of the concrete over the buckled steel reinforcing bar.

Construction Phase

Prugar reviewed the construction work in the evening. Field notes of the changes were provided to the contractor, the school district, the building department, and the insurance company the next morning.

ESO located and marked the pattern of the existing reinforcing steel so that no reinforcing steel was inadvertently cut during demolition or drilling. Also, ESO was asked to check and tighten the temporary shores daily.

The threaded steel rods were installed and pretensioned with the turn of the nut method specified by the American Institute of Steel Construction (AISC, 1985). The bearing plates at each end of the rods were set on a non-shrink grout bed (Figure 4). Prugar was present during the setting and tensioning of the threaded rods on the first day to review the process.

Prugar checked the extent of the loose and damaged concrete removal by sounding the excavated areas with a hammer. Areas that needed more removal were marked by Prugar. Little additional concrete required marking for removal. Removal of the damaged concrete exposed additional longitudinal bars. The development length of these were re-established using the methods discussed previously and shown in Figures 8 and 9.

The new reinforcing was set in place and the shotcrete was applied to the joists. The surface of the shotcrete was cut to a smooth finish to match the existing joists. It gave the appearance of formed joists. See Figure 10.

Figure 10 Finished joists.

The bond of the shotcrete to the existing joists was tested by a laboratory using the pullout test method described by the American Society of Testing Materials (ASTM C-900, 1993 and ASTM C-42, 1994). The test consisted of coring with a 6 inch diameter hollow bit into the side of the joist. The coring extended into the existing joist. A threaded mounting plate was adhered to the core and the core pulled to failure. The first set of tests revealed a bond strength of approximately 90 psi. The failure surface occurred at a cold joint in the new shotcrete material. Although these results were much lower than expected the bond was adequate for the design. A second set of tests showed that the bond to the existing concrete was on the order of 240 psi.

The school district was anxious to paint the area and get it back into full service.. Since shotcrete is drier than concrete, no moisture related painting problems were expected. However, they were advised that the shotcrete should be allowed to cure for 28 days prior to painting. After that time small pieces of plastic sheeting were taped to the sides of the joists and then examined for condensation. No evidence of excessive moisture was observed. The school district painted the area with no moisture related problems.

Final Observations

The construction work and testing were completed approximately two months after the fire occurred. Visual examination of the completed repairs disclosed that they appeared to be in compliance with the intended design. A second review, approximately one year later, disclosed no evidence of distress with the repair. A subsequent visual review while the engineer was a visitor at the school, five years later, also disclosed no evidence of distress with the repair.

Conclusions

The utilization of reinforced shotcrete to restore the shear and bending capacities of a fire damaged concrete frame resulted in a relatively quick repair with little disruption to the daily activities of the school. Shotcrete is a practical option for repair or reinforcement of cast in place concrete joists. Perhaps there can be a study of this type of repair method to better quantify the amount of the strength restored or increased by repairs with shotcrete.

References

American Concrete Institute (ACI), (1963), ACI Standard Building Code Requirements for Reinforced Concrete (ACI 318-63), American Concrete Institute, Detroit, Michigan.

American Concrete Institute (ACI), (1971), Commentary on Building Code Requirements for Reinforced Concrete (ACI 318-71), American Concrete Institute, Detroit, Michigan. Paragraph 11.1.1 and 11.1.2.

American Concrete Institute (ACI), (1992), Building Code Requirements for Reinforced Concrete (ACI 318-89) (Revised 1992) and Commentary – ACI 318R-89 (Revised 1992), American Concrete Institute, Detroit, Michigan.

American Institute of Steel Construction (AISC), (1985), Specification for Structural Joints Using ASTM A325 or A490 Bolts, November 13, 1985. American Institute of Steel Construction, Inc. Chicago, Illinois.

American Society of Testing Materials (ASTM), (1994), C 42-94, Standard Test Method for Obtaining and Testing Drilled Cores and Sawed Beams of Concrete, American Society of Testing Materials, West Conshohocken, Pennsylvania.

American Society of Testing Materials (ASTM), (1993), C 900-93, Standard Test Method for Pullout Strength of Hardened Concrete, American Society of Testing Materials, West Conshohocken, Pennsylvania.

Board of Building Standards, (1992), The 1992 Ohio Basic Building Code with amendments through January 1994, Building Officials & Code Administrators International, Inc., Country Club Hills, Illinois.

Vincent, P.E., John F., 1993, "Comprehensive Parking Garage Restoration Requires Innovative Repairs", Concrete Repair Bulletin, November/December.

EVALUATION OF PRESERVED MATERIALS REVEALS CAUSE OF COLUMN COLLAPSE

Charles J. Russo, P.E.[1] and Glenn R. Bell, P.E.[2]

ABSTRACT

This paper describes an investigation of the partial collapse of a structural steel frame during construction of a shopping mall in December 1999. The accident occurred when an assembly of a 36 ft tall steel column and an 80 ft long girder overturned in high winds and fell to the ground. Two ironworkers attempting to make a girder-to-column connection at the top of the column died in the accident.

The investigation, as well as observations by others, showed substantial deviations in the failed column's base connection to the footing from the requirements of the structural design drawings. As fabricated, the column's baseplate was a different shape with a different configuration and fewer number of anchor bolts than shown on the design drawings. A setting grout pad, placed before erection of the column, was undersized for both the as-designed baseplate and the as-fabricated baseplate. Most significantly, the cast-in column anchor bolts were installed in the wrong configuration for the as-fabricated baseplate, and the contractor cut off these bolts and replaced them with adhesive-set remedial anchors.

A unique aspect of the investigation involved testing and inspection of the anchor bolts from the base of the failed columns as well as concrete cores, which had been removed from the footing, containing the holes in which the adhesive-set anchors had been placed. We performed laboratory tests using CAT scans, materials analysis, and microscopy to examine the anchor-bolt-to-concrete bond interface. Due to several errors in installing the anchors, the anchors had no adhesive bonding to the concrete.

The deviations in the column base connection from the requirements of the structural drawings grossly reduced the overturning capacity of the column. The accident would not have occurred if the column base had been constructed as required by the design drawings.

[1] Principal, Simpson Gumpertz & Heger Inc., 1355 Piccard Drive, Suite 220, Rockville, MD 20850; PH (301) 417-0999; FAX (301) 417-9825; email: cjrusso@sgh.com

[2] Senior Principal, Simpson Gumpertz & Heger Inc., 41 Seyon Street, Building 1, Suite 500, Waltham, MA 02453; PH (781) 907-9000; FAX (781) 907-9009; email: grbell@sgh.com

INTRODUCTION

In December 1999 two ironworkers died when a portion of a steel superstructure of a shopping mall under construction collapsed during erection. Immediately after the collapse, OSHA performed an investigation and revealed that elements of the construction deviated from the contract drawings. These deviations included, at a critical column involved in the collapse, a change in the shape of the column baseplate, a reduction in the number, size, and type of anchor bolts, and a reduction in the bearing area beneath the column. However, OSHA identified neither the technical nor procedural cause(s) of the collapse.

This paper describes a structural investigation conducted by the authors more than 3-1/2 years after the collapse using project documents, recorded witness accounts and photographs, and a combination of nondestructive and destructive testing on preserved samples collected at the time of the collapse.

THE STRUCTURE

Figure 1 shows a plan of the shopping mall highlighting a food court where the collapse occurred. The principal components of the single-story food court, as shown in Figure 2, are two 18 in. diameter, 36 ft tall steel columns that each function as the center of a 90-degree layout of radial girders. The baseplates for these columns are identified on the structural plans as 30 in. diameter circular plates with eight, circumferential, 1 in. diameter anchor bolts.

The 90-degree radial girders are 36 in. deep and span 80 ft from the central columns to portions of the adjacent structure. Two additional orthogonal beams frame into each central column. The structural plans show a detail for the beam-to-column connections for the four radial girders and two orthogonal beams. The detail uses through plates to form a crucifix for the orthogonal beams, and welded shear tabs for the radial girders. The girders and the beams are bolted to the connector plates. This connection was not completely detailed on the structural drawings.

The General Notes on the structural drawings state that the connections are generally schematic and intended to define the spatial relationship of the framed members and show a feasible method of making the connection. Any connection not shown or not completely detailed on the structural drawings was to be designed by a registered professional engineer retained by the fabricator.

The shop drawings for the subject elements of the steel structure bear the review stamps of the Structural Engineer of Record (SER) and are marked either "Approved" or "Approved with Comments". On the shop drawings, the beam-and-girder-to-column connections are detailed with a series of plates in the form of a crucifix, similar to that shown on the structural design drawings. The column cap plate is listed in the bill of materials as an 18 in. square plate. In review comments the baseplate for the column is listed in the bill of materials as a 30 in. square plate. There is no detail for the anchor bolt pattern on the baseplate. The SER circled the listings for the cap plate and baseplate and wrote "round" next to them.

CONSTRUCTION DEVIATIONS

During construction of the spread footings for the central columns of the food court, the concrete subcontractor incorrectly installed cast-in anchor bolts in a configuration required for an 18 in. square baseplate used for the typical columns of the food court. Workers discovered this deviation after the concrete subcontractor had completed the installation of grout leveling beds for the steel columns. The subcontractor used an 18 in. square template to set the required elevation for the bases of the central columns and placed grout beneath the template to form a rectangular 18 in. square leveling bed.

In response a request for information submitted the contractor, which is not specifically associated with the incorrect anchor bolt pattern for the central columns of the food court, the SER issued a sketch showing a method to repair anchor bolts that were broken off at the footing by drilling in new anchors for the baseplates. The method given at a missing anchor bolt is to (1) locate new anchor bolt within 2-1/2 in. of the existing bolt hole, (2) drill a new hole into the concrete substrate with a minimum 8 in. embedment, (3) flame cut the baseplate for a new bolt hole (maximum 1-1/4 in. diameter), and (4) install an epoxy-bonded, ¾ in. diameter threaded rod with 36 ksi yield stress.

At the central column involved in the collapse the general contractor and the concrete subcontractor elected to remedy the incorrect anchor bolt pattern by cutting the cast-in anchors at the top of the grout leveling bed and installing epoxy-bonded remedial anchors at the appropriate pattern for the 30 in. square baseplate of the central columns. In as much as the baseplates of the central columns were substantially larger than the typical columns, these remedial anchors fell outside the already-installed grout leveling bed. Figure 3 is an illustration of the baseplate configuration with the remedial anchors and the grout leveling bed.

THE COLLAPSE

At the time of the collapse, employees of the structural steel erector had installed two pieces of steel: one of the 18 in. diameter columns and one of the 80 ft girders between this column and the already erected main portion of the building (Photo 1). The following summarizes witness accounts of the activities:

- Other employees had voiced concern about the column, with a 30 in. square baseplate being erected onto an 18 in. square concrete grout bed above the top of the footing (Photo 2).
- The erector placed the column onto the grout bed and remedial anchor bolts, and placed nuts on the bolts to hold the column in place. On one of the bolts

approximately 20 washers were placed over the baseplate before the nut was installed.

- The erector placed the first of the 80 ft girders between the 18 in. diameter central column and the already erected main portion of the building.
- Upon hoisting a second girder to be installed perpendicular to the first girder, the erectors realized that the square cap plate on the 18 in. diameter column and the flanges of the girders interfered with each other and would not permit direct connection. The crane operator held off the second girder to allow the ironworkers to make field modifications to the erected steel.
- One ironworker tied himself off to the first girder and proceeded to torch cut the steel. A second ironworker remained in a manlift adjacent to the column and held off one end of the second girder being suspended by the crane.
- The first girder and column toppled over, striking the manlift, and the two ironworkers fell to the ground.
- There are no witness reports of incidental contact between the second girder and the column.
- The crane operator reported that just before the failure he observed a gap between the nut and the stack of washers on one of the anchor bolts. While communicating with others on the ground concerning his observation, the erected assembly collapsed.

INITIAL FIELD INVESTIGATIONS BY OTHERS

The Occupational Safety and Health Administration investigated the collapse and revealed that elements of the construction varied from the contract drawings at the critical central columns. These deviations included a change in the shape of the column baseplate, a reduction in the number, size, and type of anchor bolts, and a reduction in the bearing area beneath the column. OSHA's inspection report stated that the general contractor or representatives of the general contractor changed the design of the steel column that collapsed. OSHA issued a Citation and Notification of Penalty to the general contractor with two citation items. The first citation was for exposing the steel erectors "to a collapse of the structural steel as a result of a change in the anchor bolts." The second citation was issued for replacing anchor bolts "without a competent person evaluation of the revised installation procedure and compliance with manufacturer's recommendations."

Relevant information contained in the OSHA report included:

- The contractor installed the remedial, epoxy anchor bolts approximately 1 to 2 weeks before the placement of the column.
- There did not appear to be any epoxy on the bolts that had pulled out of the concrete when the column fell over.
- The original design plan for the column had been changed at some time between design and erection.
- On the day of the accident, the winds were reported at approximately 20 to 30 mph.

In a separate inspection report, OSHA stated that it found no violations related to the accident on the part of the structural steel erectors.

Four days after the collapse, a testing agency retained by the developer conducted a pullout test on one of the remedial epoxy-bonded anchors of the other 18 in. diameter central column. The adhesive anchor began to move vertically at an applied load of 5,110 lbs, and was unable to resist additional load. The testing agency did not describe the condition of the extracted adhesive anchor, and no photographs are included in their report.

The general contractor retained and cataloged the four anchor bolts from the column involved in the collapse (Photo 3) and the anchor bolt from the pullout test on a companion column. The general contractor also extracted and retained large diameter cores of the foundation of collapsed column that included the drilled holes for the remedial anchors.

COLLAPSE INVESTIGATION

Our collapse investigation included the following tasks:

- a review of the project documentation
- a review of the investigations of others contemporaneous with the collapse
- an independent structural analysis of the loads on the partially erected system at the time of the collapse, the overturning resistance of the various base conditions, and the stability of the erected components
- independent nondestructive and destructive testing on the retained anchor bolts and concrete cores from the foundation

STRUCTURAL ANALYSIS

Loads Acting at the Time of Collapse

We calculated lateral wind pressures on the erected components using drag coefficients for the member shapes and applicable provisions of ASCE 7: Minimum Design Loads for Buildings and Other Structures. Based on weather records from an airport near the construction site we considered wind speeds between 25 mph and 35 mph.

Load combinations include (1) self weight of the steel components, a 200 lb allowance for the ironworker who was torching the flanges to eliminate the interference between components, and the lateral pressure from a 25 mph wind, and (2) self weight, 200 lb allowance, and a 35 mph wind. The resulting overturning moments at the base of the central column are 16 ft-kips for the 25 mph load combination and 30 ft-kips for the 35 mph load combination.

Overturning Resistance of Base Conditions

We calculated the ultimate resistance to base overturning of the 18 in. diameter column with a 30 in. square baseplate and four anchor bolts for the following bearing conditions:

Case	Grout Bearing	Anchor Bolts	Anchor Pullout Strength (lb.)	Base Connection Strength (ft.- k)
1	Full (30 in.)	A307 cast-in anchors per original design	20,000	100
2	Full (30 in.)	Epoxy set anchors at manufacturer's rated ultimate strength	19,200	100
3	Partial (18 in.)	Epoxy set anchors at manufacturer's rated ultimate strength	19,200	75
4	Partial (18 in.)	Epoxy set anchor strength as tested at companion column	5,100	25

Stability of Erected Components

We evaluated the stability of the girder/column assembly during erection using a finite-element computer program. The analysis showed that, so long as the column or column baseplate do not fail, the 80 ft long girder is stable in the erected assembly with a factor of safety against lateral buckling of about 2.3 for both the 18 in. partial bearing and 30 in. full bearing base conditions.

LABORATORY ANALYSIS

Anchor Bolts

Our testing and examination of the five anchor bolts consisted of:

- general visual and microscopic observations
- extraction of samples of residue taken from the bolts and testing of such samples by Fourier Transform Infrared Spectroscopy (FTIR)
- extraction of samples of residue taken from the bolts and high-powered microscopic examination of grain mounts prepared from those samples
- measurements of bolt diameter

Cores

Our testing and examination of the four cores consisted of:

- Computerized Axial Tomography (CAT) scanning on the cores to identify drill hole orientation and diameter with depth
- splitting open of the cores to examine the drill hole and extracting samples for further testing
- extraction of samples of materials taken from the cores and testing of such samples by Fourier Transform Infrared Spectroscopy (FTIR)
- extraction of samples of materials taken from the cores and high-powered microscopic examination of grain mounts prepared from those samples
- extraction of samples of material taken from the cores and testing of such samples by Scanning Electron Microscopy (SEM) and Energy Dispersive X-ray Spectroscopy (EDS)
- physical measurements of hole diameter

Diametric Fit of Holes and Bolts

We used computerized Axial Tomography (CAT) scans on the specimen concrete cores to characterize the holes present inside the cores. The CAT scan equipment is a medical diagnostic scanner, Picker IQ Premier CAT Scanner, SN 1436, with documented spatial linearity deviations from unity of less than 0.7%.

Each specimen was marked with a lead wire mounted on the side of the core at the exposed end. This wire is dense to the x-rays and appears on the images as a reference mark. There are approximately 50 transverse images every 0.5 cm from below the bottom of the hole to the top face of the core.

We calculated the diameters of the holes at various depths using the images from the CAT scan. Scanning at specific depths avoided interpolation artifacts. The scan provided the angle of the hole offset from vertical and the diameter of the imaged hole.

The hole diameters measured by the CAT scans range from 0.69 to 0.75 in. Measurements were to the hole edge where the CAT scan image pixels indicate a change to high density (i.e., the edge of the concrete). Diameters to the inside surface of the concrete dust layers would be somewhat smaller (nominally about 0.02 in.) depending on the thickness of the dust layer.

The measurements on cleaned surfaces of benchmark holes made with an 11/16 in. drill bit obtained from the adhesive manufacturer yielded diameters of 0.702 to 0.727 in. This compares favorably to the 0.69 to 0.75 in. measured by CAT scan on the project cores. The project correspondence file includes a receipt from the manufacturer of the adhesive system to the concrete subcontractor for an 11/16 in. masonry drill bit. Installation instructions for the adhesive system require a drill bit of 13/16 in. diameter for the 3/4 in. nominal diameter anchors used on the project.

The upper thread measurements for the four bolts from the collapsed column range from 0.718 in. to 0.738 in., with a mean of 0.732 in. Comparing these thread measurements to the 0.69 to 0.75 in. measured diameters in the lower parts of the holes indicates the bolts would have had a tight fit on insertion, especially in the lower parts of the holes. Concrete dust coatings will make the fit even tighter.

The anchor bolts have varying degrees of damage on the bottom threads that is consistent with a forced insertion. This agrees with our "tight-fit" thread and hole measurements.

Core Splitting Technique

We used a four-step method to open the concrete cores containing drilled holes. First, obtain images of the core internal holes using CAT scans (transverse X-ray image slices every 0.5 cm from just below the bottom of the hole to the top face of the core). Using computer post-processing, combine these slice images into a three-dimensional image composite, processed to yield longitudinal images of the hole at orientations parallel and perpendicular to the longitudinal plane of the hole. In preparation for the imaging, a location mark placed on the side of the core allows location of the position of the images on the core afterwards.

Second, we determined the angle of offset from vertical for each hole, and the orientation of the hole dip angle relative to the location mark. We transcribed the hole angles onto the core surface, and determined the location of the plane that contained the hole and was perpendicular to the top face of the core. Marking this plane on the sides of the core guided the saw kerfs along the core sides. This also allowed determination of the maximum depth of the saw kerfs that would be less than the depth of the hole from the sides of the cores.

Third, mount the cores, with the open hole end protected with duct tape, on the saw table, and wet cut the marked saw kerfs. Each cylinder receives two cuts, both in the same plane with the hole, and on opposing sides of the core.

Fourth, place the cores on their sides on a hard level surface, and inside a large plastic bag. Orient the cylinder such that the plane of saw kerfs and hole is vertical. Place a cold chisel into the upward oriented kerf, and strike the chisel with a firm blow from a mallet. We split the core open along the kerf - hole boundary, exposing the hole (Photo 4). We collect any loose debris in the plastic bag, and photograph the location of debris as the hole is opened.

Characterization of Material in Drill Holes and on Anchor Bolts

Based on our observations of materials in the holes of the split cores and on project-based knowledge of the history of installation of the anchors, the failure, and extraction of the cores, we concluded that the likely sources of the material in the drill holes, from bottom to top, are as follows (see Figure 4):

- Compacted concrete drill dust: Dust from the drilling of the bolt holes, incompletely cleaned before insertion of adhesive.
- Adhesive: Adhesive injected in the hole during anchor bolt installation. The FTIR signature matched example specimens of the epoxy adhesive we prepared in our laboratory.
- Soil/aggregate/concrete debris: Site materials that fell or washed into the holes, in large measure or in whole, after the bolts withdrew during the failure.

The material on the sides of the shafts of the embedded ends of all five anchors bolts is concrete dust (Photo 5). There was no adhesive on the sides of the drill holes above the layers of the adhesive plugs or on the sides of the shafts of the anchor bolts. There were traces of adhesive on the bottom tips of two of the embedded anchor bolts.

Insertion Depth and Anchor Nut Cinch-Down

With the cores split open we examined the depth of insertion of each of the four anchor bolts at the collapsed column, and from this, analyzed whether or not the bolts were embedded sufficiently far to allow the anchor nuts on each bolt to have been cinched down to the top of the baseplate. Except where otherwise noted, the cinch-down analysis is based on the presence of one washer under the head of the anchor bolt nut. "Available" insertion depth means the insertion depth that would be achieved if the anchor bolt was inserted, so that the bottom of the bolt met the top of the adhesive plug.

At one of the anchor bolts on the tension side of the baseplate, examination of the anchor bolt at the installed advanced location of the nut and post-failure photographs showing a tape measure placed against the anchor bolt indicate about 5-7/8 in. of embedment into the concrete. The available depth for embedment is about 7-3/16 in.

At the second anchor bolt on the tension side of the baseplate, physical evidence on the anchor bolt indicates a 6-3/4 in. depth of bolt embedment. In addition, photos taken of this anchor bolt by others after the failure show 6-3/4 in. of residue on the sides of the bolt. The available embedment based on core measurements is 6-3/4 in.

Lack of Cleanliness of Drill Holes

The bottom of the drill holes were filled with 13/16 in. to 1-3/16 in. of concrete dust. In addition, the inside surfaces of the drill holes, even in the depths of the holes filled with adhesive plugs, contained concrete dust. These observations indicate the installer(s) did not properly clean the holes prior to injection of the adhesive. In his deposition, an employee for the concrete subcontractor testified that he brushed the holes out but did not blow them out. This is not in accordance with the manufacturer's installation instructions, which require blowing and brushing.

Lack of Adhesive Bonding on Anchors

The analysis of the materials on embedded parts of the shafts (including the lower threaded portion) of the five anchor bolts revealed no adhesive. On two bolts there were traces of adhesive on the bottom bolt tips.

The lack of adhesive bonding on the bolt sides is due principally to the holes being drilled to the wrong diameter. There was a tight fit between the outer diameter of the threaded portion of the bolts and the inside diameter of the holes. The lack of annular space between the bolt and hole walls left no room for adhesive to flow up around the bolt when the bolt was inserted in the hole. In addition, too little adhesive was placed in the hole. The manufacturer's instruction for installing the adhesive-set anchors into concrete requires the drill holes to be filled approximately half full of adhesive before inserting the bolt. We found far less adhesive in the cores. During insertion of the anchor bolts in the epoxy system, the installer should ensure that adhesive flows to the top of the hole. Such a check would have revealed the problems with adhesive bonding on the bolts.

Withdrawal Strength of Anchor Bolts

Based on the observations and materials testing reported herein we concluded that adhesive did not contribute to the withdrawal strength of any of the four anchor bolts at the collapsed column or to the single anchor bolt subjected to the pullout test at the companion column. Whatever pullout resistance the bolts had was derived from friction between the bolt and the sides of the core hole, owing to the tight fit between the bolt and the hole.

We carefully compared the condition of the embedded ends of the anchor bolt from the pull test at the companion column, and the embedded ends of anchors from the tensile side of the collapsed column. The condition of the threads on one of the tensile bolts, in terms of the type and extent of concrete debris in the threads and the nature and degree of damage to the threads, was similar. Thus, we concluded that one of the tensile bolts had similar pullout resistance to the pull-tested bolt at the companion column (5,110 lb).

The second tensile bolt of the collapsed column was different, however. Most of the debris on the embedded end of the rod was mud-like, and there was relatively little concrete debris. Moreover, there was considerably less thread damage on the embedded end of this second bolt than the pull-tested bolt on the companion column or the other tensile bolt on the collapsed column. From this we concluded that the second tensile bolt at the collapsed column pulled out at substantially less load than the pull-tested bolt.

Overturning Resistance of Column Base

Considering the partial-bearing baseplate (18 in.) and two anchors on the tension side of the connection effective at the tested pullout value of 5,100 lb, we calculated a resistance of 25 ft-k. Recognizing the lesser pullout resistance of one of the tensile anchors, as described above, we prepared another calculation based on the assumption of one anchor effective at 5,100 lb. This calculation results in an overturning resistance of 16 ft-k. Thus we conclude that the actual resistance of the column base was in the range of 16 to 25 ft-k, probably near the middle part of that range.

The calculated overturning moments from our analysis, as discussed earlier, are 16 ft-kips for the 25 mph load combination and 30 ft-kips for the 35 mph load combination.

CONCLUSIONS

From our investigation described above we drew the following conclusions:

- The collapse most likely initiated when the central 18 in. diameter column overturned under the action of wind, incidental impact, or a combination of wind and incidental impact.
- The anchor bolts at central column had no significant adhesive bond into the concrete footing. The small resistance they had relative to the expected resistance of properly installed anchors was from unreliable friction between the bolt shafts and the concrete drill holes. This understrength is the critical factor in reducing the overturning resistance and creating the potential for failure.
- The lack of adhesive bonding of the anchor bolts at the collapsed column was caused by improper installation of the anchors. Most significantly, an 11/16 in. diameter drill bit was used for the 3/4 in. bolts, resulting in a tight fit of the bolt in the hole with no annular space between the bolt and the hole for the adhesive to flow. The manufacturer's literature for the epoxy adhesive system requires a 13/16 in. drill bit for the 3/4 in. diameter bolts.
- Other deviations in the installation from adhesive manufacturer's instructions include failure to properly clean the drill hole, use of too little adhesive in the hole, and failure to ensure that, on insertion of the bolts, the adhesive completely filled the holes.
- The change in baseplate bearing from a full 30 in. bearing to a partial 18 in. bearing represent a 25% reduction in resistance to overturning of the column. However, if the adhesive anchor bolts obtained their rated capacity, this reduction would not represent a diminution in overturning resistance sufficient to cause the collapse.

REFERENCES

American Society of Civil Engineers (ASCE). ANSI/ASCE 7-Minimum Design Loads for Buildings and Other Structures, ASCE, Reston, Virginia.

ILLUSTRATIONS

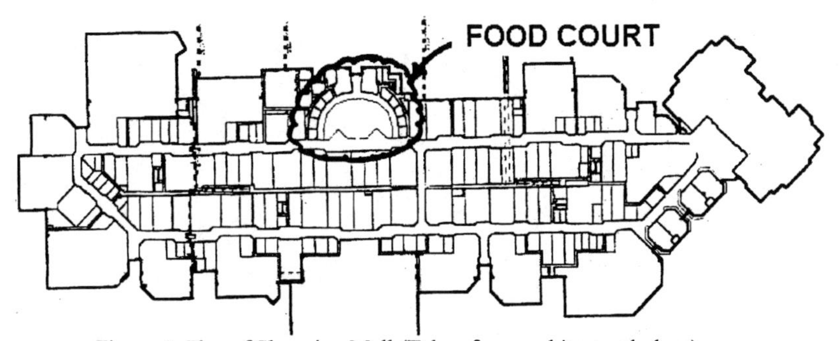

Figure 1: Plan of Shopping Mall (Taken from architectural plans)

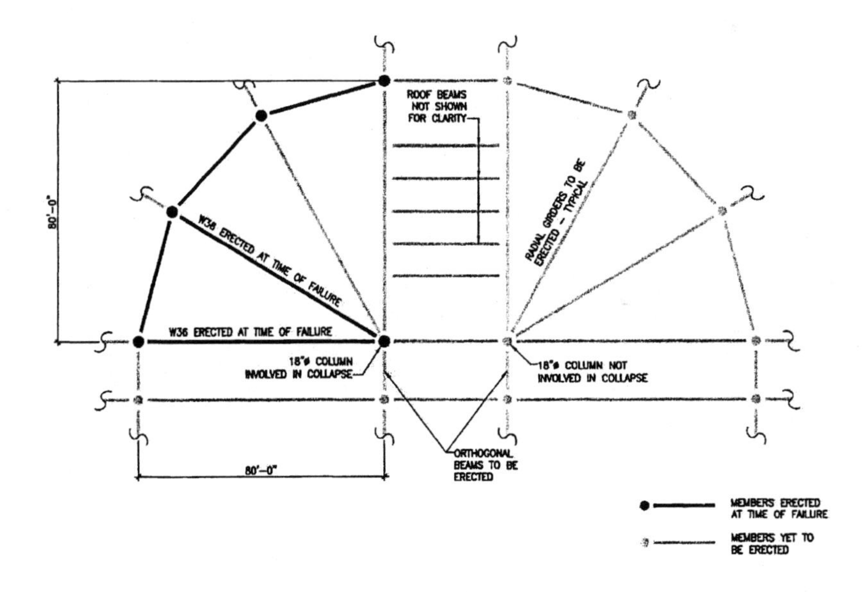

Figure 2: Partial Plan of Roof Structure at Collapsed Area
(Taken from structural plans)

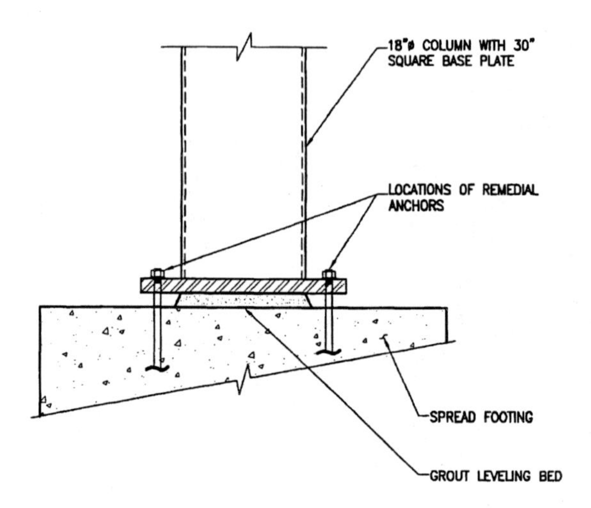

Figure 3: As-built Column Base Condition

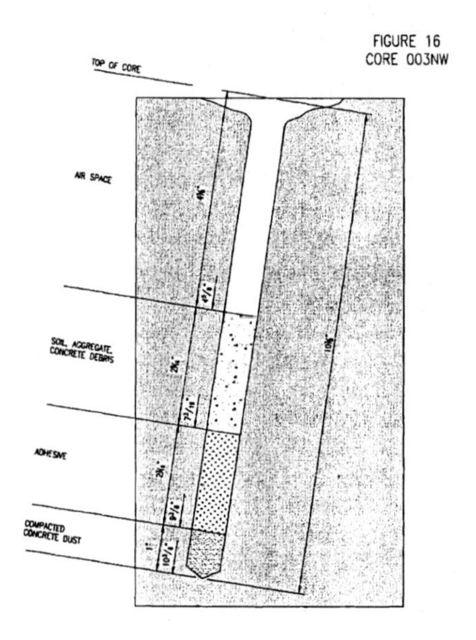

Figure 4: Materials Found in Core 003NW

Photo 1: Scene of Collapse (Source: General Contractor)

Photo 2: Column Base After Collapse (Source: General Contractor)

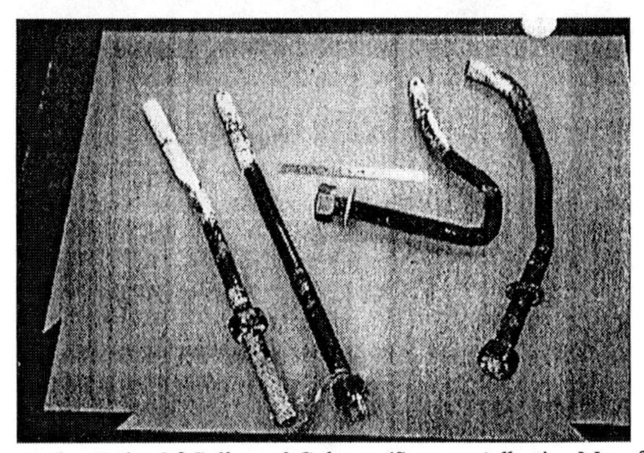

Photo 3: Anchor Bolts Of Collapsed Column (Source: Adhesive Manufacturer)

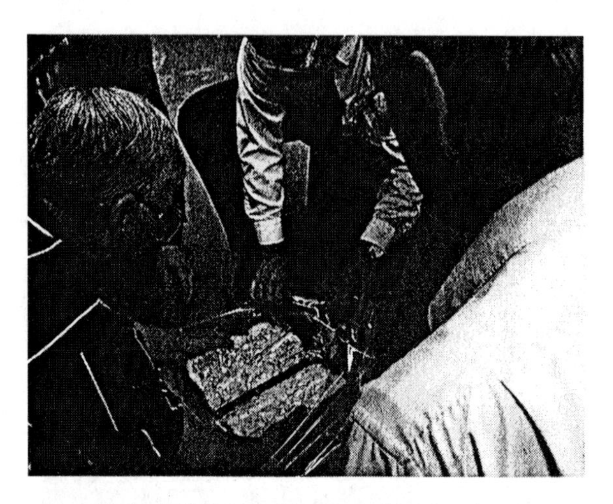

Photo 4: Splitting of Concrete Core Specimen
(Source: Simpson Gumpertz & Heger Inc.)

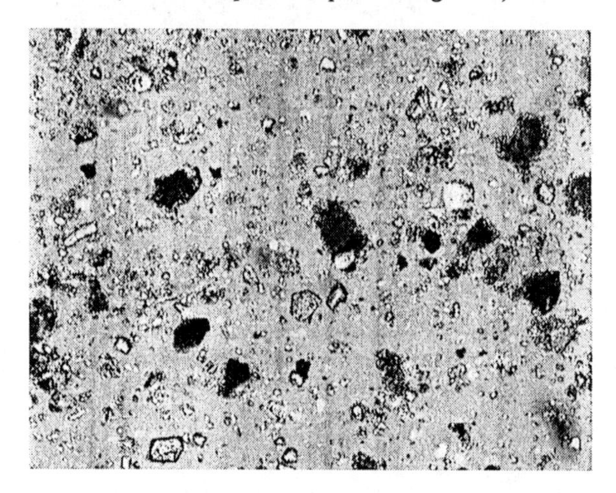

Photo 5: Example of White Material from Anchor Bolt Thread Showing Predominately Hydrated Cement Paste, Aggregate Particles, and Minor Amounts of Soil.
(Source: Simpson Gumpertz & Heger Inc.)

Structural Evaluation and Repair of Internally Damaged Concrete

Ethan C. Dodge,[1] and Matthew R. Sherman[2]

[1]Staff Engineer, Simpson Gumpertz & Heger Inc., 41 Seyon Street, Bldg. 1, Suite 500, Waltham, MA 02453

[2]Senior Project Manager, Simpson Gumpertz & Heger Inc., 41 Seyon Street, Bldg. 1, Suite 500, Waltham, MA 02453

Abstract

This paper presents an overview of the effective use of non-destructive testing at two evaluation and repair projects involving concrete structures with internal defects. Frequently, visual inspection is insufficient to locate the extent of known or suspected damage in concrete, such as in the case of concrete undergoing internal distress, deteriorated bond of overlays, underside damage on concrete placed on grade or on stay-in-place forms, or from poor concrete consolidation. This paper presents the application of Impulse Response (IR) testing to locate internal planar fractures and micro-cracking in an elevated viaduct and to locate areas of poor concrete consolidation and horizontal cold joints in a structural slab. This paper also presents an overview of the repair methods and explains how IR testing can be used to measure the success of the repairs.

Background

The IR test method was first used for testing of deep foundations circa 1970; its application has since grown to include concrete structures of all types. The basic operating principle involves striking the concrete surface with a rubber-tipped instrumented hammer and measuring the response of the structure to the recorded impact using a velocity transducer (geophone). After the time-histories of the hammer blow and measured velocities are transformed into the frequency domain using a Fast Fourier Transform (FFT) algorithm, the test data is standardized to account for the varying strengths of the hammer blows by dividing the magnitude and frequency of the structural response by the magnitude and frequency content of the mechanical input. The fundamental parameter of the test method is referred to as the average mobility given in units of (m/s /N), which is essentially the structural output divided by the force input. The average mobility can be viewed as a measure of the response of the structure as a whole to a known stimulus (Malhotra and Carino,

2004). Additional information can also be gained from the slope of the average mobility plot, "mobility slope", over varying frequencies.

Introduction

We present two studies of the use of the IR test method and supplemental concrete coring to locate and assess internal deterioration or flaws that could not otherwise have been evaluated. The subject structures include an elevated concrete and steel beam train viaduct, and a 2-ft-thick (0.6 m) structural slab placed in two horizontal lifts with a long delay between the lifts. The structures cover a range of concrete deterioration issues that are not apparent during a visual condition survey. The approaches used for evaluation differ slightly in each case, but they rely on the following basic procedures typically used to evaluate concrete structures with the potential for hidden defects:

- Review design calculations and structural drawings to develop an understanding of the designer's intent and the structure's behavior.
- Gather information from involved parties to reconstruct what is known regarding the cause and location of the potential problem areas.
- Make visual observations, conduct non-destructive testing, and remove concrete samples.
- Correlate the condition of the core samples to visual observations and non-destructive test results for calibration and verification.
- Evaluate the test data in conjunction with the physical samples to determine the condition of the structure and the effectiveness of the test program.

Investigation Case Studies

Case 1 - Train Viaduct. The multi-span elevated train viaduct was constructed circa 1912 using lightly-reinforced concrete panels placed into a grid of built-up steel floor beams and main support girders (Photo 1).

Photo 1. Overview of train viaduct.

The support piers are spaced at approximately 48 ft along the length of the 29 ft wide viaduct. The concrete panels are typically 7 ft-6 inches long, vary from 4 ft to 6 ft-6 inches in width, and vary from 10 inches to 14 inches in thickness. Prior to performing scheduled maintenance a structural condition assessment was performed. The assessment involved visual inspection and hammer sounding to help identify areas with potential concrete deterioration and to choose the locations of concrete core samples. The concrete core samples showed that the concrete was not air entrained, contained reactive aggregates, and was in varying states of deterioration due primarily to Alkali Silicate Reaction (ASR). Early attempts to define the extent of concrete deterioration resulted in the removal of 46 concrete cores from two spans of the structure, but did not reliably define the extent of the actual concrete deterioration.

Following these attempts to define the extent of concrete deterioration for repairs using coring, IR tests were performed on each concrete panel to better define the extent of deterioration so that a comprehensive rehabilitation program could be developed. The IR test used a grid of test points beginning approximately 6 inches (15 cm) from the edges and spaced at approximately 1 foot (30 cm) on center in the middle of the panels. The core samples were classified as "good", "partial", and "bad" based on observations during the drilling process and on visual inspections in the laboratory (Photo 2), for comparison to the IR test results.

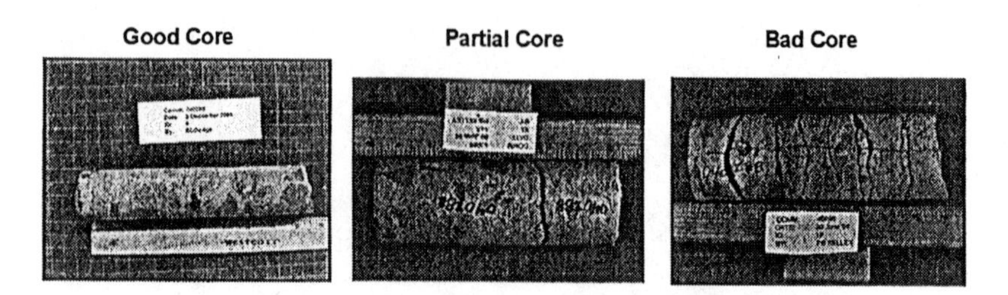

Photo 2. Classification of concrete core samples.

The "good" cores showed no signs of ASR and were removed intact. The "partial" cores showed some signs of internal cracking and were removed in 1 or 2 sections with little evidence of fracture planes in the drilled holes. The "bad" cores were removed in multiple sections and showed significant signs of ASR and fracture planes in the drilled holes. The "partial" cores typically were most clearly indicated by the average mobility values, the "bad" cores were typically most clearly indicated by the mobility slope values. All of the core conditions were closely indicated by the mathematical product of the average mobility and the mobility slope parameters ("mobility times slope") in Figure 1.

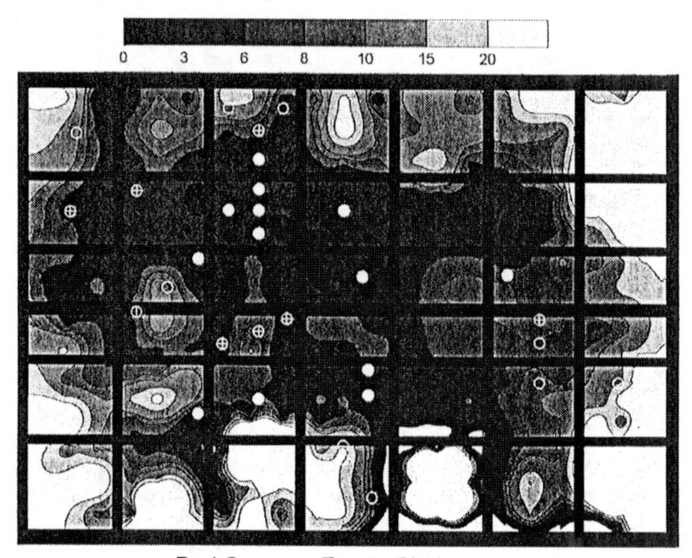

Figure 1. Mobility times slope IR (m/s /N) values and core classifications for a single span of the structure.

The IR results and core classifications show that the concrete in areas with "mobility times slope" values greater than 6 (m/s /N) was damaged to some degree and that the concrete in areas with "mobility times slope" values greater than 10 (m/s /N) was severely deteriorated.

Case 2 - Poor Consolidation Between Concrete Lifts. A circular structural concrete roof slab was constructed with up to a 10-hour delay between concrete lifts. The roof slab is heavily reinforced, approximately 2-ft–thick (0.6m), and is cast integrally with several mechanical structures. The upper concrete lift was placed directly onto the hardened lower lifts of varying elevation (Photo 3).

Photo 3. Variable elevation hardened lower lift of concrete.

Document review and conversations with involved parties regarding the construction of the structure revealed that there were three potential concrete conditions that required investigation: sound concrete, areas with potential defects, and areas with likely internal defects. The initial investigation used IR tests spaced at 2 feet on center over approximately 60% of the slab surface selected to cover the range of expected concrete conditions. Based on a preliminary interpretation of the IR data, we selected 18 core locations to correlate the existing condition to the range of observed IR values. The core samples were classified as "good", "partial", and "poor" based on observations during the drilling process and visual inspections in the laboratory. A "good" core showed no signs of voiding or cold joints (Photo 4).

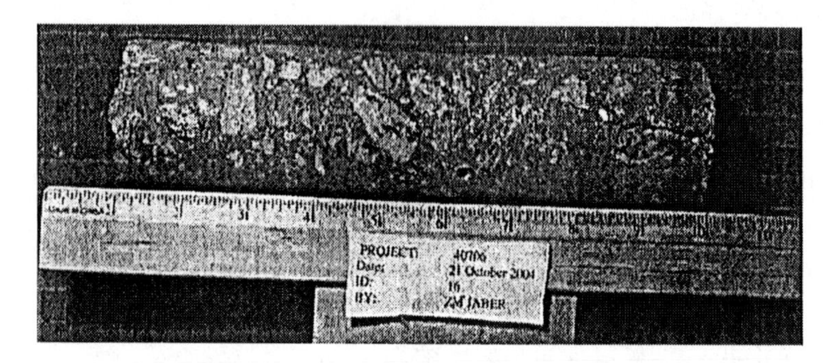

Photo 4. Concrete core classified as "good".

A "partial" core showed minor voids through a low percentage of the core cross-section and no cold joints (Photo 5).

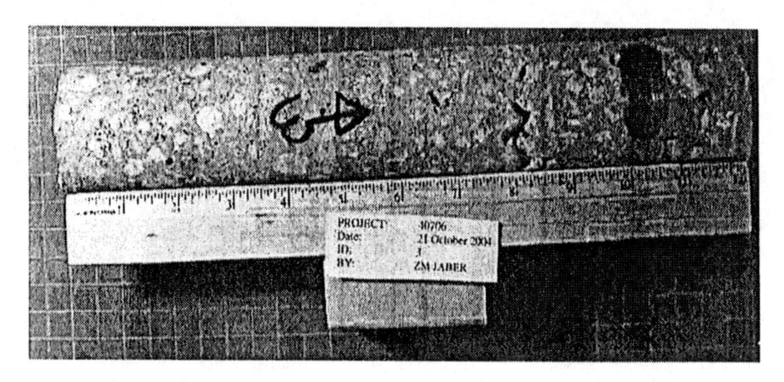

Photo 5. Concrete core classified as "partial".

A "poor" core showed large voids over the majority of the core cross-section or cold joints (Photo 6).

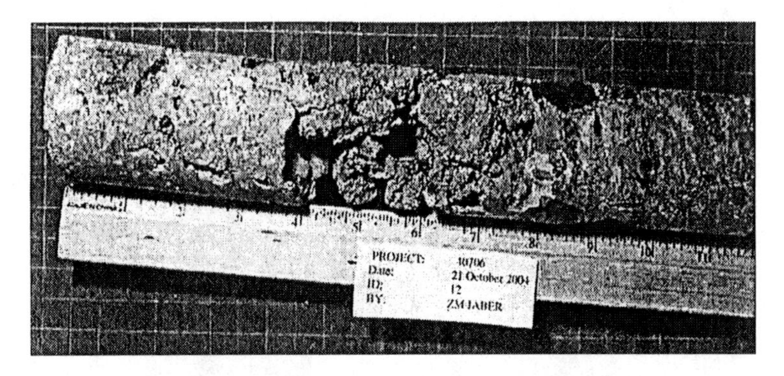

Photo 6. Concrete core classified as "poor".

As at the viaduct, based on the calibration coring the average mobility values were separated into categories. Although the same test method was used, the different structural configurations of the two structures and different nature of the internal defects caused them to respond differently. As such, at the roof slab, the average mobility was a better indicator of the internal defects, and it was used to classify the concrete condition as follows:

- Average mobility values less than 2.25 typically correspond to areas with core samples classified as "good".
- Average mobility values between 2.25 and 4.0 typically correspond to areas with core samples classified as 'partial".
- Average mobility values greater than 4.0 typically correspond to areas with core samples classified as "poor".

The core locations and classifications are presented with the average mobility values from the IR testing in Figure 2.

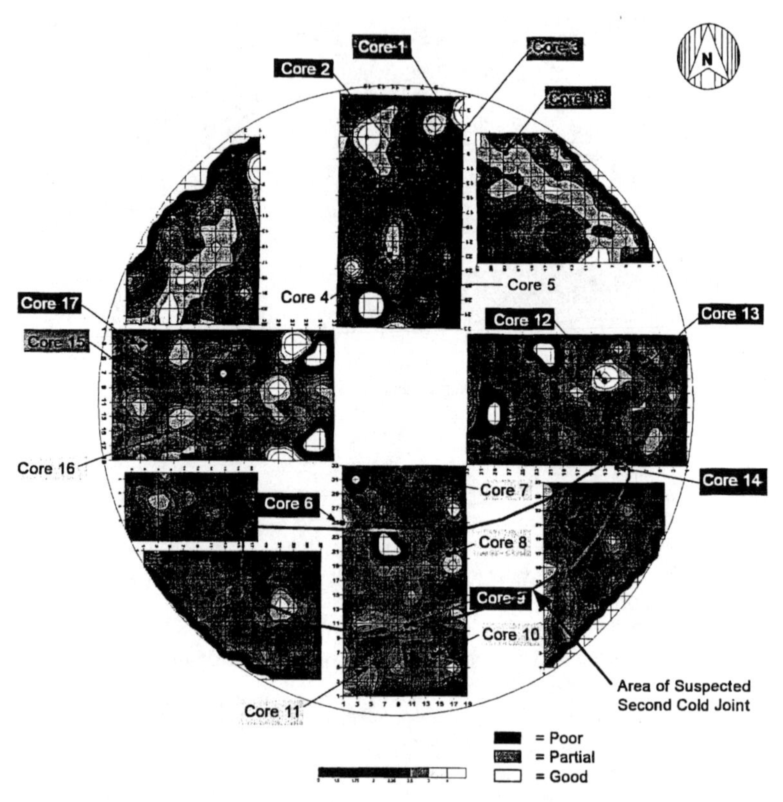

Figure 4: Pelham WWTP Dome Investigation
Suspected Second Cold Joint

Figure 2. Average mobility values and core classifications.

Concrete Repairs

After the IR test results were used to define the scope of the internal concrete distress, repairs were made to the structure and the IR tests were repeated to evaluate the effectiveness of the repairs.

Case 1 - Train Viaduct. An epoxy injection technique was used to attempt to repair the internal fracturing with the epoxy injected through holes drilled to approximately 80% of the section depth in the areas with "mobility times slope" values greater than 6 (m/s /N) using the same pattern as the IR test points. A pilot study was requested; however the relatively small size of the project did not justify a costly mobilization of the specialty contractor and the entire project proceeded under close monitoring.

Areas with "mobility times slope" values between 6 and 10 did not allow for significant volumes of epoxy to be injected in any tested locations. This indicates that the cracks observed in the cores in these areas were insufficiently open or were discontinuous.

The areas with a "mobility times slope" values greater than 10 allowed for effective epoxy penetration in approximately 30%, of the areas as shown by a comparison of the pre- and post-injection IR testing (Figure 3) and by post-injection coring.

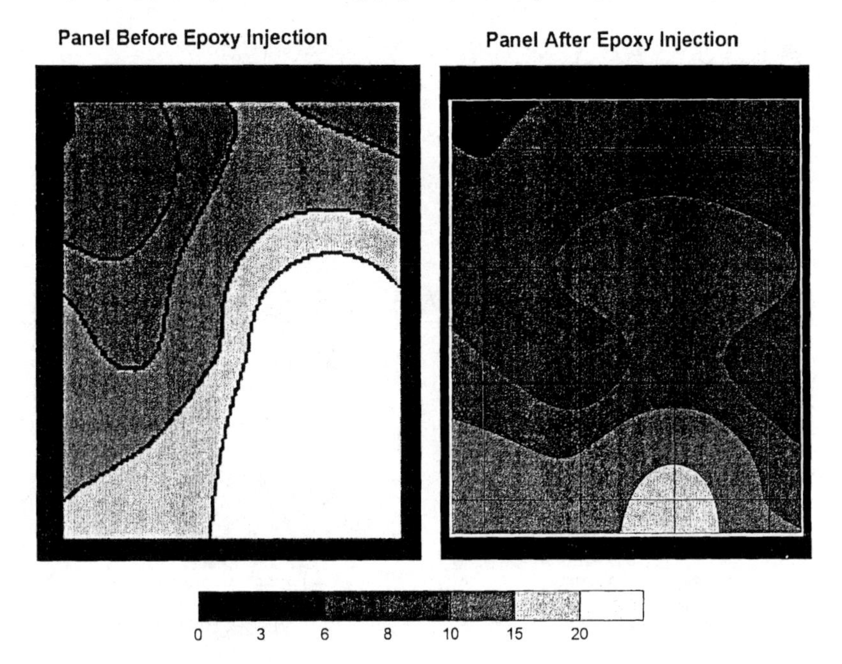

Figure 3. Comparison of mobility times slope values.

The areas with "mobility times slope" values greater than 10 that did not show significant changes in IR results were pinned together using threaded rods and large washers where shallow delaminations were present.

In general, the injection of repair material into the concrete deteriorated from internal reactions at this structure was generally only effective at filling the large, continuous voids and cracks, with unopened and discontinuous fractures remaining unfilled. The overall project was deemed acceptable because the concrete panels were stabilized and entire viaduct received a new waterproofing system which will minimize future deterioration from the ASR reaction.

Case 2 - Poor Consolidation between Concrete Lifts. After the IR results and the core data showed widespread areas of poor consolidation and separation between concrete layers a pilot study was conducted to test the repair concept of "pinning" the concrete layers together using double leg "hair-pin" anchors and injecting epoxy into poorly-consolidated areas to prevent water migration and to address long term durability concerns. IR testing and coring after the pilot study indicated that the proposed repair was feasible, after which a second pilot study was performed using a modified anchor design and improved epoxy injection equipment and procedures.

The IR test results show that the average mobility values are significantly lower after the anchor installations and epoxy injections (Figure 4). Even after successful repair, the slab-epoxy composite structure responds differently than solid concrete as evidenced by areas of the elevated average mobility values that have adequate epoxy filling of poorly consolidated zones (Photo 7).

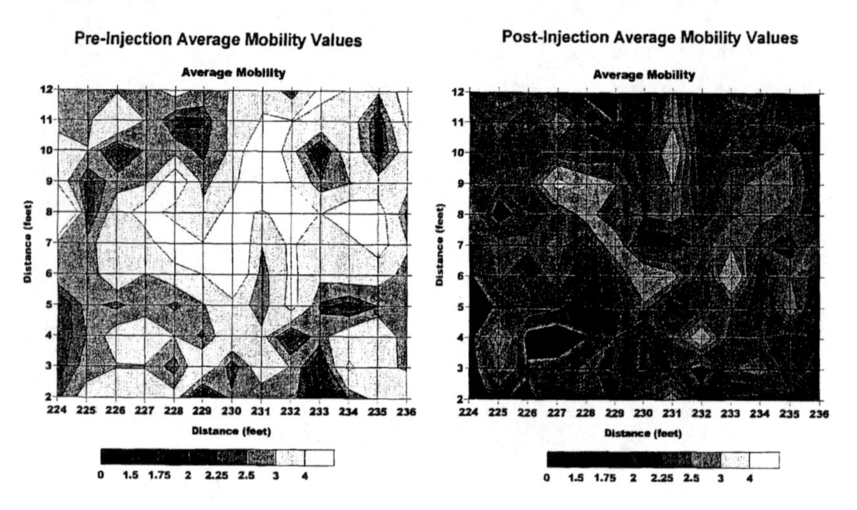

Figure 4. Comparison of IR results before and after repairs.

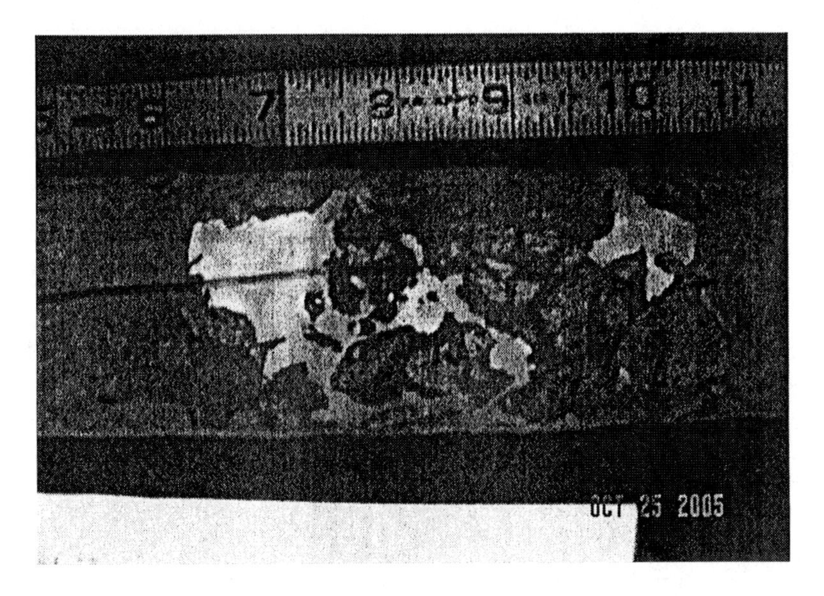

Photo 7. Epoxy penetration into poorly-consolidated areas.

The results from the final repair methodology showed that the repair material had effectively penetrated into the cold joints and into the poorly consolidated areas of the concrete, and that the anchors were capable of supporting the design load based on pullout testing.

Conclusions

The use of IR testing at these two structures with two different internal distresses provided information that could not have otherwise have been obtained. As shown by these two studies, it is important to note that the structures must be evaluated independently when using the IR non-destructive technique because IR results are significantly influenced by the characteristics and type of internal defect, structural dimensions, concrete type, and the support conditions. With proper calibration testing (typically coring) the IR testing allowed the extents of damage to be accurately determined and the effectiveness of the concrete repairs quantified.

References

Malhotra, V. M. and Carino, N. J. (2004). *Handbook on nondestructive testing of concrete*, CRC Press, Boca Raton, FL.

Kyoto protocol application in Italy: a renewed role of IRT for testing buildings

E. Rosina

B.E.S.T. Dept., Polytechnic of Milan, V. Ponzio 31, 20133 Milano – Italy
Tel. +39 02 23994150; Fax +39 02 23996020; e-mail: elisabetta.rosina@polimi.it

Abstract

Kyoto protocol proclaims major targets for the next decades: saving energy uses and not renewable resources, such as Cultural Heritage. In Italy the large amount of old and historic buildings requires to articulate different approaches in order to reach these goals. Priority to historic buildings preservation will ensure to also reach the energy savings. Planned conservation is based on the effective and real evaluation of residual performances of building elements and building system. Suitable diagnostics use images from different spectral bands to obtain information regarding damage distribution and materials. Among images analysis techniques, Infrared Thermography (IRT) has a prominent role because of the variety of applications on buildings and sourrondings, feasibility, real time results and non destructivity. Integration with quantitative tests and microclimatic monitoring allows to obtain necessary input for preservation project and maintenance.

Key words

Energy use, IRT, gravimetric tests, building techniques, masonry, historic buildings, evaporation, condensation, frescoes, planned conservation

Introduction

Assessment of heat loss in buildings has became a necessity which can not be postponed, according to the perspective of worldwide politics, which are increasingly focused on energy savings and on a conscious use of resources. Among them, cultural heritage received a high priority for their safeguard. In

Kyoto protocol, cultural heritage is meant as ethnic-cultural heritage of nations, prior to historic-artistic monuments. According to this kind of approach, preservation of built environment and its valorization require an effective protocol for diagnostic, which should be based on early detection by techniques releasing real time results, and effective for the high amount of buildings that have to be kept under control. Operators require non destructive tests to be suitable as much for preliminary and qualitative inspection, as for advanced diagnostics, for the application on those zones resulted critical from a preliminary scanning.

This paper shows the application of Infrared Thermography (IRT), as a fundamental technique for the certification of energy use of traditional and contemporary buildings, and monitoring for final scientific report in planned preservation of historic buildings.

1. Italian regulation

In Italy, Kyoto protocol accomplishment started with intervention on a specific sector policy. Particularly, intervention was focused on performing and managing building properties, which is responsible for 40% of total use of energy.

Italian government claimed a mandatory regulation to enforce the European directive 2002/91/CE about energy earning in buildings. According to the rule in force, Regions issue specific laws to control the enforcement dispositions at council level. Italian Regions especially favour any kind of energy savings (heating systems, thermal insulation).

Regarding design of new buildings, regulation provides incentives and penalizations: the increase of building gross surface up to 5% is the benefit for technologies using renewable sources and for passive solar buildings. Designers, professionals and contractors are presecutable in case of lower energey profile of building rather than the requirement, and for using highly energy dissipating technologies and materials.

New and existing buildings have to be classified within categories, depending on the energy use by surface: tax burdens correspond to any class of energy use, increasing the penalization with the energy use.

2. Rules application

European Regulation of Energy Earn of Buildings pinpoints the Energy Certification of Buildings as an effective tool for reducing the energy use in buildings. Energy certification has a high impact on all the stakeholders of the sector: contractors, professionals, public administrations. Reduction of the energy use affects the choice of the designer, for materials and execution techniques, and it becomes a major factor in the production phase.

Regarding already existing buildings, verification of congruity of building typology and energy use is an effective boost for fitting HVAC plants and refurbishment of the enveloping structures.

Stakeholders with a crucial role are the companies responsible for certification of compliance of the project, building testers and, starting form 2006, actors releasing energy certification.

Research agencies are in charge of production of steady procedures for testing and measuring heat loss. Their aim is to guarantee that the results obtained from regulated tests are comparable and reliable. Private companies and research agencies provide non destructive testing services.

Building envelope and system building-plants are indicators for classifying buildings versus their energy use. Quantification of the first indicator is a ratio of required heating (by year), second is the use of primary energy for thermal usage (heating, hot water). Regulation states the periodic control of heating plants, expecially if they are older than 15 years.

3. Consistency of building properties in Italy

Table 1. Residential property by period of construction (elaboration of the Census from Istat data at 2001)

TYPE OF BUILDING	(%)
Historical property (pre 1919)	17,6
Property 40 or more years older (1919-1960)	25
Property constructed between 1961-71	22,6
Property constructed between 1972-1981	18,3
Property constructed between 1982-1991	9,3
Property increase 1991-1999	7,1

Some questions and discussion can be drawn from regulation and current classification of building properties:

- How many historic buildings (17.6%) and properties of 40 or more years old (25%) have a HVAC plant? Among the furnished ones, how long lasts the average weekly heating? For example, most of the historic properties that spread into the countryside, where costs of soil is lower than in the large towns, have scarcely a prolonged use of heating during the day, or even during the week.
- Which of the above listed properties have historic elements (window frames, roof structures and covering, etc) that do not fit with the present requirements for energy use?

- **And at last, in case of "historic plants", which standard is to be considered in order to evaluate their efficiency?**
- **How is it possible to take in account the residual performances of historic elements fitting modern standards?**

All of these questions seems to point out a gordian knot of philosophy of energy use in historic buildings. A milestone research has been carried out in the last years by Polytechnic and Region of Lombardy to face these and further questions regarding knowledge and management of historic buildings massive properties, vernacular architecture and built environment. "Planned conservation", it is the theme of the research, and it moves from the analysis of the building system, to specification of the technological elements, to evaluate the residual performance of each element. The total evaluation does not come from simple addiction of elements' performances, but it is a holistic estimation which takes in account the connections among elements in a wider scale. Aging and all the modifications that occurred in time are factors which impact the system and generate new connection and adjustment that can supply and support a renewed balance. Beyond a organic vision of the building, the strategy of Planned conservation systemic analysis is aimed to materials conservation of the elements, for warranting the authenticity of the handwork. The specificty of the conservative approach has priority to the performances approach (as usually proposed for new buildings). In fact, the registration of the perfomances allows to evaluate the buildings performaces independently on the standard model.

This strategy fits perfectly the aims of present regulation, because lasting materials in situ is the first undisputable step towards conservation of Cultural Heritage for which we are responsible. More than the calculation of energy loss for each element, a global energy balance of the building's surfaces is suitable for historic fabric.

The role of IRT in the energy certification and the evaluation of the thermohygrometric balance of buildings

The scenario is composed of mandatory regulations, which urge a prompt process to produce regulatory diagnostic. Most of existing tests have already consolidated procedures in their specific field of application, but they require integration in references towards the energy balance assessment.

The evaluation of energy balance includes some particular tests that are necessary to energy certification and to put trials buildings on envelopes.

In modern building, steady triage includes non destructive testing for mapping thermal anomalies on the surfaces, direct tests add for probing materials, and for analyzing the thermal characteristics of structures (for

example measures of heat flux across masonry). Mathematical models are effective for studying the thermal behaviour and the heat transfer across the structures.

Regarding historic buildings, both the building techniques and materials (high thickness masonry, wood and ceramic materials for covering, limited openings, favorable orientation in the site, etc) restrain most of the heat loss. Nevertheless, some thermal bridges can be detected also in the sound structures and in those areas where damage and modification occurred accomplished beyond any energy requirements. Moreover, many historic buildings do not have HVAC system, and even where the heating plant exists, it is "on" only few hours a week.

At last, but this is the first criterion for assessing old structures, the value of the historic property allows to define new standards for its expected performances, according to planned preservation philosophy. More than ever, in planned preservation thermohygrometrical triage are necessary for reliable assessment of residual performance of the elements and system, and to evaluate the balance conditions with the environment. Nature of the tests is mainly non destructive, avoiding any disruption of materials. For that most of the analysis are based on multi-spectral images, plus monitoring microclimatic and ambient conditions. These tests are principally addressed to measure the thermohygrometrical balance of the building surface in the ambient, more than the mere localization of the heating loss.

Among most used techniques, Infrared Thermography (IRT) has a prominent role, because of its feasibility on the field (obtaining results comparable with the lab ones), versatility, speed execution, and the decades of documented experiences in the energy branch and in border fields as restoration, civil engineering, and industrial.

In the arena of tests for the energy certification of buildings, IRT has the advantages to be completely non destructive, as it is a telemetric test, fast execution on wide surface, it does not require the interruption of activities inside the buildings. A first level of qualitative information is immediately released; nevertheless, a complete analysis requires some hours/days of elaboration.

IRT is a telemetric technique, that maps emitted radiation into a two-dimensional image representative of distribution of the temperature on the emitting surface. Anomalies of thermal characteristics of elements constituting the external envelope of a building are evident as a temperature variation on the surface. The temperature distribution can be used for detecting thermal irregularities due to insulation defects, for examples, and/or air leak in the elements of the building envelope. IRT s a valid application on contemporary and historic buildings as well.

In recent decades there has been much IRT research in Italy, both in the laboratory and in the field at buildings important to Italy's cultural heritage.

Thermal anomalies have been shown to be related to the local thermal characteristics of the materials and to the defects of the examined elements .

IRT investigation permits researchers to gather information about the location, shape, material characteristics, and state of decay of building elements and systems. Moreover, the measures of surface temperature, of the heating flow across the section of the structures, the map of any thermal loss and microclimatic monitoring allows to evaluate any risk area in addition to prevent thermal imbalance and dispersion.

Study cases

1) Contemporary architecture: Office building, Milano, evaluation of heat loss on courtain walls

Tests regarded the IR Thermographic scanning of the exterior elevations, by passive approach, (according to UNI 9252/88 procedure). The second step of assessment, consisted of the IRT scanning of 6 windows at forth level, from inside. The preliminary scanning of the exteriors allowed to select the critical zones around these windows.

Thermal analysis permitted to evaluate the temperature on those areas in the windowframes and the exterior façade which appeared as heating loss. Technical documentation (projects drawings) and visual inspection confirmed the hypothesis of thermal bridges.

Thermal analysis pointed out many thermal anomalies between the ventilated façade and the courtain walls. In particular, on the stone slab, IRT measurements revealed a maximum thermal gradient 6.1°C between points B, C, I; those points corresponds to major heating loss (emissivity = 0.92 for opaque, porous, white surface of stone slabs). Points H, L, M are at the same level of previous one but they do not correspond to such high thermal loss. Between points B,C, I and F,G, thermal gradient is higher (7.6°C). Points E and K indicate further variation of temperature of the stone slabs at balance with ambient (fig. 1-2).

The investigation allowed to survey all the heating loss on the Southern and Eastern facades. Major thermal anomalies resulted at the last level, between the windows and the stone slabs. The comparison between average temperature on sound areas (for example, lateral bands) and defects allowed to quantify thermal gradients. Fewer thermal anomalies resulted on Southern façade. Inside, the higher thermal loss is around the windowframes, and it is about 2°C.

2) Old but not ancient buildings, economic buildings in Bovisio Masciago: heat loss and damage of the plaster

Aim of the investigation is the measurement of thermo-hygrometrical state of masonry, its building technique and employed materials, and microclimate conditions in some apartments, where scattered spots stained the plasters. Visual inspection allowed to detect major damage at the basis of the interior walls, as well as along the first level floor on the exterior façade.

IRT and gravimetric tests were performed after some weeks of rain, in order to measure the maximum of water content inside the structures. A preliminary thermography scanning on the whole envelope of the buildings allowed to locate the major thermal anomalies (fig. 4). According to their location, geometry and connection with the damage, operators collected some samples from the surface and inside the masonry, by drilling. Results of gravimetric tests indicated water content for each sample. This procedure joins map of humid surfaces on the whole extention of the plaster, and the quantitative measurements of water content, without disrupting and wasting materials. In fact the number of samples is very reduced, because results of only few samples are significative of the part of surface at the same temperature. Inside, in the apartments at ground floor, microclimatic monitoring pointed out a high RH (connected with the damage) where the boundary wall joins the ceiling.

Moreover, investigation results showed the paths of water coming in to the structures (pipes leakages). A further step of analysis consisted on the investigation of the building techniques by fiberscope. IRT showed large thermal bridge where the exterior damage was most spread (at first floor). Fiberscope results showed that the long horizontal thermal discontuinity is due to the presence of a concrete beam, without insulation, directly plastered with stucco.

Lack of insulation of the concrete beam causes a thermal imbalance of the structure. The calculation of transmittance across the section of the brick masonry and concrete, allows to quantify the different thermal capacity of the two materials.
Moreover, the exterior stucco can not be adhesive to the concrete substrate: that's the cause of the documented delamination, and it is a certain information for further intervention. A simple refurbishment of stucco can not be an effective and long lasting intervention. A mandatory requirement is the insulation of the boundary wall, on the whole surface, in order to decrease the thermal gradient, which is responsible for the damage inside and outside.

3.1) Historic Buildings: Istituto Musicale Donizetti, Bergamo: evaluation of heat loss on traditional masonry

The 17th century building is the site of the music school of the town since the 19th century. Major refurbishment after the 2ndWW inserted the present

heating plant (radiators). Thermal bridges clearly appear at IRT, despite of the high thickness of the pebbles masonry (more than 40 cm). In addition, in the thermograms, minor leakage from the wooden windowframes appears. Considering the small thermal range of images (7,2°-11,5°C), thermal gradient of the areas below the windows has lower values than thermal loss across the window glass (see fig. 6).

3.2) Historic Buildings: S. Stefano, Lentate, energy balance of a piece of urban fabric

The small church, 14th century, is nearby to the monastery. The interior of the church is completely frescoed (14-15th centuries). Monastery, completely refurbished, is a school and residence, while the church is seldom open, and it does not have any heating plant. By assessment, major damage resulted on Southern frescoed wall, where salt efflorescence, delaminations, lacks od plaster are spread up to 1-1.5 m up to the floor. IRT and gravimetric tests allowed to assess rising damp and localize water infiltration. Microclimatic conditions resulted steady, and sufficiently safe for optimal preservation of frescos, except for winter season.

Infact, IRT shows almost homogeneous temperatures of the walls (fig. 8-11) during the good seasons, despite of the presence of rising damp at the bottom.

At winter conditions, the heating of the adjunct building generates a heat flow across the southern masonry, as fig. 12 shows. The increase of temperature on the damp areas causes major evaporation, which is responsible for the higher damage of frescos. The heat flowing from the masonry contributes to generate convective fluxes inside the church, altering the balance between microclimatic conditions and surfaces.

From the perspective of energy use, a new heating plant in the church could balance the heat loss across the southern wall; nevertheless this criterion is not suitable for the best preservation of the surfaces. In fact, assuming that the whole building was built with the same building technique and materials, further thermal exchange will occur across the other walls, towards the

environment, increasing the risk of condensation on the colder surfaces (Northern elevation, vault, opening walled up etc).
In a few words: the natural balance of the surfaces could be destroyed in a few days of heating. Moreover, the re-use will be an occasional use (liturgical use, conference hall, concert hall): a continuos heating will have a high cost, while the occasional heating will cause very high and sharp variations of the microclimatic conditions.

Project has to take in account all these issues, designing an insulation intervention on the Southern wall (from the monastery side), intercepting rising damp and infiltration, preserving microclimatic condition untill water infiltration will be dryed on. For that a de-humidifier is suitable to smoothly decrease RH of the air.

Conclusions

The new regulation after Kyoto agreement at worldwide level, brings to a one-shot occasion to focus on a system of effective triage for heat loss in buildings.

On the other side, it is necessary that this triage takes in account the differences between traditional, historic and contemporary buildings, and their current uses and plants. IRT is the most suitable technique for the analysis process of building envelopes and thermohygrometrical balance between surfaces, structures and ambient, because of its versatility and its proper integration with other tests (quantitative). IRT complies both preliminary and advanced phases of diagnostic in the field, as steady experiences and procedures permit to read out thermal anomalies on the surfaces.

References

Croce S. (2003), "Introduzione generale alle linee guida per i documenti tecnici del Pano di conservazione." ***La Conservazione Programmata del Patrimonio Storico Architettonico: linee guida per il piano di conservazione e consuntivo scientifico*, (vol. 25-30)**
Guerini, Milano

Della Torre S. (2004), "Process innovation in preservation of Built Cultural Heritage: tests on Palazzo te in Mantua and other study cases." ***Cultural Heritage and the Politecnico di Milano***
Rivista del Politecnico di Milano, n 8

Della Torre S. (2003), "Piano di manutenzione e consuntivo scientifico nella legislazione sui lavori pubblici." ***La Conservazione Programmata del Patrimonio Storico Architettonico: linee guida per il piano di conservazione e consuntivo scientifico,*** **(vol. 25-30)**
Guerini, Milano

Rosina E. (2004), "Oltre la percezione, architettura all'infrarosso"
Alinea, Firenze

Convegno Edilizia Sostenibile e Risparmio Energetico. Le strategie di incentivazione degli Enti Locali ed i vantaggi economici ed ambientali. (2004). ***Il progetto European Solar Building Exhibition Altener marketing di case solari con standard di basso consumo energetico,*** **Salsedo B.**
Rome, Italy

Conference The Human Being and the City. Towards a Human and Sustainable Development. (2005). ***Voluntary Guidelines And Sustainable Building Code,*** **Maiellaro N.**
Naples, Italy

Tavolo Energia e Ambiente, Provincia di Milano. (2005). ***Certificazione energetica degli edifici,*** **Dall'O G.**
Milan. Italy

ART 2005 Conference (2005). ***Early detection and monitoring procedures by means of multispectral image analysis,*** **Della Torre S., Rosina E., Catalano M., Faliva C., Suardi G., Sansonetti A., Toniolo L.**
Lecce, Italy

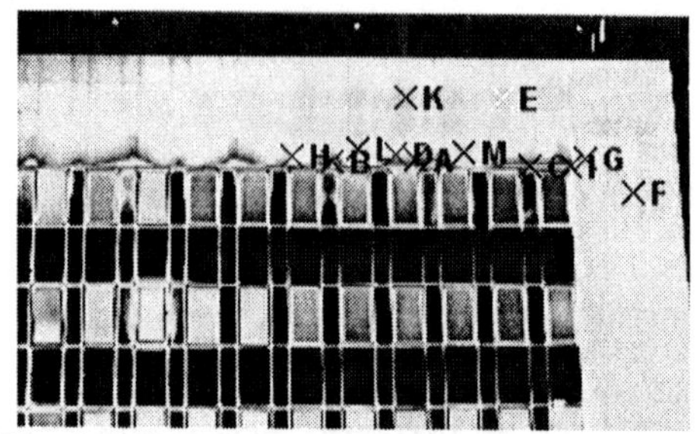

Fig. 1, Office building, thermogram of eastern facade (thermal range -4°C/3.5°C)

Tab. n 2, Temperature of points A-M (stone slabs), ambient Tem. 1.5°C

Punto	T °C
A	3,0
B	5,1
C	6,6
D	1,6
E	0,2
F	-1,0
G	-0,5
H	1,8
I	5,6
K	-0,3
L	0,5
M	0,8

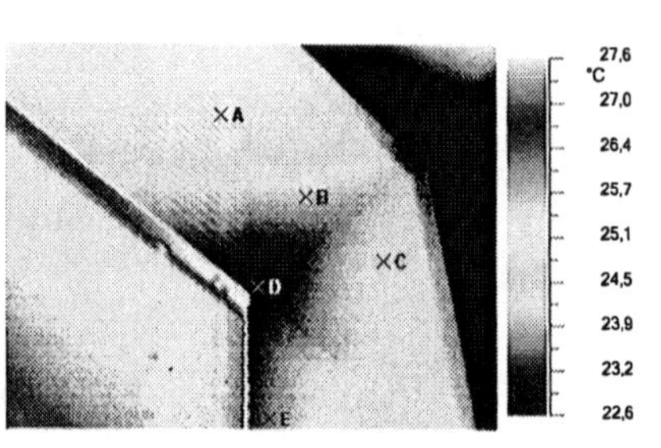

Fig. 2, Office Building, windows in room n 4025; fig. 3, thermogram of the right window in room 4025

Tab. n 3, temperature of points A-E

Punto	T °C	Emis.	Ta °C
A	25,6	0,85	23,6
B	24,7	0,85	23,6
C	25,2	0,85	23,6
D	22,6	0,85	23,6
E	24,1	0,85	23,6

Fig. 3. Bovisio Masciago, BV1 buildings, Eastern elevation

Fig. 4. Bovisio Masciago, thermograms of Eastern elevation (thermal range 17.7-22.5°C)

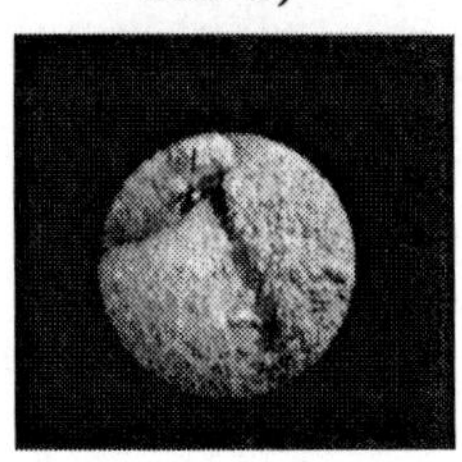

fig. 5. Fiberscope image from then Eastern facade

Fig. 6. Istituto Donizettti, thermograms of Southern and northern elevations, thermal range 7,2°-11,5°C;

Fig. 7. S. Stefano in Lentate, facade and Northern elevation

Fig. 8, Southern wall of the apse; fig. 9, mosaic of thermograms of the Southern wall of the apse (thermal range 8.1-12.3°C). Consider that the higher thermal gradient resulted on the eastern corner, due to sun irradiation

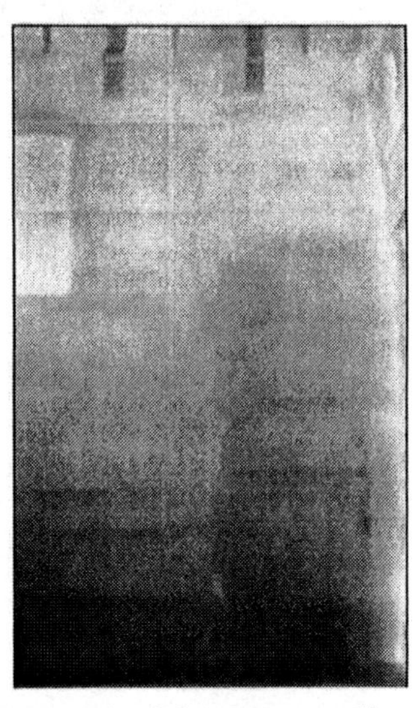

Fig. 10, nave, Southern elevation; fig. 11, mosaic of thermograms (thermal range 8.1-12.3°C)

Fig. 12. Mosaic of thermograms of the Southern elevation in the nave, during the heating of the adjunct building

The Impact of Market Demands on Residential Post-Tensioned Foundation Design: An Ethical Dilemma

Bart B. Barrett, B.S., P.E.[1]
Kerry S. Lee, M.B.A., P.E., M. ASCE[2]
Erik L. Nelson, Ph.D., P.E., M. ASCE[3]

Abstract

In 2005, a lawsuit in excess of 6 million dollars was brought to trial pertaining to 31 single-family residences in Arlington, Texas. The basis for the lawsuit was distress related to foundation movement. An extensive forensic investigation was performed on each of the structures, and the results of the investigation indicated improper construction and inadequate design of the post-tensioned slab-on-grade foundation systems. Testimony given during the legal proceedings indicated that the foundation design methodologies used deviated from the governing Post-Tensioning Institute (PTI) Design Manual. This paper summarizes the construction and design defects observed during the forensic investigation and the reasons why industry standards were not followed in the design and construction of the foundation systems. The market pressure for the least expensive residential foundations is driving design engineers to compromise their ethics and ignore industry standards.

Introduction

This paper will review a significant number of foundations from a subdivision in North Texas with a focus on the correlation between the financial restraints imposed upon engineers and the quality of engineering produced. The pressure from the residential building industry to provide lower cost products has impacted post-tensioned foundations due to poor design and lack of quality control.

[1] Senior Project Director, Nelson Architectural Engineers, Inc. 2755 Border Lake Road, Suite 103, Apopka, Florida 32703; bbarrett@architecturalengineers.com, phone 877-850-8765

[2] Director of Engineering, Nelson Architectural Engineers, Inc. 2740 Dallas Parkway, Suite 220, Plano, Texas 75093; klee@architecturalengineers.com, phone 469-429-9000

[3] President, Nelson Architectural Engineers, Inc. 2740 Dallas Parkway, Suite 220, Plano, Texas 75093; enelson@architecturalengineers.com, phone 469-429-9000

This market pressure for the least expensive foundations is driving the design engineers to compromise their ethics and ignore industry standards resulting in inadequate foundation designs, poor quality control of construction, and flexible foundations that perform below acceptable limits.

Background

Between March and September 2002, forensic investigations of 31 single-family residential structures were performed. The structures were located in a subdivision of 218 homes in Arlington, Texas and ranged in size from approximately 2200 to 2900 square feet including living and non-living areas. Prior to construction, the subdivision was divided into three sections for geotechnical exploration. Geotechnical data for the site indicated that the soil consisted of expansive clays with Potential Vertical Rise (PVR) of the soils ranging from 2.0" to 4.75". The post-tensioning foundation design parameters provided by the geotechnical engineer of record for soils in the subject section of the subdivision are indicated in Table 1:

Table 1: Post-tensioning Foundation Design Parameters

	e_m (feet)	y_m (inches)
Center Lift	5.0	3.8
Edge Lift	4.5	2.4

A forensic investigation strategy was developed to test the quality of construction, the adequacy of foundation designs, and other influences that could affect the foundation performance. The investigation strategy was extensive including distress mapping, relative elevation surveys of the foundation systems, tendon mapping, testing of post-tensioning strand tension, reviewing as-built dimensions of slabs and grade beams, testing compressive strength of slab cores, reviewing foundation designs, and plumbing testing. The forensic investigations were conducted following the methodologies presented in the Texas Board of Professional Engineers Residential Foundation Committee Policy Advisory 09-98-A, and the Texas Section of the American Society of Civil Engineers (ASCE), "Guidelines for the Evaluation and Repair of Residential Foundations - Version 1".

Distress

Observed distress was mapped on floor plans for each of the structures reviewed. The distress mapping was analyzed to determine patterns of distress in the structures and to determine correlation with the foundation movement topography. The majority of the distress indicated edge lift conditions typical of expansive soil movement. Other distress patterns indicated edge drop and center lift conditions.

Observed distress included cracking and separations at the interior and exterior architectural finishes, inoperable doors/windows, fractures and/or separations in the attic framing, and fractures in the concrete slabs. Several of the fractures at the exterior were observed to open at the bottom of the wall and taper towards the top, which is evident of edge lift. Fractures at some locations were also observed to open at the top and taper to the bottom, which is indicative of edge drop or center lift conditions. Relative elevation surveys of the foundations further indicated generally excessive movement in the structures.

Construction

Nonconformance to foundation plans and specifications was observed at each of the structures reviewed. The nonconformance included weaker concrete material than specified, excessive variations in slab thickness and grade beam depths, and poor placement of slab tendons. In addition, the structures were reviewed for other construction items that could affect the performance of the foundation systems, i.e., roof guttering systems, lot grading, and stressing of post-tensioning tendons.

The specifications for the foundation systems conformed to typical industry standards including construction of a 4" thick slab with 10" wide grade beams of varying depth (24" to 30"). The specified 28-day compressive concrete strength was 3000 psi, and tendons were specified to be ½" diameter, 270 ksi, 7-wire strands.

Concrete Strength

Concrete cores were taken from the slab at each structure and tested in accordance with ASTM C 42. The results from the testing indicated at least one core below 75% of the specified 28-day compressive strength at five of the structures. Two of the structures had compressive strength averages below 85% of the 28-day compressive strength.

Slab and Grade Beam Depths

The concrete cores were measured to determine as-built slab thicknesses. The thicknesses varied from 3 1/4" to 9 3/4" (81 to 225% of the specified thickness). In addition, grade beams were exposed at four areas of each structure in order to measure the depths of the grade beams. The percent variation of the as-built to the design depths of the grade beams ranged from -21% (20 1/2" as-built, 26" design) to +13% (31 1/2" as-built, 28" design).

Tendon Placement

The foundation plans specified that slab tendons should be placed at the center of the 4" slab. The 1996 PTI Design Manual allows for a construction tolerance of $\pm$ 1/2" for placement of tendons. With construction tolerances, the concrete cover should range from 1 1/2" to 2 1/2". The observed cover at the subject structures ranged from 0.157" to 6.1". The 6.1" value was the greatest value that the equipment could read. It is possible that the multiple 6.1" readings (7 readings) were actually greater than 6.1".

Additional Construction Items

The finished site grading at eleven (11) of the structures produced negative drainage (slopes toward the structure). A review of the foundation topographies indicated a correlation between the negative drainage and the foundation movement at five (5) of the structures. In addition to the poor drainage, absence of rain guttering was observed at some of the structures. However, it was difficult to determine which structures had been impacted from the lack of rain guttering systems since rain guttering had been installed at various times as a remedial action. Piers had also been installed as a remedial action at a portion of the structures. The foundation systems that had been piered still indicated significant foundation distortion (L/349 to L/124).

Testing of Post-Tensioning Tendons

To determine if the strands were stressed properly during the original construction of the foundations, "Lift-off" tests were performed at each structure, typically involving four tendons. The live ends of the strands were pulled by means of a hydraulic jack until the anchor was unseated. A pressure gauge was monitored during the testing process and values were recorded at the time of the anchor unseating. Under-stressing of some tendons was observed when compared to expected after stress loss values from the PTI manual.

Design

Available foundation designs of the observed structures were analyzed for conformance with PTI design criteria. The analysis was performed using PTI software, PTISlab, and was based on the foundation plans and the geotechnical design parameters contained in the original geotechnical report for the development of the site. As required by the 1996 PTI Design Manual, one (1) to three (3) design rectangles were analyzed as necessary to properly design for the configuration of the foundation. Deflection analysis was based on criteria specified in the 1996 PTI Design Manual.

Due to the irregular geometric layout of some of the foundations, simplifying, conservative assumptions were made for analytical purposes. For example, discontinuous grade beams on the interior were conservatively treated as continuous in determining the quantity of long and short beams. In some cases, these assumptions may have contributed additional strength to the foundations that was not actually present in the design, which gave benefit to the design engineer.

Information was available to analyze foundation designs for 24 of the 31 structures. Analysis of all of the foundation designs reviewed indicated overstress in the edge lift condition that did not meet design criteria put forth in 1996 PTI Design Manual. Overstress ranged up to 105.3% in bending, 59.9% in shear, and 100.2% in deflection.

Foundation designs reviewed were prepared by three different Professional Engineers licensed in the State of Texas. For the purposes of this paper, the engineers that were involved will be referred to as Engineer Alpha, Engineer Beta, and Engineer Gamma. In addition, there were seven different architectural floor plans with minor variations in any given floor plan. There were some similarities in the foundation designs provided by the engineers (slab thickness, beam widths, etc.); however, the designs of similar floor plans varied between the engineers in beam spacing, beam depth, and number of tendons used in the slabs and beams.

Engineer Alpha

Engineer Alpha provided the majority of the foundation designs for the reviewed structures. Analysis of the foundation designs prepared by Engineer Alpha indicated significant overstress; however, the overstress was generally lower than the other engineers (up to 39.1% in bending, 41.7% in shear, and 60.7% in deflection). There were significant variations observed in Engineer Alpha's foundation designs for structures with the same floor plan, even though the geotechnical design parameters were unchanged. For one particular design (Floor Plan A), the tendons used to reinforce the beams were reduced by 5 tendons in a time span of less than four months. Further, the same design was again reduced by 3 more beam tendons eight days later.

The relative elevation survey of the foundation systems for these Floor Plan A plans indicated more out-of-levelness in the foundations as the number of beam tendons decreased. The out-of-levelness for the foundations with 28 total beam tendons, 23 total beam tendons, and 20 total beam tendons was measured to be 2 1/8", 3 3/8", and 4 3/8", respectively. Representations of the change in out-of-levelness in Floor Plan A are provided in the following Figures 1, 2, and 3:

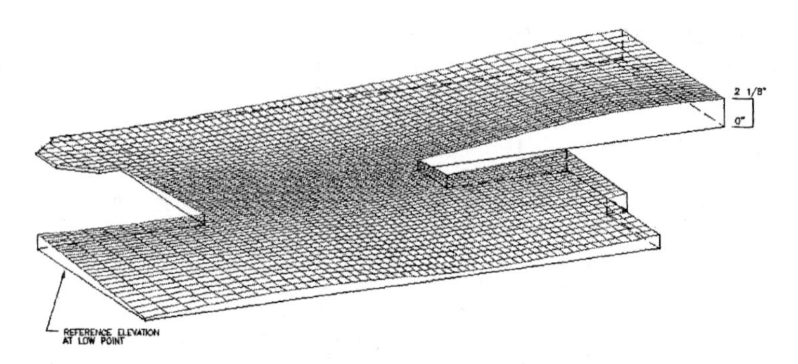

Date of Design	**02/24/98**
Total Number of Beam Tendons	**28 Tendons**
Total Out-of-Levelness	**2 1/8"**

Figure 1. Three Dimensional Representation of Total Out-of-Levelness Floor Plan A (28 Total Beam Tendons)

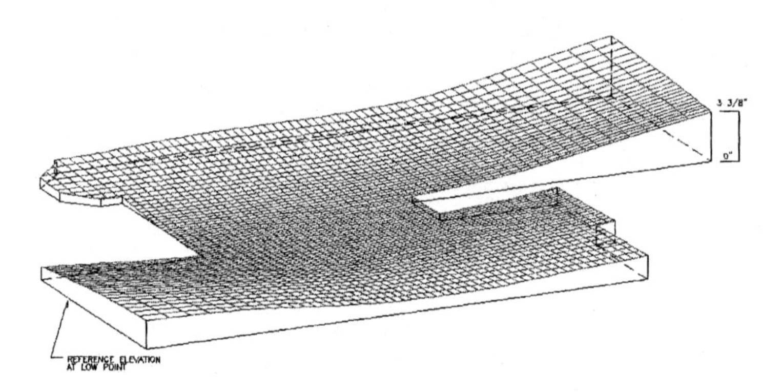

Date of Design	**06/12/98**
Total Number of Beam Tendons	**23 Tendons**
Total Out-of-Levelness	**3 3/8"**

Figure 2. Three Dimensional Representation of Total Out-of-Levelness Floor Plan A (23 Total Beam Tendons)

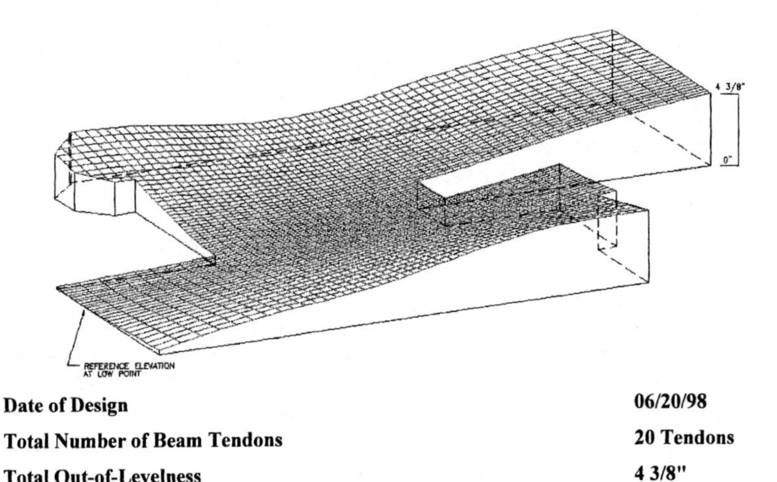

Date of Design	**06/20/98**
Total Number of Beam Tendons	**20 Tendons**
Total Out-of-Levelness	**4 3/8"**

Figure 3. Three Dimensional Representation of Total Out-of-Levelness Floor Plan A (20 Total Beam Tendons)

The output from the software used by Engineer Alpha was available for review. The output indicated that only one design rectangle was used for "H" and "Z" shaped foundations, which is in direct conflict with the 1996 PTI Design Manual. Further, Engineer Alpha's design output indicated beam depths greater than those specified in the foundation plans. The variation between the beam depth shown on the foundation plans and the deeper depth suggested by the calculations was significant, ranging from 2.63" to 5.99".

Engineer Beta

Engineer Beta prepared foundation designs for 11 of the foundations reviewed. Comparison of the foundation designs prepared by Engineer Beta to those prepared by Engineer Alpha revealed deeper grade beams by an increment of 2" on average. However, the designs prepared by Engineer Beta consisted of greater beam spacing and did not include double tendons in any of the grade beams. When comparing the foundation designs for similar floor plans between Engineer Beta and Engineer Alpha, there was a 50% reduction of post-tensioning force in the grade beams due to the reduction in the number of beam strands. An analysis of the designs by Engineer Beta using PTISlab indicated higher overstress than those foundation designs by Engineer Alpha (up to 80.2% in bending, 31.8% in shear, and 94.7% in deflection).

A graphic presentation has been prepared to demonstrate the differences in beam layout between Engineers Alpha and Beta. The grade beam locations were highlighted in a floor plan used by both engineers (Floor Plan B), and the highlighted grade beam locations from the plan designed by Engineer Beta was overlain on top of the highlighted beam locations from the design by Engineer Alpha. The graphic presentation is provided in Figure 4; the darker beams are representative of the design by Engineer Beta. Note that there are fewer short direction grade beams in the design by Engineer Beta.

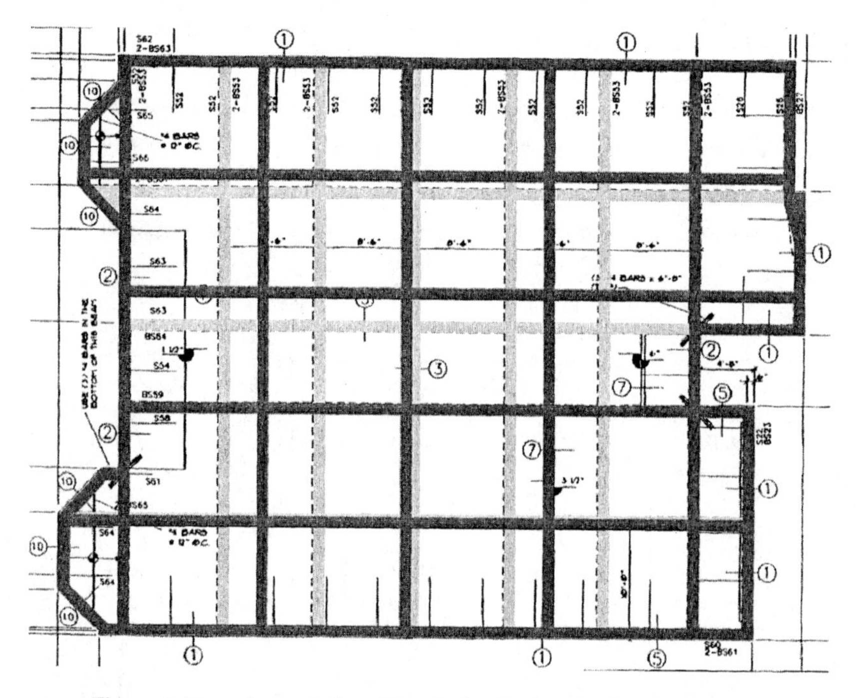

Figure 4. Beam Layouts from Foundation Designs for Floor Plan B

It is interesting to note that Engineer Beta had the deepest grade beams of the three engineers, and, for this sole reason, was not brought in as a party to the litigation. However, due to the use of fewer beams and less reinforcing tendons, the designs performed by Engineer Beta actually failed with much higher design overstress than Engineer Alpha.

Engineer Gamma

One foundation reviewed by our office was designed by Engineer Gamma. The design included single strand beams that were approximately 4" shallower in depth than the designs by Engineer Alpha. As expected, the analysis of the design indicated more overstress than the foundation designs prepared by Engineer Alpha. In fact, it exhibited the highest overstress of the foundations analyzed (105.3% in bending, 59.9% in shear, and 100.2% in deflection).

Engineering Conduct

The review of the foundation designs provided in the subdivision raises questions as to the design processes that have been used in the residential foundation industry. With clear methods indicated in both the Uniform/International Building Code and the PTI Design Method, how can engineers with the same geotechnical data produce such varied results and results that consistently fall so far below the documented industry standards?

Testimony regarding the design methodology used in designing the foundations in this case provided insight into what degree the engineers were involved with or excluded from the construction process and how that impacted the foundation design and construction from start to finish.

Design Process

The design process included a "pre-engineering" of foundations for prototype floor plans based on information provided by the builder. Then information was sent by the builders to the engineers indicating the address, lot and block numbers, and type floor plan. The engineer would then apply site-specific information to the "pre-engineered" foundation plan and issue a sealed letter stating that the foundation had been designed based on the appropriate code and geotechnical report. It was indicated by Engineer Alpha that the total design and production time related to foundation design was approximately three and a half hours and time related to review of drawings and calculations was approximately one half hour.

Testimony by Engineer Alpha indicated that in residential foundation design it is common for the builder to provide a typical floor plan so that the engineer can produce a typical foundation plan for pricing and comparison to designs provided by other engineers. The implication is that the lowest cost design determines which engineer gets the project.

Engineer Alpha did not use the PTI Design Manual method preferring to use proprietary software developed in-house, which had not been peer reviewed. However, the output of the software did not provide the same results as industry accepted software or hand calculations of the procedure put forth in the PTI Design Manual. It should be noted that requests to obtain software were denied; so the exact calculation method performed by Engineer Alpha is not known. However, the results of the calculations by Engineer Alpha do not agree with the calculations performed in accordance with typical industry software.

Pre-Pour Inspections

Pre-pour inspections were not performed by the design engineers since the builder did not request the inspections. It was indicated that pre-pour inspections had been discussed with the builder by Engineer Alpha and that it was presented as a line item under additional services along with an associated fee. Engineer Alpha estimated that a pre-pour inspection would probably take an hour including travel time and further stated that the fee associated with the pre-pour inspection was $75.

As can be seen from the results of this investigation, variations commonly occur between actual construction and the design documents often to the detriment of the performance of the foundations. The purpose of the pre-pour inspections is to ensure that the construction in the field conforms to the original intent of the design documents and that a higher quality product is constructed. Engineers typically review slab and beam depths and tendon placement, the very construction parameters that were documented to be outside of construction tolerance in this case.

Design Costs

According to Engineer Alpha, the fee associated with the design of the foundations in a subdivision would include a fee of $150 for each floor plan and another $50 fee for each house using the same floor plan. The profit from these fees was indicated to be approximately $30 per plan. The design fees were indicated by Engineer Gamma to be $100 per house.

Engineer Alpha added that the total cost for production of foundation designs and a pre-pour site inspection have increased to about $250 since the time of the charges indicated, which was around 1997 and 1998.

Design Divergence

During the time of investigation, Engineer Alpha prepared a spreadsheet of foundations in the subdivision that were designed by their firm. One column in the spreadsheet indicated whether the calculations originally performed by their firm met the allowable design values (overstress, deflection, etc.) as specified by the Post-Tensioning Institute (PTI). By his own review of the foundation design calculations, Engineer Alpha indicated that some of his calculations did not meet the specifications in the PTI Design Manual.

When Engineer Alpha was asked why he did not use multiple rectangles as specified in the PTI Design Manual, he indicated that he chooses not to follow the methodology in the PTI Design Manual, presumably to produce a less expensive foundation and to reduce design time. In fact, the rectangle used by Engineer Alpha typically was not the controlling rectangle, and the design by Engineer Alpha produced beam depths that were shallower and had fewer tendons than an adequate design. The design process by Engineer Alpha resulted in less expensive, less stiff foundation systems that performed poorly.

Testimony by Engineer Gamma indicated that at the time of foundation design the builder had indicated that the geotechnical engineer was still working with the numbers and that they would be about the same as another phase in the subdivision. Engineer Gamma indicated that the builder provided a geotechnical report without PTI design parameters, e_m and y_m values. Since no supplemental information was provided, Engineer Gamma produced his own parameters for the design of the foundation systems.

During the forensic investigation of the homes in the subdivision, Engineer Gamma reviewed his calculations with the PTI design parameters provided by the geotechnical engineer and indicated that his design did not meet the criteria specified in the PTI Design Manual. Engineer Gamma provided some insight into what was driving the design process to diverge from industry standards when he said, "all they [builders] want to do is save money" and "all they care about is money." He further indicated the builder's disregard for engineering codes and standards, stating "they don't care how you got there."

Conclusion

This paper has reviewed a significant number of foundations from a subdivision (31 out of 218) in North Texas with a focus on the correlation between the financial restraints imposed upon engineers and the quality of engineering produced. The pressure from the residential building industry to provide lower cost products has impacted post-tensioned foundations due to poor design and lack of quality control.

Testimony from the Engineers of Record for the foundation designs revealed that engineers have ignored building codes and industry standards in order to provide foundations that use less design time, less material, and less labor, which in effect cost less. The engineers indicated that they did not follow steps specified in the PTI Design Manual, choosing not to use the formulas provided and not to use multiple rectangles for "Z" and "H" shaped foundations. In addition to deviating from industry standards for design, site visits that are typically performed to ensure quality control during the construction phase were not part of the base pricing but were priced on an "a la carte" basis giving builders the opportunity to choose not to include these site visits as part of the design package. Testimony from the engineers indicated that most of these site visits were not performed by the engineers of record or by representatives acting on their behalf.

The residential market pressure for the least expensive foundations is driving the design engineers to compromise their ethics and ignore industry standards. The results are inadequate foundations, poor quality control of construction, and flexible foundations that perform below acceptable limits. The total loss due to the foundation movement in the 31 sites reviewed was claimed in excess of 6 million dollars, and a jury held the builder, the originator of the litigation, responsible for 80% of the liability and held Engineer Alpha responsible for 20% of the liability.

It is clear to the authors that the builder and the engineers involved in this litigation were responsible for the loss; however, the engineering ethics and professional responsibility that governs the professional engineers holds them to a higher standard. To knowingly give into the economic pressures applied by the residential building industry is inexcusable.

References

International Code Council (2000), "International Residential Building Code", 2000 Edition.

International Conference of Building Officials (1997), "Uniform Building Code", 1997 Edition.

Post-Tensioning Institute (PTI), "Design and Construction of Post-Tensioned Slabs-on-Ground," 1996 Edition.

Texas Board of Professional Engineers Residential Foundation Committee, "Policy Advisory 09-98-A, Regarding Design, Evaluation and Repair of Residential Foundations", September 11, 1998.

The Texas Section of the American Society of Civil Engineers (ASCE), "Guidelines for the Evaluation and Repair of Residential Foundations - Version 1", 2003.

An Update on Problems Associated with Metal-Plate-Connected Wood-Roof Trusses

Leonard J. Morse-Fortier Ph.D., P.E.[1]

[1]Member ASCE, Senior Project Manager, Simpson Gumpertz & Heger Inc., 41 Seyon Street, Building #1, Suite 500, Waltham, MA 02453; PH (781) 907-9370

Abstract

Our practice in forensic engineering includes wood structures. Over time, we have received many calls about buildings with metal-plate-connected wood-roof trusses (trusses). Trusses are used extensively, both within the housing industry and in commercial buildings, and their engineering is fine tuned and precise. At the Second Forensic Congress in 2000, and based upon experiences designing and investigating the use of engineered wood trusses, I presented a view of the problems I saw. Many of those problems grew out of the difference between how trusses actually work and how the builders using them understood them. Other problems seemed to grow directly out of the nature of the design and construction industry and in the way trusses and their design fit into the existing contracting process. This paper provides an update on the use of manufactured trusses, changes in legislation and codes that affect their design and use, and problems that remain.

Trusses are finely engineered, using sophisticated design software. They may replace numerous, complex, and redundant elements that traditionally formed part of a wood-frame roof structure. While contractors embrace the improved performance and reduction in cost that trusses provide, these contractors may not fully understand the challenges posed by using trusses. The results of this misunderstanding can be costly and potentially dangerous. Trusses may collapse in groups or overall. The collapse may occur during construction or soon after, under heavy load, or under severe wind conditions. Overall, trusses and the buildings that exploit them have found their way into trouble fairly often. In 2000, I discussed my experiences and tried to explain why and how problems occur. In addition, I proposed some steps to reduce the risks associated with using trusses and noted that efforts are ongoing within the Wood Truss Council of America (WTCA) and at the Truss Plate Institute (TPI) to address these problems. Six years later, it seems appropriate to review what has transpired since then.

The WTCA and TPI have published an extensive document on the proper handling, installing, and bracing of trusses to better guide the construction phase of their use

[BCSI-1-03]. This document runs to 136 pages, provides exhaustive guidance to those responsible for placing trusses, and should prevent problems during construotion. However, construction-phase failures remain a problem.

The WTCA and TPI also publish a document outlining and clarifying their view that the building designer is responsible for the design of a truss system's permanent bracing [WTCA 1-1995]. While this document addresses a serious problem in the design and performance of trusses, it does not remove a fundamental paradox that lies within the building engineering design process.

Introduction

The metal-plate-connected wood truss industry dates from 1955 with J. Calvin Jureit's invention of the "Gang-Nail" connector plate (Meeks 1979). In the 50 years since, the design and manufacture of wood trusses has grown into a multibillion dollar industry. This growth owes a great deal to the aggressive sale and marketing of trusses, but it owes as much to their strength, economy, and utility. Wood trusses have allowed wood-frame construction to evolve and expand. With trusses spanning greater distances, their use allows larger enclosed spaces. In the last 25 years, as business has expanded beyond the relatively modest scale of residential applications, wood trusses have occasionally failed (Kagan 1993). These failures include those that occur during construction, when builders have failed to provide adequate temporary bracing for the trusses as they were placing them. Other failures have occurred on fully completed buildings where the design and/or installation of permanent bracing were inadequate to provide proper stability to the truss members that are loaded in compression.

WTCA and TPI publish BCSI 1-03 – Guide to Good Practice for Handling, Installing & Bracing of Metal Plate Connected Wood Trusses (WTC/TPI 2003). This document is comprehensive and should, if followed properly, prevent accidents during construction. Our practice has included investigating several truss failures that occurred during construction, with the most recent example in 2005. That investigation serves as an example in the following discussion of problems that still occur during construction.

For completed buildings, truss failures most often result from improper or incomplete installation of permanent web bracing. In WTCA 1-1995, WTCA and TPI assign responsibility for the design of permanent bracing for truss web members to the building's overall designer of record. While this assignment makes the responsibility for this important project requirement completely clear, it ignores the general sequence of design, bid, and contract award. Also, it requires that design work be done that can affect a project's final cost after the contract price has been set. Finally, it assumes that the building designer is capable of designing the permanent bracing. For many classes of buildings, an architect or builder may serve as the building designer and these professionals may not have the expertise to design permanent bracing.

The following sections deal first with problems that still arise during construction, review improvements and limitations, and recommend possible strategies for further reducing the number of these problems. A second section reviews the problems associated with design and installation of permanent bracing. In addition to technical challenges, there remains a fundamental problem in the ordering and assignment of responsibilities for this critical element of truss design and performance.

The design of trusses usually occurs after the building is designed. The truss fabricator may be filing a subcontract bid to one of several general contractors competing for the project. In this context, the desired outcome may be a low bid, and the truss designer may optimize trusses individually at the expense of making their permanent bracing simple and affordable. In this case, the actual cost of bracing may remain hidden until after the contract has been awarded. Alternatively, the overall roof truss system and project cost may benefit from spending more money on individual trusses so that they form an overall system that is simpler and cheaper to brace. In this case, the merits of a higher truss bid may not be understood, and the bid may fail. The following sections describe this paradox further and recommend ways to overcome the current limitations.

Construction Problems and Solutions

In an oversimplification, wood roof trusses replace conventionally framed rafters and ceiling joists. Each truss is a single component that replaces a combination of components. Their use allows greater speed in building and results in a shorter time to close in. This generally saves money as well as time, but their use also constitutes a real change in building method.

For the typical wood-frame builders, the use of trusses appears to be a logical extension of their framing experience. Often, a framer may already have experience installing simple gable trusses at modest spans. However, when spans get long, or if trusses have parallel chords, the framer's experience may not be enough to insure safe truss placement. The challenges posed by using trusses are clearly implied by BCSI 1-03 with its extensive documentation of measures necessary to successfully and safely handle, install, and brace them. BCSI 1-03 is provided as part of the truss package and is intended to reduce the likelihood of problems associated with using trusses. However, the guide is only good if it is followed. This document is long (136 pages), deals complex problems, and may not be completely understood. It is likely rare that the guide is followed completely, or even well.

During late summer 2005, we were called to a building under construction to help sort out a partial collapse of the roof trusses. Trusses had toppled over, a few failed, and several were left leaning against a demising wall. The Town Building Official had issued a stop work order, so the Contractor was in a bind. Initially, we were told that work had been forced to a premature halt on a rainy and windy afternoon. A sudden thunderstorm then caused the collapse. The truss collapse pattern included several trusses leaning against one another, and these had clearly toppled from north to south. After pointing out that thunderstorm winds generally blow from southwest to northeast, the project site superintendent acknowledged that maybe the trusses fell because the framing subcontractor had not followed directions for safely installing and bracing them. Their spans ranged from 36 to 52 ft.

The cost of the crane used to lift the trusses was a primary concern. This concern led the framers to ignore the need to properly brace and secure the trusses as they were being lifted into place. If the framers had a larger crew, some of the personnel could have been properly bracing the trusses while other crew members helped place them. As it was, there was a crew of two, and both individuals were working with the crane operator to place the trusses as rapidly as possible. If they had remained lucky – and of course they insisted that this is the way they always install trusses – the expensive crane would have driven away in enough time to allow the framers to return to the task of sheathing and bracing. Although their luck with the trusses did not hold, no one was injured. Luck was at least partially on their side.

For this project, the town's building official felt uneasy about his own ability to monitor the project site. The town hired a local engineering firm to observe the framing subcontractor and to review the work. Although the engineer had been on site earlier in the day, there had been no input regarding the progress of the construction or the means and methods being employed to install the trusses. The absence of any input from the engineer likely derives from the respective – and respected – roles that engineers and contractors play. The contractor has complete authority over construction means and methods. Here, however, the framing subcontractor's crew was inexperienced and was not following the guidelines provided to them with the truss package. It is not clear that the engineer observing the truss placement was substantially more experienced than the framers, but the presence and silence of the engineer may have been interpreted as tacit approval of how the framers were installing the trusses. WTCA and TPI recommend that a professional engineer be consulted when installing long-span trusses – typically for spans of 60 ft or greater [BCSI 1-03]. Truss spans on this project were shorter than 60 ft and so would not have necessarily required that an engineer be present. However, an experienced engineer could have helped here, but the engineer would have to be working directly for the contractor and would have to have the authority to intervene as necessary to maintain safety.

Although the framing subcontractor was generally ignoring most of the recommendations published in BCSI 1-03, the only knowledgeable party who could have intervened was the General Contractor's site superintendent. When asked about this, he noted that the framing subcontractor acts somewhat autonomously. The framer's bid price is based upon an expectation that the work will proceed following the framer's means and methods. If the site super insists that the framers follow the guidelines of BCSI 1-03, for example, the framers may claim that the General Contractor is impeding their ability to work at their expected rate of productivity, thereby costing them money. When the result of this insistence is a safe installation of the trusses, there is no objective measure of how much of the work that was done to ensure safety was actually necessary and how much was more than required. Paradoxically, we can only be sure of how much work corresponds to how much safety when a failure occurs. In addition to its other recommendations, WTCA should insist that every framing subcontract that calls for placing wood trusses include language specifically requiring that their installation be supervised by a registered professional engineer or follow strictly the guidelines of BCSI 1-03.

In response to the problem this project presented, we worked out a plan to remove the very few damaged trusses to right the ones that were simply leaning over but remained undamaged, and we developed a sequence for stabilizing and bracing the trusses. The following day, I met with the contractor's site superintendent and a wide range of representatives, including the architect, the owners' representative, the town building official, and his engineer/observer. My role in this meeting was to convey that I knew something about how to safely place and secure trusses and that I would interact as much as necessary to ensure that when work resumed, it would be done according to my recommendations. The building official rescinded the stop work order.

I returned the third day, shared my diagrams with the framers, and explained how they should sequence the work and how they should brace as they went along. Although the framers nodded their ascent to everything we discussed, when I returned the fourth day, they had only followed about half of what they had agreed to. Frustration and anger did little to correct the situation. To this day, I remain grateful that even by doing only some of what I recommended, the framers' luck held long enough to replace the broken trusses, straighten and brace the undamaged ones, and finish the work without further incident or injury. Luck, however, is unreliable. Any structural engineer involved with a project where trusses will be used must find ways within the project documents to force the issue of safe practice by the framers. Even with a Code requirement, and language in the contract, however, oversight remains necessary, and a site visit that coincides with the truss installation seems a good idea. A quick conversation with the site superintendent may be required to insure that safe practices are followed, but if the contract language is there to support it, insisting that these guidelines be followed should not pose a contractual problem even if it doesn't win friends among the framers. If you take this on, be forewarned that preventing a failure or construction accident inevitably goes unnoticed and unrewarded.

Truss Design and the Paradox of Design for Permanent Bracing

When trusses fail in a completed building, the problem is most often a lack of proper permanent bracing for those truss members that are loaded in compression and which require lateral bracing to resist buckling. In our practice, we once investigated a truss failure where truss plates had been improperly placed. In that case, snow drifted from the windward slope to the leeward slope of a simple gable roof. The resulting drift created a significant unbalanced load, web members loaded in tension pulled out of their plated connection, and the trusses collapsed. However, in over a decade of practice, this example remains unique. Failure usually results from inadequate bracing.

WTCA and TPI assign the responsibility for the design of permanent bracing to the building designer – either the architect or structural engineer [WTCA 1-1995]. Over the past few years, WTCA and TPI have worked to incorporate this assignment into various codes. There is language in the Florida State Building Code as well as the International Building Code (IBC) reflecting this assignment of responsibility. While this effort removes any ambiguity regarding who is responsible for permanent bracing, putting the building designer in charge of permanent bracing design creates a nearly intractable problem.

Especially in a public project, the trusses are likely to be provided as part of a subcontract bid. As such, the building designer will specify the overall building geometry, will specify the loads that the trusses must sustain, and will generally include a performance specification to cover their design and installation. For many large classes of buildings – from one and two-family houses to single-story commercial buildings of a certain size – the building designer may not be qualified to design the permanent bracing. In that case, the building designer may be relying upon the performance specification and the truss provider to meet the specification's requirements. In any case, at the time contract documents are prepared, the building designer will not know the precise geometry of the trusses that various fabricators and designers might propose. Even if qualified to design permanent truss bracing, the building designer cannot design permanent bracing for trusses that have not yet been designed. Consequently, the contract bid documents will not and cannot include a design for permanent bracing.

As part of the bidding process, the general contractor will usually solicit bids from one or more truss providers. If the bid package includes all materials for installing and bracing the trusses, then the provider may anticipate the need for permanent bracing. An experienced provider may even realize that there may be significant improvement in the overall truss system performance and real economy in its installation if the particular design of each truss is altered to facilitate permanent bracing. However, by assigning the design of the permanent bracing to the building designer, the truss provider may perceive no advantage in considering the welfare of the project overall. Further, without a specific requirement for providing the overall truss system, the provider may rightly perceive that his bid will unlikely be the lowest if he is proposing to provide anything more than a set of lowest-cost trusses. Consequently, the provider will most likely submit a bid based upon providing trusses that meet the building's geometry and load carrying requirements for the lowest possible cost.

Truss design synthesizes information about building loads and geometry with fabricator inventory, practice, and truss-designer preferences. The geometry of each truss is based upon the building's architecture (spans, slopes, hips, valleys, etc.) but also reflects the preferences of the truss designer for truss type, web configuration, etc. The truss layout is usually proposed by the truss designer, while the design of each truss may reflect collaboration between the truss designer and the truss fabricator. Within each truss, long members that must resist compression forces usually require lateral bracing to prevent buckling. Therefore, the design of each truss also dictates the requirements for permanent bracing. As the location and frequency of bracing stems from the design of the trusses themselves, the truss designer is the logical choice for design of the permanent bracing.

Each truss member that requires bracing exerts a force on the brace element, these forces may accumulate if several braced members are tied together, and the resultant brace force has serious implications for the structure that supports the trusses. Additionally, the trusses function as part of the overall building structure, transferring wind and seismic loads into the rest of the building. As the building designer must engineer the structure to accept loads from the trusses and to resist truss brace forces, the building designer is the logical choice to design the permanent bracing. Clearly, the work must be shared, even if the responsibility must be assigned to a single entity. The natural interaction among floor truss systems, the roof-truss system, and the remainder of the building structure demand that there be collaboration among the design engineers.

This interdependence among design issues and their implications requires a level of cooperation among designers that conflicts with the traditional isolation of the truss-design process. The truss designer will only know how the building below can support and brace the roof trusses if the building designer provides that information. The building designer will only know what demands the trusses will make on the rest of the building if the truss designer provides that information. Overall, the complex interrelationships between individual trusses and the whole building have implications for the cost and construction of both. When the effort required to design the truss system and its interface with the rest of the structure is postponed until after the bids are in, the validity of those bids is at the mercy of the delayed design process.

Using load and fabricator inventory information, the truss designer may run several variations to arrive at an optimum design truss by truss. When required, the truss designer or fabricator hires a truss engineer of record, and the truss designs are stamped, usually at the last stage in the design review process. If there are changes in design loads, fabricator inventory, or errors detected in other aspects of the design (dimensions for example), then the truss designs may warrant modification even after the truss engineer of record has "signed off" on the designs. When the truss engineer of record is not part of the design and fabrication process, there may be problems with information feedback and review. Overall, the truss fabricator must track carefully how information flows from design to fabrication. Further, if a change in fabricator inventory justifies a design change, all other information (load, geometry, and special provisions) must be retained in the new design.

Proper permanent-bracing design requires clear communication between truss designer and the building designer. The building designer must know what brace forces the structure must sustain. The truss designer needs to know where the building structure can most easily sustain the brace forces. Without clear communication, the building designer will not likely plan correctly for the truss braces. The WTCA (WTCA 1-1995) assigns responsibility for the design of permanent bracing to the building designer. In practice, there must be collaboration between the building designer and the truss designer. Further, the building designer must be capable of designing the permanent bracing or at least understanding how to work with the truss designer to meet that goal.

If the building designer is a contractor or architect, they may not have the knowledge to accomplish this.

Example Conflict and Proposed Resolution

Conflict is a natural part of the building and contracting businesses, but seems inevitable in bracing design. Even if the truss designer and building designer communicate effectively, proper bracing design lies either within a natural gap in – or a natural overlap of – design responsibilities, and this overlap may lead to confusion about which designer "owns" the liability for the bracing and its attachment to the building structure. Further, the cost of bracing the trusses depends considerably on the design of the trusses and their bracing.

Bracing seems to be provided by the truss fabricator in the form of the TPI standards, and by the brace-point locations indicated on the truss drawings. Where a long series of adjacent trusses are all of the same design and geometry, the alignment of web members often facilitates bracing. However, where the details of the truss design vary from truss to truss, brace points move around and bracing becomes very difficult. If the truss designer makes compromises to facilitate bracing, trusses may become more expensive – falling short of the optimum otherwise achievable. In this case, the truss designer is at odds with the truss fabricator / provider who is trying to meet their original bid price. However, if the truss designer optimizes each truss individually, with no regard to facilitating the bracing, the contractor / builder will likely incur additional (and probably unexpected) expense to brace the trusses. In one case, truss web members line up perfectly, allowing a few continuous braces to stabilize whole rows of web elements. The builder braces all corresponding web members with relatively few braces, and the brace forces are resisted systematically through diagonals or by direct connection to brace points on the building below. In a contrasting case, the webs requiring bracing do not line up, continuous bracing is impossible, and the contractor must brace each member individually, possibly with a 2x6 attached to create T-section columns of each affected member. If the contractor's bid price covered the trusses only, and carried a small fraction of that cost for bracing, the additional bracing cost may represent a substantial loss.

When truss designer and building designer do collaborate to create a roof system out of the trusses, the line between their responsibilities may blur but the project will benefit overall. The obvious solution is to communicate ahead of time with the contractor about truss design, placement, and permanent bracing. This communication should begin with the invitation to bid, there should be contract language to communicate the relevant issues, and there may even be a justification for changing the design-bid-build process to select a truss provider and fabricator up front. In a public project, this last suggestion may not be possible. However, it is important to communicate early and often precisely what must happen and when, in order to coordinate the truss and bracing designs to assure a successful project.

Conclusions and Recommendations

WTCA and TPI have published a comprehensive guide to handling, installing, and bracing trusses [BCSI 1-03]. This document, if followed, should significantly reduce the likelihood of a truss collapse during construction. However, there is anecdotal evidence that this guide is not always followed. It is long, it deals with a complex engineering subject, and its audience may not be able to understand and assimilate its contents. Its provisions should become part of the contract language, possibly called out within the overall truss specification, or alternative language should be included that insists upon an engineer's involvement in truss installation. Although project site safety is the responsibility of the general contractor, framing subcontractors may resent apparent interference with their normal means and method of truss installation, however flawed. By mandating adherence to BCSI 1-03, this problem may be solved. However, the length and complexity of BCSI 1-03, together with the number of failures that still occur during construction strongly suggest that there is something special about trusses. Just as steel construction now requires four anchor bolts at the base of each column, there may one day be codified requirements for truss placement if the inherent challenges are not worked out within the community of designers, fabricators, and installers.

WTCA and TPI have assigned responsibility for the design of permanent bracing to the building designer. Although this assignment creates and promulgates problems of its own, when the building designer is a structural engineer, they should embrace the opportunity to participate in the complete design of the building. The building designer should be prepared to collaborate with the truss designer and the contract language should insist upon this collaboration. Among other requirements, contract language should allow the building designer to dictate that the design of trusses follow protocols that make permanent bracing as easy and cost-effective as possible. While this may well lead to an increase in the cost of the trusses themselves, it should result in a lower overall project cost, and it should make it easier for the contactor to install the bracing without incurring unusual and unexpected costs. The truss providers must know to expect this before they submit competitive bids – bids that the fabricator will be unable to meet if the bids are based upon lowest-cost trusses. Further, the overall contract documents should require and budget for the additional design work required after the bids have been submitted and the contract awarded.

At present, WTCA and TPI have been working to improve truss performance and safety both during construction and in finished buildings. However, problems remain, and all practicing structural engineers and building designers must play an active role in seeing that trusses are handled, installed, and braced properly and permanently.

References

Cabler, Steve (1999). "Protect yourself before the accident happens," *WOODWORDS*, Wood Truss Council of America (WCTA), (http://www.woodtruss.com/) (Jan./Feb. 1999).

Kagan, Harvey A. (1993). "Common causes of collapse of metal-plate-connected wood roof trusses," *Journal of Performance of Constructed Facilities*, ASCE, 7(4).

McMartin, K.C., Quaile, A.T., Keenan, F. J. (1984). "Strength and structural safety of long-span light wood roof trusses." *Canadian Journal of Civil Engineering*, 11(4), 978-992

Meeks, John E. (1979). "Industrial profile of the metal-plate-connected wood truss industry." *Proceedings - metal plate wood truss conference, EPRS proceedings 79-28.*

Rojiani, K.B., and Tarbell, K.A. (1985). "Analysis of the reliability of wood roof trusses." *Vol. 1 proceedings - International conference on structural safety and reliability*, Kobe, Japan.

Truss Plate Institute (1976), "Bracing wood roof trusses: commentary and recommendations." BWT-76. Madison, Wisconsin.

Truss Plate Institute (1985). "Design specification for metal-plate-connected wood trusses", (1985), TPI-85, Madison, Wisconsin.

Truss Plate Institute (1991). "Commentary and recommendations for handling, installing, and bracing metal-plate-connected wood trusses", HIB-91, Madison, Wisconsin.

Wood Truss Council of America (WCTA) (1995). "Commentary to WTCA 1-1995 standard responsibilities in the design process involving metal-plate-connected wood trusses." *Commentary to WTCA 1-1995*, Madison, Wisconsin

Wood Truss Council of America (WTCA) (1995). "Standard practice for metal-plate-connected wood truss design responsibilities." WCTA 1-1995, Madison, Wisconsin.

Wood Truss Council of America (WCTA) (2002). *Metal plate connected wood truss handbook*, 3rd ed. (WTCA-2002). Madison, Wisconsin.

Wood Truss Council of America (WTCA) and Truss Plate Institute (TPI) (2003). "*Building component safety information, BCSI 1-03, Guide to good practice for handling, installing & bracing of metal plate connected wood trusses.*" Madison, Wisconsin.

Roof Collapse: Forensic Uplift Failure Analysis

Erik L. Nelson, Ph.D., P.E., M. ASCE
Deepak Ahuja, M.S., P.E., M. ASCE
Stewart M. Verhulst, M.S., P.E., M. ASCE
Erin Criste, M.S., M. ASCE

ABSTRACT

Many factors affect the performance of structural roof framing, and if deficient components exist, the structural integrity is compromised. When a roof system is improperly designed, failure may result from under-design regarding net uplift pressures. Today's commonly used lightweight roofing products (EPDM, poly-isocyanurate) have made net uplift loads a more critical design load, and in some instances, the controlling case. In particular, a commercial warehouse building was under-designed for net uplift pressures, which in conjunction with unclear bridging spacing requirements per Steel Joist Institute (SJI) requirements, resulted in a roof collapse during a storm event. The net uplift design load for the steel joist roofing system should have been higher than what was specified on construction drawings. Additionally, a lack of clarity in the SJI requirement for joist bottom chord bridging resulted in excessive bridging spacing, which lessened the capacity of the roof framing considering uplift. Consideration of the lightweight roofing materials in the joist design and clarity in SJI uplift tables would have prevented the roof collapse.

INTRODUCTION

A roof collapse occurred at a commercial warehouse building during a storm event. The structure included an open warehouse space at the interior, with interior demising walls. The foundation consisted of a conventionally reinforced concrete slab-on-grade with perimeter piers and interior footings. The structure was indicated as being 540,000 square feet. The interior construction and framing consisted of steel joists and girders at the roof framing and concrete tilt-up wall panels at the perimeter walls. The structure was built in 1996.

The roof was indicated as a mechanically attached, single-ply EPDM membrane over 1 1/2" Isocyanurate Foam. The roof deck was a 1 1/2" deep, 22-gage wide rib painted metal deck. The typical roof joists were 26K9 (K-series) joists with three (3) rows of horizontal bridging for the top chord, four (4) rows of horizontal bridging for the bottom chord, and one (1) row of X-bridging. The joists were approximately 50' in length, spaced at 6'-3" on-center and spanned between column bays in a 50' by 50' grid.

The available information regarding the structure and the storm event included storm data, design documents, and shop drawings for the framing. The net uplift load used for the design of the joists was listed on the design drawings as 10 psf.

OBSERVATIONS AT WAREHOUSE

Observations of the structure were made and items including damaged members and inadequate connections were noted. The following is a summary of a few of the items observed:

The top chord of an original joist was observed to not be straight (i.e. it appears to have been displaced out of the plane of the joist). Additionally, some lateral movement was evident at the bottom chord, occurring near mid-span of the joist between the bridging locations. The bridging for the steel roof joists consisted of both horizontal bridging (top and bottom) and cross bridging (also denoted as "X-bridging"). Generally, the typical connection at the X-bridging was a bolted connection to an angle plate welded to the joist, and the typical connection for the horizontal bridging to the joists was specified as a fillet weld.

Measurements of the spacing between bridging locations along the bottom joist chords were taken. At the original joists, a typical bottom chord bridging spacing of 12'-8" to 12'-9" was observed across the mid-span of the joists. Typically, the bottom chord bridging was located at the end panel points of the joists and was included as part of the four (4) bottom chord bridging locations.

Based on observations, a typical failure mode for the joists was buckling at the bottom joist chords, near mid-span of the joists. Buckling failures of the end web members of the joists were also observed, suggesting multiple or combined failure modes due to uplift. Refer to Figure 1 and Figure 2.

METEOROLOGICAL REPORTS

Several meteorological reports regarding the storm occurring at the site were available. Based on the reports received, severe thunderstorms occurred with wind gusts of 75-80 mph. The maximum reported wind gust is indicated as 77 mph-recorded at an airport approximately 3 miles from the site.

Figure 1. Buckling at joist bottom chord (Photo courtesy of Roof Technical Services, Inc.).

Figure 2. Buckling failure at end web (Photo courtesy of Roof Technical Services, Inc.).

One of the meteorological reports indicated that the storm at the site was a rotating supercell thunderstorm (mesocyclone). It was reported that a supercell thunderstorm is the most violent and forceful classified storm and that these types of storms commonly have "intense micro burst updrafts and associated downdrafts." Furthermore, the meteorological reports noted that supercell thunderstorms are the type which most frequently produce tornadoes and that the peak wind gusts would likely have been even higher if a tornado was produced by the storm.

DISCUSSION OF WIND LOADS AT THE SITE

The applicable Building Code for the design of the structure was the 1991 Uniform Building Code (UBC).

Based on Figure No. 23-1 in the 1991 UBC, the design wind speed for the site is 70 mph. This is based on a "fastest-mile" wind speed criteria, which is partially defined in the UBC as, "the highest sustained average wind speed based on the time required for a mile-long sample of air to pass a fixed point". It should be noted that the fastest-mile wind speed criteria was also used by the 1994 and 1997 editions of the UBC.

More recent standards and codes, including recent ASCE 7 standards and the 2000 and 2003 editions of the International Building Code (IBC), use similar parameters for determining wind pressures. However, in these standards and codes, other factors such as site topography and wind gusts are used more explicitly in the determination of wind pressures. Also, these more recent standards use a peak gust wind speed rather than the fastest-mile wind speed.

The roof dead load for the warehouse structure was calculated to be only 5.32 psf, including the self-weight of the joist framing. Due to the geometry of the warehouse structure, it is considered as an "Open Structure" per the UBC for determination of wind pressures. Open structures generally have higher uplift pressures due to wind than structures which are not "open". In this case, the design net uplift pressure at the field of the roof is approximately 60% higher for an open condition compared to a not-open condition.

The uplift loads for the subject warehouse roof were calculated in accordance with the 1991 UBC. Based on the tributary area of the joists, they were considered as "elements and components" regarding wind uplift loading per the 1991 UBC. The gross and net uplift pressures are included in Table 1 below (the "field" of the roof refers to the main roof area and "discontinuities" refer to the areas of the roof where architectural features result in increased uplift load – such as near the eaves of the roof). The calculated net uplift exceeds the 10 psf indicated on the construction drawings by more than 75%.

Table 1. Wind Uplift Loading.

Location	Gross Uplift (psf)	Net Uplift (psf)
Discontinuities	-24.46	-19.14
Field	-22.83	-17.51

For the purposes of comparison, the change in the gross and net uplift wind pressures for incremental changes in the wind speed are indicated in Table 2. The values indicated in Table 2 are calculated for elements and components in the field of the roof for the structure, using the method of the 1991 UBC.

Table 2. Incremental Wind Uplift Loading.

Wind Speed (mph)	q_s	Gross Uplift (psf)	Net Uplift
70	12.6	-22.83	-17.51
80	16.4	-29.71	-24.39
90	20.8	-37.68	-32.36
100	25.6	-46.38	-41.06
110	31	-56.16	-50.84

As indicated in Table 2, gross wind pressures increase as a square of the wind speed (using the "fastest mile" speed per the 1991 UBC). Therefore, wind speeds in excess of the design wind speed of 70 mph could cause significant increases in the net uplift wind pressures on the roof and roof framing.

FACTORS OF SAFETY

In structural engineering design, "Factors of Safety" (FS) are employed to account for unknown conditions, variability in materials, inherent design assumptions, construction deficiencies, and to provide for the safety of the public. Considering Allowable Stress Design (ASD) of steel structures, a FS=1.67 is used for tension members and beams and a FS=23/12=1.92 is used for typical long compression members (those which perform as column members, etc.).

Generally, a Factor of Safety is not a reserve capacity, and cannot be used as such during the design or construction of a structure. The Factor of Safety is a minimum design requirement as established by the applicable building code and applicable structural codes and standards.

While SJI does require a 1.65 factor of safety in the design, the actual factor of safety with regard to compression is the 23/12 factor (FS=1.92, as noted above) applied to the Euler buckling formula. It should be noted that, for the joist designs performed, a 1/3 increase was included for the allowable stresses due to wind load per Section A5.2 of the AISC Specifications (AISC, 1989). Therefore, the actual factor of safety for the compression design of the joist members was about 1.44 (23/12 divided by 4/3 for the wind stress increase).

As noted above, the proper net uplift design load for the joists at the roof of the warehouse was 17.51 psf, which is about 75% higher than the design load for which the joists were actually designed (10 psf). This 75% increase in uplift pressure would therefore exceed the Factor of Safety of 44%, leading to a likely failure for wind speeds approaching the design wind load.

STEEL JOIST DESIGN AND BRIDGING REQUIREMENTS

It is further indicated that the bridging shall conform to the SJI specifications. The Structural Plan notes for the warehouse structure indicate that, "Steel Joists shall be braced by horizontal and/or diagonal bridging as required by the Steel Joist Institute."

Additionally, the applicable building code for the project (1991 UBC) includes the SJI Specifications as a UBC Standard.

The SJI Specifications indicate 2 types of bridging: horizontal bridging and diagonal bridging (also called X-bridging). The SJI Specifications state that,

> *Horizontal bridging shall consist of two continuous horizontal steel members, one attached to the top chord and the other attached to the bottom chord.*

Also, regarding the amount and spacing of bridging, the SJI Specifications state,

> *In no case shall the number of rows of bridging be less than shown in the bridging table. Spaces between rows shall be approximately uniform. See section 5.11 for bridging required for uplift forces.*

Section 5.11 of the SJI Specifications discusses uplift provisions for steel joists and is included here for reference:

> *5.11 UPLIFT*
>
> *Where uplift forces due to wind are a design requirement, these forces must be indicated on the contract drawings in terms of net pounds per square foot. When these forces are specified, they must be considered in design of joists and/or bridging. A single line of **bottom chord** bridging must be provided near the first bottom chord panel points whenever uplift due to wind forces is a design consideration.**
>
> **For further reference, refer to Steel Joist Institute Technical Digest #6, "Structural Design of Steel Joist Roofs to Resist Uplift Loads."*

Based on the bridging table included in the SJI Specifications; for the 26K9 joists indicated in the joist shop drawings, 4 rows of bridging are required for spans from 46' to 59'. Therefore, the maximum spacing for bridging at the joist would be 11'-9" (59'/5 spaces). This approach for determining the spacing limitation for bottom chord bridging is also indicated in SJI Technical Digest No. 6. As noted above, SJI Technical Digest No. 6 is specifically referred to for further reference by Section "5.11 Uplift" of the SJI Specifications. The SJI Specifications do not list which version of SJI Technical Digest No. 6 to follow. It is the authors' opinion that the joists should have been designed according to the most current version of the Technical Digest at that time. Based on our discussion with SJI, at the time of the design and construction of the structure, the most current SJI Technical Digest No. 6 was the 1994 version.

The steel joist shop drawings for the steel joist roof framing indicate three (3) rows of horizontal top chord bridging and four (4) rows of horizontal bottom chord bridging (including bridging at each end panel point), in addition to the one (1) row of X-bridging. The X-bridging serves as bridging for both the top and bottom joist chords and is indicated at one of the equally-spaced top chord bridging locations nearest the mid-span of the joists. Therefore, there were four (4) rows total of top chord bridging and five (5) rows total of bottom chord bridging. As noted above, the roof joists were typically 50'-0" in length.

Based on the measured geometry of the joists at the site (including the location of the end panel points) and based on the X-bridging being placed at one of the equally-spaced top chord bridging locations, a total of 6 rows of bottom chord bridging would be required for compliance with the SJI maximum spacing limitation of 11'-9". The bridging indicated in the shop drawings and the bridging layout observed at the site have typical spacings between points of bottom chord bridging which are in excess of the spacing limitations of the SJI Specifications (12'-9" and 12'-8" vs. 11'-9"). If the proper net uplift pressure had been used, the spacing of the bottom chord bridging would not have exceeded the SJI limitations.

The bridging layout on the steel drawings and the layout observed at the site did not conform with Section 5.4 of the SJI Standard Specifications because the bridging spacings are not approximately uniform. During site visits, the bridging spacing at a typical original joist was measured to be 12'-8" at one side of the X-bridging and 6'-8" on the other side. Thus, the bridging spacings vary by up to 90% along a single joist.

As noted above, failures of joist end web members were observed. Under normal gravity load, these members are tension members; however, load reversal occurs when net uplift loads control the design. Therefore, these members are in compression under net uplift conditions. For the particular joists at the subject warehouse, net uplift was the governing design condition for the joist end webs.

It should be noted that the roof joist calculations indicated a Kl/r ratio of 185.2 and a K factor of 0.8 for the end web members. Based on these values, the (l/r) ratio for the end web members on the joists was 231.5. Section 4.3 of the SJI Specifications defines the maximum allowable slenderness ratios (defined as l/r) for use in K-series steel joists as follows:

Top chord interior panels	*90*
Top chord end panels	*120*
Compression members other than top chord	*200*
Tension members	*240*

In a case when a joist is to resist a net uplift, all diagonal members, bottom chord and top chord members shall be in compression in at least one of the load cases. In fact, the governing load case for the design of these end web members was the uplift condition, where they are in compression. Based on this criteria, the limiting l/r ratio for the end web member in compression, as indicated by the SJI Specifications, would be 200. This is exceeded by the actual l/r of 231.5.

However, the SJI Technical Digest No. 6, uses the tension member criteria of 240 for a limiting slenderness ratio of an end web member. Additionally, as indicated above, the SJI Technical Digest also uses an effective length factor of K = 0.8 for the calculation of allowable compressive stress in the member. As noted, the Kl/r ratio is 185.2, which is less than the slenderness ratio of 200 indicated in the SJI Specifications. This issue appears to be an ambiguity between the SJI Specifcations and SJI Technical Digest No. 6.

ANALYSIS

The roof joists were analyzed for joist capacity considering different failure modes. As noted above, the specific joist failure modes observed at the joists included compression failures (buckling) of the bottom chords near mid-span.

It appears that the bottom chords of the failed joists buckled laterally – for the purposes of this discussion, it will be considered as buckling about the y-y axis. The design calculations for the joists were available for review and they indicated an allowable L_{yy} (allowable bridging spacing) of 14'-2" for the stress in the bottom chord. This allowable bridging spacing was calculated using the 1/3 stress increase in the allowable bottom chord stress and 10 psf net uplift loading as indicated on the design documents. This is an important reference point when considering the effect of the inadequate net uplift design load on the joist design.

The capacity of the joists, considering the failure mode at the bottom chord, is presented in Table 3. The capacity is indicated in terms of the net uplift pressure (on the joists) for different bridging spacings. Table 3 includes the capacities based on the allowable load, the permitted 1/3 stress increase per AISC, and the Euler buckling load (without the buckling safety factor). The bottom chord capacity was determined for a range of bridging spacings.

Table 3. Bottom Chord Net Uplift Capacities (psf).

Bridging Spacing	Allowable Capacity Pressure – No Stress Increase	Allowable Capacity Pressure – 1/3 Stress Increase	Euler Buckling Capacity Pressure
11'-9" [1]	10.91 psf	14.55 psf	20.92 psf
12-6"	9.64 psf	12.86 psf	18.48 psf
12'-8"	9.39 psf	12.51 psf	17.99 psf
12'-9" [2]	9.27 psf	12.36 psf	17.76 psf
13'-0"	8.92 psf	11.89 psf	17.09 psf
13'-2"	8.69 psf	11.59 psf	16.66 psf
14'	7.69 psf	10.25 psf	14.73 psf
14'-2" [3]	7:51 psf	10.01 psf	14.39 psf
14'-6"	7.17 psf	9.55 psf	13.73 psf

Notes: 1) Maximum allowable bottom chord bridging spacing per SJI.
2) Maximum measured bottom chord bridging spacing at the site.
3) Maximum allowable bottom chord bridging spacing per design.

Table 3 indicates the increase in bottom chord capacity as the bridging spacing decreases. The net uplift pressure which should have been used for the roof design was 17.51 psf, which exceeds all of the allowable values listed in Table 3. Also, this proper net uplift exceeds the capacity (no factor of safety) of the joists if they had a bridging spacing of 14'-2", further indicating that the Factor of Safety for the joist design was eclipsed by the use of the improper design load. Finally, Table 3 indicates that, if the joists had been designed and constructed in conformance to the SJI specifications, the actual capacity of the bottom chord would have exceeded the proper design load. The actual capacity of the joists would have increased 18% if the bridging layout had conformed with the SJI Specifications (20.92 psf for 11'-9" spacing vs. 17.76 psf for 12'-9" spacing observed).

This illustrates the effect of improper design loading and excessive joist bottom chord bridging spacing on the actual capacity of the joist for wind uplift. Any Factor of Safety in the joist design was eclipsed by the combination of the mis-calculated design load and the failure to comply with SJI standards for the spacing of the bottom chord bridging. Of course, the use of a non-conservative design load may result in failure, irrespective of the SJI standards. However, designing a bridging layout which complies with the SJI standard can only increase the capacity.

CHANGES TO SJI SPECIFICATIONS

In the most recent standard specifications for K-series joists, dated 2003 and effective 2005, SJI has made changes, including clarification of top chord bridging and bottom chord bridging requirements. As part of this clarification, SJI requires that the number of rows of bottom chord bridging not be less than the number of rows of top chord bridging. The bottom chord bridging spacing will also be required such that the bottom chord complies with the slenderness requirements of SJI and any specified strength requirements. The language regarding bridging has been further clarified to distinguish between the bottom chord and top chord bridging, noting that they may be spaced independently.

It is the opinion of the authors that these changes implemented by SJI are helpful in clarifying the top and bottom chord bridging requirements. However, some ambiguity remains; including the determination of the governing slenderness ratio for a bottom chord member and an end web member if uplift controls the design. The use of l/r also remains in the SJI standard, which causes some confusion due to the use of Kl/r in the Technical Digest as noted above. The ambiguous "approximately uniform" spacing requirement has been removed from the 2003 SJI specifications for K-series joists.

CONCLUSIONS

The net uplift design load for the joists was inadequate and the design load should have been about 75% higher. As noted in this report, the roof systems selected for the original construction were uniquely light. This should have been considered in the design process regarding wind uplift. As noted, the Factor of Safety for the joist design was eclipsed by the use of the improper design load for uplift and by the failure to comply with the proper SJI bridging requirements for the bottom chord bridging.

The misuse of the SJI Specifications regarding the bridging spacing apparently is the result of misunderstanding of the SJI Specifications for the joists. Ambiguity in the SJI Specifications, such as calling for "approximately uniform" spacing of the bridging and a failure to explicitly state that the bottom chord bridging is also subject to maximum spacing requirements has contributed to the misunderstanding.

Additionally, there is ambiguity between the SJI specifications and the SJI Technical Digest regarding the proper slenderness ratio for the end web members. This also requires clarification to prevent further misunderstanding. In the most recent standard specifications for K-series joists, SJI has made changes regarding bridging. Specific changes include clarification of top chord bridging and bottom chord bridging requirements. It is the opinion of the authors that these changes by SJI are helpful in clarifying the top and bottom chord bridging requirements, although some ambiguity remains.

In the case of the subject warehouse, the failure to comply with SJI lessened the capacity of the joists in uplift. These joists ultimately failed in a violent manner. An increase in the joist capacities for uplift could have prevented or lessened this failure. The actual capacity of the joists would have increased 18% if the bridging layout had conformed with the SJI specifications. This illustrates the role of Factors of Safety and minimum standards (such as those by SJI) in the arena of public safety.

REFERENCES

American Institute of Steel Construction (AISC). (1989). *Manual of Steel Construction – Allowable Stress Design*, Ninth Edition, Second Revision, Chicago, Illinois

American Society of Civil Engineers (ASCE), (2002). *SEI/ASCE 7-02 Minimum Design Loads for Buildings and Other Structures*, Reston, Virginia.

International Conference of Building Officials (ICBO). (1991). *Uniform Building Code*, 1991 Edition, Whittier, California.

International Code Council, Inc. (ICC). (2000). *International Building Code*, 2000 Edition, Whittier, California.

International Code Council, Inc. (ICC). (2003). *International Building Code*, 2003 Edition, Whittier, California.

Salmon, Charles G. and Johnson, John E. (1996). *Steel Structures – Design and Behavior*, Fourth Edition, Harper Collins Publishers, Inc.

Steel Joist institute (SJI). (1989). *Standard Specifications for Open Web Steel Joists, K-Series*, Adopted by the Steel Joist Institute November 4, 1985; Revised to November 15, 1989, Myrtle Beach, South Carolina.

Steel Joist Institute (SJI). (1994). *Technical Digest No. 6 – Design of Steel Joist Roofs To Resist Uplift Loads*, 1994 Edition, Myrtle Beach, South Carolina.

Steel Joist Institute (SJI). (2003). *Technical Digest No. 6 – Design of Steel Joist Roofs To Resist Uplift Loads*, 2003 Edition, Myrtle Beach, South Carolina.

Steel Joist institute (SJI). (2003). *Standard Specifications for Open Web Steel Joists, K-Series*, Adopted by the Steel Joist Institute November 4, 1985; Revised to November 10, 2003, Effective March 1, 2005, Myrtle Beach, South Carolina.

Steel Joist Institute (SJI). (2004). "Minutes of Engineering Practice Committee Meeting", February 9-10, 2004.

Claims and Forensic Engineering in Tunneling

Wolfgang Roth - URS Los Angeles, CA
(formerly Dames & Moore – Principal Geotechnical Engineer, CM JV Parsons-Dillingham)

Anthony Stirbys – PTG Pasadena, CA
(formerly Dames & Moore – Geotechnical Engineering Manager, CM JV Parsons-Dillingham)

ABSTRACT: This paper discusses a series of tunneling mishaps and related claims, which occurred during construction of the Hollywood segments of the LA Metro Red Line project. Forensic engineering was used as an important tool for evaluating likely failure causes in an effort to evaluate the merit of the construction claims, or lack there of. Since some of the failure causes remain controversial to this day, little has been published, and valuable lessons which could be learned are in danger of being forgotten. Even though all legal aspects eventually were settled, these events also created an atmosphere of public anxiety about tunneling in Los Angeles, which weakened the political support for future tunnel projects in the region.

1.0 INTRODUCTION

1.1 The Role of Forensic Engineering

When it comes to identifying the cause of underground-construction failures, owners and contractors are usually on opposite sides. Where contractors claim differing site conditions, owners often see inadequate construction means and methods or noncompliance with design specifications. In order to decide which is which, it is necessary to evaluate the most likely failure cause. This is where forensic engineering comes in.

This paper discusses a series of tunneling mishaps and related claims, which occurred during construction of the Hollywood segments of the Los Angeles Metro Red Line project. Since some of the failures remain controversial to this day, little has been published to date, and valuable lessons which could be learned are in danger of being forgotten. Even though the legal aspects are all settled by now, these events also created an atmosphere of public anxiety about tunneling in Los Angeles, which weakened the political support for future tunnel projects in the region.

1.2 Project Overview

The 17-mile long Metro Red Line was under construction from 1984 through 1999 (Fig.1). Segment 1 was opened in two phases from Union Station to Mc.Arthur Park, and then to Wilshire/Western in 1993 and 1996, respectively. Segment 2 followed to Hollywood/Vine in 1999, and Segment 3 to North Hollywood in 2000.

Except for a short hard-rock section through the Santa Monica Mountains, subsurface conditions ranged from alluvium to relatively soft silt-, clay-, and sandstones of the Puente formation. Throughout most of the alignment, groundwater was well below tunnel invert, and excavation was accomplished with 22-foot diameter, open-face digger shields with one or two breasting tables. Temporary support was by expanded, pre-cast concrete segments; and the final liner was cast in place.

2.0 PILLAR FAILURE, BARNSDALL SHAFT

Both the Hollywood Blvd. and Vermont Ave twin tunnels were mined from a starter shaft at Barnsdall Park. The contractor chose to construct a shaft significantly smaller than the available space allowed for in the contract documents. So in order to be able to excavate all 4 headings simultaneously, he chose to hand-mine four 72-foot long starter adits (Fig. 2).

Ground conditions consisted of Puente formation siltstone with an average unconfined compressive strength (UCS) of 90 psi. The Geotechnical Design Summary Report (GDSR - also referred to as Geotechnical Baseline Report, or GBR) described the shaft location as a transition zone from fresh to weathered siltstone, where strength values could be expected to be significantly less than average.

The 100-foot deep shaft excavation was supported by 48-inch cast-in-place tangent piles with multiple levels of tiebacks (Fig.3). The four adits were excavated simultaneously with headings followed by bench excavations. Headings were supported by steel arches and wall beams, and timber lagging was backfilled with concrete. In excavating the benches, vertical steel ribs were placed under the steel arches and wall beams. Backfilling behind the lagging of the bench excavation was sporadic at best.

2.1 PILLAR FAILURE

After advancing all 4 bench excavations about 50 feet, the tangent piles between the tunnel portals showed signs of axial-compression failure. Forty-five degree shear planes developed simultaneously at both the west and east walls, and the ground surface above the portals settled 3 to 5 inches. The shaft was immediately evacuated, but an engineering assessment soon concluded that there was no eminent danger of collapse. Even though the narrow, 13-foot wide siltstone pillars between the adits had failed, load transfer had taken place in the form of soil arching across adjacent adits. This process had dragged down the tangent piles, thereby exceeding their axial-load capacity.

Remedial measures included the installation of additional tiebacks for the tangent piles adjacent to the tunnel portals (Fig. 4). In addition, the pillars were laterally confined by backfilling behind the lagging and installing cross ties between adjacent adits as can be seen in Figure 2.

2.2 DIFFERING-SITE CONDITIONS (DSC) CLAIM

While there was eventual agreement about the cause and mechanism of the pillar failures, the contractor claimed entitlement of adjustment in time and money on the basis of any, or all, of the following:

1. Type-I DSC, because nothing in the contract documents indicated that the pillars would be a construction problem;

2. Type-II DSC, because the soil condition causing the problem was of an unusual character not inherent in the work;
3. Defective specifications, because the problem was not identified anywhere in the GDSR or in the Plans and Specifications; and/or
4. The contractor's submittal was approved by the owner without raising concerns about the stability of the pillars.

2.3 DRB RECOMMENDATION

The claim was brought before a Dispute Resolution Board (DRB), which found the first three points without merit. However, it put 25% liability on the owner, because the contractor's submittal had been approved by the design consultant without recognizing its engineering inadequacy. The most pertinent points raised by the DRB are summarized below:

1. There was no reason for the pillar problem to be discussed in the GDSR, because the contract documents assumed tunneling to start without pre-mined adits. The latter had been proposed as a change by the contractor;
2. There was nothing unusual about the soil conditions, because they were not materially different from the characteristics described in the GDSR;
3. The starter adits had not been part of the original contract documents and, hence, there was no reason for showing them on any plans or addressing them in any specifications;
4. Based on the strength of the siltstone provided in the GDSR, the DRB found that the adit support had been grossly under-designed and, thus, was bound to fail; and
5. While the owner's approval of contractor submittals does not relieve the contractor of responsibility for his design, this should not entirely insulate the owner from some share of responsibility.

Further aggravating the situation was the testimony of the owner's engineer, who had reviewed and approved the contractor submittal. He stated that, based on his experience and perception, he had concluded that the strength of the Puente siltstone was much higher than stated in the GDSR and, therefore, he had no concern about the adequacy of the pillar.

2.4 LESSONS LEARNED

Contrary to popular opinion, and in spite of cleverly worded legal disclaimers, the review of contractor submittals inevitably involves some sharing of responsibility. Rather than trying to avoid liability by taking a hands-off approach in reviewing tunneling means and methods, owners and their consultants would be well advised to maintain an active dialogue with the contractor. More specifically, they should not shy away from providing constructive feedback, including discussing alternative approaches with the contractor.

3.0 HARD-ROCK CLAIM, VERMONT TUNNELS

Leaving the Barnsdall shaft to the east and turning south along Vermont Ave, the digger shields were working in rather competent Puente formation siltstone. A short distance into the curved alignment, the leading shield began to deform (vertical ovalling) to the point where tunneling had to be stopped in order to repair/ strengthen the shield. Thanks to the competent ground conditions, it was possible to repair the shield underground by hand-mining around it, and jacking it back to its original shape.

3.1 DSC CLAIM

Claiming that the shield had been damaged by ground conditions harder/stronger than predicted by the GDSR, the contractor filed a Type-I DSC claim. Specifically, he presented test data from block samples obtained during tunneling, which showed UCS data ranging from 280 to 510 psi (Fig. 5).

Even though these values did not exceed the maximum baseline value of USC=600 psi mentioned in the GDSR, they exceeded the range of actual lab-test data listed in a table of the same report.

The owner rejected the claim as having no merit, since the GDSR had established a clear 600-psi baseline for bidding purposes. Furthermore, it was the owner's position that the ovalling of the shield had been due to steering difficulties entirely unrelated to the strength of the siltstone. In fact, rock-strength data had not even been a factor in the contractor's shield design.

3.2 GDSR STATEMENTS

The GDSR described the Puente formation siltstone as "*hard to dense soil, ...with local presence of hard sand/siltstone ranging from less than a foot to several feet thick....*" Furthermore, the report referred to a table with 48 lab-test results showing UCS values ranging from 11 to 276 psi, with a mean of 89 psi. Both the qualitative description and the lab-test data contrasted with the following statement, which was given without further explanation: *"For the purpose of this bid, the contractor should assume that UCS values can reach up to 600 psi."*

This statement was based on experience from a previous claim on Red Line Segment 1. There, block samples taken from the tunnel face turned out to be much stronger than the lab-test data reported in the GDSR. Based on a testing program conducted as part of this earlier claim, it had been concluded that UCS data from testing disturbed drive-samples should be multiplied by a factor of 2 to 3 in order to reflect more realistic in-situ values.

In preparing the Vermont tunnel GDSR, the question of whether or not to mention this previous experience was the subject of lively debates. In the end, the owner's design consultant recommended against it, because it would have amounted to an admission that mistakes had been made in the past.

3.3 CONTROVERSIAL DRB FINDINGS

Much to the surprise of almost everyone involved in this case, the DRB decided in favor of the contractor. In the opinion of the DRB, by not disclosing the reason for the discrepancy of the UCS data, the owner had withheld vital information which could have assisted the contractor in preparing his bid.

In the owner's opinion, the DRB's finding defeated the very purpose of providing a geotechnical baseline. It was argued that the objective of including the 600-psi statement for bidding purposes was not to provide "accurate" soil data, but a baseline in the very sense of the word. Depending on how much risk the owner wants to take, it should be his right to include as much conservatism in the baseline (and pay for it by receiving higher bids) as he deems appropriate.

In the end, the DRB's recommendation was rejected by the owner, and the case continued with several years of legal hearings. It was finally settled as part of an overall package deal involving a long string of suits and counter suits surrounding the B251 contract.

3.4 Lessons Learned

While the purpose of a GDSR (or GBR) is to establish well defined bid assumptions, it should also reflect actual conditions as closely as possible. Introducing excessive conservatism, by quoting seemingly unrealistic soil data, is counterproductive. Such inflated baselines are tantamount to denying a contractor the right of using his engineering know-how and experience in preparing his bid.

If actual test data are different from those given as a baseline, the contractor is likely to choose the interpretation which suits him best. While there may be good reasons for such differences, in the absence of any explanation provided, the contractor is likely to prevail in a DSC claim.

4.0 RUNNING GROUND, EDGEMONT AVE

Leaving the Barnsdall shaft and advancing west along Hollywood Blvd, the tunnels leave the weathered Puente siltstone and enter saturated alluvium (Fig. 6). Even though the GDSR predicted the groundwater table to be about 20 feet above tunnel crown, the contractor decided to push ahead without dewatering. He argued that the silty and clayey interbeds within the alluvium would make dewatering difficult and not very effective. Counting on these same interbeds to provide sufficient standup time, he thought it possible to control the open tunnel face with two breasting tables. Since the performance-based specifications lacked any contractual requirement for dewatering, the contractor's submittal was eventually approved with the usual disclaimer of not assuming any liability.

4.1 Tunnel-Face Collapse in Running Ground

As soon as the leading shield pushed out of the siltstone into saturated alluvium, running-ground conditions caused the tunnel heading to collapse (Fig. 7). Even though it took more than 100 cubic yards of lean mix to fill the resulting underground cavity, a relatively thick cohesive soil layer above the groundwater table had prevented the cave-in to progress to the street surface some 80 feet above the tunnel crown. Luckily, with little surface expression of the cave-in, this incident didn't even make the news. However, its impact on the construction schedule was quite significant. Both shields had to be stopped for 6 months while the contractor lowered the groundwater table below the tunnel-invert elevation in the 1-mile long saturated portion of the alignment ahead.

4.2 DSC CLAIM

The contractor requested a Type-I DSC change order in the amount of $12 mill for costs associated with the dewatering effort and construction delays. In support of this request, he submitted data from an observation well monitored after the cave-in, which showed the groundwater table to be about 10 feet higher than had been predicted in the GDSR. Faced with this evidence, the owner initially was considering approval of the change order, but eventually decided against it. The resulting claim was never resolved on its own merit. Rather, it was settled many years later as part of a package agreement ending a long drawn-out string of suits and counter suits involving many other issues related to the B-251 contract.

4.3 LESSONS LEARNED

Though it was quite clear that the tunnel face would have collapsed, even with the groundwater table "only" 20 feet above tunnel crown (as stated in the GDSR), the claim would have had a reasonable chance to be decided in favor of the contractor. Groundwater conditions described in the GDSR should have taken into account seasonal and other long-term fluctuations, rather than merely reporting the groundwater table observed at the time of the site investigation taking place long before the start of construction.

In any case, this incident could have been avoided had the contract contained prescriptive-type specifications requiring the obvious, namely lowering the groundwater table below tunnel invert. In the absence of such requirements, the owner and its engineers were reluctant to force the contractor to dewater, fearing to get stuck with the bill. Allowing the contractor to proceed was thought to get the owner off the hook without relieving the contractor of his liability.

5.0 TUNNEL LINER FAILURE, HUDSON AVE

Temporary tunnel liners were erected in rings consisting of 4 pre-cast concrete segments. After clearing the tunnel shield, these rings were expanded and locked in place with hardwood wedges in the 10- and 2 o'clock positions (Fig.8). The wedges had been approved, at the request of the contractor, in place of originally specified steel screw jacks.

A day after the leading shield had passed Hudson Avenue, water was observed seeping through the tunnel liner. This, even though no water had been encountered at this location during tunnel excavation the day before. Suspected water sources included broken utility lines or perched groundwater above the tunnel, which could have seeped through tunneling-induced cracks of the cohesive layers confining them.

Ground settlements, which initially were about 2 inches, steadily increased over the next 3 weeks, but finally stabilized after reaching 4 inches. Both shields were on hold during this time. Finally, because adjacent structures were not believed to be in any danger, the trailing shield was given the green light to pass the critical location without implementing any ground improvement measures. When the shield reached the seepage-affected location, however, the first tunnel nearly collapsed as the hardwood wedges in the 10-o'clock-position were crushed and crown segments dropped up to 7 inches (Figure 9). As street settlements reached 11 inches, Hollywood Boulevard was closed, and the crown segments were supported by vertical steel posts (Figure 10).

5.1 DSC CLAIM

The contractor submitted a request for change order asserting that the wedge failures had been caused by collapsible Young Alluvium not anticipated in the GDSR. The owner denied this request based on test results from soil samples taken at the failure location. The amount of hydro-compaction observed upon wetting these samples simply was not large enough for these soils to be classified "collapsible."

When the change-order request became a claim, an investigation was launched into the probable cause of failure. Because the same hardwood wedges had been used previously for many miles of tunneling without incident, the question was "why did they fail at this particular location?" To solve this puzzle, a failure analysis was performed by way of numerically simulating the sequence of events leading up to the crushing of the wedges (Roth, et al, 2001).

5.2 FAILURE ANALYSIS

The analysis results indicated that even a small amount of hydro-compaction, if triggered by wetting after the tunnel liner had been expanded and locked in place, caused significant stress increase in the tunnel liner (Fig. 11). With liner compression forces already having been elevated due to wetting before the arrival of the trailing shield, the additional stress increase due to excavating the second tunnel, finally, lead to the crushing of the wedges.

Complicating this case was the fact that the compressive strength of the hard wood wedges had been less than half of the specified value. In fact, the analysis results showed that the wedges would have survived had they been as strong as specified. Since this finding appeared to implicate just about all parties involved in this dispute, the case was never resolved on its own merit. It was later included in a "package settlement" which ended years of suits and counter suits involving this and many other disputes in connection with Contract B251.

5.3 LESSONS LEARNED

Had the hardwood wedges failed in spite of meeting strength specifications, the contractor could have taken advantage of the fact that the GDSR was silent on the subject of hydro-compaction. The owner's argument that a slight tendency towards hydro-compaction should have been expected for Young Alluvium may have been difficult to prevail in the eyes of a DRB or judge.

6.0 GROUTING CLAIM, LANKERSHIM BLVD

The roughly 2-mile long Lankershim Blvd tunnels were excavated in ground conditions dominated by Young Alluvium. These soils consisted of randomly inter-bedded silts, sands, and gravels with occasional cobble nests. In the absence of groundwater, the contractor selected open-face digger shields (Fig. 12) with two breasting tables to excavate the 22-foot diameter twin tunnels. Expanded pre-cast concrete segments were used for temporary support. The tunnels were driven from north to south starting at a shaft near Lankershim and Magnolia.

As soon as the first shield left a short break-out zone stabilized by chemical grouting, the contractor encountered difficulties with caving of the tunnel face below the lower breasting table. Sand and gravel lenses encountered at this elevation tended to stand vertically at first, but then suddenly collapsed in the form of 1- to 2-foot thick, near-vertical slabs separating and falling away from the face. Caving then advanced upward to the crown and above, ahead of the cutting edge of the shield (Fig. 13).

The problem was caused by the oversized digger (Fig. 14), which could not be fully retracted and, hence, tended to undermine the face below the breasting tables. Concerns about this digger had been voiced even before the machines were installed underground. The contractor, however, opted to leave the diggers unchanged and add fixed hoods instead. This measure turned out to be ineffective, and caving continued to advance above the crown.

6.1 Request for Change Order

After periodic shut downs and directions given to the contractor to improve his means and methods, it came to a total impasse. The contractor requested a change order for implementing ground improvement, but this request was denied. In the interest of proceeding with the project, the owner finally directed the contractor to stabilize the entire alignment by chemical grouting from the surface (Fig. 15).

A conditional change order in the amount of $20 mill was issued for this work, with the owner reserving the right to challenge it later. The owner argued that the ground conditions had been presented realistically in the GDSR and, therefore, should have been taken into account in preparing the bid. I.e., the contractor either should have chosen tunneling means and methods capable of handling the ground or, alternatively, improve the soil as needed for his equipment to function properly. In fact, the owner's argument continued, optional grouting had been included in the specifications for this very reason.

The owner's interpretation above was disputed by the contractor, who argued that the specifications called for intermittent grouting from the tunnel face, but did not include a large-scale grouting program to be implemented from the surface as the owner had directed them to do.

6.2 DRB Recommendation

In hearings before the DRB, the owner requested that the conditional change order be retracted and the payment for grouting returned. The contractor, on the other hand, requested payment of additional $40 mill for "owner-caused delays."

The DRB recommended that the $20 mill paid for grouting was justified, but the additional $40 mill were not. Specifically, the DRB agreed with the contractor's opinion that the owner, by directing grouting be performed from the surface, had requested additional work not provided for in the contract specifications. However, the DRB refused to blame the owner for the substantial construction delays which had accumulated due to the frequent shutdowns and related drawn-out arguments and bickering, which had plagued this job for many months.

6.3 Legal Settlement

The contractor rejected the DRB's recommendation and went to court. After an expensive legal battle drawn out over several years, the case was settled. The owner agreed to pay an additional $7 mill on top of the $20 mill already paid for grouting.

6.4 Lessons Learned

Even though ground conditions were described quite realistically in the GDSR, one could argue that the tendency of sudden/brittle failure by "slabbing" could have been specifically pointed out. Also, the defective shield design could have been rejected early on in the submittal stage. Alternatively, the need for grouting could have been made a condition for approving the shield submittal. Above all, however, there was an unfortunate failure to communicate between the various parties.

7.0 CONCLUSIONS

The examples discussed above, as well as experience from other Los Angeles tunnel projects (i.e. Roth & Kamine, 1997), suggest that owners often end up paying for the downside risk which low bidders are forced to take on performance-based contracts. Given today's availability of hi-tech tunneling technology and an increasingly competitive tunnel-construction market, the old adage of "don't tell a contractor how to do his job," may just have outlived its usefulness.

While a trend towards more prescriptive tunneling specifications has already begun, the will and determination to strictly enforce such specifications is still lacking with many owners. But for owners to assume more control over construction risks is crucial for the success of future tunnel projects. After all, construction mishaps on public-works projects not only affect adjacent properties, but also undermine public trust and political support. A contractor may be held responsible for fixing the former, but he can't do anything about the latter.

8.0 REFERENCES

Roth, W. H. & Haber Kamine M. (1997), *Two Construction Claims on the NORS Project*, Proceedings, 23rd World Tunnel Congress 197, Vienna, Austria, pp 745-746.

Roth, W.H., Dawson E.M., Nesarajah S., (2001), *Numerical Modeling in Forensic Engineering,* Proceedings 10th Int. Conf. Computer Methods and Advances in Geomechanics; Tucson, Arizona.

9.0 FIGURES

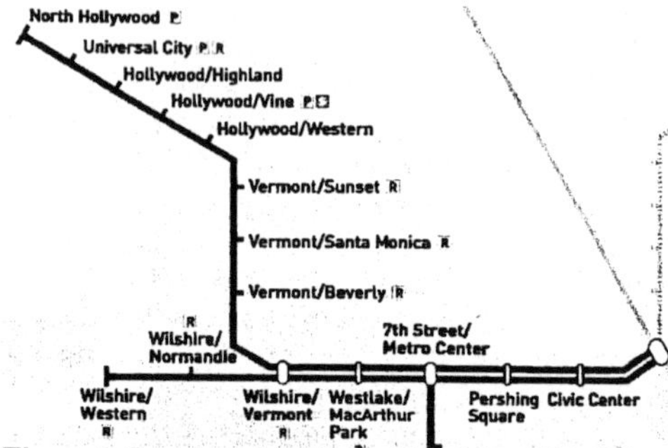

Figure 1. Los Angeles Metro Red Line alignment.

Figure 2. Hand-minded starter adit at Barnsdall shaft.

Figure 3. Barnsdall shaft supported by concrete tangent piles.

FIGURES CONTINUED

Figure 4. Reinforcement of failed pillar at Barnsdall shaft, west wall.

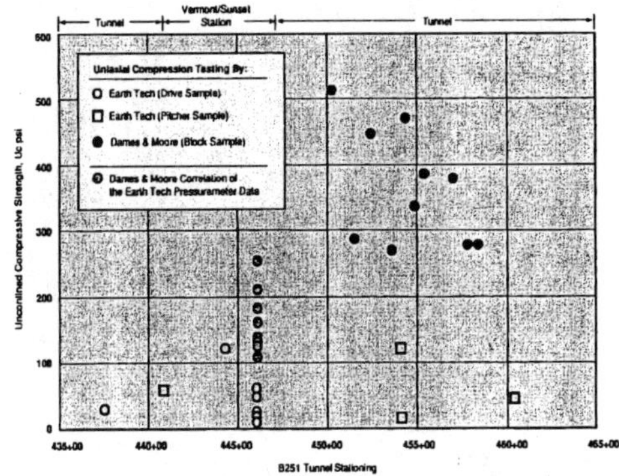

Figure 5. Unconfined compressive strength of siltstone - open symbols are lab-test data listed in the GDSR; full symbols represent test data from block samples taken during tunneling.

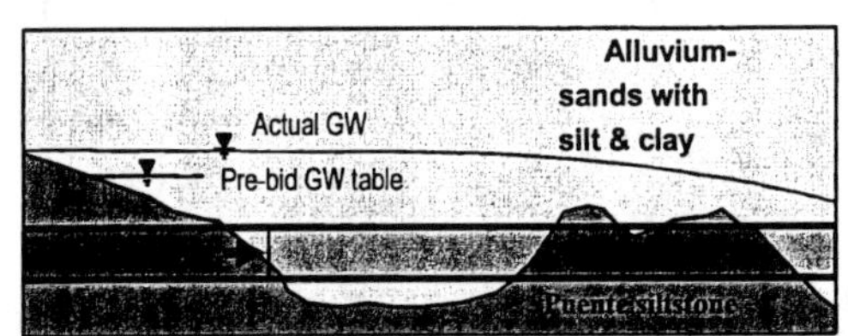

Figure 6. Subsurface conditions near Edgemont Avenue (not to scale); the black arrow marks the collapsed tunnel face.

FIGURES CONTINUED

Figure 7. Boarded up tunnel face following the collapse due to running ground at Edgemont Ave.

Figure 8. Hardwood wedges in the 10- and 2-o'clock positions of the expanded pre-cast concrete liner.

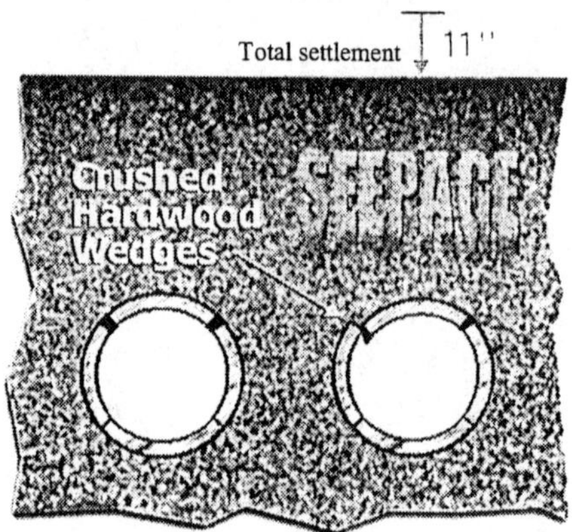

Figure 9. The hardwood wedges in the 10-o'clock position of the outbound tunnel crushed when the trailing shield passed.

FIGURES CONTINUED

Figure 10. Steel posts supporting the failed temporary liner.

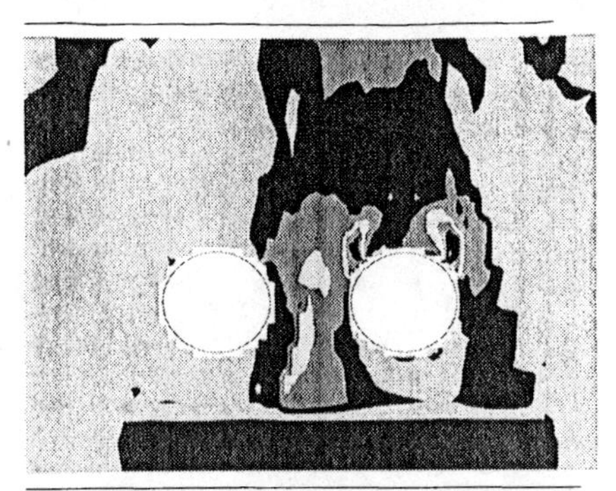

Figure 11. Volumetric-strain contours showing pattern of wetting-induced hydro-compaction as predicted by the numerical model; darker areas indicate more compaction.

Figure 12. Digger shield with 2 breasting tables, used for Lankershim Blvd tunnels.

FIGURES CONTINUED

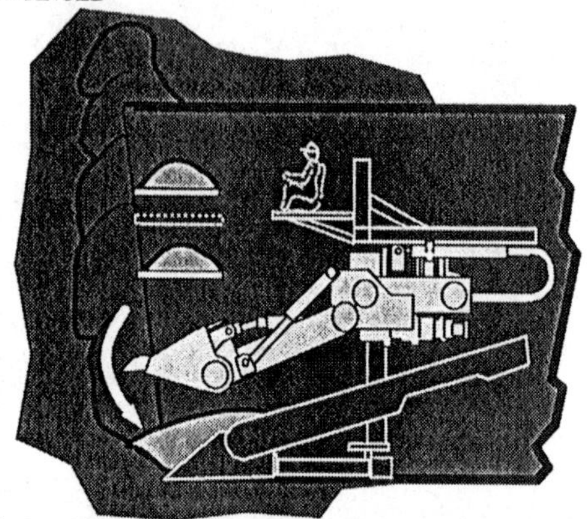

Figure 13. Caving below the breasting tables resulted in the development of chimneys above the crown.

Figure 14. The over-sized digger could not be retracted inside the cutting edge of the shield.

Figure 15. Grouting from the surface along Lankershim Blvd.

Investigation and Repairs to Damaged Duck Creek Culvert

Terry M. Sullivan, P.E., Jeremiah R. Nichols, P.E., Steven J. Smith, Ph.D., P.E., Steven Gebler, P.E., Honggang Cao, P.E., Michael G. Carfagno, P.E. and Larry P. DeRoo, P.E.

The Duck Creek Phase III flood protection project presented the Corps of Engineers' Louisville District with numerous difficult challenges in the design, contracting and construction arenas. The purpose of this project was to lower flood elevations by eliminating an oxbow bend in Duck Creek, an urban stream located on the east side of Cincinnati, Ohio. The proposed long-span culvert, designed to carry flood flows around the existing creek channel bottleneck, was oriented parallel to and in several locations was nearly underneath the centerline of an adjacent, relocated roadway (Figure 1). The culvert alignment included two significant horizontal curves in order to snake around existing high-tension overhead power line towers. The culvert also had to be constructed with very little cover, presenting a challenge to the precast culvert designer and manufacturer including limited space within the contractor's work limits for the storage of overburden material from the culvert excavation.

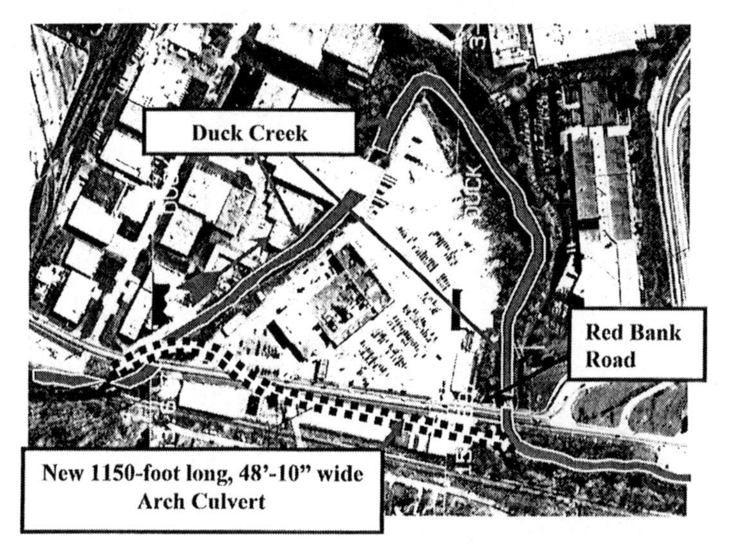

Figure 1: The Duck Creek culvert eliminates an oxbow in an urban stream, and is constructed in a congested site bounded by a railroad track and an existing highway.

During studies prior to development of final plans, the Corps of Engineers proposed a structure composed of triple cast-in-place 18-foot box culverts to enclose an approximately 1200-foot long stretch of Duck Creek along a complex curved alignment. In the contract construction drawings, the Corps of Engineers presented a standard 48'-0" wide Con/Span® precast arch segment. This reduced the overall excavation width by approximately 11-feet, which created a significant material savings and also allowed for the elimination of over 300 feet of sheet piling (a savings of approximately $1 million) needed in the original box culvert scheme to protect adjacent railroad tracks.

The Corps left bidders with the option to provide an "or approved equal" design by a competing manufacturer. The specifications further presented the requirement to construct a precast concrete arch type culvert with a minimum clear opening area of 557.50 square feet. The construction contract was awarded to Ahern & Associates of Springfield, Ohio on May 30, 2002. Once detailed shop drawing development started after contract award, the prime contractor found that overhead clearance issues with the new Red Bank Road subgrade were going to create conflicts with the standard 48-foot Con/Span® section. The extensive length of the conduit allowed a custom cross-section to be designed that optimized the balance between the required hydraulic performance, the overhead clearance and its structural efficiency. Thus, Ahern selected a prototype long span culvert design developed by Con/Span®.

The Corps of Engineers designed the cast-in-place base slab and knee walls upon which the precast culvert elements were to be placed, using design loads from Con/Span®. BridgeTek fabricated the 161 arch segments, each 49'-8" wide, 11-inches thick and approximately seven feet long, at their Milton, Kentucky outdoor precasting facility. A typical cross-section of the culvert is shown in Figure 2.

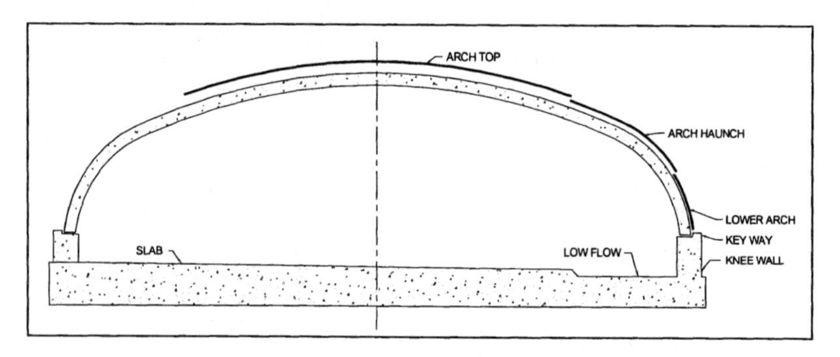

Figure 2: The Duck Creek culvert typical section. The precast concrete arch elements are placed on top of the knee walls, which were cast monolithically as part of the cast-in-place base slab.

The construction of the culvert proceeded with Ahern performing a sequential operation. Because of slope stability concerns with the adjacent railroad track, the contractor was required to limit his excavation to a maximum length of about 150 feet. Therefore the base slab was constructed in segments generally between 100 and 140 feet long, with the majority being 126 feet. For each base slab segment, Ahern performed an excavation of the overburden, prepared the subgrade, cast the base slab, placed 14 to 18 precast arch elements (see Figure 4) and backfilled the arch with imported select granular fill. Ahern constructed the base slab and installed the precast arch culvert elements in a timely and highly satisfactory manner. Installation of as many 18 arch segments was routinely performed in a single long day. Each base slab and arch placement cycle took approximately one to 1½-months.

In late July, 2004 Ahern's quality control manager noted some spall damage on the inside of the culvert. Ahern immediately informed the Corps of Engineers and representatives from Con/Span®. A preliminary check showed that a large overburden stockpile had inadvertently been placed over the top of the culvert in an area that had been completed and backfilled several months prior (Figure 3). The Corps and Ahern reviewed their archives for site photographs from the previous year to see if any other stockpiles had been placed over the culvert. The photos showed several distinct stockpiles that covered about 400 linear feet of the 1150 foot long culvert. It was inferred by the Corps that the stockpile had simply been moved as required so the contractor could continue to proceed with other work on site. Aerial photos of the stockpile were taken and a field survey was completed. The maximum stockpile height was approximately 15 feet, with the maximum height over the culvert of about 12 feet. The culvert had been designed by Con/Span® for a maximum dead load of four feet of earth.

Ahern procured the failure investigation services of CTLGroup and coordinated a thorough and detailed inspection of the entire culvert Con/Span®. Findings from this investigation were used by the parties to design and construct repairs, restoring the structural performance of the arch.

STRUCTURAL INVESTIGATION OVERVIEW

A preliminary damage survey conducted by Con/Span® provided a summary of the type and extent of visible damage (Figure 4). This drawing also identifies the numbering system for the arch units, which will also be used for location identification in this paper.

A detailed site investigation followed that considered five potential types of structural distress resulted from the overloading: 1) Tension cracking at the interior surface of the top of the arch; 2) Tension cracking at the exterior surface of the haunch of the arch; 3) Shear failure at the base of the arch in the form of delamination of the precast concrete from the steel reinforcement; 4) Knee wall damage via one of three modes of failure 4a) Complete failure of the knee wall due to shear cracking through the entire section; 4b) Localized failure of the key way; and

4c) Regions of transition from complete knee wall to key way failure;.and 5) Tension cracking at the top surface of the slab in the vicinity of the knee wall.

The investigation used nondestructive testing (NDT) in addition to conventional visual investigation techniques and coring. The NDT methods included: Ultrasonic Pulse Velocity (UPV), Impact Echo (IE), Impulse Response (IR) and Ground Penetrating Radar (GPR). All of the above techniques are included in the American Concrete Institute Report, ACI 228.2R-98 "Nondestructive Test Methods for Evaluation of Concrete in Structures."

INSPECTION

Table 1 identifies the locations of application for the various inspection methods employed during the site investigation.

Conventional

A general visual inspection was conducted on the interior of units 70 – 120 and exterior of units 97-105. The knee wall and slab were also inspected at these locations. This inspection included observations of cracks in the slab, knee wall and arch, separation between the knee wall and arch, and uninstrumented hammer sounding of the arch. Detailed crack maps were developed for selected regions as identified in Table 1.

Ultrasonic Pulse Velocity (UPV)

The UPV method was used to evaluate the depth of cracks in the base slab. The tests were conducted primarily on two cracks that ran parallel and in proximity to the west knee wall. These two cracks were deemed to be among the most significant from the visual inspection, with widths up to 0.015 in. The test area ranged from arch segments numbers 98 to 107. The UPV method uses two transducers, one placed on each side of a crack. The transmission of an ultrasonic pulse from one transducer to the other can be correlated to the depth of the crack. The transducers were positioned at 8 in. and/or 12 in. from each side from the crack. Prior to the test, the reference pulse velocity was estimated on a region of solid concrete using the indirect transmission method. Three tests were conducted at a path length of 16, 24, and 32 in., respectively. The velocity was calculated as the mean of the three results.

Figure 3: One of several distinct overburden stockpiles placed over the culvert.

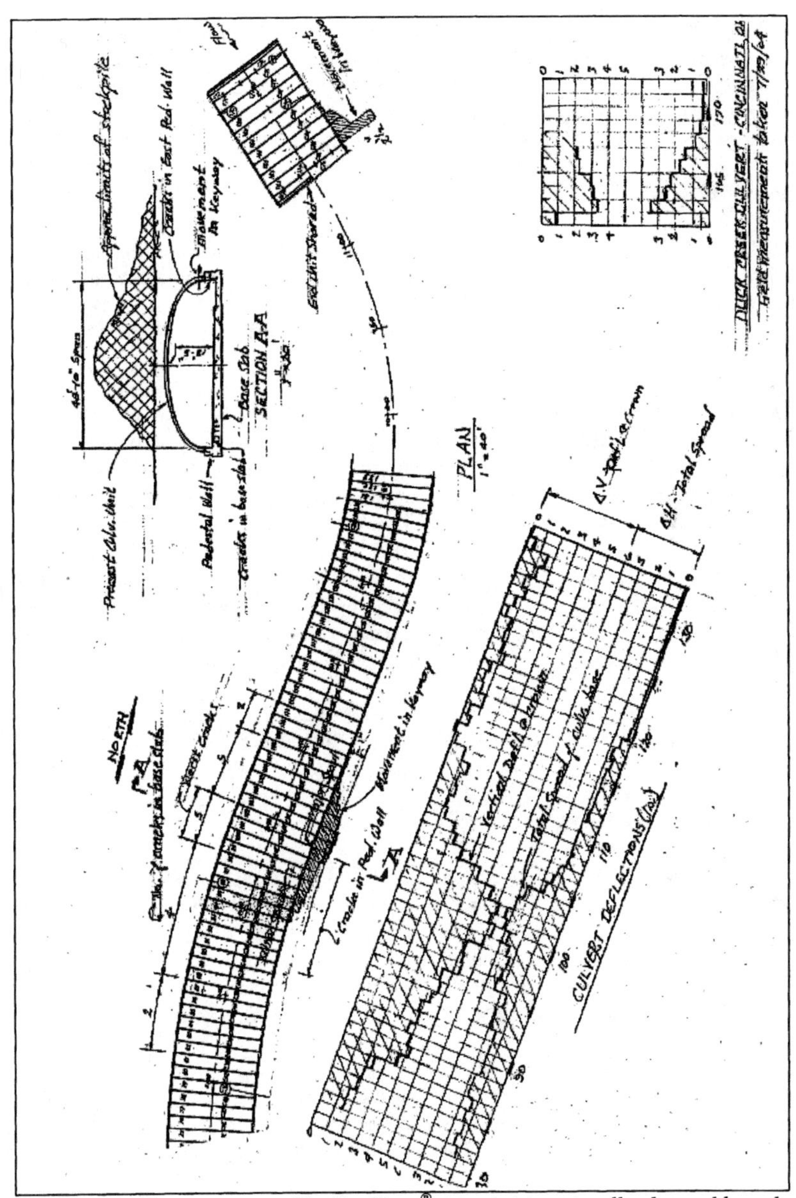

Figure 4: Field notes prepared by Con/Span® *summarizing visually observable and measurable damage to culvert precast elements, base slab and knee walls.*

Based on the results of UPV tests at the slab cracks near the west wall, the maximum depth was estimated as 13 in. while the majority of the crack depths were within approximately 5 to 10 in. Two core samples, taken at the points of maximum UPV readings, showed the crack depths of 8 and 9 in. This indicated that the UPV readings provided a conservative measure of the crack depth.

Impact Echo (IE)

The IE method was used to evaluate the presence of delamination in the knee wall and near the base of arch segments. The method uses an instrumented hammer to generate a stress wave in the tested element. A reflection of this stress wave is generated from the back face of the element as well as from any significant internal discontinuity.

The test was performed at the intersections of a 2 ft grid on the arch units, from the base of the unit to a height of approximately 9 ft. The density of reinforcing steel in the arch units added a degree of complexity to the interpretation of the signals. For example, the IE data gave a weak indication of an internal discontinuity in the region of unit 110. However, a core from this unit revealed the concrete to be undamaged.

Analysis of the IE test results on the arch elements showed no significant internal delamination at the areas tested, implying that delamination was limited to the units with visible damage on the interior surface (units 102 to 105). Unit 120 was used for reference measurements of a visually undamaged, 11-in. thick arch unit.

The IE test was found to be ineffective when testing the knee wall due to its relatively large thickness and the associated large attenuation of the transmitted stress wave.

Impulse Response (IR)

The IR method was used to evaluate structural uniformity and the presence of internal damage or defects in the arch units and knee wall. The IR method is similar to the IE method, but uses a higher energy impact and measures the structural response of the tested element, rather than the transmission of the stress wave. The IR method was not capable of detecting the relatively superficial effect of failure of the key way and was also sensitive to variability in the stiffness provided by the fill material.

On the knee walls, the IR tests were conducted from the interior face on a 1 ft horizontal by 0.5 ft vertical grid. On the arch segments, the tests were conducted from the interior face at a 1 ft by 1 ft grid up to 9 ft high from the base.

For a continuous section of knee wall with sound concrete, average mobility values were less than 2. Values greater than 4 indicate concrete anomalies within the wall. The east knee wall at units 90 through 95 exhibited very high average mobility values in the upper and mid sections (greater than 10). These areas correspond well with visual evidence of shear cracking in the knee wall in this area. Units 98 through 100 show similar high mobility values in their upper and mid sections, agreeing with the observation of shear cracking in this area. Units 101 through 103 showed relatively high average mobility values (~4) around the vertical construction joints and in the uppermost panel zone. This also confirmed the observations made on the exterior wall in this area, with the observation from the IR results that the damage was considerably reduced here compared with the two areas described above. All other knee wall panels tested, including those locations on the west knee wall, gave IR average mobility results between 1.5 and 3, indicating sound concrete. Figure 5 shows sample IR data.

The IR did not indicate any significant damage in the arch outside of the areas visually identified as delaminated.

Impulse Radar (GPR)

The GPR survey was used to evaluate the locations of reinforcing bars to facilitate the coring operation and avoid damaging the reinforcement.

Coring

Cores were extracted from multiple locations in the slab, knee wall and arch with nominal diameters including 2, 3 and 4 in. The purpose of the cores was to verify limits of damage indicated by the other inspection methods.

Where cores contained unexpected cracks, the core holes were examined to determine whether the cracking was due to distress during the coring process. Cores taken in the east knee wall identified limits of shear failure and those from the base of the arch identified limits of delamination. Cores from the top region of the arch and from the slab indicated that the tensile cracks in these regions did not compromise the structural integrity of the section. The majority of the cracks in the slab propagate around the aggregate, implying they occurred soon after the concrete was cast, while the concrete was still relatively weak compared to the aggregate, and are likely due to thermal differences and drying shrinkage associated with placing large quantities of concrete.

CONDITION ASSESSMENT

1. Tension cracking at the interior surface of the top of the arch: The interior of the arch exhibited minor hairline cracking based on crack mapping of selected interior and exterior units and cores taken from unit 102. The cracks were completely closed implying that the arch geometry was elastically restored from the overloading and the reinforcing steel likely did not yield.
2. Tension cracking at the exterior surface of the haunch of the arch: Cracking on the exterior of the units was considered for application of sealant relative to

corrosive agents from road salts, though the rebar is epoxy coated, and the cracking appeared consistent with expectations for a cracked-section design.

3. Shear failure at the base of the arch in the form of delamination of the precast concrete from the steel reinforcement: Cores at units 101, 102, 104, 106 and 110 combined with the IR and IE data indicated that damage was limited to units 102 to 105. The interior surface of units 102 through 104 exhibited significant delamination of the outer region of concrete with some spalling extending to unit 105 and 112. Figure 6 shows the delamination in unit 104 before and after the spalled concrete had been removed, revealing the damage to be limited to the outer layer of concrete. The interior layer of rebar in unit 102 had been plastically deformed along the failure plane in the concrete. It was deemed that this steel could be bent back into alignment and reused, given the compressive stress regime in this region of the arch.

4. Knee Wall: Separation between the arch units and the interior grout pack of the knee wall was documented in units 98 through 112. This implied a failure of the key (Figure 7). Cores at units 92 and 94, visual observations and the IR data indicated complete failure of the knee wall in units 90 to 95 and 98 to 100 (Figure 8). All other units within the ranges 90 through 112 had key way or transitional damage due to displacement of the arch base (based on visual observations).

5. Tension cracking at the top surface of the slab in the vicinity of the knee wall: Cores, combined with the UPV data, indicated that the cracking in the slab was primarily of the sort that would be expected from casting mass quantities of concrete (temperature and shrinkage cracks). The cracks develop into a flexural pattern within approximately 10 ft of the east and west walls at units 90 through 104. Cracks in these regions may have developed under the action of gravity loads transferred from the knee wall during the overload. However, cores taken through cracks in these regions also had indications of being generated by thermal effects (the cracks were very tight and propagated through the aggregate). The cracks were generally less than 0.015 in., and most were hairline with no offset, implying no yielding of the reinforcing steel from tensile or shear strains.

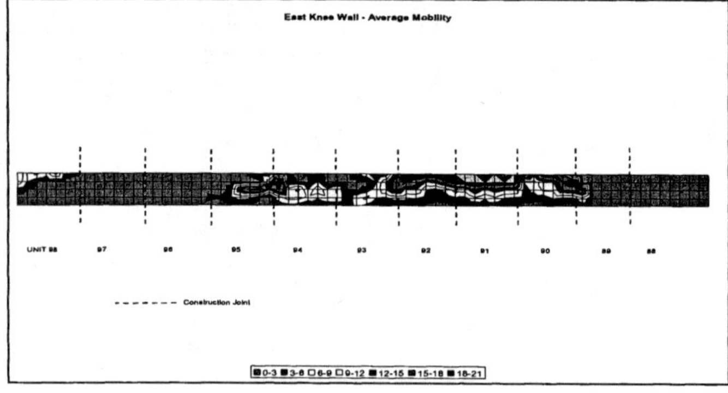

Figure 5: Sample Impulse Response data from the east Knee Wall.

Table 1: Inspection method locations (IR and IE locations apply to East side, except as noted)

UNIT	Crack Map				IR		IE		UPV	Coring
(#)	Knee	Slab	Arch	Ext[1]	Arch	Knee	Arch	Knee	(Slab)	(Location)
88						X				East Knee Wall / Slab
89						X				
90	X	X	X		X	X				
91	X	X	X		X	X	X			
92	X	X	X		X	X	X			East Knee Wall
93	X	X	X		X	X	X			
94	X	X	X		X	X	X			East Knee Wall
95	X	X	X			X				East Slab
96	X	X	X			X				East Knee Wall
97	X	X	X			X				
98	X	X	X	X		X			X	
99	X	X	X	X	X	X	X		X	
100	X	X	X	X	X	X	X		X	
101	X	X	X	X	West	X	West		X	East Arch Base
102	X	X	X	X	West	X	West		X	7 locations [2]
103	X	X	X			X			X	
104	X	X	X			X			X	West Slab, East Arch Base
105	X	X	X			X			X	
106	X	X	X			X			X	East Arch Base
107	X	X	X			X			X	West Slab, East Slab
108	X	X	X		X	X				
109		X	X		X	X				
110		X	X		X	X	X			East Arch Base
111		X			X	X	X			
112		X			X	X				
113						X				
114						X				East Knee Wall
115										
116										
117										
118										
119						X				
120					X	X	X			

[1] Exposed exterior east side of the arch

[2] 1) West Arch Haunch, 2) East Arch Haunch, 3) Crown, 4) West Arch Base, 5) East Arch Base, 6) West Knee Wall, 7) East Slab (at low flow taper)

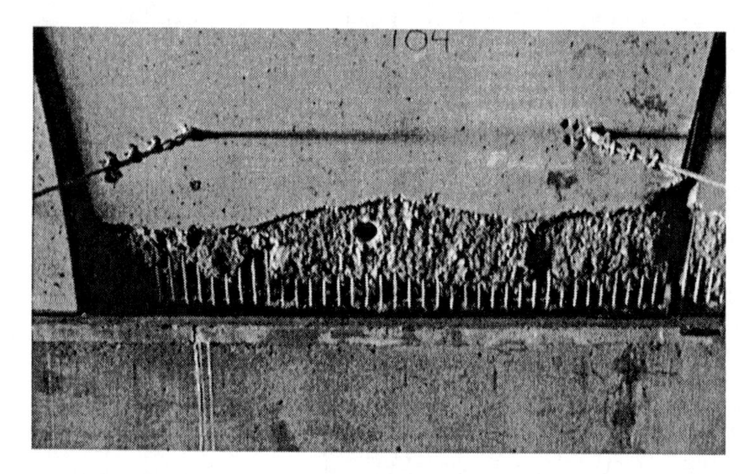

Figure 6: Damage to unit 104 at the interior surface of the east base with and without spalled concrete removed (exploratory core hole also visible).

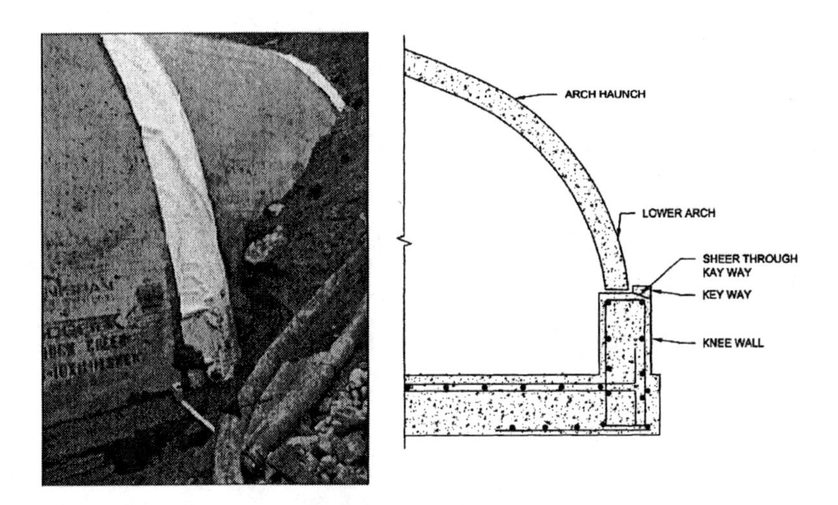

Figure 7: Keyway failure.

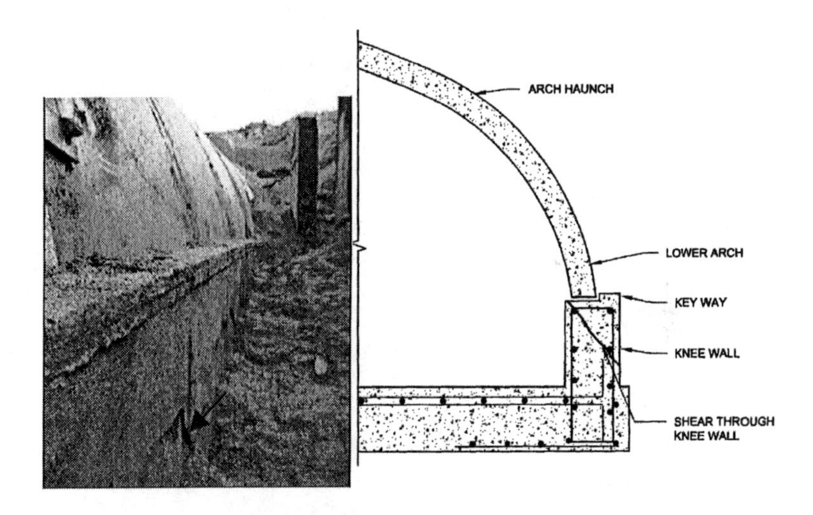

Figure 8: Knee wall shear failure.

REPAIR PLAN

Once the results of the Field Investigations were summarized, a repair plan was formulated jointly by Ahern, Con/Span® and the Corps. Some of the arch unit legs required partial or full depth spall repairs; the arch units that increased in span by lateral spreading would have to be jacked back to their original span; cracks on the top of the arch segments were to be sealed to prevent future corrosion of the reinforcement; the foundation knee wall required one of four different repairs, depending on the type and severity of the damage; and the foundation base slab required sealing of cracks in its top surface to insure durability.

The first step at each location was to provide the temporary shoring (sheeting) required to maintain traffic on the temporary road. The arch units were then shored up vertically and cable ties were installed (Figure 9) to keep the legs from spreading due to loss of support from the side fill. Earth fill was then removed evenly on each side of the precast arch units to minimize any unbalanced loading. The required concrete removals were then made based on the extent of damage and the required repair (Figure 10). The arch unit span was then adjusted by a combination of jacking the arch units up to bring the leg in and by tensioning the cable ties to get the arch back as close to the original span of 48ft-10in as possible. The knee wall repairs were then completed with the arch segments in place (Figure 11).

Once the knee wall repair had cured the arch unit leg spall repairs were made by applying shotcrete to the prepared surfaces. A close watch on the weather forecast was kept during the repairs, as a heavy rain event had the possibility of bringing heavy flows and large debris through the culvert. With the arch units up on vertical shoring, they would be particularly vulnerable to the types of debris (large trees, concrete chunks, discarded appliances etc.) that had occasionally been witnessed to be transported by Duck Creek during flood events. In the event of a heavy rain, secondary shoring was planned as a fallback, though this was never actually required as Mother Nature cooperated.

Figure 9: The arch units were shored up vertically and cable ties were installed to keep the legs from spreading due to loss of support from the side fill. This photo shows where a damaged section of the knee wall has been removed.

The complete knee wall repair, consisting of concrete removal, reinforcing placement, formwork, and concrete placement and curing, was completed in a two to three day window depending on the number of arch unit sections repaired at one time. Sealing of cracks on the exterior of the arch surface was also performed during this phase for units 100 through 105.

Figure 10: After the backfill was removed, demolition and replacement of the damaged portions of the kneewall proceeded. The contractor drilled holes and used hydraulic splitters and jackhammers to demolish the existing concrete. Splitter holes are shown.

CONCLUSIONS

The local sponsors, the Corps of Engineers, the general contractor (Ahern & Associates), the culvert designer (Con/Span®) and the forensic consultant (CTLGroup) all have a strong interest in understanding this overloading event, the damage investigation and repair operations. More importantly all of these parties hope to help prevent similar overloading events from occurring on other projects in the future.

There were a variety of contributing factors that led to the overloading. The congested nature of the jobsite, with temporary routings of heavy urban traffic on one side and an active freight railroad on the other, combined with the curving alignment of the large culvert, probably contributed to both the contractor's and the government inspectors' lack of attention to loading issues. The project site was also relatively tight and provided little space for temporary stockpiling of overburden. The fact that several overloading events occurred over a period of months in numerous different locations leads to the conclusion that the project's designers, contractors and inspectors likely had never really considered such a scenario as a realistic issue.

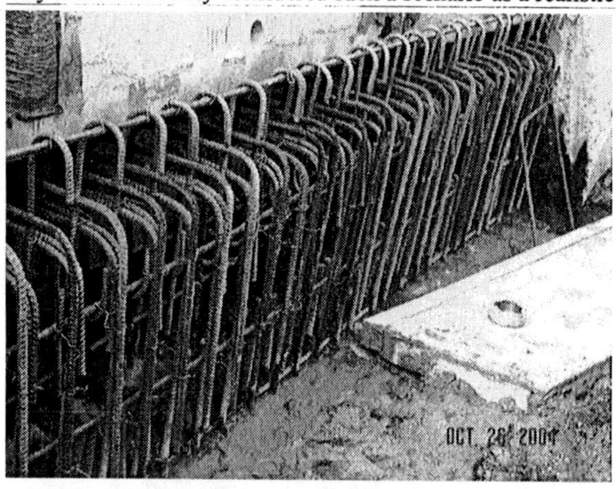

Figure 11: New reinforcement was added in the knee walls and especially in the keyway to provided added shear resistance in locations where the arch segments spread.

Despite inclusion in the specification that "The load case reviewed for this design is HS20-44 loading with 2ft-0in to 4ft-0in of fill." it is likely that the people in the field for both the contractor and the Corps of Engineers did not have such design issues foremost in their minds in their dedication to complete this very complex and challenging construction project. Because of the repetitive nature of the construction of this very long culvert, both the Corps' designers and the culvert's designers were not frequent visitors to the project site after initial construction issues were resolved. The overloads occurred long after the construction activities related to the culvert had become "routine."

In the future, discussion about potential construction overloading issues will be better highlighted on the plans and in the specifications, and discussed more clearly and boldly in instructions to field personnel. Designers of Record need to visit the project site on a more frequent basis, even if constructability issues are not arising. For the culvert designers, the lessons are more complex. In retrospect, the Corps-designed knee walls could have been designed with more shear reinforcement to guard against a greater potential variation in loads. The keyway shown on the Corps contract documents was a detail recommended by Con/Span® based on many prior arch culvert installations; all parties now agree that it should have been designed as a more robust structural keyway, as this culvert section was a true arch, and the thrust loads were somewhat higher than those generated by a conventional Con/Span® arch installation. The precast Con/Span® elements survived largely intact, demonstrating the great structural load-carrying capability of a true arch. The culvert designers in the future would wisely provide more clarity on their shop drawings regarding what constitutes an allowable construction load.

The initial field damage assessment conducted by Con/Span® provided a detailed overview of the types and extents of damage. This assessment serves as a reminder that a significant amount of structural information can be gleaned from conventional methods when applied carefully and thoughtfully. The concealed damage required more intensive inspection methods. The use of advanced NDT methods calibrated with coring provided an efficient method for determining accurate identification of damage types and locations allowing the design of tailored repairs and development of efficient repair construction staging.

The repairs to the culvert were completed successfully in 2005, and the entire project, including the culvert, adjacent flood walls and the roadway relocation, have now been completed. The repairs are expected to restore long-term durability and serviceability. The completed project is a tribute to the vision and hard work of the designers, consultants and constructors (see Figure 12). The culvert will reduce flooding as a benefit for the local sponsors for many years to come.

Figure 12: The completed culvert is almost invisible to the surrounding community.

Field Comparison of NDE Methods for Tunnel Condition Assessment

Paul A. Bosela[1], M.ASCE, Sathaporn Lek-udom[2], Satya Mullangi[3], and Norbert Delatte[4], M.ASCE

ABSTRACT

Several different nondestructive evaluation (NDE) techniques were used to detect delaminations/flaws in a reinforced concrete tunnel liner. This case study involved three areas located in different parts of an abandoned streetcar/subway tunnel in Cleveland, Ohio. The tests were performed during the Summer of 2002. Four methods were used, and a comparison was made of the testing time and extent of deteriorated area detected. The Impact-Echo method provided the maximum amount of flaw detection, followed by Ground Penetrating Radar (GPR), Rotary Percussion (RP) and Visual Inspection (VI), respectively. The GPR and RP methods provided the minimum testing times. Based upon this study, the GPR method, along with a sophisticated data/image processing technique, has the potential for practical rapid NDE of subway tunnels.

[1] Professor and Chair, Department of Civil and Environmental Engineering, Cleveland State University, 2121 Euclid Avenue, Cleveland, OH 44115-2214

[2] Graduate Assistant, Department of Civil and Environmental Engineering, Cleveland State University, 2121 Euclid Avenue, Cleveland, OH 44115-2214

[3] Graduate Student, Department of Civil and Environmental Engineering, Cleveland State University, 2121 Euclid Avenue, Cleveland, OH 44115-2214

[4] Associate Professor, Department of Civil and Environmental Engineering, Cleveland State University, 2121 Euclid Avenue, Cleveland, OH 44115-2214

INTRODUCTION

The 1997 "Synthesis of Transit Practice 23" (Russell and Gilmore, 1997) reported that tunnel leaks or ground water intrusion were the major problems leading to the structural deterioration of tunnels, and that there was a need for active research contributing to the development of standard practices for tunnel inspections. The Cleveland tunnel (Watson and Wolfs, 1981) was selected for study using various nondestructive testing methods, including Visual Inspection, Rotary Percussion, Impact-Echo, and Ground Penetrating Radar. The objective of the study was to evaluate the capabilities and limitations of each method, based on the delaminations/flaws detected and testing time required. Core sampling was not permitted, so physical verification of the results was not possible.

The main tunnel wall (Test Area 1) is a 12 in. (305 mm) thick reinforced concrete slab, which has a near face longitudinal reinforcement spacing of approximately 24 in. (610 mm) and a near face transverse reinforcement spacing of about 5 in. (127 mm). The Pedestrian Underpass Extension floor (Test Area 2) is a 13 in. (330 mm) thick reinforced concrete slab having top and bottom rebar grids, with longitudinal spacing of 15 in. (381 mm) and transverse rebar spacing of 12 in. (305 mm). The Pedestrian Underpass Extension wall (Test Area 3) is a 12 in. (305 mm) thick reinforced concrete slab, with 17 in. (432 mm) spaced near face and far face longitudinal rebar and 5 in. (127 mm) spaced near face and far face transverse rebar. This concrete section is also covered with a 4-1/2 in. (114.3 mm) thick glazed brick liner. The rebar sizes ranged from ½ in. Number 4 (12.7 mm) to 5/8 in. Number 5 (16 mm) and concrete cover from 1 in. (25 mm) to 2 in. (50 mm), respectively.

Due to constraints on time and personnel, the study was focused on three subareas of the tunnel, including Test Areas 1, 2 and 3. All four methods were used for test areas 1 and 2. Testing of Area 3, which had concrete walls and a glazed brick liner, was not completed.

FIELD TESTING AND RESULTS

Visual Inspection (VI) Method

A well-trained engineer may be able to differentiate between the various types of cracking encountered during a visual inspection. Material deterioration is normally indicated by surface cracking and spalling. Mapping of cracks (Pollock et al, 1981) also provides a valuable diagnostic tool for determining causes of deterioration. By the VI method, the condition of the structure was reviewed for indication of problems, including cracking, surface distress, water leakage, structural movement, metal corrosion, and staining.

In 2001, the Nondestructive Evaluation Validation Center (NDEVC) research group studied the reliability of visual inspection for bridges (Grabeal et al., 2001), by having teams of inspectors perform routine and in-depth inspections of the same set of bridges, and comparing the variance of their results.. Routine bridge inspections, characterized by visual observations done from a distance, are used to assess physical and functional conditions of the bridge. The inspector will establish a single condition rating numbers for the bridge deck, substructure and superstructure, based upon the severity of deterioration and extent of districution. In-depth inspections, on the other hand, are close-up or hands-on type inspections of the various elements. They are intended to identify deficiencies that cannot normally be detected from a distance. The study indicated that in most cases even in-depth inspections of bridges rely primarily on visual inspection. The study also revealed that there were aspects of bridge inspection that need significant improvement. For routine inspections, evaluation of the statistical data indicated that the condition ratings, element-level inspection results, inspection notes, and photographs were found to have significant variability. Of greatest importance was the amount of variability found in the assignment of condition ratings. For in-depth inspections, VI might not yield any findings beyond those noted during a routine inspection. The results of the deck-delamination survey conducted during this investigation indicated that this type of inspection did not consistently provide accurate results.

Since rapid screening of subway tunnels is also typically done by visual inspection, it was the first inspection method used. Grid points were laid out on the test surface as shown in Figure 1. The 112 points, which cover a 20.16 square meter (217 sf) area, were created using 16 vertical lines and 7 horizontal lines, with spacings of 0.60 (23.6 in.) and 0.30 m (11.8 in), respectively.

In the Test Area 1, two cracks in the vertical surface and exposed rebar were found as shown in Figure 1. No major flaws were found on the Test Area 2 surface, only a hairline crack resulting from a resurfacing process. The testing times for each of these areas was less than one hour. Most of the testing time was spent on writing down comments and sketching the items detected.

Rotary Percussion (RP) and Impact-Echo (IE) methods

According to ASTM D4580 (ASTM, 2002), the RP method uses a simple gear and wheel instrument to detect the delamination on both horizontal and vertical surfaces. The concept of this method is analogous to those of the Electro-Mechanical Sounding Device and the Chain Drag methods. A dull or hollow sound is created when the rotary percussion device strikes the delaminated area while rolling across the surface. The method is effective for locating gross near surface delaminations and for detecting delamination in concrete where the surface is not covered by other materials.

The concept of the IE method is analogous to that of other sounding methods, except a receiving transducer is used to detect the sound echo from the subsurface discontinuities such as cracks, voids, and layer interfaces. The stress waves, which are generated using various sizes of steel balls ranging from 5-12.5 mm (0.2-0.5 in.), travel through the layered media and are reflected back to a broad band low frequency transducer placed on the test surface (Sansalone and Streett, 1997). The signal is then recorded and processed using a field computer with data acquisition software. According to ASTM C1383-98 (ASTM, 1998) P-wave (Primary wave) speed is measured and a thickness of the slab can be calculated using the formula;

$$T = \frac{0.96C_P}{2\,f_T} \qquad (1)$$

where

T	=	thickness of the concrete slab (mm).
C_P	=	primary or compression wave speed of concrete (m/s);
f_T	=	frequency corresponding to the thickness of the concrete slab (kHz);
P	=	primary wave;

In general, the criteria and limitations for detectable flaws are based on the interrelation between the lateral flaw size, depth of that flaw, and wave length (Sansalone and Streett, 1997). Studies have shown that accurate results may be obtained using IE method to assess the condition of underground structures (Lin and Sansalone, 1994a, Lin and Sansalone, 1994b, FHWA, 1997).

The same grid setup shown in Figure 1 was used for these test methods as well. IE data analysis and interpretation used signal processing software, which transformed the time domain signal (waveform) into the frequency domain. In time domain, it is possible to exclude an invalid waveform (caused by loss of transducer contact with the surface), from the valid one as shown in Figures 2a and 2b, respectively. Waveforms obtained from Test Areas 1 and 2, as shown in Figures 2c and 2e, indicated delaminations. Although these waveforms looked similar to those obtained from the solid section, they possessed different dominant frequencies. In Figures 2d and 2f , frequencies of 6.35 kHz and 6.84 kHz indicate delaminations at test grid point F5 in Test Area 1 and at test grid point D10 in Test Area 2, respectively. These values were compared to the solid (thickness) frequencies of 5.86 kHz and 6.36 kHz with allowable error of 6%, which can be calculated using Eq.(1), for solid sections in Test Areas 1 and 2 respectively. The rest of the field waveforms were analyzed and interpreted and their results are presented in Figure 3.

Results obtained using the RP method were compared to those from the IE method. Since the IE method can detect deeper delaminations and anomalies than RP, it yielded a higher percentage of deteriorated areas for both cases. In addition, the thickness of each slab can be estimated using the measured P-wave speeds which were 3,579 m/s (13,770 ft/s) and 4,197 m/s (11,742 ft/s) for Test Areas 1 and 2, respectively.

The IE method took up to 25 times the testing time used by the RP method due to the fact that the sensitive equipment had to be moved carefully in a dark tunnel (Test Area 1) and the rough, dusty surface area at each grid point of the Test

Area l had to be cleaned for each measurement. Also, testing the vertical area (Test Area 1) took almost twice the time, when compared to testing a horizontal ground slab of similar size (Test Area 2). The RP device took only a fraction of an hour to perform the test regardless of whether the testing surface was vertical or horizontal. The only problem encountered while running the RP device was the excessive echo generated within the Test Area 2.

Ground Penetrating Radar (GPR) Method

The GPR method has been used extensively for rapid scanning of bridge decks for about a decade (Saarenketo and Scullion, 2000). A more complete survey discussed the use of GPR for nondestructive evaluation of road and bridge pavements, noting that air-coupled antenna systems were more popular than the ground-coupled system used in this study (Longstreet, 2003). A case study of an N.Y. tunnel inspection indicated that use of GPR was feasible (Tranbarger, 1985).

GPR systems are classified as either ground-coupled or air-coupled systems. Ground-coupled systems must remain in physical contact with the surface in order to collect data, whereas air-coupled systems do not require the physical contact.. A ground-coupled antenna system was chosen for this study based upon the necessary penetration depth and resolution and location of targets. For this system, the two antennas, tranmitter and receiver, operated within a high frequency range (0.5 – 1.5 GHz). A two or three cycle transmitted pulse propagates through the host medium and is reflected back to the receiving antenna. The travel time of that pulse is recorded and displayed in the viewing screen. The amplitude of the signal depends on the dielectric properties of the media as well as the performance of the equipment (Annan, 1996, Annan, 2002). This antenna system will also generate direct air waves, which can cause an interference in the recorded signal (Sensors and Software, 1999a).

At the time the GPR survey was planned, the target defect was assumed to be a deteriorated layer, i.e., delamination zone near the rebar, rebar grid, and concrete/soil interface. In a subway tunnel, the air gap may be filled with water when the ground water table is high. The time window was set up as 10 ns and an estimated wave speed of 0.10 m/s (0.33 ft/s) was used. The data was collected every 0.025 m (1 in.) horizontally and sampled every 0.1 ns. Again, the grid setup for this test is similar to that used in the previous sections. Since high water content and temperature have an adverse effect on concrete (Halabe et al., 1993), a protimeter was used to measure the surface moisture content of the concrete. In summer, the temperature measured in the tunnel was about 25 degrees Celcius (77 degrees Fahrenheit). The measured surface moisture content is shown in Figure 4.

Seven longitudinal survey lines (Line_x0-Line_x6) and sixteen transverse survey lines (Line_y0-Line_y15) were used for Test Areas 1 and 2 as shown in Figure 5. The resulting 112 grid points covered the same area as the other NDE methods. Only the GPR method was used for Test Area 3. No grid points were set up, and data was collected using one longitudinal survey line for this Test Area.

The collected data were analyzed and interpreted using a commercially available image processing software package (Sensors and Software, 1999b, Sensors and Software, 2002). In general, GPR data processing consists of data acquisition, recording and play back, data editing, basic processing, advanced data processing, and visual or interpretation processing (Sensors and Software, 1999c). This study used two sets of image processing techiques to enhance the reconstructed data images. The first set using Auto Gain (AG), which exponentially amplifies each echo in waveform, was applied to process the raw data. The other set used included No Gain (NG), Migration (M), and Background Subtraction (BS). This processing set reduced the effect due to the long tail of a parabolic-shaped pattern across the subsurface profile, partially removed direct air wave pulse and surface echoes, and enhanced the near surface or early time events (echoes) from the presence of rebar, surface opening cracks, high moisture content areas, and delaminations. These two sets of image processing techiques were used to process raw data collected from Test Areas 1 and 2, but raw data collected from Test Area 3 was processed using the first set and second set without BS.

Section plots, which consist of a series of traces or Radar waveforms, were used to analyze and interpret the processed data. The analysis consists of two steps as follows.

1. Identify the concrete/soil interface using post-processed images of longitudinal and tranverse subsurface profiles. If there exist echoes from this interface, flat layers of black and white would appear at the travel time corresponding to the location of the interface, which is somewhere between 6.0 – 7.0 nanoseconds (ns), based upon the estimated travel time corresponding to 305 mm (12 inch) concrete, which is about 6.0 ns using 0.10 m/ns for radar wave speed propagation in concrete.

2. Identify the locations of rebar and presence of surface opening cracks, high moisture content areas, and delaminations. If there is a surface opening crack, a vertical wedge like pattern would appear in the subsurface profile. Since a delaminated area close to rebar causes distortion to the smooth parabolic-shaped pattern formed by a series of echoes from rebar, it is possible to distinguish this event from those caused by high moisture content areas, which normally slow down the radar wave propagation speed but do not distort the shape of rebar echoes.

Interpretation of data was based on location, orientation, and degree of variation of each identified area. One sample of radar waveform and post-processed data obtained from survey Line_x3's and Line_y7's, which were located near the horizontal and vertical centerlines of Test Areas 1 and 2, is discussed in the following paragraphs to illustrate the GPR data analysis and interpretation scheme for this study.

An example of radar waveform obtained using a ground-coupled radar system is shown in Figure 6a. This waveform consists of four expotentially decaying echoes, which were from direct air wave and surface echo, rebar, and concrete/soil interface as shown in Figure 6b. The transmitted pulse is not an ideal pulse. Instead, it is a two or three cycle oscillatory pulse having a duration time of about 2 ns. The near face rebar grid with concrete cover less than 76 mm (3 inches), in general, causes overlapping of early time echoes and a multiple reflection or ringing phenomenon.

Examples of post-processed data from Test Area 1 are shown in Figure 7. Unless structural information was provided, it became difficult to indicate the location of the concrete/soil interface because echoes from unexpected reinforcing steel appeared in many locations in the subsurface profiles as shown in Figures 7a and 7c. The image of a thick dark layer, appeared at 6.0 ns in Figure 7a and between 6.0 – 7.0 ns in Figure 7c. Locations of near face rebar can be determined from Figures 7b and 7d, whereas those of far face rebar can be determined from Figures 7a and 7c. Surface cracks, between 1.8 – 2.4 m (6 – 8 feet) and 5.5 – 7.3 m (18 – 24 feet), are marked in Figure 7b using three solid circles. Another variation, marked by the second solid circle from left, indicates a possible deteriorated area beneath the surface.

Analysis and interpretation of data obtained from Test Area 2 was similar. The location of concrete/soil interface is between 6.0 – 7.0 ns as shown in Figures 8a and 8c. No flaws were found in this test area. There are two likely false indicators, which may cause misinterpretation of post-processed data image. The first one is the arch located at 9 feet and 6.5 nanoseconds in Figure 8a. Rather than being a deeper lying rebar, it may simply be a reflection from the closely-spaced rebar layer located above. The other is that the varying location of the near face transverse rebar locations, marked using a solid ellipse in Figure 8b, may be due to a varying surface moisture content of the concrete at that location, which affects the signal wave speed, rather than a variance in the actual rebar depth. Figure 8d shows locations of the near face longitudinal rebar locations as well as near face transverse rebar, which is indicated by a dark layer at about 2.0 ns.

Two examples of post-processed data images obtained from Test Area 3 are presented in Figure 9. The information of interest includes the glazed brick liner/concrete interface and the near face transverse rebar. The glazed brick liner and the near face transverse rebar can be easily verified from Figure 9a, but locations of those features are shown enhanced in Figure 9b.

COMPARISON AND DISCUSSION OF RESULTS

Results obtained using GPR and other methods for Test Area 1 and 2 are shown in Figures 10 and 11, respectively. In order to compare results, percentages of deteriorated areas found using a single method, the same deteriorated areas found using pairs of methods with the corresponding conditional probabilities, and using all methods together are summarized in Table 1.

Conditional probability may be defined as "the likelihood of one ovent occurring given that another event occurs" (p. 12, Bray and Stanley, 1997). In the context of NDE, conditional probability refers to the overlap between methods, e.g. the probability that IE will locate a defect discovered using GPR. Methods that can only find surface defects (VI, for example) would be expected to have a low conditional probability with GPR, unless the defect was visible on the surface and also extended deeply into the concrete.

The IE method gives the maximum percentages of deteriorated areas (61 and 16), whereas the VI method gives the minimum percentages (10 and 0) for Test Areas 1 and 2. Conditional probability obtained using IE with GPR gives a maximum value of 0.20 and those obtained using two combinations, VI with IE and RP with IE, give the same minimum value of 0.08 for Test Area 1. Using all four methods together and every pair combination of methods, except VI with GPR, gives a conditional probability value of zero for Test Area 2. It is interesting to note that the GPR method gives the maximum average value of conditional probability (0.17), whereas the IE method gives the minimum one (0.12), and that the others give the same value of 0.13 for Test Area 1.

The RP method took the least testing time, while the IE method required the most for both test areas as shown in Table 2. Testing times used by VI and GPR methods are 30 and 25 minutes for Test Area 1 and 25 and 20 minutes for Test Area 2. None of the testing times for VI, RP, and GPR methods exceeded 30 minutes. However, the IE method required more than 4 hours. The VI and IE testing times for Test Area 1 were as much as twice of those for Test Area 2.

CONCLUSIONS AND RECOMMENDATIONS

VI, as a classical standard method, requires little testing time but is not useful for subsurface damage. Significant variability is caused by human and environmental factors. RP also requires little testing time and can detect some shallow delaminated areas. IE takes the most testing time but is the most effective for detecting deteriorated areas. This method requires a considerable amount of time for test preparation, validation of data, data analysis and interpretation. The equipment is delicated and had difficulties while testing Test Area 1 due to a loss of transducer contact with the surface.

GPR, a newer NDE technology, reduces testing time and is more effective at detecting deteriorated areas than RP but less effective than IE. The field time for GPR testing was much less than that required for IE. However, GPR data analysis and interpretation are still somewhat subjective. The common problem found from the results obtained using those methods is the mislocation of the deteriorated area, caused by user interpretation and limitations of the equipment. Cores were not obtained; thus physical confirmation of deterioration present could not be made with certainty. However, a previous study was performed as the basis for the restoration and renovation of a commercial building with an interior multi-level parking facility (National Terminal Warehouse, Cleveland, Ohio), using both the RP and IE methods. Damaged concrete detected by the NDE survey was verified, removed and repaired or replaced, as necessary. The NDE survey proved to be reliable and effective. Unfortunately, at this time the results of that study have not been released for publication by the client. Using the four methods together provides the highest probability of detecting deteriorated areas. If possible, two destructive test methods, combined with core samples and chemical analysis of the structural materials, should be used for confirmation of results.

None of the methods used provides a panacea for rapid screening of a subway tunnel. The IE method is able to locate the most defects, including those deep within the concrete that cannot be seen by VI or RP, but the field testing is extremely time consuming, making it unacceptable for rapid screening of large areas. The VI, RP and GPR methods all have similar, relatively rapid field observation times, but have limitations in the defects observed. The VI method only discovers visible surface defects. RP finds near surface delaminations, but cannot detect deeper delaminations. GPR finds most of the defects detected by VI and RP, many of the deeper defects and other pertinent information, such as location of rebar, slab thickness, etc., but may miss some of the near surface delaminations due to surface airwave event interference. Based upon the testing time and amount of defects detected, GPR has a high potential for rapid screening of subway tunnels, and further study and development is needed, particularly with respect to data interpretation and interference filtering.

ACKNOWLEDGEMENTS

Support for this research was provided by the USDOT Federal Transit Administration, through the Great Cities Universities Coalition Transportation Initiative. Matching funds were provided by UAB, CCNY, CSU, UIC, and UC. The authors also wish to thank the Cuyahoga County Engineer's office for their assistance, including the use of the tunnel, and former graduate student Franz Buce for his assistance in the field testing. Opinions expressed in this paper are solely those of the authors and not of the research sponsors.

REFERENCES

ASTM (1998). *ASTM. C1383-98: Standard test method for measuring the P-wave speed and the thickness of concrete plates using the Impact-Echo method.* ASTM, West Conshohocken, Pa.

ASTM (2002). *ASTM. D4580-02: Standard practice for measuring delaminations in concrete bridge decks by sounding.* ASTM, West Conshohocken, Pa.

Annan, A.P. (1996). "Transmission Dispersion and GPR." *JEEG*, vol.0, no.2, January 1996, pp. 125-136.

Annan, A.P. (2002). "GPR Signal Amplitude Charaterization and Recording Dynamic Range." *Sensors & Software, Inc.*, 2002.

Bray, D.E., and Stanley, R.K. (1997). "Nondestructive Evaluation: A Tool in Design, Manufacturing, and Service," CRC Press, Boca Raton, Florida.

Federal Highway Administration (FHWA). (1997). "Concrete Testing Puts Subway Riders at Ease." *Publication No. FHWA-SA-96-045*, Washington, D.C.

Grabeal, Benjamin A., Phares, Brent M., Rolander, Dennis D., and Washer Glenn A. (2001). "Reliability of Visual Bridge Inspection." *Public Roads*, vol.64, March/April 2001.

Halabe, U.B., A. Sotoodehnia, K.R., Maser and E.A., Kausel. (1993). "Modeling the Electromagnetic Properties of Concrete." *American Concrete Institute (ACI) Materials Journal*, vol.90, no.6, November-December 1993, pp. 552-563.

Lin, J.M. and Sansalone, M. (1994a). "Impact-Echo Response of Hollow Cylindrical Concrete Structures Surrounded by Soil and Rock: Part I-Numerical studies." *Geotechnical Testing Journal*, GTJODJ, vol.17, no.2, June 1994, pp. 207-219.

Lin, J.M. and Sansalone, M. (1994b). "Impact-Echo Response of Hollow Cylindrical Concrete Structures Surrounded by Soil and Rock: Part II-Experimental studies." *Geotechnical Testing Journal*, GTJODJ, vol.17, no.2, June 1994, pp. 220-226.

Longstreet, P. William (2003). SHRP PRODUCT 2015: Ground-Penetrating Radar for Bridge Deck Evaluations. Concrete Assessment and Rehabilitation, AASHTO Innovative Highway Technologies.
http://leadstates.tamu.edu/car/shrp_products/2015.stm. Accessed July 15, 2003.

Pollock, D.J., Kay, E.A. and Fookes, P.G. (1981) "Crack Mapping for Investigation of Middle East Concrete." *Concrete*, vol.15, no.5, May 1981, pp. 12-18.

Russell, H.A., and Gilmore, J. (1997). "Inspection Policy and Procedures for Transit Tunnels and Underground Structures." *Transit Cooperative Research Program Synthesis 23.* National Academy Press, Washington, D.C.

Saarenketo, T. and Scullion, T. (2000). "Road Evaluation With Ground Penetrating Radar." *Journal of Applied Geophysics*, vol.43, pp. 119-138.

Sansalone, M.J. and Streett , W.B. (1997). *Impact-Echo, Nondestructive Evaluation of Concrete and Masonry: Field Testing.* Bullbrier Press, Ithaca, N.Y.

Sensors and Software. (1999a). *Ground Penetrating Radar Survey Design: 1992-1999*. Sensors & Software, Inc.

Sensors and Software. (1999b). *Ekko_3D user's guide version 2.0: 1999*. Sensors & Software, Inc.

Sensors and Software. (1999c). *Practical Processing of GPR Data: 1999*. Sensors & Software, Inc.

Sensors and Software. (2002). *Ekko_Mapper user's guide version 2.0: 2002*. Sensors & Software, Inc.

Tranbarger, O. (1985). "FM Radar for Inspecting Brick and Concrete Tunnels." *Materials Evaluation*, vol.43, September 1985, pp. 1254-1261.

Watson, S.R., and Wolfs, J.R. (1981). *Bridges of Metropolitan Cleveland: The Detroit-Superior High Level Bridge (1918)*. USA.

TABLE 1 Percentages of deteriorated areas found using single method and percentages of the same deteriorated areas found using pairs of methods and all methods together

TEST AREA	VI	RP	IE	GPR	VI AND RP	VI AND IE	VI AND GPR	RP AND IE	RP AND GPR	IE AND GPR	ALL METHODS
1	10	18	61	43	4 (0.17)[1]	5 (0.08)	5 (0.15)	6 (0.08)	6 (0.15)	15 (0.20)	4
2	0	1	16	9	0 (0)[1]	0 (0)[1]	0 (NC)	0 (0)[1]	0 (0)[1]	4 (0)[1]	0

[1] Conditional Probability
[2] Undetermined Conditional Probability

TABLE 2 Testing times required by VI, RP, IE, and GPR methods

TEST	VI	RP	IE	GPR	TOTAL
1	30 minutes	20 minutes	8:35 hours	25 minutes	9:50 hours
2	25 minutes	10 minutes	4:45 hours	20 minutes	5:40 hours
Total	55 minutes	30 minutes	13:20 hours	45 minutes	15:30

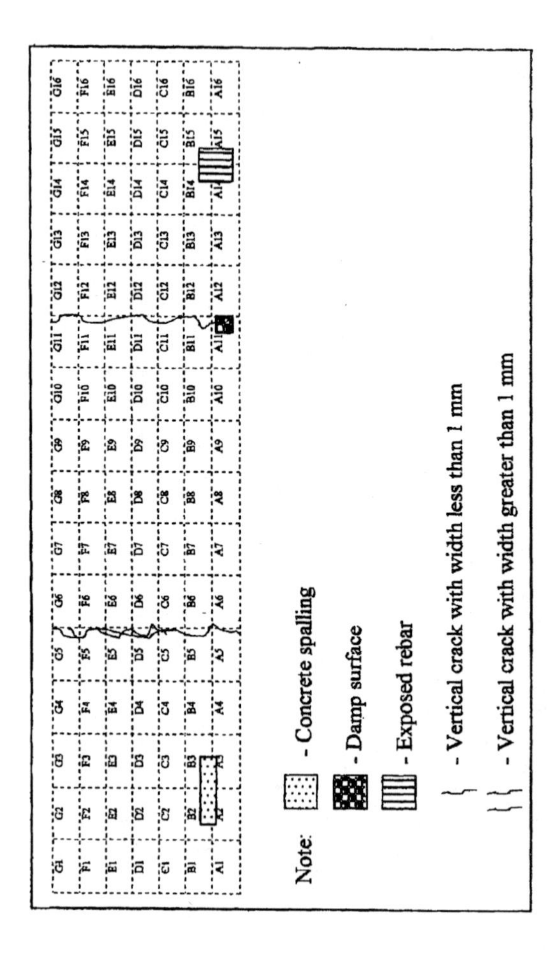

FIGURE 1 Surface flaws detected using VI method for Test Area 1.

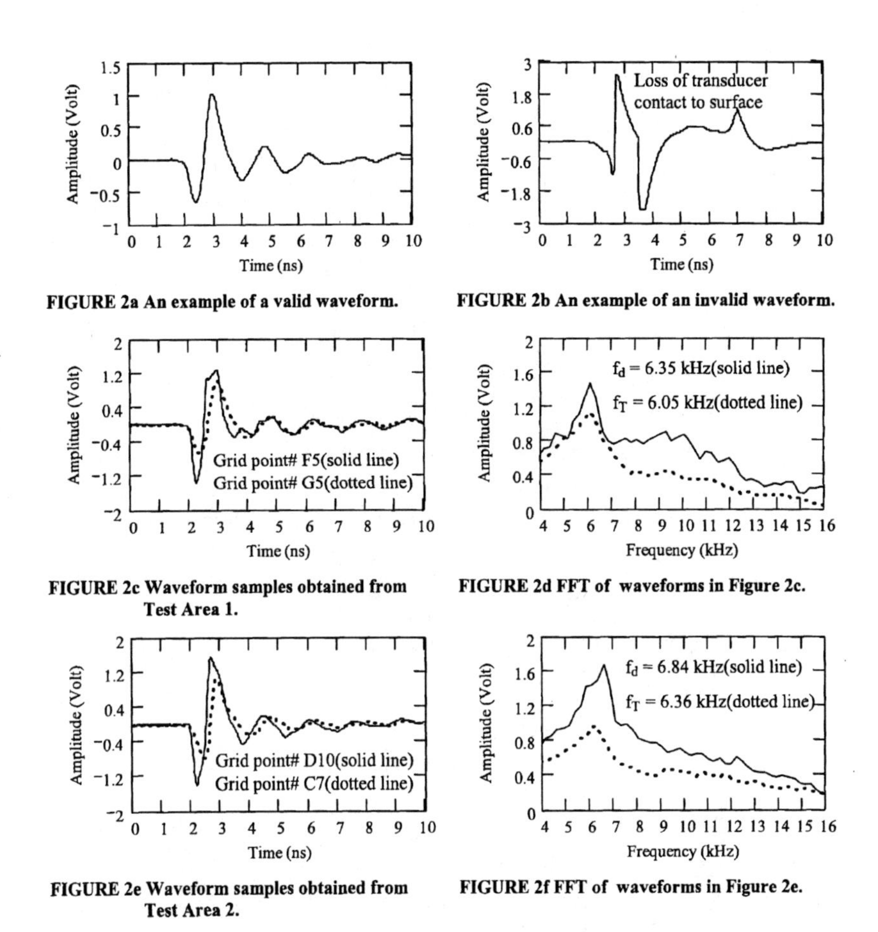

FIGURE 2a An example of a valid waveform.

FIGURE 2b An example of an invalid waveform.

FIGURE 2c Waveform samples obtained from Test Area 1.

FIGURE 2d FFT of waveforms in Figure 2c.

FIGURE 2e Waveform samples obtained from Test Area 2.

FIGURE 2f FFT of waveforms in Figure 2e.

FIGURE 2 Examples of IE waveforms and amplitude-spectrum plots.

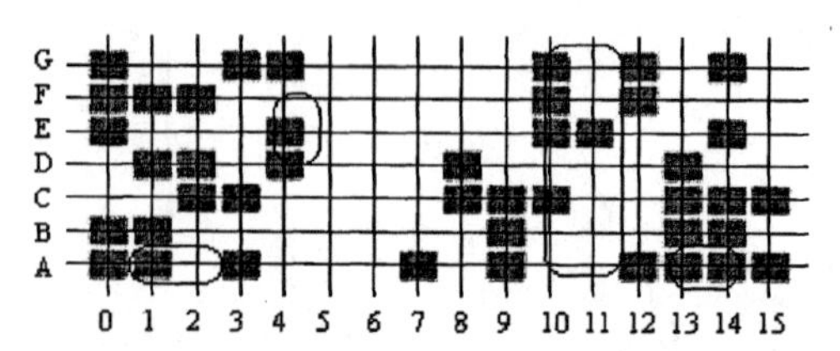

FIGURE 3a Deteriorated areas obtained from Test Area 1 using RP and IE methods.

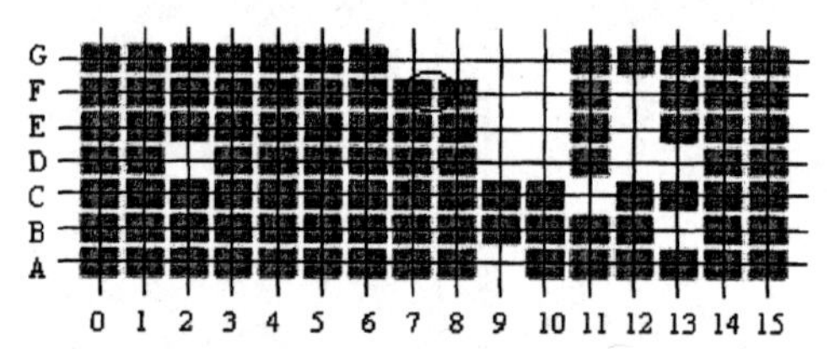

FIGURE 3b Deteriorated areas obtained from Test Area 2 using RP and IE methods.

Legends:

- Delamination found using RP mehod

- Flaw detected using IE method

- No flaw detected using IE method

FIGURE 3 Deteriorated areas obtained from Test Areas 1 and 2 using RP and IE methods.

18	18	19	19	19	20	20	20	19	19	19	22	22	22	19	19
18	18	19	19	19	20	20	20	19	19	19	22	22	22	19	19
20	20	19	19	19	19	19	19	21	21	21	22	22	22	22	22
20	20	19	19	19	19	19	19	21	21	21	22	22	22	22	22
20	20	19	19	19	19	19	19	21	21	21	22	22	22	22	22
18	18	27	27	27	19	19	19	21	21	21	22	22	22	21	21
18	18	27	27	27	19	19	19	21	21	21	22	22	22	21	21

FIGURE 4a Surface moisture content (%) of Test Area 1.

27	27	27	27	27	25	25	25	25	25	25	25	25	25	25	25
27	27	27	27	27	25	25	25	25	25	25	25	25	25	25	25
27	27	27	27	27	45	45	45	45	45	45	25	25	25	25	25
27	27	27	27	27	45	45	45	45	45	45	25	25	25	25	25
27	27	27	27	27	45	45	45	45	45	45	25	25	25	25	25
27	27	27	27	27	27	27	27	27	27	27	25	25	25	25	25
27	27	27	27	27	27	27	27	27	27	27	25	25	25	25	25

FIGURE 4b Surface moisture content (%) of Test Area 2.

Legend:

- Surface moisture content (%) greater than average values

FIGURE 4 Percentages of surface moisture content of Test Areas 1 and 2.

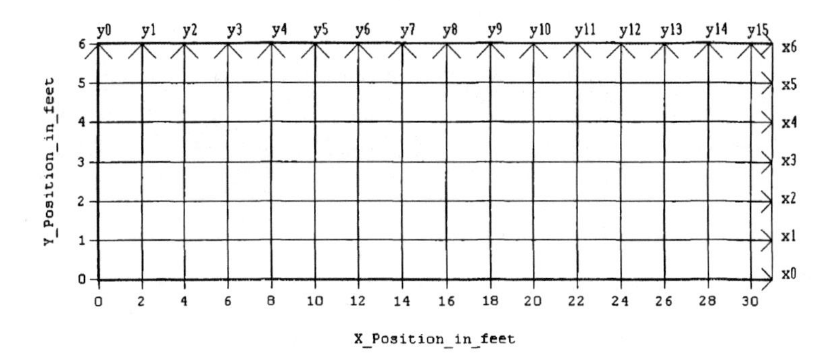

FIGURE 5 GPR survey lines used for collecting data from Test Areas 1 and 2 (1 ft. = 305 mm).

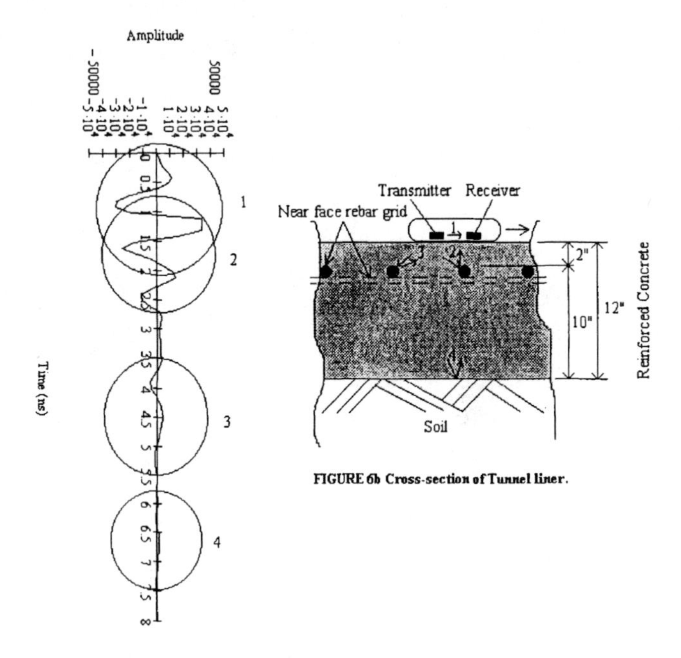

FIGURE 6a Waveform obtained using ground-coupled radar system.

FIGURE 6b Cross-section of Tunnel liner.

FIGURE 6 An example of GPR waveform obtained using ground-coupled radar system and cross section of tunnel liner (1 inch = 25 mm).

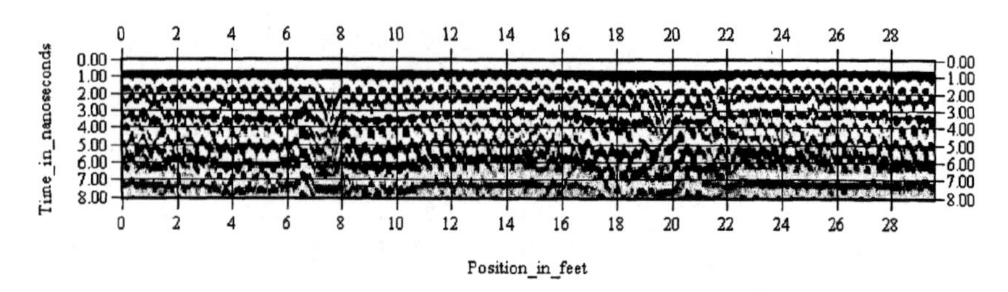

FIGURE 7a Post-processed and reconstructed subsurface profile from survey Line_x3 of Test Area 1 using Auto Gain.

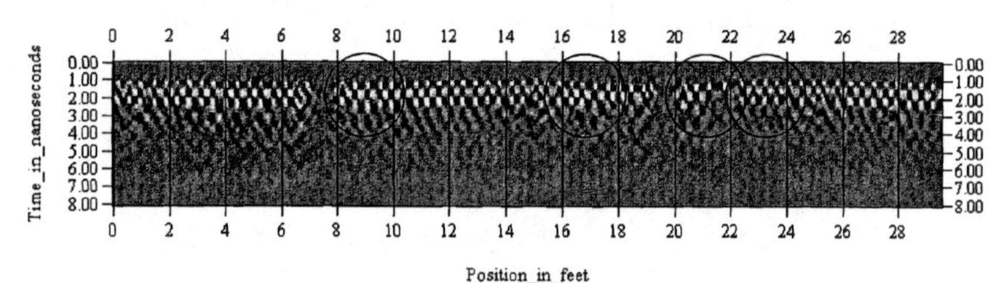

FIGURE 7b Post-processed and reconstructed subsurface profile from survey Line_x3 of Test Area 1 using No Gain, Migration, and Background Subtraction.

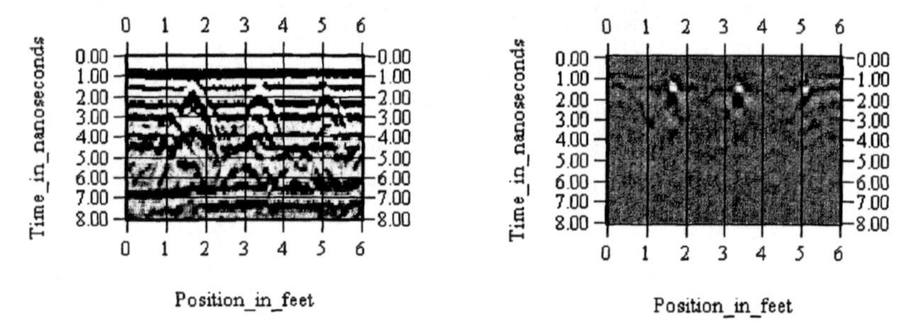

FIGURE 7c Post-processed and reconstructed
reconstructed
subsurface profile from survey Line_y7 of
Line_y7 of
Test Area 1 using Auto Gain.
Migration, and

FIGURE 7d Post-processed and

subsurface profile from survey

Test Area 1 using No Gain,

Background Subtraction.

FIGURE 7 Examples of post-processed and reconstructed subsurface profiles obtained from Test Area 1 (1 ft. = 305 mm).

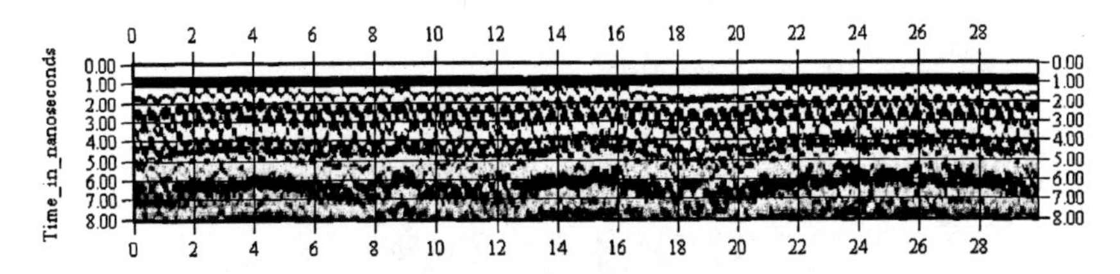

FIGURE 8a Post-processed and reconstructed subsurface profile from survey Line_x3 of Test Area 2 using Auto Gain.

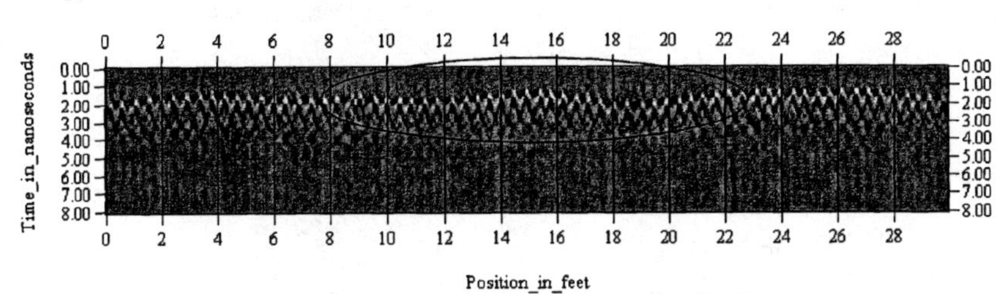

FIGURE 8b Post-processed and reconstructed subsurface profile from survey Line_x3 of Test Area 2 using No Gain, Migration, and Background Subtraction.

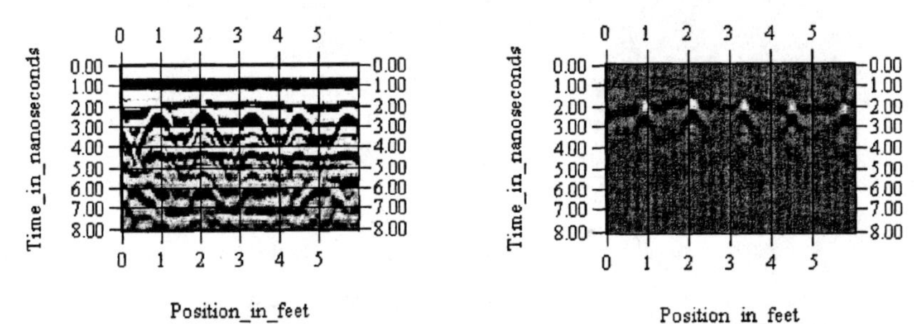

FIGURE 8c Post-processed and reconstructed reconstructed
subsurface profile from survey Line_y7 of Line_y7 of
Test Area 2 using Auto Gain. Migration, and

FIGURE 8d Post-processed and
subsurface profile from survey
Test Area 2 using No Gain,
Background Subtraction.

FIGURE 8 Examples of post-processed and reconstructed subsurface profiles obtained from Test Area 2 (1 ft. = 305 mm).

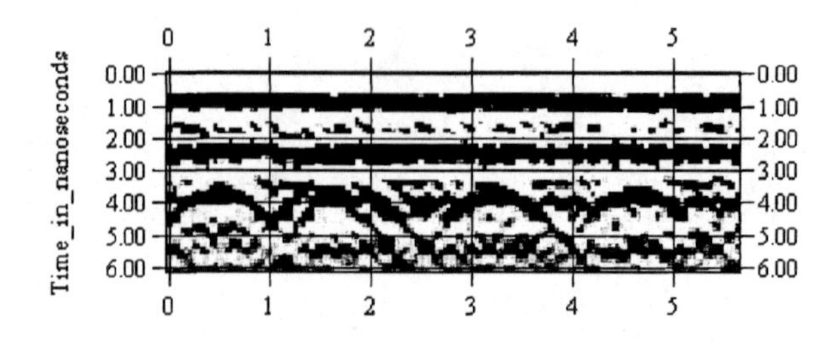

FIGURE 9a Post-processed and reconstructed subsurface profile from longitudinal survey line at center of Test Area 3 using Auto Gain.

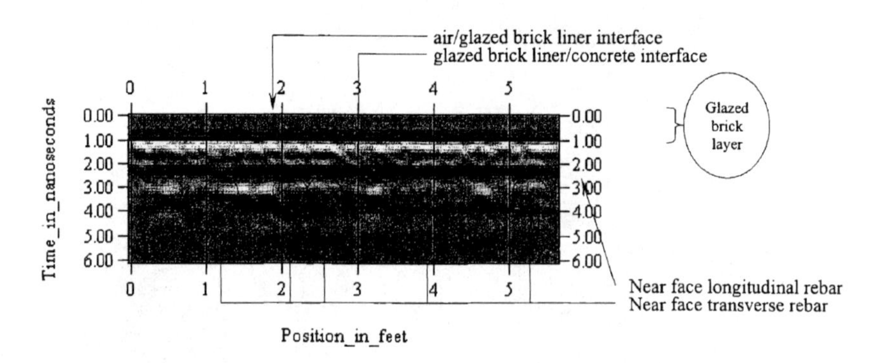

FIGURE 9b Post-processed and reconstructed subsurface profile from longitudinal survey line at center of Test Area 3 using No Gain and Migration.

FIGURE 9 Examples of post-processed and reconstructed subsurface profiles obtained from Test Area 3 (1 ft. = 305 mm).

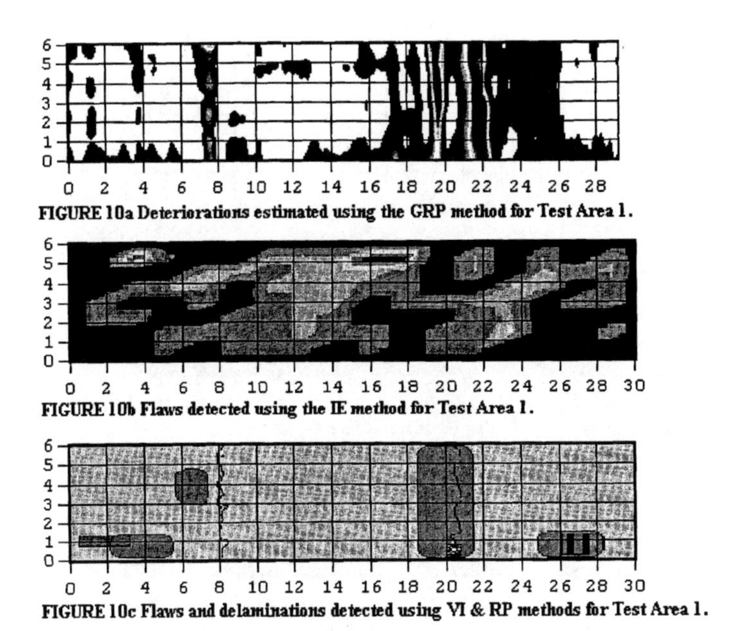

FIGURE 10a Deteriorations estimated using the GRP method for Test Area 1.

FIGURE 10b Flaws detected using the IE method for Test Area 1.

FIGURE 10c Flaws and delaminations detected using VI & RP methods for Test Area 1.

FIGURE 10 Visual map for the results obtained using GPR, IE, VI, and RP methods for Test Area 1.

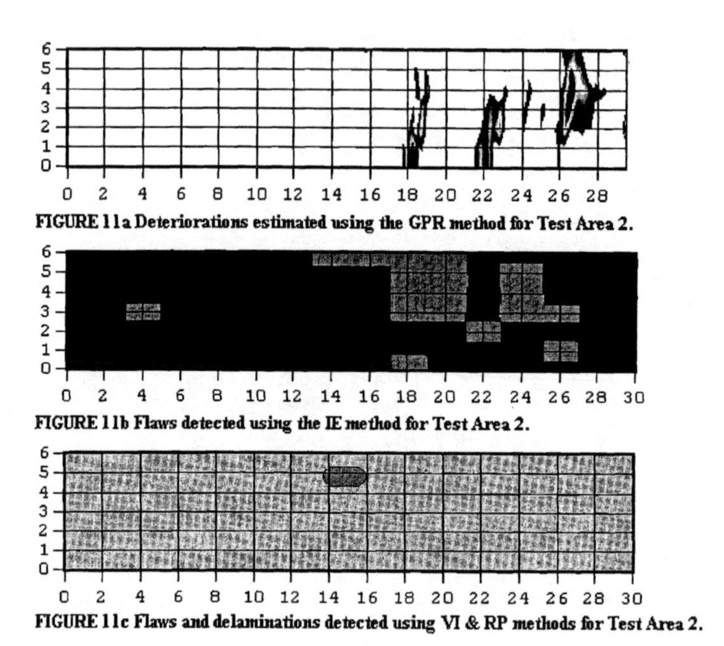

FIGURE 11a Deteriorations estimated using the GPR method for Test Area 2.

FIGURE 11b Flaws detected using the IE method for Test Area 2.

FIGURE 11c Flaws and delaminations detected using VI & RP methods for Test Area 2.

FIGURE 11 Visual map for the results obtained using GPR, IE, VI, and RP methods for Test Area 2.

Subject Index

Page number refers to the first page of paper

Author Index

Page number refers to the first page of paper